Introduction to Quantum Mechanics

Second Edition

David J. Griffiths
Reed College

PEARSON

Pearson Prentice Hall® is a trademark of Pearson Education, Inc.

ISBN 978-81-7758-230-7

10 9 8

Published by Dorling Kindersley (India) Pvt. Ltd., licensees of Pearson Education in South Asia.

Head Office: 7th Floor, Knowledge Boulevard, A-8(A), Sector-62, Noida 201309, UP, India.
Registered Office: 11, Community Centre, Panchsheel Park, New Delhi 110 017, India.

Printed in India by Anubha Printers.

CONTENTS

PREFACE

Unlike Newton's mechanics, or Maxwell's electrodynamics, or Einstein's relativity, quantum theory was not created—or even definitively packaged—by one individual, and it retains to this day some of the scars of its exhilarating but traumatic youth. There is no general consensus as to what its fundamental principles are, how it should be taught, or what it really "means." Every competent physicist can "do" quantum mechanics, but the stories we tell ourselves about what we are doing are as various as the tales of Scheherazade, and almost as implausible. Niels Bohr said, "If you are not confused by quantum physics then you haven't really understood it"; Richard Feynman remarked, "I think I can safely say that nobody understands quantum mechanics."

The purpose of this book is to teach you how to *do* quantum mechanics. Apart from some essential background in Chapter 1, the deeper quasi-philosophical questions are saved for the end. I do not believe one can intelligently discuss what quantum mechanics *means* until one has a firm sense of what quantum mechanics *does*. But if you absolutely cannot wait, by all means read the Afterword immediately following Chapter 1.

Not only is quantum theory conceptually rich, it is also technically difficult, and exact solutions to all but the most artificial textbook examples are few and far between. It is therefore essential to develop special techniques for attacking more realistic problems. Accordingly, this book is divided into two parts;[1] Part I covers the basic theory, and Part II assembles an arsenal of approximation schemes, with illustrative applications. Although it is important to keep the two parts *logically* separate, it is not necessary to study the material in the order presented here. Some

[1]This structure was inspired by David Park's classic text, *Introduction to the Quantum Theory*, 3rd ed., McGraw-Hill, New York (1992).

instructors, for example, may wish to treat time-independent perturbation theory immediately after Chapter 2.

This book is intended for a one-semester or one-year course at the junior or senior level. A one-semester course will have to concentrate mainly on Part I; a full-year course should have room for supplementary material beyond Part II. The reader must be familiar with the rudiments of linear algebra (as summarized in the Appendix), complex numbers, and calculus up through partial derivatives; some acquaintance with Fourier analysis and the Dirac delta function would help. Elementary classical mechanics is essential, of course, and a little electrodynamics would be useful in places. As always, the more physics and math you know the easier it will be, and the more you will get out of your study. But I would like to emphasize that quantum mechanics is not, in my view, something that flows smoothly and naturally from earlier theories. On the contrary, it represents an abrupt and revolutionary departure from classical ideas, calling forth a wholly new and radically counterintuitive way of thinking about the world. That, indeed, is what makes it such a fascinating subject.

At first glance, this book may strike you as forbiddingly mathematical. We encounter Legendre, Hermite, and Laguerre polynomials, spherical harmonics, Bessel, Neumann, and Hankel functions, Airy functions, and even the Riemann zeta function—not to mention Fourier transforms, Hilbert spaces, hermitian operators, Clebsch-Gordan coefficients, and Lagrange multipliers. Is all this baggage really necessary? Perhaps not, but physics is like carpentry: Using the right tool makes the job *easier*, not more difficult, and teaching quantum mechanics without the appropriate mathematical equipment is like asking the student to dig a foundation with a screwdriver. (On the other hand, it can be tedious and diverting if the instructor feels obliged to give elaborate lessons on the proper use of each tool. My own instinct is to hand the students shovels and tell them to start digging. They may develop blisters at first, but I still think this is the most efficient and exciting way to learn.) At any rate, I can assure you that there is no *deep* mathematics in this book, and if you run into something unfamiliar, and you don't find my explanation adequate, by all means *ask* someone about it, or look it up. There are many good books on mathematical methods—I particularly recommend Mary Boas, *Mathematical Methods in the Physical Sciences*, 2nd ed., Wiley, New York (1983), or George Arfken and Hans-Jurgen Weber, *Mathematical Methods for Physicists*, 5th ed., Academic Press, Orlando (2000). But whatever you do, don't let the mathematics—which, for us, is only a *tool*—interfere with the physics.

Several readers have noted that there are fewer worked examples in this book than is customary, and that some important material is relegated to the problems. This is no accident. I don't believe you can learn quantum mechanics without doing many exercises for yourself. Instructors should of course go over as many problems in class as time allows, but students should be warned that this is not a subject about which *any*one has natural intuitions—you're developing a whole new set of muscles here, and there is simply no substitute for calisthenics. Mark Semon suggested that I offer a "Michelin Guide" to the problems, with varying numbers of stars to indicate the level of difficulty and importance. This seemed like a good idea (though, like the quality of a restaurant, the significance of a problem is partly a matter of taste); I have adopted the following rating scheme:

* an *essential* problem that every reader should study;
** a somewhat more difficult or more peripheral problem;
*** an unusually challenging problem, that may take over an hour.

(No stars at all means fast food: OK if you're hungry, but not very nourishing.) Most of the one-star problems appear at the end of the relevant section; most of the three-star problems are at the end of the chapter. A solution manual is available (to instructors only) from the publisher.

In preparing the second edition I have tried to retain as much as possible the spirit of the first. The only wholesale change is Chapter 3, which was much too long and diverting; it has been completely rewritten, with the background material on finite-dimensional vector spaces (a subject with which most students at this level are already comfortable) relegated to the Appendix. I have added some examples in Chapter 2 (and fixed the awkward definition of raising and lowering operators for the harmonic oscillator). In later chapters I have made as few changes as I could, even preserving the numbering of problems and equations, where possible. The treatment is streamlined in places (a better introduction to angular momentum in Chapter 4, for instance, a simpler proof of the adiabatic theorem in Chapter 10, and a new section on partial wave phase shifts in Chapter 11). Inevitably, the second edition is a bit longer than the first, which I regret, but I hope it is cleaner and more accessible.

I have benefited from the comments and advice of many colleagues, who read the original manuscript, pointed out weaknesses (or errors) in the first edition, suggested improvements in the presentation, and supplied interesting problems. I would like to thank in particular P. K. Aravind (Worcester Polytech), Greg Benesh (Baylor), David Boness (Seattle), Burt Brody (Bard), Ash Carter (Drew), Edward Chang (Massachusetts), Peter Collings (Swarthmore), Richard Crandall (Reed), Jeff Dunham (Middlebury), Greg Elliott (Puget Sound), John Essick (Reed), Gregg Franklin (Carnegie Mellon), Henry Greenside (Duke), Paul Haines (Dartmouth), J. R. Huddle (Navy), Larry Hunter (Amherst), David Kaplan (Washington), Alex Kuzmich (Georgia Tech), Peter Leung (Portland State), Tony Liss (Illinois), Jeffry Mallow (Chicago Loyola), James McTavish (Liverpool), James Nearing (Miami), Johnny Powell (Reed), Krishna Rajagopal (MIT), Brian Raue (Florida International), Robert Reynolds (Reed), Keith Riles (Michigan), Mark Semon (Bates), Herschel Snodgrass (Lewis and Clark), John Taylor (Colorado), Stavros Theodorakis (Cyprus), A. S. Tremsin (Berkeley), Dan Velleman (Amherst), Nicholas Wheeler (Reed), Scott Willenbrock (Illinois), William Wootters (Williams), Sam Wurzel (Brown), and Jens Zorn (Michigan).

Introduction to Quantum Mechanics

PART I THEORY

CHAPTER 1

THE WAVE FUNCTION

1.1 THE SCHRODINGER EQUATION

Imagine a particle of mass m, constrained to move along the x-axis, subject to some specified force $F(x,t)$ (Figure 1.1). The program of *classical* mechanics is to determine the position of the particle at any given time: $x(t)$. Once we know that, we can figure out the velocity ($v = dx/dt$), the momentum ($p = mv$), the kinetic energy ($T = (1/2)mv^2$), or any other dynamical variable of interest. And how do we go about determining $x(t)$? We apply Newton's second law: $F = ma$. (For *conservative* systems—the only kind we shall consider, and, fortunately, the only kind that *occur* at the microscopic level—the force can be expressed as the derivative of a potential energy function,[1] $F = -\partial V/\partial x$, and Newton's law reads $m\, d^2x/dt^2 = -\partial V/\partial x$.) This, together with appropriate initial conditions (typically the position and velocity at $t = 0$), determines $x(t)$.

Quantum mechanics approaches this same problem quite differently. In this case what we're looking for is the particle's **wave function**, $\Psi(x,t)$, and we get it by solving the **Schrödinger equation**:

$$\boxed{i\hbar\frac{\partial\Psi}{\partial t} = -\frac{\hbar^2}{2m}\frac{\partial^2\Psi}{\partial x^2} + V\Psi.} \qquad [1.1]$$

[1]Magnetic forces are an exception, but let's not worry about them just yet. By the way, we shall assume throughout this book that the motion is nonrelativistic ($v \ll c$).

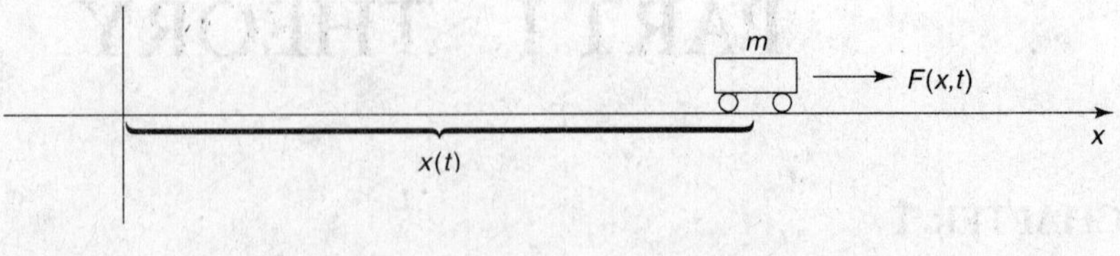

FIGURE 1.1: A "particle" constrained to move in one dimension under the influence of a specified force.

Here i is the square root of -1, and $\hbar$ is Planck's constant—or rather, his *original* constant (h) divided by 2π:

$$\hbar = \frac{h}{2\pi} = 1.054572 \times 10^{-34}\text{J s}. \quad [1.2]$$

The Schrödinger equation plays a role logically analogous to Newton's second law: Given suitable initial conditions (typically, $\Psi(x, 0)$), the Schrödinger equation determines $\Psi(x, t)$ for all future time, just as, in classical mechanics, Newton's law determines $x(t)$ for all future time.[2]

1.2 THE STATISTICAL INTERPRETATION

But what exactly *is* this "wave function," and what does it do for you once you've *got* it? After all, a particle, by its nature, is localized at a point, whereas the wave function (as its name suggests) is spread out in space (it's a function of x, for any given time t). How can such an object represent the state of a *particle*? The answer is provided by Born's **statistical interpretation** of the wave function, which says that $|\Psi(x, t)|^2$ gives the *probability* of finding the particle at point x, at time t—or, more precisely,[3]

$$\int_a^b |\Psi(x, t)|^2\, dx = \left\{ \begin{array}{l} \text{probability of finding the particle} \\ \text{between } a \text{ and } b, \text{ at time } t. \end{array} \right\} \quad [1.3]$$

Probability is the *area* under the graph of $|\Psi|^2$. For the wave function in Figure 1.2, you would be quite likely to find the particle in the vicinity of point A, where $|\Psi|^2$ is large, and relatively *un*likely to find it near point B.

[2]For a delightful first-hand account of the origins of the Schrödinger equation see the article by Felix Bloch in *Physics Today*, December 1976.

[3]The wave function itself is complex, but $|\Psi|^2 = \Psi^*\Psi$ (where Ψ^* is the complex conjugate of Ψ) is real and nonnegative—as a probability, of course, *must* be.

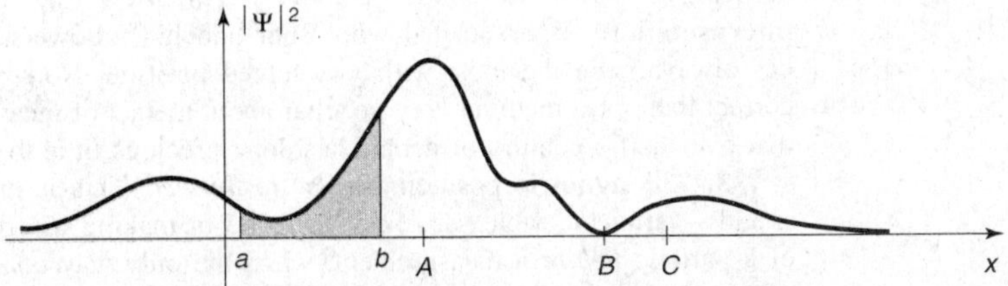

FIGURE 1.2: A typical wave function. The shaded area represents the probability of finding the particle between a and b. The particle would be relatively likely to be found near A, and unlikely to be found near B.

The statistical interpretation introduces a kind of **indeterminacy** into quantum mechanics, for even if you know everything the theory has to tell you about the particle (to wit: its wave function), still you cannot predict with certainty the outcome of a simple experiment to measure its position—all quantum mechanics has to offer is *statistical* information about the *possible* results. This indeterminacy has been profoundly disturbing to physicists and philosophers alike, and it is natural to wonder whether it is a fact of nature, or a defect in the theory.

Suppose I *do* measure the position of the particle, and I find it to be at point C.[4] *Question:* Where was the particle just *before* I made the measurement? There are three plausible answers to this question, and they serve to characterize the main schools of thought regarding quantum indeterminacy:

1. The **realist** position: *The particle was at C.* This certainly seems like a sensible response, and it is the one Einstein advocated. Note, however, that if this is true then quantum mechanics is an *incomplete* theory, since the particle *really was* at C, and yet quantum mechanics was unable to tell us so. To the realist, indeterminacy is not a fact of nature, but a reflection of our ignorance. As d'Espagnat put it, "the position of the particle was never indeterminate, but was merely unknown to the experimenter."[5] Evidently Ψ is not the whole story—some additional information (known as a **hidden variable**) is needed to provide a complete description of the particle.

2. The **orthodox** position: *The particle wasn't really anywhere.* It was the act of measurement that forced the particle to "take a stand" (though how and why it decided on the point C we dare not ask). Jordan said it most starkly: "Observations not only *disturb* what is to be measured, they *produce* it ... We *compel* (the

[4]Of course, no measuring instrument is perfectly precise; what I *mean* is that the particle was found *in the vicinity* of C, to within the tolerance of the equipment.

[5]Bernard d'Espagnat, "The Quantum Theory and Reality" (Scientific American, November 1979, p. 165).

particle) to assume a definite position."[6] This view (the so-called **Copenhagen interpretation**), is associated with Bohr and his followers. Among physicists it has always been the most widely accepted position. Note, however, that if it is correct there is something very peculiar about the act of measurement—something that over half a century of debate has done precious little to illuminate.

3. The **agnostic** position: *Refuse to answer.* This is not quite as silly as it sounds—after all, what sense can there be in making assertions about the status of a particle *before* a measurement, when the only way of knowing whether you were right is precisely to conduct a measurement, in which case what you get is no longer "before the measurement?" It is metaphysics (in the pejorative sense of the word) to worry about something that cannot, by its nature, be tested. Pauli said: "One should no more rack one's brain about the problem of whether something one cannot know anything about exists all the same, than about the ancient question of how many angels are able to sit on the point of a needle."[7] For decades this was the "fall-back" position of most physicists: They'd try to sell you the orthodox answer, but if you were persistent they'd retreat to the agnostic response, and terminate the conversation.

Until fairly recently, all three positions (realist, orthodox,, and agnostic) had their partisans. But in 1964 John Bell astonished the physics community by showing that it makes an *observable* difference whether the particle had a precise (though unknown) position prior to the measurement, or not. Bell's discovery effectively eliminated agnosticism as a viable option, and made it an *experimental* question whether 1 or 2 is the correct choice. I'll return to this story at the end of the book, when you will be in a better position to appreciate Bell's argument; for now, suffice it to say that the experiments have decisively confirmed the orthodox interpretation:[8] A particle simply does not *have* a precise position prior to measurement, any more than the ripples on a pond do; it is the measurement process that insists on one particular number, and thereby in a sense *creates* the specific result, limited only by the statistical weighting imposed by the wave function.

What if I made a *second* measurement, *immediately* after the first? Would I get C again, or does the act of measurement cough up some completely new number each time? On this question everyone is in agreement: A repeated measurement (on the same particle) must return the same value. Indeed, it would be tough to prove that the particle was really found at C in the first instance, if this could not be confirmed by immediate repetition of the measurement. How does the orthodox

---

[6] Quoted in a lovely article by N. David Mermin, "Is the moon there when nobody looks?" (Physics Today, April 1985, p. 38).

[7] Quoted by Mermin (footnote 6), p. 40.

[8] This statement is a little too strong: There remain a few theoretical and experimental loopholes, some of which I shall discuss in the Afterword. There exist viable nonlocal hidden variable theories (notably David Bohm's), and other formulations (such as the **many worlds** interpretation) that do not fit cleanly into any of my three categories. But I think it is wise, at least from a pedagogical point of view, to adopt a clear and coherent platform at this stage, and worry about the alternatives later.

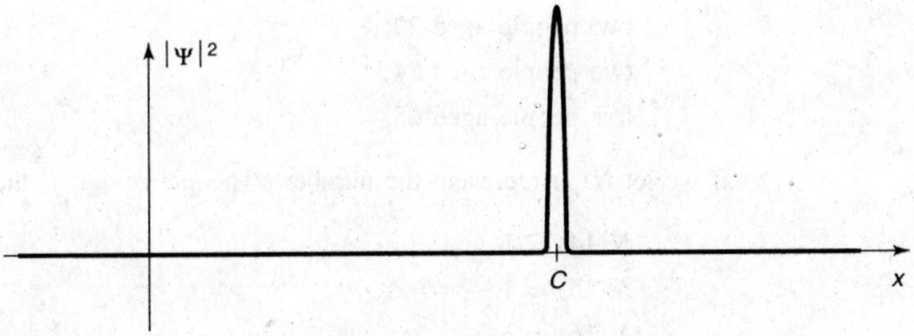

FIGURE 1.3: Collapse of the wave function: graph of $|\Psi|^2$ immediately *after* a measurement has found the particle at point C.

interpretation account for the fact that the second measurement is bound to yield the value C? Evidently the first measurement radically alters the wave function, so that it is now sharply peaked about C (Figure 1.3). We say that the wave function **collapses**, upon measurement, to a spike at the point C (it soon spreads out again, in accordance with the Schrödinger equation, so the second measurement must be made quickly). There are, then, two entirely distinct kinds of physical processes: "ordinary" ones, in which the wave function evolves in a leisurely fashion under the Schrödinger equation, and "measurements," in which Ψ suddenly and discontinuously collapses.[9]

1.3 PROBABILITY

1.3.1 Discrete Variables

Because of the statistical interpretation, probability plays a central role in quantum mechanics, so I digress now for a brief discussion of probability theory. It is mainly a question of introducing some notation and terminology, and I shall do it in the context of a simple example.

Imagine a room containing fourteen people, whose ages are as follows:

one person aged 14,

one person aged 15,

three people aged 16,

[9]The role of measurement in quantum mechanics is so critical and so bizarre that you may well be wondering what precisely *constitutes* a measurement. Does it have to do with the interaction between a microscopic (quantum) system and a macroscopic (classical) measuring apparatus (as Bohr insisted), or is it characterized by the leaving of a permanent "record" (as Heisenberg claimed), or does it involve the intervention of a conscious "observer" (as Wigner proposed)? I'll return to this thorny issue in the Afterword; for the moment let's take the naive view: A measurement is the kind of thing that a scientist does in the laboratory, with rulers, stopwatches, Geiger counters, and so on.

two people aged 22,
two people aged 24,
five people aged 25.

If we let $N(j)$ represent the number of people of age j, then

$$N(14) = 1,$$
$$N(15) = 1,$$
$$N(16) = 3,$$
$$N(22) = 2,$$
$$N(24) = 2,$$
$$N(25) = 5,$$

while $N(17)$, for instance, is zero. The *total* number of people in the room is

$$N = \sum_{j=0}^{\infty} N(j). \quad [1.4]$$

(In the example, of course, $N = 14$.) Figure 1.4 is a histogram of the data. The following are some questions one might ask about this distribution.

Question 1. If you selected one individual at random from this group, what is the **probability** that this person's age would be 15? *Answer:* One chance in 14, since there are 14 possible choices, all equally likely, of whom only one has that particular age. If $P(j)$ is the probability of getting age j, then $P(14) = 1/14$, $P(15) = 1/14$, $P(16) = 3/14$, and so on. In general,

$$P(j) = \frac{N(j)}{N}. \quad [1.5]$$

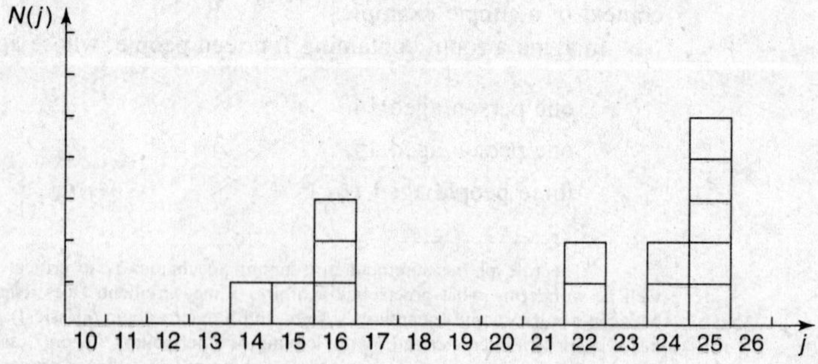

FIGURE 1.4: Histogram showing the number of people, $N(j)$, with age j, for the distribution in Section 1.3.1.

Notice that the probability of getting *either* 14 *or* 15 is the *sum* of the individual probabilities (in this case, 1/7). In particular, the sum of *all* the probabilities is 1—you're *certain* to get *some* age:

$$\sum_{j=0}^{\infty} P(j) = 1. \qquad [1.6]$$

Question 2. What is the **most probable** age? *Answer:* 25, obviously; five people share this age, whereas at most three have any other age. In general, the most probable j is the j for which $P(j)$ is a maximum.

Question 3. What is the **median** age? *Answer:* 23, for 7 people are younger than 23, and 7 are older. (In general, the median is that value of j such that the probability of getting a larger result is the same as the probability of getting a smaller result.)

Question 4. What is the **average** (or **mean**) age? *Answer:*

$$\frac{(14) + (15) + 3(16) + 2(22) + 2(24) + 5(25)}{14} = \frac{294}{14} = 21.$$

In general, the average value of j (which we shall write thus: $\langle j \rangle$) is

$$\langle j \rangle = \frac{\sum j N(j)}{N} = \sum_{j=0}^{\infty} j P(j). \qquad [1.7]$$

Notice that there need not be anyone with the average age or the median age—in this example nobody happens to be 21 or 23. In quantum mechanics the average is usually the quantity of interest; in that context it has come to be called the **expectation value**. It's a misleading term, since it suggests that this is the outcome you would be most likely to get if you made a single measurement (*that* would be the *most probable value*, not the average value)—but I'm afraid we're stuck with it.

Question 5. What is the average of the *squares* of the ages? *Answer:* You could get $14^2 = 196$, with probability 1/14, or $15^2 = 225$, with probability 1/14, or $16^2 = 256$, with probability 3/14, and so on. The average, then, is

$$\langle j^2 \rangle = \sum_{j=0}^{\infty} j^2 P(j). \qquad [1.8]$$

In general, the average value of some *function* of j is given by

$$\boxed{\langle f(j) \rangle = \sum_{j=0}^{\infty} f(j) P(j).} \qquad [1.9]$$

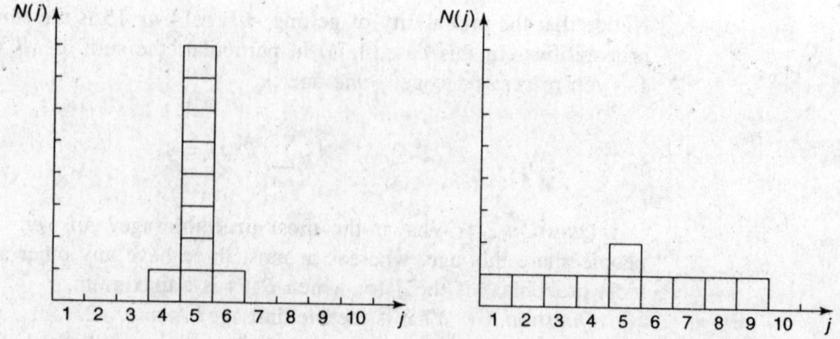

FIGURE 1.5: Two histograms with the same median, same average, and same most probable value, but different standard deviations.

(Equations 1.6, 1.7, and 1.8 are, if you like, special cases of this formula.) *Beware:* The average of the squares, $\langle j^2\rangle$, is *not* equal, in general, to the square of the average, $\langle j\rangle^2$. For instance, if the room contains just two babies, aged 1 and 3, then $\langle x^2\rangle = 5$, but $\langle x\rangle^2 = 4$.

Now, there is a conspicuous difference between the two histograms in Figure 1.5, even though they have the same median, the same average, the same most probable value, and the same number of elements: The first is sharply peaked about the average value, whereas the second is broad and flat. (The first might represent the age profile for students in a big-city classroom, the second, perhaps, a rural one-room schoolhouse.) We need a numerical measure of the amount of "spread" in a distribution, with respect to the average. The most obvious way to do this would be to find out how far each individual deviates from the average,

$$\Delta j = j - \langle j\rangle, \qquad [1.10]$$

and compute the average of Δj. Trouble is, of course, that you get *zero*, since, by the nature of the average, Δj is as often negative as positive:

$$\langle \Delta j\rangle = \sum (j - \langle j\rangle)P(j) = \sum jP(j) - \langle j\rangle \sum P(j)$$
$$= \langle j\rangle - \langle j\rangle = 0.$$

(Note that $\langle j\rangle$ is constant—it does not change as you go from one member of the sample to another—so it can be taken outside the summation.) To avoid this irritating problem you might decide to average the *absolute value* of Δj. But absolute values are nasty to work with; instead, we get around the sign problem by *squaring* before averaging:

$$\sigma^2 \equiv \langle (\Delta j)^2\rangle. \qquad [1.11]$$

This quantity is known as the **variance** of the distribution; σ itself (the square root of the average of the square of the deviation from the average—gulp!) is called the **standard deviation**. The latter is the customary measure of the spread about $\langle j \rangle$.

There is a useful little theorem on variances:

$$\sigma^2 = \langle (\Delta j)^2 \rangle = \sum (\Delta j)^2 P(j) = \sum (j - \langle j \rangle)^2 P(j)$$
$$= \sum (j^2 - 2j\langle j \rangle + \langle j \rangle^2) P(j)$$
$$= \sum j^2 P(j) - 2\langle j \rangle \sum j P(j) + \langle j \rangle^2 \sum P(j)$$
$$= \langle j^2 \rangle - 2\langle j \rangle \langle j \rangle + \langle j \rangle^2 = \langle j^2 \rangle - \langle j \rangle^2.$$

Taking the square root, the standard deviation itself can be written as

$$\sigma = \sqrt{\langle j^2 \rangle - \langle j \rangle^2}. \tag{1.12}$$

In practice, this is a much faster way to get σ: Simply calculate $\langle j^2 \rangle$ and $\langle j \rangle^2$, subtract, and take the square root. Incidentally, I warned you a moment ago that $\langle j^2 \rangle$ is not, in general, equal to $\langle j \rangle^2$. Since σ^2 is plainly nonnegative (from its definition in Equation 1.11), Equation 1.12 implies that

$$\langle j^2 \rangle \geq \langle j \rangle^2, \tag{1.13}$$

and the two are equal only when $\sigma = 0$, which is to say, for distributions with no spread at all (every member having the same value).

1.3.2 Continuous Variables

So far, I have assumed that we are dealing with a *discrete* variable—that is, one that can take on only certain isolated values (in the example, j had to be an integer, since I gave ages only in years). But it is simple enough to generalize to *continuous* distributions. If I select a random person off the street, the probability that her age is *precisely* 16 years, 4 hours, 27 minutes, and 3.333 ... seconds is *zero*. The only sensible thing to speak about is the probability that her age lies in some *interval*—say, between 16 and 17. If the interval is sufficiently short, this probability is *proportional to the length of the interval*. For example, the chance that her age is between 16 and 16 plus *two* days is presumably twice the probability that it is between 16 and 16 plus *one* day. (Unless, I suppose, there was some extraordinary baby boom 16 years ago, on exactly that day—in which case we have simply chosen an interval too long for the rule to apply. If the baby boom lasted six hours, we'll take intervals of a second or less, to be on the safe side. Technically, we're talking about *infinitesimal* intervals.) Thus

$$\left\{\begin{array}{l}\text{probability that an individual (chosen}\\ \text{at random) lies between } x \text{ and } (x+dx)\end{array}\right\} = \rho(x)\,dx. \qquad [1.14]$$

The proportionality factor, $\rho(x)$, is often loosely called "the probability of getting x," but this is sloppy language; a better term is **probability density**. The probability that x lies between a and b (a *finite* interval) is given by the integral of $\rho(x)$:

$$P_{ab} = \int_a^b \rho(x)\,dx, \qquad [1.15]$$

and the rules we deduced for discrete distributions translate in the obvious way:

$$1 = \int_{-\infty}^{+\infty} \rho(x)\,dx, \qquad [1.16]$$

$$\langle x \rangle = \int_{-\infty}^{+\infty} x\rho(x)\,dx, \qquad [1.17]$$

$$\langle f(x) \rangle = \int_{-\infty}^{+\infty} f(x)\rho(x)\,dx, \qquad [1.18]$$

$$\sigma^2 \equiv \langle (\Delta x)^2 \rangle = \langle x^2 \rangle - \langle x \rangle^2. \qquad [1.19]$$

Example 1.1 Suppose I drop a rock off a cliff of height h. As it falls, I snap a million photographs, at random intervals. On each picture I measure the distance the rock has fallen. *Question:* What is the *average* of all these distances? That is to say, what is the *time average* of the distance traveled?[10]

Solution: The rock starts out at rest, and picks up speed as it falls; it spends more time near the top, so the average distance must be less than $h/2$. Ignoring air resistance, the distance x at time t is

$$x(t) = \frac{1}{2}gt^2.$$

The velocity is $dx/dt = gt$, and the total flight time is $T = \sqrt{2h/g}$. The probability that the camera flashes in the interval dt is dt/T, so the probability that a given

[10] A statistician will complain that I am confusing the average of a *finite sample* (a million, in this case) with the "true" average (over the whole continuum). This can be an awkward problem for the experimentalist, especially when the sample size is small, but here I am only concerned, of course, with the true average, to which the sample average is presumably a good approximation.

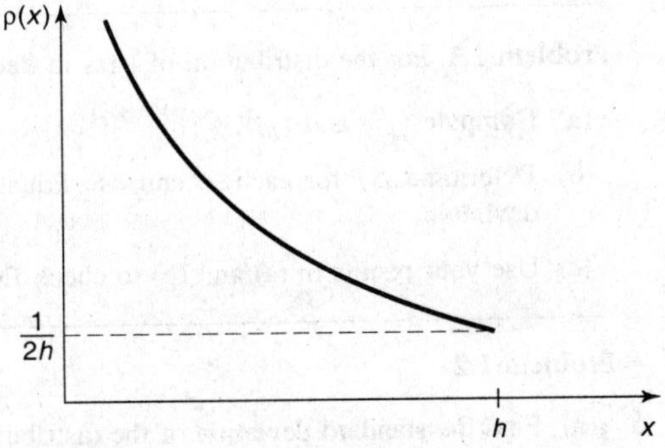

FIGURE 1.6: **The probability density in Example 1.1: $\rho(x) = 1/(2\sqrt{hx})$.**

photograph shows a distance in the corresponding range dx is

$$\frac{dt}{T} = \frac{dx}{gt}\sqrt{\frac{g}{2h}} = \frac{1}{2\sqrt{hx}}\,dx.$$

Evidently the probability *density* (Equation 1.14) is

$$\rho(x) = \frac{1}{2\sqrt{hx}}, \qquad (0 \leq x \leq h)$$

(outside this range, of course, the probability density is zero).

We can check this result, using Equation 1.16:

$$\int_0^h \frac{1}{2\sqrt{hx}}\,dx = \frac{1}{2\sqrt{h}}\left(2x^{1/2}\right)\Big|_0^h = 1.$$

The *average* distance (Equation 1.17) is

$$\langle x\rangle = \int_0^h x\frac{1}{2\sqrt{hx}}\,dx = \frac{1}{2\sqrt{h}}\left(\frac{2}{3}x^{3/2}\right)\Big|_0^h = \frac{h}{3},$$

which is somewhat less than $h/2$, as anticipated.

Figure 1.6 shows the graph of $\rho(x)$. Notice that a probability *density* can be infinite, though probability itself (the *integral* of ρ) must of course be finite (indeed, less than or equal to 1).

∗Problem 1.1 For the distribution of ages in Section 1.3.1:

(a) Compute $\langle j^2 \rangle$ and $\langle j \rangle^2$.

(b) Determine Δj for each j, and use Equation 1.11 to compute the standard deviation.

(c) Use your results in (a) and (b) to check Equation 1.12.

Problem 1.2

(a) Find the standard deviation of the distribution in Example 1.1.

(b) What is the probability that a photograph, selected at random, would show a distance x more than one standard deviation away from the average?

∗Problem 1.3 Consider the **gaussian** distribution

$$\rho(x) = Ae^{-\lambda(x-a)^2},$$

where A, a, and λ are positive real constants. (Look up any integrals you need.)

(a) Use Equation 1.16 to determine A.

(b) Find $\langle x \rangle$, $\langle x^2 \rangle$, and σ.

(c) Sketch the graph of $\rho(x)$.

1.4 NORMALIZATION

We return now to the statistical interpretation of the wave function (Equation 1.3), which says that $|\Psi(x, t)|^2$ is the probability density for finding the particle at point x, at time t. It follows (Equation 1.16) that the integral of $|\Psi|^2$ must be 1 (the particle's got to be *some*where):

$$\boxed{\int_{-\infty}^{+\infty} |\Psi(x, t)|^2 \, dx = 1.} \qquad [1.20]$$

Without this, the statistical interpretation would be nonsense.

However, this requirement should disturb you: After all, the wave function is supposed to be determined by the Schrödinger equation—we can't go imposing an extraneous condition on Ψ without checking that the two are consistent. Well, a glance at Equation 1.1 reveals that if $\Psi(x, t)$ is a solution, so too is $A\Psi(x, t)$, where A is any (complex) constant. What we must do, then, is pick this undetermined multiplicative factor so as to ensure that Equation 1.20 is satisfied. This process is called **normalizing** the wave function. For some solutions to the Schrödinger equation the integral is *infinite*; in that case *no* multiplicative factor is going to make it 1. The same goes for the trivial solution $\Psi = 0$. Such **non-normalizable** solutions cannot represent particles, and must be rejected. Physically realizable states correspond to the **square-integrable** solutions to Schrödinger's equation.[11]

But wait a minute! Suppose I have normalized the wave function at time $t = 0$. How do I know that it will *stay* normalized, as time goes on, and Ψ evolves? (You can't keep *re*normalizing the wave function, for then A becomes a function of t, and you no longer have a solution to the Schrödinger equation.) Fortunately, the Schrödinger equation has the remarkable property that it automatically preserves the normalization of the wave function—without this crucial feature the Schrödinger equation would be incompatible with the statistical interpretation, and the whole theory would crumble.

This is important, so we'd better pause for a careful proof. To begin with,

$$\frac{d}{dt}\int_{-\infty}^{+\infty} |\Psi(x,t)|^2\, dx = \int_{-\infty}^{+\infty} \frac{\partial}{\partial t}|\Psi(x,t)|^2\, dx. \qquad [1.21]$$

(Note that the *integral* is a function only of t, so I use a *total* derivative (d/dt) in the first expression, but the *integrand* is a function of x as well as t, so it's a *partial* derivative ($\partial/\partial t$) in the second one.) By the product rule,

$$\frac{\partial}{\partial t}|\Psi|^2 = \frac{\partial}{\partial t}(\Psi^*\Psi) = \Psi^*\frac{\partial \Psi}{\partial t} + \frac{\partial \Psi^*}{\partial t}\Psi. \qquad [1.22]$$

Now the Schrödinger equation says that

$$\frac{\partial \Psi}{\partial t} = \frac{i\hbar}{2m}\frac{\partial^2 \Psi}{\partial x^2} - \frac{i}{\hbar}V\Psi, \qquad [1.23]$$

and hence also (taking the complex conjugate of Equation 1.23)

$$\frac{\partial \Psi^*}{\partial t} = -\frac{i\hbar}{2m}\frac{\partial^2 \Psi^*}{\partial x^2} + \frac{i}{\hbar}V\Psi^*, \qquad [1.24]$$

so

$$\frac{\partial}{\partial t}|\Psi|^2 = \frac{i\hbar}{2m}\left(\Psi^*\frac{\partial^2 \Psi}{\partial x^2} - \frac{\partial^2 \Psi^*}{\partial x^2}\Psi\right) = \frac{\partial}{\partial x}\left[\frac{i\hbar}{2m}\left(\Psi^*\frac{\partial \Psi}{\partial x} - \frac{\partial \Psi^*}{\partial x}\Psi\right)\right] \qquad [1.25]$$

---

[11] Evidently $\Psi(x, t)$ must go to zero faster than $1/\sqrt{|x|}$, as $|x| \to \infty$. Incidentally, normalization only fixes the *modulus* of A; the *phase* remains undetermined. However, as we shall see, the latter carries no physical significance anyway.

The integral in Equation 1.21 can now be evaluated explicitly:

$$\frac{d}{dt}\int_{-\infty}^{+\infty}|\Psi(x,t)|^2\,dx = \frac{i\hbar}{2m}\left(\Psi^*\frac{\partial\Psi}{\partial x} - \frac{\partial\Psi^*}{\partial x}\Psi\right)\Bigg|_{-\infty}^{+\infty}. \quad [1.26]$$

But $\Psi(x,t)$ must go to zero as x goes to $(\pm)$ infinity—otherwise the wave function would not be normalizable.[12] It follows that

$$\frac{d}{dt}\int_{-\infty}^{+\infty}|\Psi(x,t)|^2\,dx = 0, \quad [1.27]$$

and hence that the integral is *constant* (independent of time); if Ψ is normalized at $t = 0$, it *stays* normalized for all future time. QED

Problem 1.4 At time $t = 0$ a particle is represented by the wave function

$$\Psi(x,0) = \begin{cases} A\dfrac{x}{a}, & \text{if } 0 \le x \le a, \\ A\dfrac{(b-x)}{(b-a)}, & \text{if } a \le x \le b, \\ 0, & \text{otherwise,} \end{cases}$$

where A, a, and b are constants.

(a) Normalize Ψ (that is, find A, in terms of a and b).

(b) Sketch $\Psi(x, 0)$, as a function of x.

(c) Where is the particle most likely to be found, at $t = 0$?

(d) What is the probability of finding the particle to the left of a? Check your result in the limiting cases $b = a$ and $b = 2a$.

(e) What is the expectation value of x?

***Problem 1.5** Consider the wave function

$$\Psi(x,t) = Ae^{-\lambda|x|}e^{-i\omega t},$$

where A, λ, and ω are positive real constants. (We'll see in Chapter 2 what potential (V) actually produces such a wave function.)

(a) Normalize Ψ.

(b) Determine the expectation values of x and x^2

---

[12] A good mathematician can supply you with pathological counterexamples, but they do not arise in physics; for us the wave function *always* goes to zero at infinity.

(c) Find the standard deviation of x. Sketch the graph of $|\Psi|^2$, as a function of x, and mark the points $(\langle x\rangle + \sigma)$ and $(\langle x\rangle - \sigma)$, to illustrate the sense in which σ represents the "spread" in x. What is the probability that the particle would be found outside this range?

1.5 MOMENTUM

For a particle in state Ψ, the expectation value of x is

$$\boxed{\langle x\rangle = \int_{-\infty}^{+\infty} x|\Psi(x,t)|^2\,dx.} \tag{1.28}$$

What exactly does this mean? It emphatically does *not* mean that if you measure the position of one particle over and over again, $\int x|\Psi|^2 dx$ is the average of the results you'll get. On the contrary: The first measurement (whose outcome is indeterminate) will collapse the wave function to a spike at the value actually obtained, and the subsequent measurements (if they're performed quickly) will simply repeat that same result. Rather, $\langle x\rangle$ is the average of measurements performed on particles *all in the state* Ψ, which means that either you must find some way of returning the particle to its original state after each measurement, or else you have to prepare a whole **ensemble** of particles, each in the same state Ψ, and measure the positions of all of them: $\langle x\rangle$ is the average of *these* results. (I like to picture a row of bottles on a shelf, each containing a particle in the state Ψ (relative to the center of the bottle). A graduate student with a ruler is assigned to each bottle, and at a signal they all measure the positions of their respective particles. We then construct a histogram of the results, which should match $|\Psi|^2$, and compute the average, which should agree with $\langle x\rangle$. (Of course, since we're only using a finite sample, we can't expect perfect agreement, but the more bottles we use, the closer we ought to come.)) In short, *the expectation value is the average of repeated measurements on an ensemble of identically prepared systems*, not the average of repeated measurements on one and the same system.

Now, as time goes on, $\langle x\rangle$ will change (because of the time dependence of Ψ), and we might be interested in knowing how fast it moves. Referring to Equations 1.25 and 1.28, we see that[13]

$$\frac{d\langle x\rangle}{dt} = \int x\frac{\partial}{\partial t}|\Psi|^2\,dx = \frac{i\hbar}{2m}\int x\frac{\partial}{\partial x}\left(\Psi^*\frac{\partial\Psi}{\partial x} - \frac{\partial\Psi^*}{\partial x}\Psi\right)dx. \tag{1.29}$$

[13]To keep things from getting too cluttered, I'll suppress the limits of integration.

This expression can be simplified using integration-by-parts:[14]

$$\frac{d\langle x\rangle}{dt} = -\frac{i\hbar}{2m}\int\left(\Psi^*\frac{\partial\Psi}{\partial x} - \frac{\partial\Psi^*}{\partial x}\Psi\right)dx. \qquad [1.30]$$

(I used the fact that $\partial x/\partial x = 1$, and threw away the boundary term, on the ground that Ψ goes to zero at ($\pm$) infinity.) Performing another integration by parts, on the second term, we conclude:

$$\frac{d\langle x\rangle}{dt} = -\frac{i\hbar}{m}\int \Psi^*\frac{\partial\Psi}{\partial x}dx. \qquad [1.31]$$

What are we to make of this result? Note that we're talking about the "velocity" of the *expectation* value of x, which is not the same thing as the velocity of the *particle*. Nothing we have seen so far would enable us to calculate the velocity of a particle. It's not even clear what velocity *means* in quantum mechanics: If the particle doesn't have a determinate position (prior to measurement), neither does it have a well-defined velocity. All we could reasonably ask for is the *probability* of getting a particular value. We'll see in Chapter 3 how to construct the probability density for velocity, given Ψ; for our present purposes it will suffice to postulate that the *expectation value of the velocity is equal to the time derivative of the expectation value of position*:

$$\langle v\rangle = \frac{d\langle x\rangle}{dt}. \qquad [1.32]$$

Equation 1.31 tells us, then, how to calculate $\langle v\rangle$ directly from Ψ.

Actually, it is customary to work with **momentum** ($p = mv$), rather than velocity:

$$\boxed{\langle p\rangle = m\frac{d\langle x\rangle}{dt} = -i\hbar\int\left(\Psi^*\frac{\partial\Psi}{\partial x}\right)dx.} \qquad [1.33]$$

---

[14]The product rule says that

$$\frac{d}{dx}(fg) = f\frac{dg}{dx} + \frac{df}{dx}g,$$

from which it follows that

$$\int_a^b f\frac{dg}{dx}dx = -\int_a^b \frac{df}{dx}g\,dx + fg\Big|_a^b.$$

Under the integral sign, then, you can peel a derivative off one factor in a product, and slap it onto the other one—it'll cost you a minus sign, and you'll pick up a boundary term.

Let me write the expressions for $\langle x\rangle$ and $\langle p\rangle$ in a more suggestive way:

$$\langle x\rangle = \int \Psi^*(x)\Psi\, dx, \qquad [1.34]$$

$$\langle p\rangle = \int \Psi^*\left(\frac{\hbar}{i}\frac{\partial}{\partial x}\right)\Psi\, dx. \qquad [1.35]$$

We say that the **operator**[15] x "represents" position, and the operator $(\hbar/i)(\partial/\partial x)$ "represents" momentum, in quantum mechanics; to calculate expectation values we "sandwich" the appropriate operator between Ψ^* and Ψ, and integrate.

That's cute, but what about other quantities? The fact is, *all* classical dynamical variables can be expressed in terms of position and momentum. Kinetic energy, for example, is

$$T = \frac{1}{2}mv^2 = \frac{p^2}{2m},$$

and angular momentum is

$$\mathbf{L} = \mathbf{r}\times m\mathbf{v} = \mathbf{r}\times\mathbf{p}$$

(the latter, of course, does not occur for motion in one dimension). To calculate the expectation value of *any* such quantity, $Q(x, p)$, we simply replace every p by $(\hbar/i)(\partial/\partial x)$, insert the resulting operator between Ψ^* and Ψ, and integrate:

$$\boxed{\langle Q(x,p)\rangle = \int \Psi^* Q\left(x, \frac{\hbar}{i}\frac{\partial}{\partial x}\right)\Psi\, dx.} \qquad [1.36]$$

For example, the expectation value of the kinetic energy is

$$\langle T\rangle = -\frac{\hbar^2}{2m}\int \Psi^*\frac{\partial^2\Psi}{\partial x^2}\, dx. \qquad [1.37]$$

Equation 1.36 is a recipe for computing the expectation value of any dynamical quantity, for a particle in state Ψ; it subsumes Equations 1.34 and 1.35 as special cases. I have tried in this section to make Equation 1.36 seem plausible, given Born's statistical interpretation, but the truth is that this represents such a radically new way of doing business (as compared with classical mechanics) that it's a good idea to get some practice *using* it before we come back (in Chapter 3) and put it on a firmer theoretical foundation. In the meantime, if you prefer to think of it as an *axiom*, that's fine with me.

---

[15] An "operator" is an instruction to *do something* to the function that follows it. The position operator tells you to *multiply* by x; the momentum operator tells you to *differentiate* with respect to x (and multiply the result by $-i\hbar$). In this book *all* operators will be derivatives (d/dt, d^2/dt^2, $\partial^2/\partial x\partial y$, etc.) or multipliers ($2$, i, x^2, etc.), or combinations of these.

Problem 1.6 Why can't you do integration-by-parts directly on the middle expression in Equation 1.29—pull the time derivative over onto x, note that $\partial x/\partial t = 0$, and conclude that $d\langle x\rangle/dt = 0$?

∗Problem 1.7 Calculate $d\langle p\rangle/dt$. *Answer*:

$$\frac{d\langle p\rangle}{dt} = \left\langle -\frac{\partial V}{\partial x}\right\rangle \qquad [1.38]$$

Equations 1.32 (or the first part of 1.33) and 1.38 are instances of **Ehrenfest's theorem**, which tells us that *expectation values obey classical laws.*

Problem 1.8 Suppose you add a constant V_0 to the potential energy (by "constant" I mean independent of x as well as t). In *classical* mechanics this doesn't change anything, but what about *quantum* mechanics? Show that the wave function picks up a time-dependent phase factor: $\exp(-iV_0t/\hbar)$. What effect does this have on the expectation value of a dynamical variable?

1.6 THE UNCERTAINTY PRINCIPLE

Imagine that you're holding one end of a very long rope, and you generate a wave by shaking it up and down rhythmically (Figure 1.7). If someone asked you "Precisely where *is* that wave?" you'd probably think he was a little bit nutty: The wave isn't precisely *any*where—it's spread out over 50 feet or so. On the other hand, if he asked you what its *wavelength* is, you could give him a reasonable answer: It looks like about 6 feet. By contrast, if you gave the rope a sudden jerk (Figure 1.8), you'd get a relatively narrow bump traveling down the line. This time the first question (Where precisely is the wave?) is a sensible one, and the second (What is its wavelength?) seems nutty—it isn't even vaguely periodic, so how can you assign a wavelength to it? Of course, you can draw intermediate cases, in which the wave is *fairly* well localized and the wavelength is *fairly* well defined, but there is an inescapable trade-off here: The more precise a wave's position is, the less precise is its wavelength, and vice versa.[16] A theorem in Fourier analysis makes all this rigorous, but for the moment I am only concerned with the qualitative argument.

[16] That's why a piccolo player must be right on pitch, whereas a double-bass player can afford to wear garden gloves. For the piccolo, a sixty-fourth note contains many full cycles, and the frequency (we're working in the time domain now, instead of space) is well defined, whereas for the bass, at a much lower register, the sixty-fourth note contains only a few cycles, and all you hear is a general sort of "oomph," with no very clear pitch.

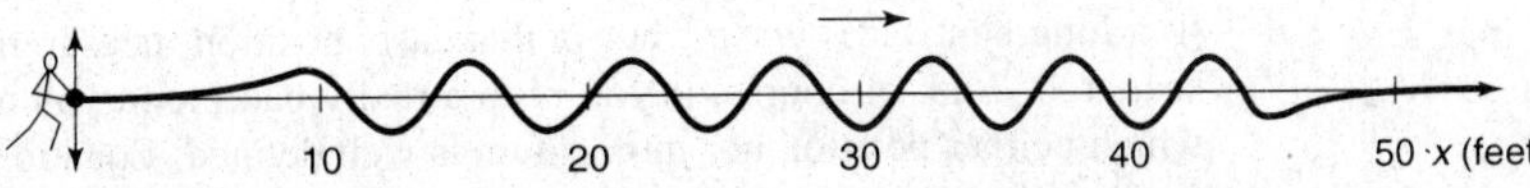

FIGURE 1.7: A wave with a (fairly) well-defined *wavelength*, but an ill-defined *position*.

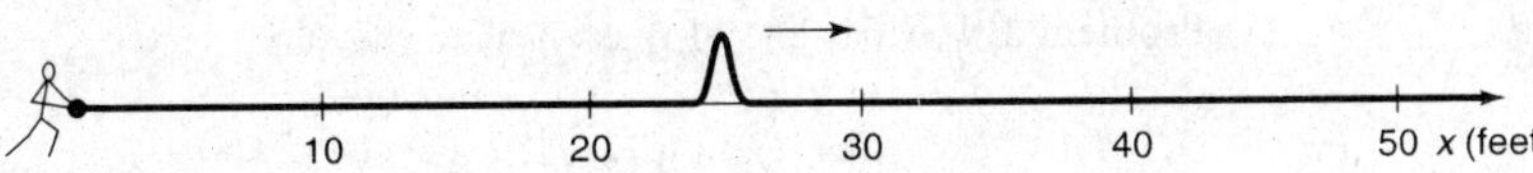

FIGURE 1.8: A wave with a (fairly) well-defined *position*, but an ill-defined *wavelength*.

This applies, of course, to *any* wave phenomenon, and hence in particular to the quantum mechanical wave function. Now the wavelength of Ψ is related to the *momentum* of the particle by the **de Broglie formula**:[17]

$$p = \frac{h}{\lambda} = \frac{2\pi\hbar}{\lambda}. \qquad [1.39]$$

Thus a spread in *wavelength* corresponds to a spread in *momentum*, and our general observation now says that the more precisely determined a particle's position is, the less precisely is its momentum. Quantitatively,

$$\boxed{\sigma_x \sigma_p \geq \frac{\hbar}{2},} \qquad [1.40]$$

where σ_x is the standard deviation in x, and σ_p is the standard deviation in p. This is Heisenberg's famous **uncertainty principle**. (We'll prove it in Chapter 3, but I wanted to mention it right away, so you can test it out on the examples in Chapter 2.)

Please understand what the uncertainty principle *means*: Like position measurements, momentum measurements yield precise answers—the "spread" here refers to the fact that measurements on identically prepared systems do not yield identical results. You can, if you want, construct a state such that repeated position measurements will be very close together (by making Ψ a localized "spike"), but you will pay a price: Momentum measurements on this state will be widely scattered. Or you can prepare a state with a reproducible momentum (by making

[17] I'll prove this in due course. Many authors take the de Broglie formula as an *axiom*, from which they then deduce the association of momentum with the operator $(\hbar/i)(\partial/\partial x)$. Although this is a conceptually cleaner approach, it involves diverting mathematical complications that I would rather save for later.

Ψ a long sinusoidal wave), but in that case, position measurements will be widely scattered. And, of course, if you're in a really bad mood you can create a state for which neither position nor momentum is well defined: Equation 1.40 is an *inequality*, and there's no limit on how *big* σ_x and σ_p can be—just make Ψ some long wiggly line with lots of bumps and potholes and no periodic structure.

∗Problem 1.9 A particle of mass m is in the state

$$\Psi(x, t) = Ae^{-a[(mx^2/\hbar)+it]},$$

where A and a are positive real constants.

(a) Find A.

(b) For what potential energy function $V(x)$ does Ψ satisfy the Schrödinger equation?

(c) Calculate the expectation values of x, x^2, p, and p^2.

(d) Find σ_x and σ_p. Is their product consistent with the uncertainty principle?

FURTHER PROBLEMS FOR CHAPTER 1

Problem 1.10 Consider the first 25 digits in the decimal expansion of π (3, 1, 4, 1, 5, 9, ...).

(a) If you selected one number at random, from this set, what are the probabilities of getting each of the 10 digits?

(b) What is the most probable digit? What is the median digit? What is the average value?

(c) Find the standard deviation for this distribution.

Problem 1.11 The needle on a broken car speedometer is free to swing, and bounces perfectly off the pins at either end, so that if you give it a flick it is equally likely to come to rest at any angle between 0 and π.

(a) What is the probability density, $\rho(\theta)$? *Hint:* $\rho(\theta)\,d\theta$ is the probability that the needle will come to rest between θ and $(\theta + d\theta)$. Graph $\rho(\theta)$ as a function of θ, from $-\pi/2$ to $3\pi/2$. (Of course, *part* of this interval is excluded, so ρ is zero there.) Make sure that the total probability is 1.

(b) Compute $\langle\theta\rangle$, $\langle\theta^2\rangle$, and σ, for this distribution.

(c) Compute $\langle\sin\theta\rangle$, $\langle\cos\theta\rangle$, and $\langle\cos^2\theta\rangle$.

Problem 1.12 We consider the same device as the previous problem, but this time we are interested in the x-coordinate of the needle point—that is, the "shadow," or "projection," of the needle on the horizontal line.

(a) What is the probability density $\rho(x)$? Graph $\rho(x)$ as a function of x, from $-2r$ to $+2r$, where r is the length of the needle. Make sure the total probability is 1. *Hint:* $\rho(x)\,dx$ is the probability that the projection lies between x and $(x + dx)$. You know (from Problem 1.11) the probability that θ is in a given range; the question is, what interval dx corresponds to the interval $d\theta$?

(b) Compute $\langle x\rangle$, $\langle x^2\rangle$, and σ, for this distribution. Explain how you could have obtained these results from part (c) of Problem 1.11.

Problem 1.13 **Buffon's needle. A needle of length l is dropped at random onto a sheet of paper ruled with parallel lines a distance l apart. What is the probability that the needle will cross a line? *Hint:* Refer to Problem 1.12.

Problem 1.14 Let $P_{ab}(t)$ be the probability of finding a particle in the range $(a < x < b)$, at time t.

(a) Show that

$$\frac{dP_{ab}}{dt} = J(a,t) - J(b,t),$$

where

$$J(x,t) \equiv \frac{i\hbar}{2m}\left(\Psi\frac{\partial\Psi^*}{\partial x} - \Psi^*\frac{\partial\Psi}{\partial x}\right).$$

What are the units of $J(x,t)$? *Comment:* J is called the **probability current**, because it tells you the rate at which probability is "flowing" past the point x. If $P_{ab}(t)$ is increasing, then more probability is flowing into the region at one end than flows out at the other.

(b) Find the probability current for the wave function in Problem 1.9. (This is not a very pithy example, I'm afraid; we'll encounter more substantial ones in due course.)

****Problem 1.15** Suppose you wanted to describe an **unstable particle**, that spontaneously disintegrates with a "lifetime" τ. In that case the total probability of finding the particle somewhere should *not* be constant, but should decrease at (say) an exponential rate:

$$P(t) \equiv \int_{-\infty}^{+\infty} |\Psi(x,t)|^2\,dx = e^{-t/\tau}.$$

A crude way of achieving this result is as follows. In Equation 1.24 we tacitly assumed that V (the potential energy) is *real*. That is certainly reasonable, but it leads to the "conservation of probability" enshrined in Equation 1.27. What if we assign to V an imaginary part:

$$V = V_0 - i\Gamma,$$

where V_0 is the true potential energy and Γ is a positive real constant?

(a) Show that (in place of Equation 1.27) we now get

$$\frac{dP}{dt} = -\frac{2\Gamma}{\hbar}P.$$

(b) Solve for $P(t)$, and find the lifetime of the particle in terms of Γ.

Problem 1.16 Show that

$$\frac{d}{dt}\int_{-\infty}^{\infty} \Psi_1^*\,\Psi_2\,dx = 0$$

for any two (normalizable) solutions to the Schrödinger equation, Ψ_1 and Ψ_2.

Problem 1.17 A particle is represented (at time $t = 0$) by the wave function

$$\Psi(x,0) = \begin{cases} A(a^2 - x^2), & \text{if } -a \le x \le +a, \\ 0, & \text{otherwise.} \end{cases}$$

(a) Determine the normalization constant A.

(b) What is the expectation value of x (at time $t = 0$)?

(c) What is the expectation value of p (at time $t = 0$)? (Note that you *cannot* get it from $p = md\langle x\rangle/dt$. Why not?)

(d) Find the expectation value of x^2.

(e) Find the expectation value of p^2.

(f) Find the uncertainty in x (σ_x).

(g) Find the uncertainty in p (σ_p).

(h) Check that your results are consistent with the uncertainty principle.

Problem 1.18 In general, quantum mechanics is relevant when the de Broglie wavelength of the particle in question (h/p) is greater than the characteristic size of the system (d). In thermal equilibrium at (Kelvin) temperature T, the average kinetic energy of a particle is

$$\frac{p^2}{2m} = \frac{3}{2} k_B T$$

(where k_B is Boltzmann's constant), so the typical de Broglie wavelength is

$$\lambda = \frac{h}{\sqrt{3mk_B T}}. \quad [1.41]$$

The purpose of this problem is to anticipate which systems will have to be treated quantum mechanically, and which can safely be described classically.

(a) **Solids.** The lattice spacing in a typical solid is around $d = 0.3$ nm. Find the temperature below which the free[18] *electrons* in a solid are quantum mechanical. Below what temperature are the *nuclei* in a solid quantum mechanical? (Use sodium as a typical case.) *Moral:* The free electrons in a solid are *always* quantum mechanical; the nuclei are almost *never* quantum mechanical. The same goes for liquids (for which the interatomic spacing is roughly the same), with the exception of helium below 4 K.

(b) **Gases.** For what temperatures are the atoms in an ideal gas at pressure P quantum mechanical? *Hint:* Use the ideal gas law ($PV = Nk_B T$) to deduce the interatomic spacing. *Answer:* $T < (1/k_B)(h^2/3m)^{3/5} P^{2/5}$. Obviously (for the gas to show quantum behavior) we want m to be as *small* as possible, and P as *large* as possible. Put in the numbers for helium at atmospheric pressure. Is hydrogen in outer space (where the interatomic spacing is about 1 cm and the temperature is 3 K) quantum mechanical?

[18] In a solid the inner electrons are attached to a particular nucleus, and for them the relevant size would be the radius of the atom. But the outermost electrons are not attached, and for them the relevant distance is the lattice spacing. This problem pertains to the *outer* electrons.

CHAPTER 2

TIME-INDEPENDENT SCHRÖDINGER EQUATION

2.1 STATIONARY STATES

In Chapter 1 we talked a lot about the wave function, and how you use it to calculate various quantities of interest. The time has come to stop procrastinating, and confront what is, logically, the prior question: How do you *get* $\Psi(x, t)$ in the *first* place? We need to solve the Schrödinger equation,

$$i\hbar\frac{\partial\Psi}{\partial t} = -\frac{\hbar^2}{2m}\frac{\partial^2\Psi}{\partial x^2} + V\Psi, \qquad [2.1]$$

for a specified potential[1] $V(x, t)$. In this chapter (and most of this book) I shall assume that V is *independent of* t. In that case the Schrödinger equation can be solved by the method of **separation of variables** (the physicist's first line of attack on any partial differential equation): We look for solutions that are simple *products*,

$$\Psi(x, t) = \psi(x)\,\varphi(t), \qquad [2.2]$$

where ψ (*lower*-case) is a function of x alone, and φ is a function of t alone. On its face, this is an absurd restriction, and we cannot hope to get more than a tiny

[1] It is tiresome to keep saying "potential energy function," so most people just call V the "potential," even though this invites occasional confusion with *electric* potential, which is actually potential energy *per unit charge*.

subset of all solutions in this way. But hang on, because the solutions we *do* obtain turn out to be of great interest. Moreover (as is typically the case with separation of variables) we will be able at the end to patch together the separable solutions in such a way as to *construct* the most general solution.

For separable solutions we have

$$\frac{\partial \Psi}{\partial t} = \psi \frac{d\varphi}{dt}, \quad \frac{\partial^2 \Psi}{\partial x^2} = \frac{d^2\psi}{dx^2}\varphi$$

(*ordinary* derivatives, now), and the Schrödinger equation reads

$$i\hbar\psi \frac{d\varphi}{dt} = -\frac{\hbar^2}{2m}\frac{d^2\psi}{dx^2}\varphi + V\psi\varphi.$$

Or, dividing through by $\psi\varphi$:

$$i\hbar\frac{1}{\varphi}\frac{d\varphi}{dt} = -\frac{\hbar^2}{2m}\frac{1}{\psi}\frac{d^2\psi}{dx^2} + V. \qquad [2.3]$$

Now, the left side is a function of t alone, and the right side is a function of x alone.[2] The only way this can possibly be true is if both sides are in fact *constant*—otherwise, by varying t, I could change the left side without touching the right side, and the two would no longer be equal. (That's a subtle but crucial argument, so if it's new to you, be sure to pause and think it through.) For reasons that will appear in a moment, we shall call the separation constant E. Then

$$i\hbar\frac{1}{\varphi}\frac{d\varphi}{dt} = E,$$

or

$$\frac{d\varphi}{dt} = -\frac{iE}{\hbar}\varphi, \qquad [2.4]$$

and

$$-\frac{\hbar^2}{2m}\frac{1}{\psi}\frac{d^2\psi}{dx^2} + V = E,$$

or

$$\boxed{-\frac{\hbar^2}{2m}\frac{d^2\psi}{dx^2} + V\psi = E\psi.} \qquad [2.5]$$

Separation of variables has turned a *partial* differential equation into two *ordinary* differential equations (Equations 2.4 and 2.5). The first of these (Equation 2.4)

[2]Note that this would *not* be true if V were a function of t as well as x.

is easy to solve (just multiply through by dt and integrate); the general solution is $C\exp(-iEt/\hbar)$, but we might as well absorb the constant C into ψ (since the quantity of interest is the product $\psi\varphi$). Then

$$\varphi(t) = e^{-iEt/\hbar}. \quad [2.6]$$

The second (Equation 2.5) is called the **time-independent Schrödinger equation**; we can go no further with it until the potential $V(x)$ is specified.

The rest of this chapter will be devoted to solving the time-independent Schrödinger equation, for a variety of simple potentials. But before I get to that you have every right to ask: *What's so great about separable solutions?* After all, *most* solutions to the (time *de*pendent) Schrödinger equation do *not* take the form $\psi(x)\varphi(t)$. I offer three answers—two of them physical, and one mathematical:

1. They are **stationary states**. Although the wave function itself,

$$\Psi(x, t) = \psi(x)e^{-iEt/\hbar}, \quad [2.7]$$

does (obviously) depend on t, the *probability density*,

$$|\Psi(x, t)|^2 = \Psi^*\Psi = \psi^* e^{+iEt/\hbar}\psi e^{-iEt/\hbar} = |\psi(x)|^2, \quad [2.8]$$

does *not*—the time-dependence cancels out.[3] The same thing happens in calculating the expectation value of any dynamical variable; Equation 1.36 reduces to

$$\langle Q(x, p)\rangle = \int \psi^* Q\left(x, \frac{\hbar}{i}\frac{d}{dx}\right)\psi\, dx. \quad [2.9]$$

Every expectation value is constant in time; we might as well drop the factor $\varphi(t)$ altogether, and simply use ψ in place of Ψ. (Indeed, it is common to refer to ψ as "the wave function," but this is sloppy language that can be dangerous, and it is important to remember that the *true* wave function always carries that exponential time-dependent factor.) In particular, $\langle x\rangle$ is constant, and hence (Equation 1.33) $\langle p\rangle = 0$. Nothing ever *happens* in a stationary state.

2. They are states of *definite total energy*. In classical mechanics, the total energy (kinetic plus potential) is called the **Hamiltonian**:

$$H(x, p) = \frac{p^2}{2m} + V(x). \quad [2.10]$$

[3]For normalizable solutions, E must be *real* (see Problem 2.1(a)).

The corresponding Hamiltonian *operator*, obtained by the canonical substitution $p \to (\hbar/i)(\partial/\partial x)$, is therefore[4]

$$\hat{H} = -\frac{\hbar^2}{2m}\frac{\partial^2}{\partial x^2} + V(x). \quad [2.11]$$

Thus the time-independent Schrödinger equation (Equation 2.5) can be written

$$\hat{H}\psi = E\psi, \quad [2.12]$$

and the expectation value of the total energy is

$$\langle H \rangle = \int \psi^* \hat{H}\psi\, dx = E\int |\psi|^2\, dx = E\int |\Psi|^2\, dx = E. \quad [2.13]$$

(Notice that the normalization of Ψ entails the normalization of ψ.) Moreover,

$$\hat{H}^2\psi = \hat{H}(\hat{H}\psi) = \hat{H}(E\psi) = E(\hat{H}\psi) = E^2\psi,$$

and hence

$$\langle H^2 \rangle = \int \psi^* \hat{H}^2\psi\, dx = E^2\int |\psi|^2\, dx = E^2.$$

So the variance of H is

$$\sigma_H^2 = \langle H^2 \rangle - \langle H \rangle^2 = E^2 - E^2 = 0. \quad [2.14]$$

But remember, if $\sigma = 0$, then every member of the sample must share the same value (the distribution has zero spread). *Conclusion:* A separable solution has the property that *every measurement of the total energy is certain to return the value* E. (That's why I chose that letter for the separation constant.)

3. The general solution is a **linear combination** of separable solutions. As we're about to discover, the time-independent Schrödinger equation (Equation 2.5) yields an infinite collection of solutions ($\psi_1(x)$, $\psi_2(x)$, $\psi_3(x)$, ...), each with its associated value of the separation constant (E_1, E_2, E_3, ...); thus there is a different wave function for each **allowed energy**:

$$\Psi_1(x,t) = \psi_1(x)e^{-iE_1t/\hbar}, \quad \Psi_2(x,t) = \psi_2(x)e^{-iE_2t/\hbar}, \quad \ldots.$$

Now (as you can easily check for yourself) the (time-*de*pendent) Schrödinger equation (Equation 2.1) has the property that any linear combination[5] of solutions

---

[4] Whenever confusion might arise, I'll put a "hat" (^) on the operator, to distinguish it from the dynamical variable it represents.

[5] A **linear combination** of the functions $f_1(z), f_2(z), \ldots$ is an expression of the form

$$f(z) = c_1 f_1(z) + c_2 f_2(z) + \cdots,$$

where $c_1, c_2, \ldots$ are any (complex) constants.

is itself a solution. Once we have found the separable solutions, then, we can immediately construct a much more general solution, of the form

$$\Psi(x,t) = \sum_{n=1}^{\infty} c_n \psi_n(x) e^{-iE_n t/\hbar}. \tag{2.15}$$

It so happens that *every* solution to the (time-dependent) Schrödinger equation can be written in this form—it is simply a matter of finding the right constants (c_1, c_2, ...) so as to fit the initial conditions for the problem at hand. You'll see in the following sections how all this works out in practice, and in Chapter 3 we'll put it into more elegant language, but the main point is this: Once you've solved the time-*in*dependent Schrödinger equation, you're essentially *done*; getting from there to the general solution of the time-*de*pendent Schrödinger equation is, in principle, simple and straightforward.

A lot has happened in the last four pages, so let me recapitulate, from a somewhat different perspective. Here's the generic problem: You're given a (time-independent) potential $V(x)$, and the starting wave function $\Psi(x, 0)$; your job is to find the wave function, $\Psi(x, t)$, for any subsequent time t. To do this you must solve the (time-dependent) Schrödinger equation (Equation 2.1). The strategy[6] is first to solve the time-*in*dependent Schrödinger equation (Equation 2.5); this yields, in general, an infinite set of solutions ($\psi_1(x)$, $\psi_2(x)$, $\psi_3(x)$, ...), each with its own associated energy (E_1, E_2, E_3, ...). To fit $\Psi(x, 0)$ you write down the general linear combination of these solutions:

$$\Psi(x,0) = \sum_{n=1}^{\infty} c_n \psi_n(x); \tag{2.16}$$

the miracle is that you can *always* match the specified initial state by appropriate choice of the constants c_1, c_2, c_3, ... To construct $\Psi(x, t)$ you simply tack onto each term its characteristic time dependence, $\exp(-iE_n t/\hbar)$:

$$\boxed{\Psi(x,t) = \sum_{n=1}^{\infty} c_n \psi_n(x) e^{-iE_n t/\hbar} = \sum_{n=1}^{\infty} c_n \Psi_n(x,t).} \tag{2.17}$$

The separable solutions themselves,

$$\Psi_n(x,t) = \psi_n(x) e^{-iE_n t/\hbar}, \tag{2.18}$$

[6]Occasionally you can solve the time-dependent Schrödinger equation without recourse to separation of variables—see, for instance, Problems 2.49 and 2.50. But such cases are extremely rare.

are *stationary* states, in the sense that all probabilities and expectation values are independent of time, but this property is emphatically *not* shared by the general solution (Equation 2.17); the energies are different, for different stationary states, and the exponentials do not cancel, when you calculate $|\Psi|^2$.

Example 2.1 Suppose a particle starts out in a linear combination of just *two* stationary states:

$$\Psi(x, 0) = c_1\psi_1(x) + c_2\psi_2(x).$$

(To keep things simple I'll assume that the constants c_n and the states $\psi_n(x)$ are *real*.) What is the wave function $\Psi(x, t)$ at subsequent times? Find the probability density, and describe its motion.

Solution: The first part is easy:

$$\Psi(x, t) = c_1\psi_1(x)e^{-iE_1t/\hbar} + c_2\psi_2(x)e^{-iE_2t/\hbar},$$

where E_1 and E_2 are the energies associated with ψ_1 and ψ_2. It follows that

$$|\Psi(x, t)|^2 = (c_1\psi_1 e^{iE_1t/\hbar} + c_2\psi_2 e^{iE_2t/\hbar})(c_1\psi_1 e^{-iE_1t/\hbar} + c_2\psi_2 e^{-iE_2t/\hbar})$$

$$= c_1^2\psi_1^2 + c_2^2\psi_2^2 + 2c_1c_2\psi_1\psi_2\cos[(E_2 - E_1)t/\hbar].$$

(I used **Euler's formula**, $\exp i\theta = \cos\theta + i\sin\theta$, to simplify the result.) Evidently the probability density *oscillates* sinusoidally, at an angular frequency $(E_2 - E_1)/\hbar$; this is certainly *not* a stationary state. But notice that it took a *linear combination* of states (with different energies) to produce motion.[7]

∗Problem 2.1 Prove the following three theorems:

(a) For normalizable solutions, the separation constant E must be *real*. *Hint:* Write E (in Equation 2.7) as $E_0 + i\Gamma$ (with E_0 and Γ real), and show that if Equation 1.20 is to hold for all t, Γ must be zero.

(b) The time-independent wave function $\psi(x)$ can always be taken to be *real* (unlike $\Psi(x, t)$, which is necessarily complex). This doesn't mean that every solution to the time-independent Schrödinger equation *is* real; what it says is that if you've got one that is *not*, it can always be expressed as a linear combination of solutions (with the same energy) that *are*. So you *might as well* stick to ψ's that are real. *Hint:* If $\psi(x)$ satisfies Equation 2.5, for a given E, so too does its complex conjugate, and hence also the real linear combinations $(\psi + \psi^*)$ and $i(\psi - \psi^*)$.

---

[7]This is nicely illustrated by an applet at the Web site http://thorin.adnc.com/~topquark/quantum/deepwellmain.html.

(c) If $V(x)$ is an **even function** (that is, $V(-x) = V(x)$) then $\psi(x)$ can always be taken to be either even or odd. *Hint:* If $\psi(x)$ satisfies Equation 2.5, for a given E, so too does $\psi(-x)$, and hence also the even and odd linear combinations $\psi(x) \pm \psi(-x)$.

∗**Problem 2.2** Show that E must exceed the minimum value of $V(x)$, for every normalizable solution to the time-independent Schrödinger equation. What is the classical analog to this statement? *Hint:* Rewrite Equation 2.5 in the form

$$\frac{d^2\psi}{dx^2} = \frac{2m}{\hbar^2}[V(x) - E]\psi;$$

if $E < V_{\min}$, then ψ and its second derivative always have the *same sign*—argue that such a function cannot be normalized.

2.2 THE INFINITE SQUARE WELL

Suppose

$$V(x) = \begin{cases} 0, & \text{if } 0 \le x \le a, \\ \infty, & \text{otherwise} \end{cases} \qquad [2.19]$$

(Figure 2.1). A particle in this potential is completely free, except at the two ends ($x = 0$ and $x = a$), where an infinite force prevents it from escaping. A classical model would be a cart on a frictionless horizontal air track, with perfectly elastic bumpers—it just keeps bouncing back and forth forever. (This potential is artificial, of course, but I urge you to treat it with respect. Despite its simplicity—or rather, precisely *because* of its simplicity—it serves as a wonderfully accessible test case for all the fancy machinery that comes later. We'll refer back to it frequently.)

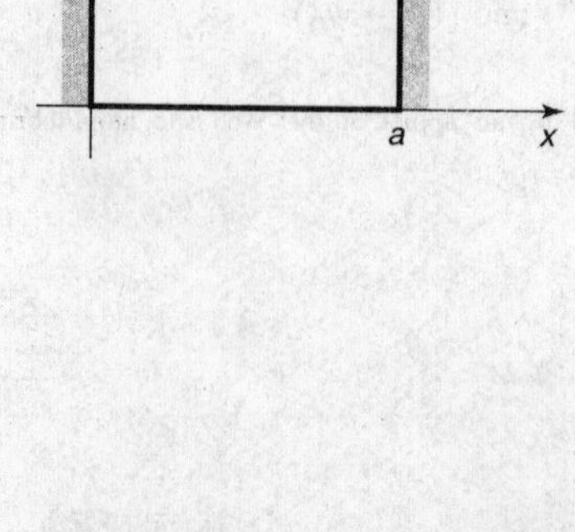

FIGURE 2.1: The infinite square well potential (Equation 2.19).

Outside the well, $\psi(x) = 0$ (the probability of finding the particle there is zero). *Inside* the well, where $V = 0$, the time-independent Schrödinger equation (Equation 2.5) reads

$$-\frac{\hbar^2}{2m}\frac{d^2\psi}{dx^2} = E\psi. \tag{2.20}$$

or

$$\frac{d^2\psi}{dx^2} = -k^2\psi, \quad \text{where } k \equiv \frac{\sqrt{2mE}}{\hbar}. \tag{2.21}$$

(By writing it in this way, I have tacitly assumed that $E \geq 0$; we know from Problem 2.2 that $E < 0$ won't work.) Equation 2.21 is the classical **simple harmonic oscillator** equation; the general solution is

$$\psi(x) = A\sin kx + B\cos kx, \tag{2.22}$$

where A and B are arbitrary constants. Typically, these constants are fixed by the **boundary conditions** of the problem. What *are* the appropriate boundary conditions for $\psi(x)$? Ordinarily, *both* ψ *and* $d\psi/dx$ *are continuous*, but where the potential goes to infinity only the first of these applies. (I'll *prove* these boundary conditions, and account for the exception when $V = \infty$, in Section 2.5; for now I hope you will trust me.)

Continuity of $\psi(x)$ requires that

$$\psi(0) = \psi(a) = 0, \tag{2.23}$$

so as to join onto the solution outside the well. What does this tell us about A and B? Well,

$$\psi(0) = A\sin 0 + B\cos 0 = B,$$

so $B = 0$, and hence

$$\psi(x) = A\sin kx. \tag{2.24}$$

Then $\psi(a) = A\sin ka$, so either $A = 0$ (in which case we're left with the trivial—non-normalizable—solution $\psi(x) = 0$), or else $\sin ka = 0$, which means that

$$ka = 0, \pm\pi, \pm 2\pi, \pm 3\pi, \ldots \tag{2.25}$$

But $k = 0$ is no good (again, that would imply $\psi(x) = 0$), and the negative solutions give nothing new, since $\sin(-\theta) = -\sin(\theta)$ and we can absorb the minus sign into A. So the *distinct* solutions are

$$k_n = \frac{n\pi}{a}, \quad \text{with } n = 1, 2, 3, \ldots \tag{2.26}$$

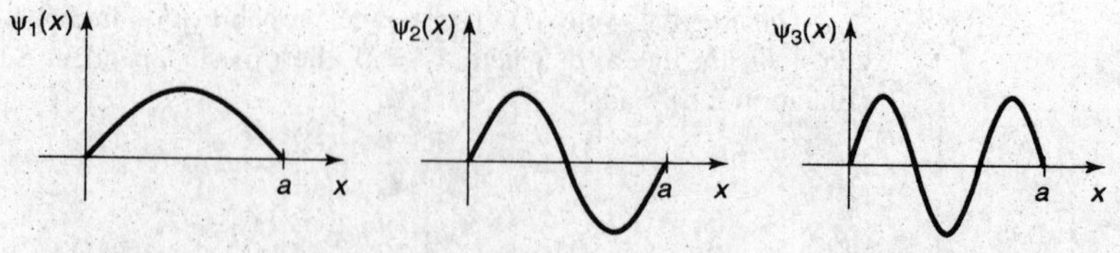

FIGURE 2.2: The first three stationary states of the infinite square well (Equation 2.28).

Curiously, the boundary condition at $x = a$ does not determine the constant A, but rather the constant k, and hence the possible values of E:

$$\boxed{E_n = \frac{\hbar^2 k_n^2}{2m} = \frac{n^2\pi^2\hbar^2}{2ma^2}.} \qquad [2.27]$$

In radical contrast to the classical case, a quantum particle in the infinite square well cannot have just *any* old energy—it has to be one of these special **allowed** values.[8] To find A, we *normalize* ψ:

$$\int_0^a |A|^2 \sin^2(kx)\,dx = |A|^2 \frac{a}{2} = 1, \quad \text{so} \quad |A|^2 = \frac{2}{a}.$$

This only determines the *magnitude* of A, but it is simplest to pick the positive real root: $A = \sqrt{2/a}$ (the phase of A carries no physical significance anyway). Inside the well, then, the solutions are

$$\boxed{\psi_n(x) = \sqrt{\frac{2}{a}} \sin\left(\frac{n\pi}{a}x\right).} \qquad [2.28]$$

As promised, the time-independent Schrödinger equation has delivered an infinite set of solutions (one for each positive integer n). The first few of these are plotted in Figure 2.2. They look just like the standing waves on a string of length a; ψ_1, which carries the lowest energy, is called the **ground state**, the others, whose energies increase in proportion to n^2, are called **excited states**. As a collection, the functions $\psi_n(x)$ have some interesting and important properties:

1. They are alternately **even** and **odd**, with respect to the center of the well: ψ_1 is even, ψ_2 is odd, ψ_3 is even, and so on.[9]

---

[8]Notice that the quantization of energy emerged as a rather technical consequence of the boundary conditions on solutions to the time-independent Schrödinger equation.

[9]To make this symmetry more apparent, some authors center the well at the origin (running it from $-a$ to $+a$). The even functions are then cosines, and the odd ones are sines. See Problem 2.36.

2. As you go up in energy, each successive state has one more **node** (zero-crossing): ψ_1 has none (the end points don't count), ψ_2 has one, ψ_3 has two, and so on.

3. They are mutually **orthogonal**, in the sense that

$$\int \psi_m(x)^* \psi_n(x)\, dx = 0, \tag{2.29}$$

whenever $m \neq n$. *Proof*:

$$\begin{aligned}
\int \psi_m(x)^* \psi_n(x)\, dx &= \frac{2}{a}\int_0^a \sin\left(\frac{m\pi}{a}x\right)\sin\left(\frac{n\pi}{a}x\right) dx \\
&= \frac{1}{a}\int_0^a \left[\cos\left(\frac{m-n}{a}\pi x\right) - \cos\left(\frac{m+n}{a}\pi x\right)\right] dx \\
&= \left\{\frac{1}{(m-n)\pi}\sin\left(\frac{m-n}{a}\pi x\right) - \frac{1}{(m+n)\pi}\sin\left(\frac{m+n}{a}\pi x\right)\right\}\Bigg|_0^a \\
&= \frac{1}{\pi}\left\{\frac{\sin[(m-n)\pi]}{(m-n)} - \frac{\sin[(m+n)\pi]}{(m+n)}\right\} = 0.
\end{aligned}$$

Note that this argument does *not* work if $m = n$. (Can you spot the point at which it fails?) In that case normalization tells us that the integral is 1. In fact, we can combine orthogonality and normalization into a single statement:[10]

$$\boxed{\int \psi_m(x)^* \psi_n(x)\, dx = \delta_{mn},} \tag{2.30}$$

where δ_{mn} (the so-called **Kronecker delta**) is defined in the usual way,

$$\delta_{mn} = \begin{cases} 0, & \text{if } m \neq n; \\ 1, & \text{if } m = n. \end{cases} \tag{2.31}$$

We say that the ψ's are **orthonormal**.

4. They are **complete**, in the sense that any *other* function, $f(x)$, can be expressed as a linear combination of them:

$$f(x) = \sum_{n=1}^{\infty} c_n \psi_n(x) = \sqrt{\frac{2}{a}} \sum_{n=1}^{\infty} c_n \sin\left(\frac{n\pi}{a}x\right). \tag{2.32}$$

---

[10] In this case the ψ's are *real*, so the * on ψ_m is unnecessary, but for future purposes it's a good idea to get in the habit of putting it there.

I'm not about to *prove* the completeness of the functions $\sin(n\pi x/a)$, but if you've studied advanced calculus you will recognize that Equation 2.32 is nothing but the **Fourier series** for $f(x)$, and the fact that "any" function can be expanded in this way is sometimes called **Dirichlet's theorem.**[11]

The coefficients c_n can be evaluated—for a given $f(x)$—by a method I call **Fourier's trick**, which beautifully exploits the orthonormality of $\{\psi_n\}$: Multiply both sides of Equation 2.32 by $\psi_m(x)^*$, and integrate.

$$\int \psi_m(x)^* f(x)\,dx = \sum_{n=1}^{\infty} c_n \int \psi_m(x)^* \psi_n(x)\,dx = \sum_{n=1}^{\infty} c_n \delta_{mn} = c_m. \qquad [2.33]$$

(Notice how the Kronecker delta kills every term in the sum except the one for which $n = m$.) Thus the nth coefficient in the expansion of $f(x)$ is[12]

$$\boxed{c_n = \int \psi_n(x)^* f(x)\,dx.} \qquad [2.34]$$

These four properties are extremely powerful, and they are not peculiar to the infinite square well. The first is true whenever the potential itself is a symmetric function; the second is universal, regardless of the shape of the potential.[13] Orthogonality is also quite general—I'll show you the proof in Chapter 3. Completeness holds for all the potentials you are likely to encounter, but the proofs tend to be nasty and laborious; I'm afraid most physicists simply *assume* completeness, and hope for the best.

The stationary states (Equation 2.18) of the infinite square well are evidently

$$\Psi_n(x,t) = \sqrt{\frac{2}{a}} \sin\left(\frac{n\pi}{a}x\right) e^{-i(n^2\pi^2\hbar/2ma^2)t}. \qquad [2.35]$$

I claimed (Equation 2.17) that the most general solution to the (time-dependent) Schrödinger equation is a linear combination of stationary states:

$$\Psi(x,t) = \sum_{n=1}^{\infty} c_n \sqrt{\frac{2}{a}} \sin\left(\frac{n\pi}{a}x\right) e^{-i(n^2\pi^2\hbar/2ma^2)t}. \qquad [2.36]$$

---

[11] See, for example, Mary Boas, *Mathematical Methods in the Physical Sciences*, 2d ed. (New York: John Wiley, 1983), p. 313; $f(x)$ can even have a finite number of finite discontinuities.

[12] It doesn't matter whether you use m or n as the "dummy index" here (as long as you are consistent on the two sides of the equation, of course); *whatever* letter you use, it just stands for "any positive integer."

[13] See, for example, John L. Powell and Bernd Crasemann, *Quantum Mechanics* (Addison-Wesley, Reading, MA, 1961), p. 126.

(If you doubt that this *is* a solution, by all means *check* it!) It remains only for me to demonstrate that I can fit any prescribed initial wave function, $\Psi(x, 0)$, by appropriate choice of the coefficients c_n:

$$\Psi(x, 0) = \sum_{n=1}^{\infty} c_n \psi_n(x).$$

The completeness of the ψ's (confirmed in this case by Dirichlet's theorem) guarantees that I can always express $\Psi(x, 0)$ in this way, and their orthonormality licenses the use of Fourier's trick to determine the actual coefficients:

$$c_n = \sqrt{\frac{2}{a}} \int_0^a \sin\left(\frac{n\pi}{a}x\right) \Psi(x, 0)\, dx. \qquad [2.37]$$

That *does* it: Given the initial wave function, $\Psi(x, 0)$, we first compute the expansion coefficients c_n, using Equation 2.37, and then plug these into Equation 2.36 to obtain $\Psi(x, t)$. Armed with the wave function, we are in a position to compute any dynamical quantities of interest, using the procedures in Chapter 1. And this same ritual applies to *any* potential—the only things that change are the functional form of the ψ's and the equation for the allowed energies.

Example 2.2 A particle in the infinite square well has the initial wave function

$$\Psi(x, 0) = Ax(a - x), \quad (0 \le x \le a),$$

for some constant A (see Figure 2.3). *Outside* the well, of course, $\Psi = 0$. Find $\Psi(x, t)$.

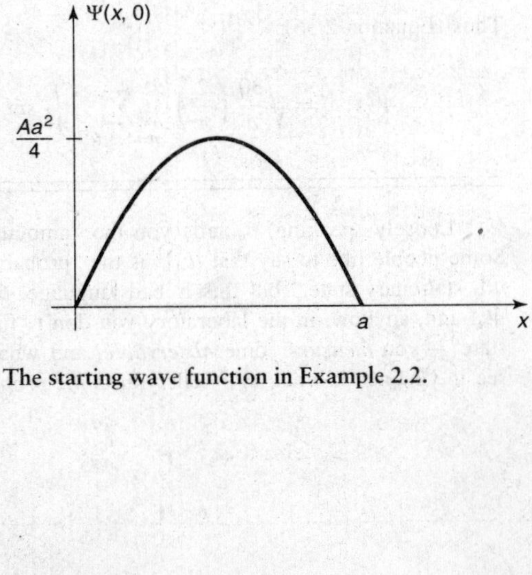

FIGURE 2.3: The starting wave function in Example 2.2.

Solution: First we need to determine A, by normalizing $\Psi(x,0)$:

$$1=\int_0^a |\Psi(x,0)|^2\,dx = |A|^2\int_0^a x^2(a-x)^2\,dx = |A|^2\frac{a^5}{30},$$

so

$$A=\sqrt{\frac{30}{a^5}}.$$

The nth coefficient is (Equation 2.37)

$$\begin{aligned}
c_n &= \sqrt{\frac{2}{a}}\int_0^a \sin\left(\frac{n\pi}{a}x\right)\sqrt{\frac{30}{a^5}}\,x(a-x)\,dx\\
&= \frac{2\sqrt{15}}{a^3}\left[a\int_0^a x\sin\left(\frac{n\pi}{a}x\right)dx-\int_0^a x^2\sin\left(\frac{n\pi}{a}x\right)dx\right]\\
&= \frac{2\sqrt{15}}{a^3}\left\{a\left[\left(\frac{a}{n\pi}\right)^2\sin\left(\frac{n\pi}{a}x\right)-\frac{ax}{n\pi}\cos\left(\frac{n\pi}{a}x\right)\right]\Bigg|_0^a\right.\\
&\quad\left.-\left[2\left(\frac{a}{n\pi}\right)^2 x\sin\left(\frac{n\pi}{a}x\right)-\frac{(n\pi x/a)^2-2}{(n\pi/a)^3}\cos\left(\frac{n\pi}{a}x\right)\right]\Bigg|_0^a\right\}\\
&= \frac{2\sqrt{15}}{a^3}\left[-\frac{a^3}{n\pi}\cos(n\pi)+a^3\frac{(n\pi)^2-2}{(n\pi)^3}\cos(n\pi)+a^3\frac{2}{(n\pi)^3}\cos(0)\right]\\
&= \frac{4\sqrt{15}}{(n\pi)^3}\left[\cos(0)-\cos(n\pi)\right]\\
&= \begin{cases} 0, & \text{if } n \text{ is even},\\ 8\sqrt{15}/(n\pi)^3, & \text{if } n \text{ is odd}.\end{cases}
\end{aligned}$$

Thus (Equation 2.36):

$$\Psi(x,t)=\sqrt{\frac{30}{a}}\left(\frac{2}{\pi}\right)^3\sum_{n=1,3,5\ldots}\frac{1}{n^3}\sin\left(\frac{n\pi}{a}x\right)e^{-in^2\pi^2\hbar t/2ma^2}$$

Loosely speaking, c_n tells you the "amount of ψ_n that is contained in Ψ." Some people like to say that $|c_n|^2$ is the "probability of finding the particle in the nth stationary state," but this is bad language; the particle is in the state Ψ, *not* Ψ_n, and, anyhow, in the laboratory you don't "find a particle to be in a particular state"—you *measure* some *observable*, and what you get is a *number*. As we'll see in Chapter 3, what $|c_n|^2$ tells you is *the probability that a measurement of the energy would yield the value* E_n (a competent measurement will always return *one* of the "allowed" values—hence the name—and $|c_n|^2$ is the probability of getting the *particular* value E_n).

Of course, the *sum* of these probabilities should be 1,

$$\boxed{\sum_{n=1}^{\infty} |c_n|^2 = 1.} \tag{2.38}$$

Indeed, this follows from the normalization of Ψ (the c_n's are independent of time, so I'm going to do the proof for $t = 0$; if this bothers you, you can easily generalize the argument to arbitrary t).

$$1 = \int |\Psi(x, 0)|^2 \, dx = \int \left(\sum_{m=1}^{\infty} c_m \psi_m(x) \right)^* \left(\sum_{n=1}^{\infty} c_n \psi_n(x) \right) dx$$

$$= \sum_{m=1}^{\infty} \sum_{n=1}^{\infty} c_m^* c_n \int \psi_m(x)^* \psi_n(x) \, dx$$

$$= \sum_{n=1}^{\infty} \sum_{m=1}^{\infty} c_m^* c_n \delta_{mn} = \sum_{n=1}^{\infty} |c_n|^2.$$

(Again, the Kronecker delta picks out the term $m = n$ in the summation over m.)

Moreover, the expectation value of the energy must be

$$\boxed{\langle H \rangle = \sum_{n=1}^{\infty} |c_n|^2 E_n,} \tag{2.39}$$

and this too can be checked directly: The time-independent Schrödinger equation (Equation 2.12) says

$$H\psi_n = E_n \psi_n, \tag{2.40}$$

so

$$\langle H \rangle = \int \Psi^* H \Psi \, dx = \int \left(\sum c_m \psi_m \right)^* H \left(\sum c_n \psi_n \right) dx$$

$$= \sum \sum c_m^* c_n E_n \int \psi_m^* \psi_n \, dx = \sum |c_n|^2 E_n.$$

Notice that the probability of getting a particular energy is independent of time, and so, *a fortiori*, is the expectation value of H. This is a manifestation of **conservation of energy** in quantum mechanics.

Example 2.3 In Example 2.2 the starting wave function (Figure 2.3) closely resembles the ground state ψ_1 (Figure 2.2). This suggests that $|c_1|^2$ should dominate, and in fact

$$|c_1|^2 = \left(\frac{8\sqrt{15}}{\pi^3}\right)^2 = 0.998555\ldots.$$

The rest of the coefficients make up the difference:[14]

$$\sum_{n=1}^{\infty}|c_n|^2 = \left(\frac{8\sqrt{15}}{\pi^3}\right)^2 \sum_{n=1,3,5,\ldots}^{\infty}\frac{1}{n^6} = 1.$$

The expectation value of the energy, in this example, is

$$\langle H\rangle = \sum_{n=1,3,5,\ldots}^{\infty}\left(\frac{8\sqrt{15}}{n^3\pi^3}\right)^2 \frac{n^2\pi^2\hbar^2}{2ma^2} = \frac{480\hbar^2}{\pi^4 ma^2}\sum_{n=1,3,5,\ldots}^{\infty}\frac{1}{n^4} = \frac{5\hbar^2}{ma^2}.$$

As one might expect, it is very close to $E_1 = \pi^2\hbar^2/2ma^2$—slightly *larger*, because of the admixture of excited states.

Problem 2.3 Show that there is no acceptable solution to the (time-independent) Schrödinger equation for the infinite square well with $E = 0$ or $E < 0$. (This is a special case of the general theorem in Problem 2.2, but this time do it by explicitly solving the Schrödinger equation, and showing that you cannot meet the boundary conditions.)

***Problem 2.4** Calculate $\langle x\rangle$, $\langle x^2\rangle$, $\langle p\rangle$, $\langle p^2\rangle$, σ_x, and σ_p, for the nth stationary state of the infinite square well. Check that the uncertainty principle is satisfied. Which state comes closest to the uncertainty limit?

***Problem 2.5** A particle in the infinite square well has as its initial wave function an even mixture of the first two stationary states:

$$\Psi(x, 0) = A[\psi_1(x) + \psi_2(x)].$$

[14]You can look up the series

$$\frac{1}{1^6} + \frac{1}{3^6} + \frac{1}{5^6} + \cdots = \frac{\pi^6}{960}$$

and

$$\frac{1}{1^4} + \frac{1}{3^4} + \frac{1}{5^4} + \cdots = \frac{\pi^4}{96}$$

in math tables, under "Sums of Reciprocal Powers" or "Riemann Zeta Function."

(a) Normalize $\Psi(x, 0)$. (That is, find A. This is very easy, if you exploit the orthonormality of ψ_1 and ψ_2. Recall that, having normalized Ψ at $t = 0$, you can rest assured that it *stays* normalized—if you doubt this, check it explicitly after doing part (b).)

(b) Find $\Psi(x, t)$ and $|\Psi(x, t)|^2$. Express the latter as a sinusoidal function of time, as in Example 2.1. To simplify the result, let $\omega \equiv \pi^2\hbar/2ma^2$.

(c) Compute $\langle x \rangle$. Notice that it oscillates in time. What is the angular frequency of the oscillation? What is the amplitude of the oscillation? (If your amplitude is greater than $a/2$, go directly to jail.)

(d) Compute $\langle p \rangle$. (As Peter Lorre would say, "Do it ze *kveek* vay, Johnny!")

(e) If you measured the energy of this particle, what values might you get, and what is the probability of getting each of them? Find the expectation value of H. How does it compare with E_1 and E_2?

Problem 2.6 Although the *overall* phase constant of the wave function is of no physical significance (it cancels out whenever you calculate a measurable quantity), the *relative* phase of the coefficients in Equation 2.17 *does* matter. For example, suppose we change the relative phase of ψ_1 and ψ_2 in Problem 2.5:

$$\Psi(x, 0) = A[\psi_1(x) + e^{i\phi}\psi_2(x)],$$

where ϕ is some constant. Find $\Psi(x, t)$, $|\Psi(x, t)|^2$, and $\langle x \rangle$, and compare your results with what you got before. Study the special cases $\phi = \pi/2$ and $\phi = \pi$. (For a graphical exploration of this problem see the applet in footnote 7.)

***Problem 2.7** A particle in the infinite square well has the initial wave function[15]

$$\Psi(x, 0) = \begin{cases} Ax, & 0 \le x \le a/2, \\ A(a - x), & a/2 \le x \le a. \end{cases}$$

(a) Sketch $\Psi(x, 0)$, and determine the constant A.

(b) Find $\Psi(x, t)$.

---

[15]There is no restriction in principle on the *shape* of the starting wave function, as long as it is normalizable. In particular, $\Psi(x, 0)$ need not have a continuous derivative—in fact, it doesn't even have to be a *continuous* function. However, if you try to calculate $\langle H \rangle$ using $\int \Psi(x, 0)^* H \Psi(x, 0)\, dx$ in such a case, you may encounter technical difficulties, because the second derivative of $\Psi(x, 0)$ is ill-defined. It works in Problem 2.9 because the discontinuities occur at the end points, where the wave function is zero anyway. In Problem 2.48 you'll see how to manage cases like Problem 2.7.

(c) What is the probability that a measurement of the energy would yield the value E_1?

(d) Find the expectation value of the energy.

Problem 2.8 A particle of mass m in the infinite square well (of width a) starts out in the left half of the well, and is (at $t = 0$) equally likely to be found at any point in that region.

(a) What is its initial wave function, $\Psi(x, 0)$? (Assume it is real. Don't forget to normalize it.)

(b) What is the probability that a measurement of the energy would yield the value $\pi^2\hbar^2/2ma^2$?

Problem 2.9 For the wave function in Example 2.2, find the expectation value of H, at time $t = 0$, the "old fashioned" way:

$$\langle H \rangle = \int \Psi(x, 0)^* \hat{H} \Psi(x, 0)\, dx.$$

Compare the result obtained in Example 2.3, using Equation 2.39. *Note:* because $\langle H \rangle$ is independent of time, there is no loss of generality in using $t = 0$.

2.3 THE HARMONIC OSCILLATOR

The paradigm for a classical harmonic oscillator is a mass m attached to a spring of force constant k. The motion is governed by **Hooke's law**,

$$F = -kx = m\frac{d^2x}{dt^2}$$

(ignoring friction), and the solution is

$$x(t) = A\sin(\omega t) + B\cos(\omega t),$$

where

$$\omega \equiv \sqrt{\frac{k}{m}} \qquad [2.41]$$

is the (angular) frequency of oscillation. The potential energy is

$$V(x) = \frac{1}{2}kx^2; \qquad [2.42]$$

its graph is a parabola.

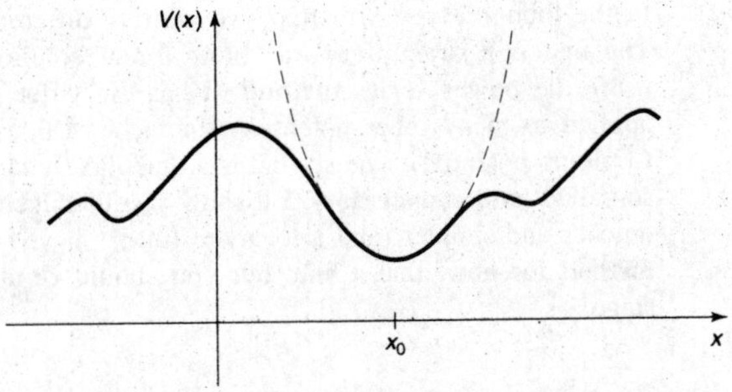

FIGURE 2.4: Parabolic approximation (dashed curve) to an arbitrary potential, in the neighborhood of a local minimum.

Of course, there's no such thing as a *perfect* harmonic oscillator—if you stretch it too far the spring is going to break, and typically Hooke's law fails long before that point is reached. But practically any potential is *approximately* parabolic, in the neighborhood of a local minimum (Figure 2.4). Formally, if we expand $V(x)$ in a **Taylor series** about the minimum:

$$V(x) = V(x_0) + V'(x_0)(x - x_0) + \frac{1}{2}V''(x_0)(x - x_0)^2 + \cdots,$$

subtract $V(x_0)$ (you can add a constant to $V(x)$ with impunity, since that doesn't change the force), recognize that $V'(x_0) = 0$ (since x_0 is a minimum), and drop the higher-order terms (which are negligible as long as $(x - x_0)$ stays small), we get

$$V(x) \cong \frac{1}{2}V''(x_0)(x - x_0)^2,$$

which describes simple harmonic oscillation (about the point x_0), with an effective spring constant $k = V''(x_0)$.[16] That's why the simple harmonic oscillator is so important: Virtually *any* oscillatory motion is approximately simple harmonic, as long as the amplitude is small.

The *quantum* problem is to solve the Schrödinger equation for the potential

$$V(x) = \frac{1}{2}m\omega^2 x^2 \qquad [2.43]$$

(it is customary to eliminate the spring constant in favor of the classical frequency, using Equation 2.41). As we have seen, it suffices to solve the time-independent Schrödinger equation:

$$-\frac{\hbar^2}{2m}\frac{d^2\psi}{dx^2} + \frac{1}{2}m\omega^2 x^2 \psi = E\psi. \qquad [2.44]$$

---

[16]Note that $V''(x_0) \geq 0$, since by assumption x_0 is a *minimum*. Only in the rare case $V''(x_0) = 0$ is the oscillation not even approximately simple harmonic.

In the literature you will find two entirely different approaches to this problem. The first is a straightforward "brute force" solution to the differential equation, using the **power series method**; it has the virtue that the same strategy can be applied to many other potentials (in fact, we'll use it in Chapter 4 to treat the Coulomb potential). The second is a diabolically clever algebraic technique, using so-called **ladder operators**. I'll show you the algebraic method first, because it is quicker and simpler (and a lot more fun);[17] if you want to skip the power series method for now, that's fine, but you should certainly plan to study it at some stage.

2.3.1 Algebraic Method

To begin with, let's rewrite Equation 2.44 in a more suggestive form:

$$\frac{1}{2m}[p^2 + (m\omega x)^2]\psi = E\psi, \tag{2.45}$$

where $p \equiv (\hbar/i)d/dx$ is, of course, the momentum operator. The basic idea is to *factor* the Hamiltonian,

$$H = \frac{1}{2m}[p^2 + (m\omega x)^2]. \tag{2.46}$$

If these were *numbers*, it would be easy:

$$u^2 + v^2 = (iu + v)(-iu + v).$$

Here, however, it's not quite so simple, because p and x are *operators*, and operators do not, in general, **commute** (xp is not the same as px). Still, this does motivate us to examine the quantities

$$\boxed{a_\pm \equiv \frac{1}{\sqrt{2\hbar m\omega}}(\mp ip + m\omega x)} \tag{2.47}$$

(the factor in front is just there to make the final results look nicer).

Well, what *is* the product $a_- a_+$?

$$a_- a_+ = \frac{1}{2\hbar m\omega}(ip + m\omega x)(-ip + m\omega x)$$

$$= \frac{1}{2\hbar m\omega}[p^2 + (m\omega x)^2 - im\omega(xp - px)].$$

---

[17]We'll encounter some of the same strategies in the theory of angular momentum (Chapter 4), and the technique generalizes to a broad class of potentials in **super-symmetric quantum mechanics** (see, for example, Richard W. Robinett, *Quantum Mechanics*, (Oxford U.P., New York, 1997), Section 14.4).

As anticipated, there's an extra term, involving $(xp - px)$. We call this the **commutator** of x and p; it is a measure of how badly they *fail* to commute. In general, the commutator of operators A and B (written with square brackets) is

$$[A, B] \equiv AB - BA. \quad [2.48]$$

In this notation,

$$a_-a_+ = \frac{1}{2\hbar m\omega}[p^2 + (m\omega x)^2] - \frac{i}{2\hbar}[x, p]. \quad [2.49]$$

We need to figure out the commutator of x and p. *Warning:* Operators are notoriously slippery to work with in the abstract, and you are bound to make mistakes unless you give them a "test function," $f(x)$, to act on. At the end you can throw away the test function, and you'll be left with an equation involving the operators alone. In the present case we have:

$$[x, p]f(x) = \left[x\frac{\hbar}{i}\frac{d}{dx}(f) - \frac{\hbar}{i}\frac{d}{dx}(xf)\right] = \frac{\hbar}{i}\left(x\frac{df}{dx} - x\frac{df}{dx} - f\right) = i\hbar f(x). \quad [2.50]$$

Dropping the test function, which has served its purpose,

$$\boxed{[x, p] = i\hbar.} \quad [2.51]$$

This lovely and ubiquitous result is known as the **canonical commutation relation.**[18]

With this, Equation 2.49 becomes

$$a_-a_+ = \frac{1}{\hbar\omega}H + \frac{1}{2}, \quad [2.52]$$

or

$$H = \hbar\omega\left(a_-a_+ - \frac{1}{2}\right). \quad [2.53]$$

Evidently the Hamiltonian does *not* factor perfectly—there's that extra $-1/2$ on the right. Notice that the ordering of a_+ and a_- is important here; the same argument, with a_+ on the left, yields

$$a_+a_- = \frac{1}{\hbar\omega}H - \frac{1}{2}. \quad [2.54]$$

In particular,

$$[a_-, a_+] = 1. \quad [2.55]$$

[18] In a deep sense all of the mysteries of quantum mechanics can be traced to the fact that position and momentum do not commute. Indeed, some authors take the canonical commutation relation as an *axiom* of the theory, and use it to *derive* $p = (\hbar/i)d/dx$.

So the Hamiltonian can equally well be written

$$H = \hbar\omega\left(a_+a_- + \frac{1}{2}\right). \qquad [2.56]$$

In terms of $a_\pm$, then, the Schrödinger equation[19] for the harmonic oscillator takes the form

$$\hbar\omega\left(a_\pm a_\mp \pm \frac{1}{2}\right)\psi = E\psi \qquad [2.57]$$

(in equations like this you read the upper signs all the way across, or else the lower signs).

Now, here comes the crucial step: I claim that *if ψ satisfies the Schrödinger equation with energy E*, (that is: $H\psi = E\psi$), *then $a_+\psi$ satisfies the Schrödinger equation with energy $(E + \hbar\omega)$*: $H(a_+\psi) = (E + \hbar\omega)(a_+\psi)$. *Proof:*

$$\begin{aligned} H(a_+\psi) &= \hbar\omega\left(a_+a_- + \frac{1}{2}\right)(a_+\psi) = \hbar\omega\left(a_+a_-a_+ + \frac{1}{2}a_+\right)\psi \\ &= \hbar\omega a_+\left(a_-a_+ + \frac{1}{2}\right)\psi = a_+\left[\hbar\omega\left(a_+a_- + 1 + \frac{1}{2}\right)\psi\right] \\ &= a_+(H + \hbar\omega)\psi = a_+(E + \hbar\omega)\psi = (E + \hbar\omega)(a_+\psi). \end{aligned}$$

(I used Equation 2.55 to replace a_-a_+ by $a_+a_- + 1$, in the second line. Notice that whereas the ordering of a_+ and a_- *does* matter, the ordering of $a_\pm$ and any *constants*—such as $\hbar$, ω, and E—does *not*; an operator commutes with any constant.)

By the same token, $a_-\psi$ is a solution with energy $(E - \hbar\omega)$:

$$\begin{aligned} H(a_-\psi) &= \hbar\omega\left(a_-a_+ - \frac{1}{2}\right)(a_-\psi) = \hbar\omega a_-\left(a_+a_- - \frac{1}{2}\right)\psi \\ &= a_-\left[\hbar\omega\left(a_-a_+ - 1 - \frac{1}{2}\right)\psi\right] = a_-(H - \hbar\omega)\psi = a_-(E - \hbar\omega)\psi \\ &= (E - \hbar\omega)(a_-\psi). \end{aligned}$$

Here, then, is a wonderful machine for generating new solutions, with higher and lower energies—if we could just find *one* solution, to get started! We call $a_\pm$ **ladder operators**, because they allow us to climb up and down in energy; a_+ is the **raising operator**, and a_- the **lowering operator**. The "ladder" of states is illustrated in Figure 2.5.

---

[19] I'm getting tired of writing "time-independent Schrödinger equation," so when it's clear from the context which one I mean, I'll just call it the "Schrödinger equation."

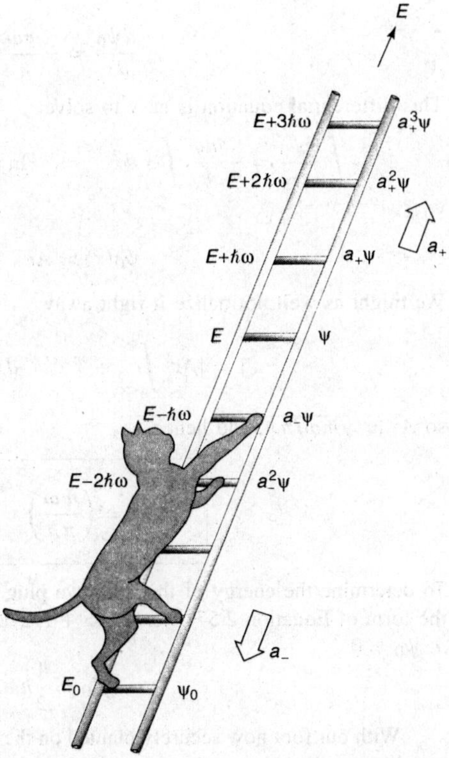

FIGURE 2.5: The "ladder" of states for the harmonic oscillator.

But wait! What if I apply the lowering operator repeatedly? Eventually I'm going to reach a state with energy less than zero, which (according to the general theorem in Problem 2.2) does not exist! At some point the machine must fail. How can that happen? We know that $a_-\psi$ is a new solution to the Schrödinger equation, but *there is no guarantee that it will be normalizable*—it might be zero, or its square-integral might be infinite. In practice it is the former: There occurs a "lowest rung" (call it ψ_0) such that

$$a_-\psi_0 = 0. \qquad [2.58]$$

We can use this to determine $\psi_0(x)$:

$$\frac{1}{\sqrt{2\hbar m\omega}}\left(\hbar\frac{d}{dx} + m\omega x\right)\psi_0 = 0,$$

or

$$\frac{d\psi_0}{dx} = -\frac{m\omega}{\hbar} x \psi_0.$$

This differential equation is easy to solve:

$$\int \frac{d\psi_0}{\psi_0} = -\frac{m\omega}{\hbar} \int x\, dx \quad \Rightarrow \quad \ln \psi_0 = -\frac{m\omega}{2\hbar} x^2 + \text{constant},$$

so

$$\psi_0(x) = A e^{-\frac{m\omega}{2\hbar} x^2}.$$

We might as well normalize it right away:

$$1 = |A|^2 \int_{-\infty}^{\infty} e^{-m\omega x^2/\hbar}\, dx = |A|^2 \sqrt{\frac{\pi\hbar}{m\omega}},$$

so $A^2 = \sqrt{m\omega/\pi\hbar}$, and hence

$$\boxed{\psi_0(x) = \left(\frac{m\omega}{\pi\hbar}\right)^{1/4} e^{-\frac{m\omega}{2\hbar} x^2}.} \qquad [2.59]$$

To determine the energy of this state we plug it into the Schrödinger equation (in the form of Equation 2.57), $\hbar\omega(a_+a_- + 1/2)\psi_0 = E_0\psi_0$, and exploit the fact that $a_-\psi_0 = 0$:

$$E_0 = \frac{1}{2}\hbar\omega. \qquad [2.60]$$

With our foot now securely planted on the bottom rung (the ground state of the quantum oscillator), we simply apply the raising operator (repeatedly) to generate the excited states,[20] increasing the energy by $\hbar\omega$ with each step:

$$\boxed{\psi_n(x) = A_n (a_+)^n \psi_0(x), \quad \text{with } E_n = \left(n + \frac{1}{2}\right)\hbar\omega,} \qquad [2.61]$$

where A_n is the normalization constant. By applying the raising operator (repeatedly) to ψ_0, then, we can (in principle) construct all[21] the stationary states of

---

[20] In the case of the harmonic oscillator it is customary, for some reason, to depart from the usual practice, and number the states starting with $n = 0$, instead of $n = 1$. Obviously, the lower limit on the sum in a formula such as Equation 2.17 should be altered accordingly.

[21] Note that we obtain *all* the (normalizable) solutions by this procedure. For if there were some *other* solution, we could generate from it a second ladder, by repeated application of the raising and lowering operators. But the bottom rung of this new ladder would have to satisfy Equation 2.58, and since that leads inexorably to Equation 2.59, the bottom rungs would be the same, and hence the two ladders would in fact be identical.

the harmonic oscillator. Meanwhile, without ever doing that explicitly, we have determined the allowed energies.

Example 2.4 Find the first excited state of the harmonic oscillator.

Solution: Using Equation 2.61,

$$\psi_1(x) = A_1 a_+ \psi_0 = \frac{A_1}{\sqrt{2\hbar m\omega}}\left(-\hbar\frac{d}{dx} + m\omega x\right)\left(\frac{m\omega}{\pi\hbar}\right)^{1/4} e^{-\frac{m\omega}{2\hbar}x^2}$$
$$= A_1\left(\frac{m\omega}{\pi\hbar}\right)^{1/4}\sqrt{\frac{2m\omega}{\hbar}}\, x e^{-\frac{m\omega}{2\hbar}x^2}. \qquad [2.62]$$

We can normalize it "by hand":

$$\int |\psi_1|^2\, dx = |A_1|^2\sqrt{\frac{m\omega}{\pi\hbar}}\left(\frac{2m\omega}{\hbar}\right)\int_{-\infty}^{\infty} x^2 e^{-\frac{m\omega}{\hbar}x^2}\, dx = |A_1|^2,$$

so, as it happens, $A_1 = 1$.

I wouldn't want to calculate ψ_{50} this way (applying the raising operator fifty times!), but never mind: In *principle* Equation 2.61 does the job—except for the normalization.

You can even get the normalization algebraically, but it takes some fancy footwork, so watch closely. We know that $a_\pm \psi_n$ is *proportional* to $\psi_{n\pm 1}$,

$$a_+\psi_n = c_n\psi_{n+1}, \quad a_-\psi_n = d_n\psi_{n-1} \qquad [2.63]$$

but what are the proportionality factors, c_n and d_n? First note that for "any"[22] functions $f(x)$ and $g(x)$,

$$\int_{-\infty}^{\infty} f^*(a_\pm g)\, dx = \int_{-\infty}^{\infty} (a_\mp f)^* g\, dx. \qquad [2.64]$$

(In the language of linear algebra, $a_\mp$ is the **hermitian conjugate** of $a_\pm$.)

Proof:

$$\int_{-\infty}^{\infty} f^*(a_\pm g)\, dx = \frac{1}{\sqrt{2\hbar m\omega}}\int_{-\infty}^{\infty} f^*\left(\mp\hbar\frac{d}{dx} + m\omega x\right) g\, dx,$$

[22]Of course, the integrals must *exist*, and this means that $f(x)$ and $g(x)$ must go to zero at $\pm\infty$.

and integration by parts takes $\int f^*(dg/dx)\,dx$ to $-\int (df/dx)^* g\,dx$ (the boundary terms vanish, for the reason indicated in footnote 22), so

$$\int_{-\infty}^{\infty} f^*(a_\pm g)\,dx = \frac{1}{\sqrt{2\hbar m\omega}}\int_{-\infty}^{\infty}\left[\left(\pm\hbar\frac{d}{dx} + m\omega x\right) f\right]^* g\,dx = \int_{-\infty}^{\infty}(a_\mp f)^* g\,dx.$$

QED

In particular,

$$\int_{-\infty}^{\infty}(a_\pm \psi_n)^*(a_\pm \psi_n)\,dx = \int_{-\infty}^{\infty}(a_\mp a_\pm \psi_n)^*\psi_n\,dx.$$

But (invoking Equations 2.57 and 2.61)

$$a_+a_-\psi_n = n\psi_n, \quad a_-a_+\psi_n = (n+1)\psi_n, \tag{2.65}$$

so

$$\int_{-\infty}^{\infty}(a_+\psi_n)^*(a_+\psi_n)\,dx = |c_n|^2\int_{-\infty}^{\infty}|\psi_{n+1}|^2\,dx = (n+1)\int_{-\infty}^{\infty}|\psi_n|^2\,dx,$$

$$\int_{-\infty}^{\infty}(a_-\psi_n)^*(a_-\psi_n)\,dx = |d_n|^2\int_{-\infty}^{\infty}|\psi_{n-1}|^2\,dx = n\int_{-\infty}^{\infty}|\psi_n|^2\,dx.$$

But since ψ_n and $\psi_{n\pm1}$ are normalized, it follows that $|c_n|^2 = n+1$ and $|d_n|^2 = n$, and hence

$$\boxed{a_+\psi_n = \sqrt{n+1}\,\psi_{n+1}, \quad a_-\psi_n = \sqrt{n}\,\psi_{n-1}.} \tag{2.66}$$

Thus

$$\psi_1 = a_+\psi_0, \quad \psi_2 = \frac{1}{\sqrt{2}}a_+\psi_1 = \frac{1}{\sqrt{2}}(a_+)^2\psi_0,$$

$$\psi_3 = \frac{1}{\sqrt{3}}a_+\psi_2 = \frac{1}{\sqrt{3\cdot 2}}(a_+)^3\psi_0, \quad \psi_4 = \frac{1}{\sqrt{4}}a_+\psi_3 = \frac{1}{\sqrt{4\cdot 3\cdot 2}}(a_+)^4\psi_0,$$

and so on. Clearly

$$\boxed{\psi_n = \frac{1}{\sqrt{n!}}(a_+)^n\psi_0,} \tag{2.67}$$

which is to say that the normalization factor in Equation 2.61 is $A_n = 1/\sqrt{n!}$ (in particular, $A_1 = 1$, confirming our result in Example 2.4).

As in the case of the infinite square well, the stationary states of the harmonic oscillator are orthogonal:

$$\int_{-\infty}^{\infty} \psi_m^* \psi_n \, dx = \delta_{mn}. \qquad [2.68]$$

This can be proved using Equation 2.65, and Equation 2.64 twice—first moving a_+ and then moving a_-:

$$\begin{aligned}\int_{-\infty}^{\infty} \psi_m^* (a_+ a_-) \psi_n \, dx &= n \int_{-\infty}^{\infty} \psi_m^* \psi_n \, dx \\ &= \int_{-\infty}^{\infty} (a_- \psi_m)^* (a_- \psi_n) \, dx = \int_{-\infty}^{\infty} (a_+ a_- \psi_m)^* \psi_n \, dx \\ &= m \int_{-\infty}^{\infty} \psi_m^* \psi_n \, dx.\end{aligned}$$

Unless $m = n$, then, $\int \psi_m^* \psi_n \, dx$ must be zero. Orthonormality means that we can again use Fourier's trick (Equation 2.34) to evaluate the coefficients, when we expand $\Psi(x, 0)$ as a linear combination of stationary states (Equation 2.16), and $|c_n|^2$ is again the probability that a measurement of the energy would yield the value E_n.

Example 2.5 Find the expectation value of the potential energy in the nth state of the harmonic oscillator.

Solution:

$$\langle V \rangle = \left\langle \frac{1}{2} m \omega^2 x^2 \right\rangle = \frac{1}{2} m \omega^2 \int_{-\infty}^{\infty} \psi_n^* x^2 \psi_n \, dx.$$

There's a beautiful device for evaluating integrals of this kind (involving powers of x or p): Use the definition (Equation 2.47) to express x and p in terms of the raising and lowering operators:

$$x = \sqrt{\frac{\hbar}{2m\omega}} (a_+ + a_-); \quad p = i \sqrt{\frac{\hbar m \omega}{2}} (a_+ - a_-). \qquad [2.69]$$

In this example we are interested in x^2:

$$x^2 = \frac{\hbar}{2m\omega} \left[(a_+)^2 + (a_+ a_-) + (a_- a_+) + (a_-)^2 \right].$$

So

$$\langle V \rangle = \frac{\hbar \omega}{4} \int \psi_n^* \left[(a_+)^2 + (a_+ a_-) + (a_- a_+) + (a_-)^2 \right] \psi_n \, dx.$$

But $(a_+)^2\psi_n$ is (apart from normalization) ψ_{n+2}, which is orthogonal to ψ_n, and the same goes for $(a_-)^2\psi_n$, which is proportional to ψ_{n-2}. So those terms drop out, and we can use Equation 2.65 to evaluate the remaining two:

$$\langle V\rangle = \frac{\hbar\omega}{4}(n + n + 1) = \frac{1}{2}\hbar\omega\left(n + \frac{1}{2}\right).$$

As it happens, the expectation value of the potential energy is exactly *half* the total (the other half, of course, is kinetic). This is a peculiarity of the harmonic oscillator, as we'll see later on.

∗Problem 2.10

(a) Construct $\psi_2(x)$.

(b) Sketch ψ_0, ψ_1, and ψ_2.

(c) Check the orthogonality of ψ_0, ψ_1, and ψ_2, by explicit integration. *Hint:* If you exploit the even-ness and odd-ness of the functions, there is really only one integral left to do.

∗Problem 2.11

(a) Compute $\langle x\rangle$, $\langle p\rangle$, $\langle x^2\rangle$, and $\langle p^2\rangle$, for the states ψ_0 (Equation 2.59) and ψ_1 (Equation 2.62), by explicit integration. *Comment:* In this and other problems involving the harmonic oscillator it simplifies matters if you introduce the variable $\xi \equiv \sqrt{m\omega/\hbar}\,x$ and the constant $\alpha \equiv (m\omega/\pi\hbar)^{1/4}$.

(b) Check the uncertainty principle for these states.

(c) Compute $\langle T\rangle$ (the average kinetic energy) and $\langle V\rangle$ (the average potential energy) for these states. (No new integration allowed!) Is their sum what you would expect?

∗Problem 2.12 Find $\langle x\rangle$, $\langle p\rangle$, $\langle x^2\rangle$, $\langle p^2\rangle$, and $\langle T\rangle$, for the nth stationary state of the harmonic oscillator, using the method of Example 2.5. Check that the uncertainty principle is satisfied.

Problem 2.13 A particle in the harmonic oscillator potential starts out in the state

$$\Psi(x, 0) = A[3\psi_0(x) + 4\psi_1(x)].$$

(a) Find A.

(b) Construct $\Psi(x, t)$ and $|\Psi(x, t)|^2$.

(c) Find $\langle x \rangle$ and $\langle p \rangle$. Don't get too excited if they oscillate at the classical frequency; what would it have been had I specified $\psi_2(x)$, instead of $\psi_1(x)$? Check that Ehrenfest's theorem (Equation 1.38) holds for this wave function.

(d) If you measured the energy of this particle, what values might you get, and with what probabilities?

Problem 2.14 A particle is in the ground state of the harmonic oscillator with classical frequency ω, when suddenly the spring constant quadruples, so $\omega' = 2\omega$, without initially changing the wave function (of course, Ψ will now *evolve* differently, because the Hamiltonian has changed). What is the probability that a measurement of the energy would still return the value $\hbar\omega/2$? What is the probability of getting $\hbar\omega$? [*Answer:* 0.943.]

2.3.2 Analytic Method

We return now to the Schrödinger equation for the harmonic oscillator,

$$-\frac{\hbar^2}{2m}\frac{d^2\psi}{dx^2} + \frac{1}{2}m\omega^2x^2\psi = E\psi, \qquad [2.70]$$

and solve it directly, by the series method. Things look a little cleaner if we introduce the dimensionless variable

$$\xi \equiv \sqrt{\frac{m\omega}{\hbar}}\,x; \qquad [2.71]$$

in terms of ξ the Schrödinger equation reads

$$\frac{d^2\psi}{d\xi^2} = (\xi^2 - K)\psi, \qquad [2.72]$$

where K is the energy, in units of $(1/2)\hbar\omega$:

$$K \equiv \frac{2E}{\hbar\omega}. \qquad [2.73]$$

Our problem is to solve Equation 2.72, and in the process obtain the "allowed" values of K (and hence of E).

To begin with, note that at very large ξ (which is to say, at very large x), ξ^2 completely dominates over the constant K, so in this regime

$$\frac{d^2\psi}{d\xi^2} \approx \xi^2\psi, \qquad [2.74]$$

which has the approximate solution (check it!)

$$\psi(\xi) \approx Ae^{-\xi^2/2} + Be^{+\xi^2/2}. \qquad [2.75]$$

The B term is clearly not normalizable (it blows up as $|x| \to \infty$); the physically acceptable solutions, then, have the asymptotic form

$$\psi(\xi) \to (\ \ )e^{-\xi^2/2}, \quad \text{at large } \xi. \tag{2.76}$$

This suggests that we "peel off" the exponential part,

$$\psi(\xi) = h(\xi)e^{-\xi^2/2}, \tag{2.77}$$

in hopes that what remains, $h(\xi)$, has a simpler functional form than $\psi(\xi)$ itself.[23] Differentiating Equation 2.77,

$$\frac{d\psi}{d\xi} = \left(\frac{dh}{d\xi} - \xi h\right) e^{-\xi^2/2},$$

and

$$\frac{d^2\psi}{d\xi^2} = \left(\frac{d^2h}{d\xi^2} - 2\xi\frac{dh}{d\xi} + (\xi^2 - 1)h\right) e^{-\xi^2/2},$$

so the Schrödinger equation (Equation 2.72) becomes

$$\frac{d^2h}{d\xi^2} - 2\xi\frac{dh}{d\xi} + (K - 1)h = 0. \tag{2.78}$$

I propose to look for solutions to Equation 2.78 in the form of *power series* in ξ:[24]

$$h(\xi) = a_0 + a_1\xi + a_2\xi^2 + \cdots = \sum_{j=0}^{\infty} a_j\xi^j. \tag{2.79}$$

Differentiating the series term by term,

$$\frac{dh}{d\xi} = a_1 + 2a_2\xi + 3a_3\xi^2 + \cdots = \sum_{j=0}^{\infty} ja_j\xi^{j-1},$$

and

$$\frac{d^2h}{d\xi^2} = 2a_2 + 2\cdot 3a_3\xi + 3\cdot 4a_4\xi^2 + \cdots = \sum_{j=0}^{\infty}(j+1)(j+2)a_{j+2}\xi^j.$$

[23] Note that although we invoked some approximations to *motivate* Equation 2.77, what follows is *exact*. The device of stripping off the asymptotic behavior is the standard first step in the power series method for solving differential equations—see, for example, Boas (footnote 11), Chapter 12.

[24] This is known as the **Frobenius method** for solving a differential equation. According to Taylor's theorem, *any* reasonably well-behaved function can be expressed as a power series, so Equation 2.79 ordinarily involves no loss of generality. For conditions on the applicability of the method, see Boas (footnote 11) or George B. Arfken and Hans-Jurgen Weber, *Mathematical Methods for Physicists*, 5th ed., Academic Press, Orlando (2000), Section 8.5.

Putting these into Equation 2.78, we find

$$\sum_{j=0}^{\infty}\left[(j+1)(j+2)a_{j+2}-2ja_j+(K-1)a_j\right]\xi^j=0. \qquad [2.80]$$

It follows (from the uniqueness of power series expansions[25]) that the coefficient of *each power* of ξ must vanish,

$$(j+1)(j+2)a_{j+2}-2ja_j+(K-1)a_j=0,$$

and hence that

$$a_{j+2}=\frac{(2j+1-K)}{(j+1)(j+2)}a_j. \qquad [2.81]$$

This **recursion formula** is entirely equivalent to the Schrödinger equation. Starting with a_0, it generates all the even-numbered coefficients:

$$a_2=\frac{(1-K)}{2}a_0, \quad a_4=\frac{(5-K)}{12}a_2=\frac{(5-K)(1-K)}{24}a_0, \quad \ldots,$$

and starting with a_1, it generates the odd coefficients:

$$a_3=\frac{(3-K)}{6}a_1, \quad a_5=\frac{(7-K)}{20}a_3=\frac{(7-K)(3-K)}{120}a_1, \quad \ldots.$$

We write the complete solution as

$$h(\xi)=h_{\text{even}}(\xi)+h_{\text{odd}}(\xi), \qquad [2.82]$$

where

$$h_{\text{even}}(\xi)\equiv a_0+a_2\xi^2+a_4\xi^4+\cdots$$

is an even function of ξ, built on a_0, and

$$h_{\text{odd}}(\xi)\equiv a_1\xi+a_3\xi^3+a_5\xi^5+\cdots$$

is an odd function, built on a_1. Thus Equation 2.81 determines $h(\xi)$ in terms of two arbitrary constants (a_0 and a_1)—which is just what we would expect, for a second-order differential equation.

However, not all the solutions so obtained are *normalizable*. For at very large j, the recursion formula becomes (approximately)

$$a_{j+2}\approx\frac{2}{j}a_j,$$

[25]See, for example, Arfken (footnote 24), Section 5.7.

with the (approximate) solution

$$a_j \approx \frac{C}{(j/2)!},$$

for some constant C, and this yields (at large ξ, where the higher powers dominate)

$$h(\xi) \approx C \sum \frac{1}{(j/2)!} \xi^j \approx C \sum \frac{1}{j!} \xi^{2j} \approx C e^{\xi^2}.$$

Now, if h goes like $\exp(\xi^2)$, then ψ (remember ψ?—that's what we're trying to calculate) goes like $\exp(\xi^2/2)$ (Equation 2.77), which is precisely the asymptotic behavior we *didn't* want.[26] There is only one way to wiggle out of this: For normalizable solutions *the power series must terminate*. There must occur some "highest" j (call it n), such that the recursion formula spits out $a_{n+2} = 0$ (this will truncate *either* the series h_{even} *or* the series h_{odd}; the *other* one must be zero from the start: $a_1 = 0$ if n is even, and $a_0 = 0$ if n is odd). For physically acceptable solutions, then, Equation 2.81 requires that

$$K = 2n + 1,$$

for some non-negative integer n, which is to say (referring to Equation 2.73) that the *energy* must be

$$E_n = \left(n + \frac{1}{2}\right) \hbar\omega, \quad \text{for } n = 0, 1, 2, \ldots. \qquad [2.83]$$

Thus we recover, by a completely different method, the fundamental quantization condition we found algebraically in Equation 2.61.

It seems at first rather surprising that the quantization of energy should emerge from a technical detail in the power series solution to the Schrödinger equation, but let's look at it from a different perspective. Equation 2.70 has solutions, of course, for *any* value of E (in fact, it has *two* linearly independent solutions for every E). But almost all of these solutions blow up exponentially at large x, and hence are not normalizable. Imagine, for example, using an E that is slightly *less* than one of the allowed values (say, $0.49\hbar\omega$), and plotting the solution (Figure 2.6(a)); the "tails" fly off to infinity. Now try an E slightly *larger* (say, $0.51\hbar\omega$); the "tails" now blow up in the *other* direction (Figure 2.6(b)). As you tweak the parameter in tiny increments from 0.49 to 0.51, the tails flip over when you pass through 0.5—only at *precisely* 0.5 do the tails go to zero, leaving a normalizable solution.[27]

---

[26] It's no surprise that the ill-behaved solutions are still contained in Equation 2.81; this recursion relation is equivalent to the Schrödinger equation, so it's *got* to include both the asymptotic forms we found in Equation 2.75.

[27] It is possible to set this up on a computer, and discover the allowed energies "experimentally." You might call it the **wag the dog** method: When the tail wags, you know you've just passed over an allowed value. See Problems 2.54–2.56.

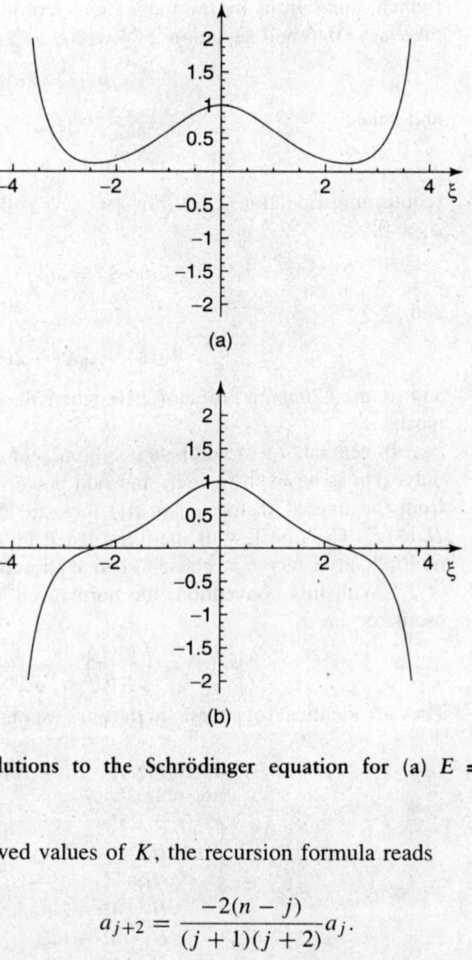

FIGURE 2.6: Solutions to the Schrödinger equation for (a) $E = 0.49\,\hbar\omega$, and (b) $E = 0.51\,\hbar\omega$.

For the allowed values of K, the recursion formula reads

$$a_{j+2} = \frac{-2(n-j)}{(j+1)(j+2)}a_j. \qquad [2.84]$$

If $n = 0$, there is only one term in the series (we must pick $a_1 = 0$ to kill h_{odd}, and $j = 0$ in Equation 2.84 yields $a_2 = 0$):

$$h_0(\xi) = a_0,$$

and hence

$$\psi_0(\xi) = a_0 e^{-\xi^2/2}$$

(which, apart from the normalization, reproduces Equation 2.59). For $n = 1$ we take $a_0 = 0$,[28] and Equation 2.84 with $j = 1$ yields $a_3 = 0$, so

$$h_1(\xi) = a_1\xi,$$

and hence

$$\psi_1(\xi) = a_1\xi e^{-\xi^2/2}$$

(confirming Equation 2.62). For $n = 2$, $j = 0$ yields $a_2 = -2a_0$, and $j = 2$ gives $a_4 = 0$, so

$$h_2(\xi) = a_0(1 - 2\xi^2),$$

and

$$\psi_2(\xi) = a_0(1 - 2\xi^2)e^{-\xi^2/2},$$

and so on. (Compare Problem 2.10, where this last result was obtained by algebraic means.)

In general, $h_n(\xi)$ will be a polynomial of degree n in ξ, involving even powers only, if n is an even integer, and odd powers only, if n is an odd integer. Apart from the overall factor (a_0 or a_1) they are the so-called **Hermite polynomials**, $H_n(\xi)$.[29] The first few of them are listed in Table 2.1. By tradition, the arbitrary multiplicative factor is chosen so that the coefficient of the highest power of ξ is 2^n. With this convention, the normalized[30] stationary states for the harmonic oscillator are

$$\psi_n(x) = \left(\frac{m\omega}{\pi\hbar}\right)^{1/4} \frac{1}{\sqrt{2^n n!}} H_n(\xi)e^{-\xi^2/2}. \qquad [2.85]$$

They are identical (of course) to the ones we obtained algebraically in Equation 2.67.

TABLE 2.1: The first few Hermite polynomials, $H_n(\xi)$.

$H_0 = 1,$
$H_1 = 2\xi,$
$H_2 = 4\xi^2 - 2,$
$H_3 = 8\xi^3 - 12\xi,$
$H_4 = 16\xi^4 - 48\xi^2 + 12,$
$H_5 = 32\xi^5 - 160\xi^3 + 120\xi.$

[28] Note that there is a completely different set of coefficients a_j for each value of n.

[29] The Hermite polynomials have been studied extensively in the mathematical literature, and there are many tools and tricks for working with them. A few of these are explored in Problem 2.17.

[30] I shall not work out the normalization constant here; if you are interested in knowing how it is done, see for example Leonard Schiff, *Quantum Mechanics*, 3rd ed., McGraw-Hill, New York (1968), Section 13.

In Figure 2.7(a) I have plotted $\psi_n(x)$ for the first few n's. The quantum oscillator is strikingly different from its classical counterpart—not only are the energies quantized, but the position distributions have some bizarre features. For instance, the probability of finding the particle outside the classically allowed range (that is, with x greater than the classical amplitude for the energy in question) is *not* zero (see Problem 2.15), and in all odd states the probability of finding the particle at the center is zero. Only at large n do we begin to see some resemblance to the classical case. In Figure 2.7(b) I have superimposed the classical position distribution on the quantum one (for $n = 100$); if you smoothed out the bumps, the two would fit pretty well (however, in the classical case we are talking about the distribution of positions over *time* for *one* oscillator, whereas in the quantum case we are talking about the distribution over an *ensemble* of identically prepared systems).[31]

Problem 2.15 In the ground state of the harmonic oscillator, what is the probability (correct to three significant digits) of finding the particle outside the classically allowed region? *Hint:* Classically, the energy of an oscillator is $E = (1/2)ka^2 = (1/2)m\omega^2 a^2$, where a is the amplitude. So the "classically allowed region" for an oscillator of energy E extends from $-\sqrt{2E/m\omega^2}$ to $+\sqrt{2E/m\omega^2}$. Look in a math table under "Normal Distribution" or "Error Function" for the numerical value of the integral.

Problem 2.16 Use the recursion formula (Equation 2.84) to work out $H_5(\xi)$ and $H_6(\xi)$. Invoke the convention that the coefficient of the highest power of ξ is 2^n to fix the overall constant.

∗∗**Problem 2.17** In this problem we explore some of the more useful theorems (stated without proof) involving Hermite polynomials.

(a) The **Rodrigues formula** says that

$$H_n(\xi) = (-1)^n e^{\xi^2} \left(\frac{d}{d\xi}\right)^n e^{-\xi^2}. \qquad [2.86]$$

Use it to derive H_3 and H_4.

(b) The following recursion relation gives you H_{n+1} in terms of the two preceding Hermite polynomials:

$$H_{n+1}(\xi) = 2\xi H_n(\xi) - 2n H_{n-1}(\xi). \qquad [2.87]$$

Use it, together with your answer in (a), to obtain H_5 and H_6.

---

[31] The parallel is perhaps more direct if you interpret the classical distribution as an ensemble of oscillators all with the same energy, but with random starting times.

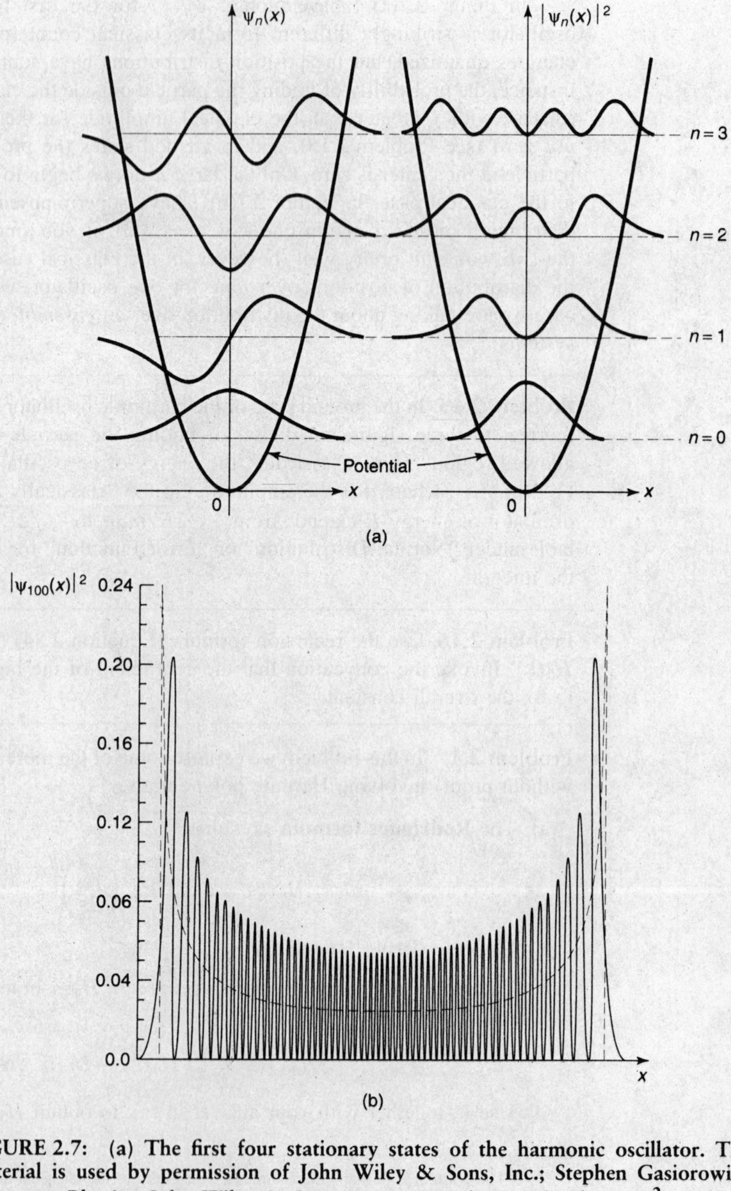

FIGURE 2.7: (a) The first four stationary states of the harmonic oscillator. This material is used by permission of John Wiley & Sons, Inc.; Stephen Gasiorowicz, *Quantum Physics*, John Wiley & Sons, Inc., 1974. (b) Graph of $|\psi_{100}|^2$, with the classical distribution (dashed curve) superimposed.

(c) If you differentiate an nth-order polynomial, you get a polynomial of order $(n-1)$. For the Hermite polynomials, in fact,

$$\frac{dH_n}{d\xi} = 2nH_{n-1}(\xi). \tag{2.88}$$

Check this, by differentiating H_5 and H_6.

(d) $H_n(\xi)$ is the nth z-derivative, at $z = 0$, of the **generating function** $\exp(-z^2 + 2z\xi)$; or, to put it another way, it is the coefficient of $z^n/n!$ in the Taylor series expansion for this function:

$$e^{-z^2+2z\xi} = \sum_{n=0}^{\infty} \frac{z^n}{n!} H_n(\xi). \tag{2.89}$$

Use this to rederive H_0, H_1, and H_2.

2.4 THE FREE PARTICLE

We turn next to what *should* have been the simplest case of all: the free particle ($V(x) = 0$ everywhere). Classically this would just mean motion at constant velocity, but in quantum mechanics the problem is surprisingly subtle and tricky. The time-independent Schrödinger equation reads

$$-\frac{\hbar^2}{2m}\frac{d^2\psi}{dx^2} = E\psi, \tag{2.90}$$

or

$$\frac{d^2\psi}{dx^2} = -k^2\psi, \quad \text{where } k \equiv \frac{\sqrt{2mE}}{\hbar}. \tag{2.91}$$

So far, it's the same as inside the infinite square well (Equation 2.21), where the potential is also zero; this time, however, I prefer to write the general solution in exponential form (instead of sines and cosines), for reasons that will appear in due course:

$$\psi(x) = Ae^{ikx} + Be^{-ikx}. \tag{2.92}$$

Unlike the infinite square well, there are no boundary conditions to restrict the possible values of k (and hence of E); the free particle can carry *any* (positive) energy. Tacking on the standard time dependence, $\exp(-iEt/\hbar)$,

$$\Psi(x,t) = Ae^{ik(x-\frac{\hbar k}{2m}t)} + Be^{-ik(x+\frac{\hbar k}{2m}t)}. \tag{2.93}$$

Now, *any* function of x and t that depends on these variables in the special combination $(x \pm vt)$ (for some constant v) represents a wave of fixed profile, traveling in the $\mp x$-direction, at speed v. A fixed point on the waveform (for example, a maximum or a minimum) corresponds to a fixed value of the argument, and hence to x and t such that

$$x \pm vt = \text{constant}, \quad \text{or} \quad x = \mp vt + \text{constant}.$$

Since every point on the waveform is moving along with the same velocity, its *shape* doesn't change as it propagates. Thus the first term in Equation 2.93 represents a wave traveling to the *right*, and the second represents a wave (of the same energy) going to the *left*. By the way, since they only differ by the *sign* in front of k, we might as well write

$$\Psi_k(x, t) = Ae^{i(kx - \frac{\hbar k^2}{2m}t)}, \qquad [2.94]$$

and let k run negative to cover the case of waves traveling to the left:

$$k \equiv \pm\frac{\sqrt{2mE}}{\hbar}, \quad \text{with} \begin{cases} k > 0 \Rightarrow & \text{traveling to the right,} \\ k < 0 \Rightarrow & \text{traveling to the left.} \end{cases} \qquad [2.95]$$

Evidently the "stationary states" of the free particle are propagating waves; their wavelength is $\lambda = 2\pi/|k|$, and, according to the de Broglie formula (Equation 1.39), they carry momentum

$$p = \hbar k. \qquad [2.96]$$

The speed of these waves (the coefficient of t over the coefficient of x) is

$$v_{\text{quantum}} = \frac{\hbar|k|}{2m} = \sqrt{\frac{E}{2m}}. \qquad [2.97]$$

On the other hand, the *classical* speed of a free particle with energy E is given by $E = (1/2)mv^2$ (pure kinetic, since $V = 0$), so

$$v_{\text{classical}} = \sqrt{\frac{2E}{m}} = 2v_{\text{quantum}}. \qquad [2.98]$$

Apparently the quantum mechanical wave function travels at *half* the speed of the particle it is supposed to represent! We'll return to this paradox in a moment—there is an even more serious problem we need to confront first: *This wave function is not normalizable.* For

$$\int_{-\infty}^{+\infty} \Psi_k^* \Psi_k \, dx = |A|^2 \int_{-\infty}^{+\infty} dx = |A|^2(\infty). \qquad [2.99]$$

In the case of the free particle, then, the separable solutions do not represent physically realizable states. A free particle cannot exist in a stationary state; or, to put it another way, *there is no such thing as a free particle with a definite energy.*

But that doesn't mean the separable solutions are of no use to us, for they play a *mathematical* role that is entirely independent of their *physical* interpretation. The general solution to the time-dependent Schrödinger equation is still a linear combination of separable solutions (only this time it's an *integral* over the continuous variable k, instead of a *sum* over the discrete index n):

$$\Psi(x,t) = \frac{1}{\sqrt{2\pi}} \int_{-\infty}^{+\infty} \phi(k) e^{i\left(kx - \frac{\hbar k^2}{2m} t\right)} \, dk. \qquad [2.100]$$

(The quantity $1/\sqrt{2\pi}$ is factored out for convenience; what plays the role of the coefficient c_n in Equation 2.17 is the combination $(1/\sqrt{2\pi})\phi(k)\,dk$.) Now *this* wave function *can* be normalized (for appropriate $\phi(k)$). But it necessarily carries a *range* of k's, and hence a range of energies and speeds. We call it a **wave packet**.[32]

In the generic quantum problem, we are *given* $\Psi(x, 0)$, and we are asked to *find* $\Psi(x, t)$. For a free particle the solution takes the form of Equation 2.100; the only question is how to determine $\phi(k)$ so as to match the initial wave function:

$$\Psi(x,0) = \frac{1}{\sqrt{2\pi}} \int_{-\infty}^{+\infty} \phi(k) e^{ikx} \, dk. \qquad [2.101]$$

This is a classic problem in Fourier analysis; the answer is provided by **Plancherel's theorem** (see Problem 2.20):

$$f(x) = \frac{1}{\sqrt{2\pi}} \int_{-\infty}^{+\infty} F(k) e^{ikx} \, dk \Longleftrightarrow F(k) = \frac{1}{\sqrt{2\pi}} \int_{-\infty}^{+\infty} f(x) e^{-ikx} \, dx. \qquad [2.102]$$

$F(k)$ is called the **Fourier transform** of $f(x)$; $f(x)$ is the **inverse Fourier transform** of $F(k)$ (the only difference is in the sign of the exponent). There is, of course, some restriction on the allowable functions: The integrals have to *exist*.[33] For our purposes this is guaranteed by the physical requirement that $\Psi(x, 0)$ itself

---

[32]Sinusoidal waves extend out to infinity, and they are not normalizable. But *superpositions* of such waves lead to interference, which allows for localization and normalizability.

[33]The necessary and sufficient condition on $f(x)$ is that $\int_{-\infty}^{\infty} |f(x)|^2 dx$ be *finite*. (In that case $\int_{-\infty}^{\infty} |F(k)|^2 dk$ is also finite, and in fact the two integrals are equal.) See Arfken (footnote 24), Section 15.5.

be normalized. So the solution to the generic quantum problem, for the free particle, is Equation 2.100, with

$$\boxed{\phi(k) = \frac{1}{\sqrt{2\pi}} \int_{-\infty}^{+\infty} \Psi(x, 0) e^{-ikx}\, dx.} \qquad [2.103]$$

Example 2.6 A free particle, which is initially localized in the range $-a < x < a$, is released at time $t = 0$:

$$\Psi(x, 0) = \begin{cases} A, & \text{if } -a < x < a, \\ 0, & \text{otherwise,} \end{cases}$$

where A and a are positive real constants. Find $\Psi(x, t)$.

Solution: First we need to normalize $\Psi(x, 0)$:

$$1 = \int_{-\infty}^{\infty} |\Psi(x, 0)|^2\, dx = |A|^2 \int_{-a}^{a} dx = 2a|A|^2 \;\Rightarrow\; A = \frac{1}{\sqrt{2a}}.$$

Next we calculate $\phi(k)$, using Equation 2.103:

$$\phi(k) = \frac{1}{\sqrt{2\pi}} \frac{1}{\sqrt{2a}} \int_{-a}^{a} e^{-ikx}\, dx = \frac{1}{2\sqrt{\pi a}} \frac{e^{-ikx}}{-ik}\bigg|_{-a}^{a}$$

$$= \frac{1}{k\sqrt{\pi a}} \left(\frac{e^{ika} - e^{-ika}}{2i} \right) = \frac{1}{\sqrt{\pi a}} \frac{\sin(ka)}{k}.$$

Finally, we plug this back into Equation 2.100:

$$\Psi(x, t) = \frac{1}{\pi\sqrt{2a}} \int_{-\infty}^{\infty} \frac{\sin(ka)}{k} e^{i(kx - \frac{\hbar k^2}{2m} t)}\, dk. \qquad [2.104]$$

Unfortunately, this integral cannot be solved in terms of elementary functions, though it can of course be evaluated numerically (Figure 2.8). (There are, in fact, precious few cases in which the integral for $\Psi(x, t)$ (Equation 2.100) *can* be calculated explicitly; see Problem 2.22 for a particularly beautiful example.)

It is illuminating to explore the limiting cases. If a is very small, the starting wave function is a nicely localized spike (Figure 2.9(a)). In this case we can use the small angle approximation to write $\sin(ka) \approx ka$, and hence

$$\phi(k) \approx \sqrt{\frac{a}{\pi}};$$

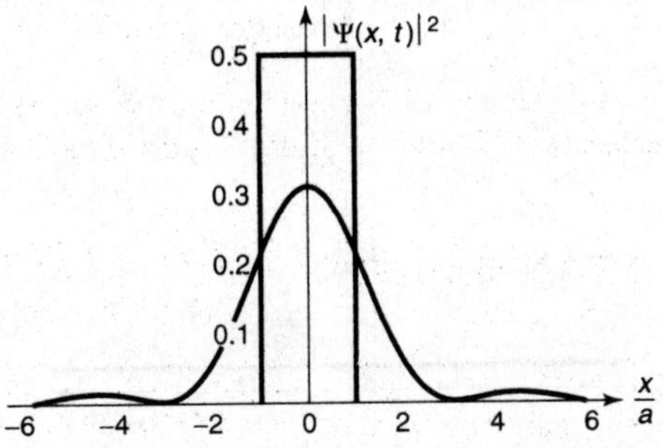

FIGURE 2.8: Graph of $|\Psi(x,t)|^2$ (Equation 2.104) at $t = 0$ (the rectangle) and at $t = ma^2/\hbar$ (the curve).

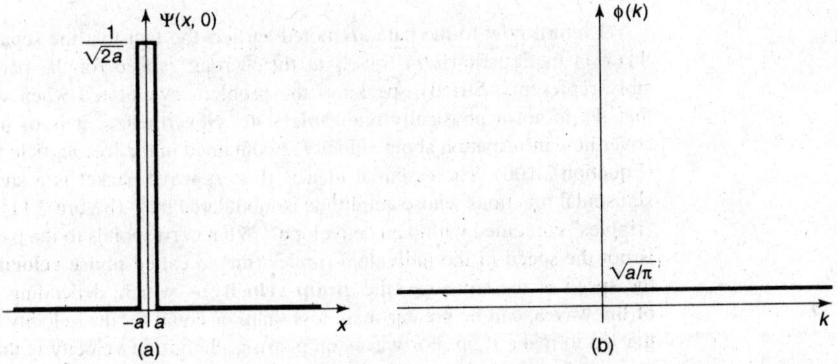

FIGURE 2.9: Example 2.6, for small a. (a) Graph of $\Psi(x, 0)$. (b) Graph of $\phi(k)$.

it's *flat*, since the k's cancelled out (Figure 2.9(b)). This is an example of the uncertainty principle: If the spread in *position* is small, the spread in *momentum* (and hence in k—see Equation 2.96) must be large. At the other extreme (*large* a) the spread in position is broad (Figure 2.10(a)) and

$$\phi(k) = \sqrt{\frac{a}{\pi}}\frac{\sin(ka)}{ka}.$$

Now, $\sin z/z$ has its maximum at $z = 0$, and drops to zero at $z = \pm\pi$ (which, in this context, means $k = \pm\pi/a$). So for large a, $\phi(k)$ is a sharp spike about $k = 0$ (Figure 2.10(b)). This time it's got a well-defined momentum but an ill-defined position.

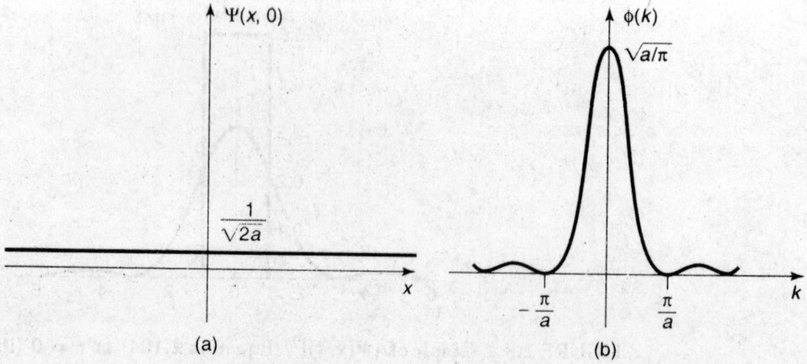

FIGURE 2.10: Example 2.6, for large a. (a) Graph of $\Psi(x, 0)$. (b) Graph of $\phi(k)$.

I return now to the paradox noted earlier: the fact that the separable solution $\Psi_k(x, t)$ in Equation 2.94 travels at the "wrong" speed for the particle it ostensibly represents. Strictly speaking, the problem evaporated when we discovered that Ψ_k is not a physically realizable state. Nevertheless, it is of interest to discover how information about velocity *is* contained in the free particle wave function (Equation 2.100). The essential idea is this: A wave packet is a superposition of sinusoidal functions whose amplitude is modulated by ϕ (Figure 2.11); it consists of "ripples" contained within an "envelope." What corresponds to the particle velocity is not the speed of the individual ripples (the so-called **phase velocity**), but rather the speed of the envelope (the **group velocity**)—which, depending on the nature of the waves, can be greater than, less than, or equal to, the velocity of the ripples that go to make it up. For waves on a string, the group velocity is the same as the phase velocity. For water waves it is one-half the phase velocity, as you may have noticed when you toss a rock into a pond (if you concentrate on a particular ripple, you will see it build up from the rear, move forward through the group, and fade away at the front, while the group as a whole propagates out at half the speed). What I need to show is that for the wave function of a free particle in quantum mechanics

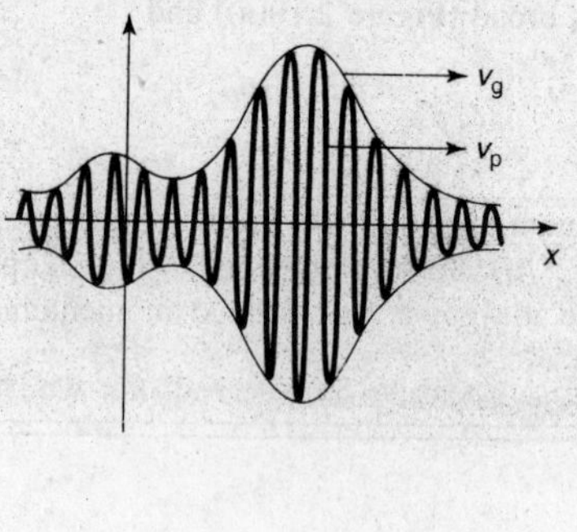

FIGURE 2.11: A wave packet. The "envelope" travels at the group velocity; the "ripples" travel at the phase velocity.

the group velocity is *twice* the phase velocity—just right to represent the classical particle speed.

The problem, then, is to determine the group velocity of a wave packet with the general form

$$\Psi(x, t) = \frac{1}{\sqrt{2\pi}} \int_{-\infty}^{+\infty} \phi(k) e^{i(kx - \omega t)}\, dk.$$

(In our case $\omega = (\hbar k^2/2m)$, but what I have to say now applies to *any* kind of wave packet, regardless of its **dispersion relation**—the formula for ω as a function of k.) Let us assume that $\phi(k)$ is narrowly peaked about some particular value k_0. (There is nothing *illegal* about a broad spread in k, but such wave packets change shape rapidly—since different components travel at different speeds—so the whole notion of a "group," with a well-defined velocity, loses its meaning.) Since the integrand is negligible except in the vicinity of k_0, we may as well Taylor-expand the function $\omega(k)$ about that point, and keep only the leading terms:

$$\omega(k) \cong \omega_0 + \omega_0'(k - k_0),$$

where ω_0' is the derivative of ω with respect to k, at the point k_0.

Changing variables from k to $s \equiv k - k_0$ (to center the integral at k_0), we have

$$\Psi(x, t) \cong \frac{1}{\sqrt{2\pi}} \int_{-\infty}^{+\infty} \phi(k_0 + s) e^{i[(k_0+s)x - (\omega_0 + \omega_0' s)t]}\, ds.$$

At $t = 0$,

$$\Psi(x, 0) = \frac{1}{\sqrt{2\pi}} \int_{-\infty}^{+\infty} \phi(k_0 + s) e^{i(k_0+s)x}\, ds,$$

and at later times

$$\Psi(x, t) \cong \frac{1}{\sqrt{2\pi}} e^{i(-\omega_0 t + k_0 \omega_0' t)} \int_{-\infty}^{+\infty} \phi(k_0 + s) e^{i(k_0+s)(x - \omega_0' t)}\, ds.$$

Except for the shift from x to $(x - \omega_0' t)$, the integral is the same as the one in $\Psi(x, 0)$. Thus

$$\Psi(x, t) \cong e^{-i(\omega_0 - k_0 \omega_0')t}\, \Psi(x - \omega_0' t, 0). \qquad [2.105]$$

Apart from the phase factor in front (which won't affect $|\Psi|^2$ in any event) the wave packet evidently moves along at a speed ω_0':

$$v_{\text{group}} = \frac{d\omega}{dk} \qquad [2.106]$$

(evaluated at $k = k_0$). This is to be contrasted with the ordinary phase velocity

$$v_{\text{phase}} = \frac{\omega}{k}. \qquad [2.107]$$

In our case, $\omega = (\hbar k^2/2m)$, so $\omega/k = (\hbar k/2m)$, whereas $d\omega/dk = (\hbar k/m)$, which is twice as great. This confirms that it is the group velocity of the wave packet, not the phase velocity of the stationary states, that matches the classical particle velocity:

$$v_{\text{classical}} = v_{\text{group}} = 2v_{\text{phase}}. \qquad [2.108]$$

Problem 2.18 Show that $[Ae^{ikx} + Be^{-ikx}]$ and $[C\cos kx + D\sin kx]$ are equivalent ways of writing the same function of x, and determine the constants C and D in terms of A and B, and vice versa. *Comment:* In quantum mechanics, when $V = 0$, the exponentials represent *traveling* waves, and are most convenient in discussing the free particle, whereas sines and cosines correspond to *standing* waves, which arise naturally in the case of the infinite square well.

Problem 2.19 Find the probability current, J (Problem 1.14) for the free particle wave function Equation 2.94. Which direction does the probability current flow?

****Problem 2.20** This problem is designed to guide you through a "proof" of Plancherel's theorem, by starting with the theory of ordinary Fourier series on a *finite* interval, and allowing that interval to expand to infinity.

(a) Dirichlet's theorem says that "any" function $f(x)$ on the interval $[-a, +a]$ can be expanded as a Fourier series:

$$f(x) = \sum_{n=0}^{\infty}[a_n \sin(n\pi x/a) + b_n \cos(n\pi x/a)].$$

Show that this can be written equivalently as

$$f(x) = \sum_{n=-\infty}^{\infty} c_n e^{in\pi x/a}.$$

What is c_n, in terms of a_n and b_n?

(b) Show (by appropriate modification of Fourier's trick) that

$$c_n = \frac{1}{2a}\int_{-a}^{+a} f(x)e^{-in\pi x/a}\,dx.$$

(c) Eliminate n and c_n in favor of the new variables $k = (n\pi/a)$ and $F(k) = \sqrt{2/\pi}\,ac_n$. Show that (a) and (b) now become

$$f(x) = \frac{1}{\sqrt{2\pi}}\sum_{n=-\infty}^{\infty} F(k)e^{ikx}\,\Delta k; \quad F(k) = \frac{1}{\sqrt{2\pi}}\int_{-a}^{+a} f(x)e^{-ikx}\,dx,$$

where Δk is the increment in k from one n to the next.

(d) Take the limit $a \to \infty$ to obtain Plancherel's theorem. *Comment:* In view of their quite different origins, it is surprising (and delightful) that the two formulas—one for $F(k)$ in terms of $f(x)$, the other for $f(x)$ in terms of $F(k)$—have such a similar structure in the limit $a \to \infty$.

Problem 2.21 A free particle has the initial wave function

$$\Psi(x, 0) = Ae^{-a|x|},$$

where A and a are positive real constants.

(a) Normalize $\Psi(x, 0)$.

(b) Find $\phi(k)$.

(c) Construct $\Psi(x, t)$, in the form of an integral.

(d) Discuss the limiting cases (a very large, and a very small).

***Problem 2.22 The gaussian wave packet.** A free particle has the initial wave function

$$\Psi(x, 0) = Ae^{-ax^2},$$

where A and a are constants (a is real and positive).

(a) Normalize $\Psi(x, 0)$.

(b) Find $\Psi(x, t)$. *Hint:* Integrals of the form

$$\int_{-\infty}^{+\infty} e^{-(ax^2+bx)}\, dx$$

can be handled by "completing the square": Let $y \equiv \sqrt{a}\,[x + (b/2a)]$, and note that $(ax^2 + bx) = y^2 - (b^2/4a)$. *Answer:*

$$\Psi(x, t) = \left(\frac{2a}{\pi}\right)^{1/4} \frac{e^{-ax^2/[1+(2i\hbar at/m)]}}{\sqrt{1 + (2i\hbar at/m)}}.$$

(c) Find $|\Psi(x, t)|^2$. Express your answer in terms of the quantity

$$w \equiv \sqrt{\frac{a}{1 + (2\hbar at/m)^2}}.$$

Sketch $|\Psi|^2$ (as a function of x) at $t = 0$, and again for some very large t. Qualitatively, what happens to $|\Psi|^2$, as time goes on?

(d) Find $\langle x\rangle$, $\langle p\rangle$, $\langle x^2\rangle$, $\langle p^2\rangle$, σ_x, and σ_p. *Partial answer:* $\langle p^2\rangle = a\hbar^2$, but it may take some algebra to reduce it to this simple form.

(e) Does the uncertainty principle hold? At what time t does the system come closest to the uncertainty limit?

2.5 THE DELTA-FUNCTION POTENTIAL

2.5.1 Bound States and Scattering States

We have encountered two very different kinds of solutions to the time-independent Schrödinger equation: For the infinite square well and the harmonic oscillator they are *normalizable*, and labeled by a *discrete index* n; for the free particle they are *non-normalizable*, and labeled by a *continuous variable* k. The former represent physically realizable states in their own right, the latter do not; but in both cases the general solution to the time-dependent Schrödinger equation is a linear combination of stationary states—for the first type this combination takes the form of a *sum* (over n), whereas for the second it is an *integral* (over k). What is the physical significance of this distinction?

In *classical* mechanics a one-dimensional time-independent potential can give rise to two rather different kinds of motion. If $V(x)$ rises higher than the particle's total energy (E) on either side (Figure 2.12(a)), then the particle is "stuck" in the potential well—it rocks back and forth between the **turning points**, but it cannot escape (unless, of course, you provide it with a source of extra energy, such as a motor, but we're not talking about that). We call this a **bound state**. If, on the other hand, E exceeds $V(x)$ on one side (or both), then the particle comes in from "infinity," slows down or speeds up under the influence of the potential, and returns to infinity (Figure 2.12(b)). (It can't get trapped in the potential unless there is some mechanism, such as friction, to *dissipate* energy, but again, we're not talking about that.) We call this a **scattering state**. Some potentials admit only bound states (for instance, the harmonic oscillator); some allow only scattering states (a potential hill with no dips in it, for example); some permit both kinds, depending on the energy of the particle.

The two kinds of solutions to the Schrödinger equation correspond precisely to bound and scattering states. The distinction is even cleaner in the quantum domain, because the phenomenon of **tunneling** (which we'll come to shortly) allows the particle to "leak" through any finite potential barrier, so the only thing that matters is the potential at infinity (Figure 2.12(c)):

$$\begin{cases} E < [V(-\infty) \text{ and } V(+\infty)] \Rightarrow & \text{bound state,} \\ E > [V(-\infty) \text{ or } V(+\infty)] \Rightarrow & \text{scattering state.} \end{cases} \quad [2.109]$$

In "real life" most potentials go to *zero* at infinity, in which case the criterion simplifies even further:

$$\begin{cases} E < 0 \Rightarrow & \text{bound state,} \\ E > 0 \Rightarrow & \text{scattering state.} \end{cases} \quad [2.110]$$

Because the infinite square well and harmonic oscillator potentials go to infinity as $x \rightarrow \pm\infty$, they admit bound states only; because the free particle potential is zero

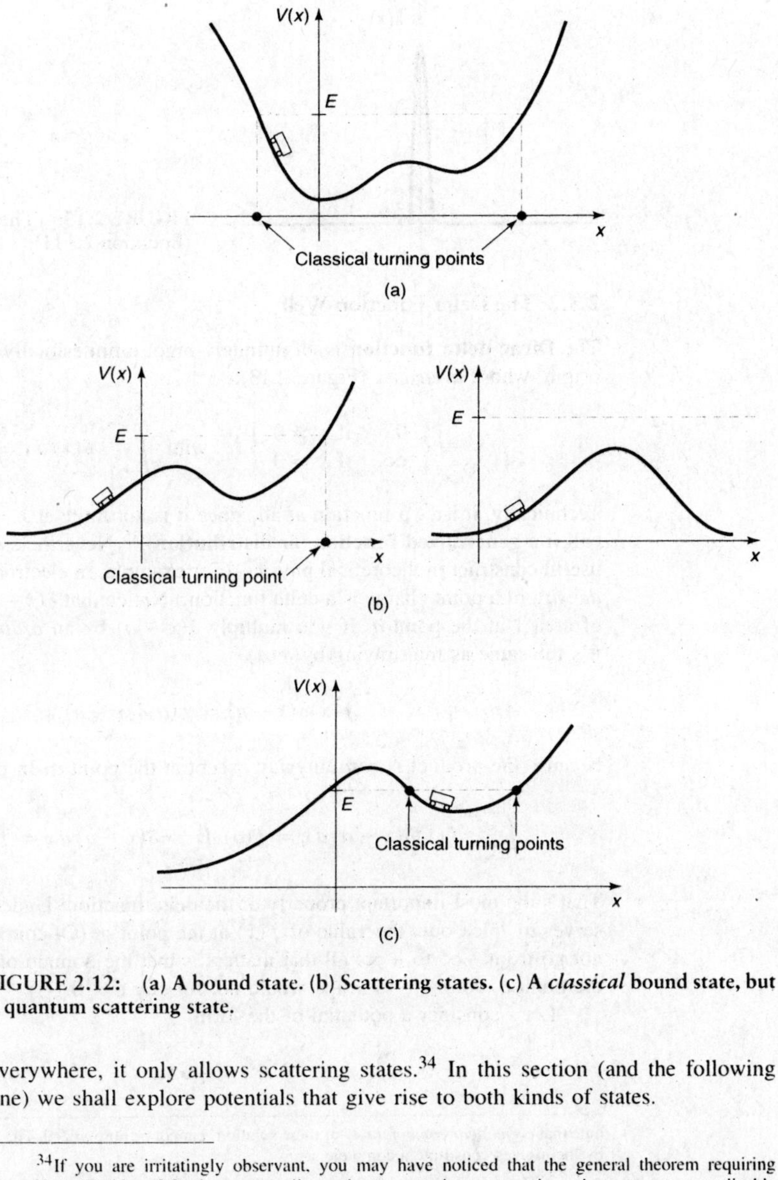

FIGURE 2.12: (a) A bound state. (b) Scattering states. (c) A *classical* bound state, but a quantum scattering state.

everywhere, it only allows scattering states.[34] In this section (and the following one) we shall explore potentials that give rise to both kinds of states.

---

[34]If you are irritatingly observant, you may have noticed that the general theorem requiring $E > V_{\min}$ (Problem 2.2) does not really apply to scattering states, since they are not normalizable anyway. If this bothers you, try solving the Schrödinger equation with $E \le 0$, for the free particle, and

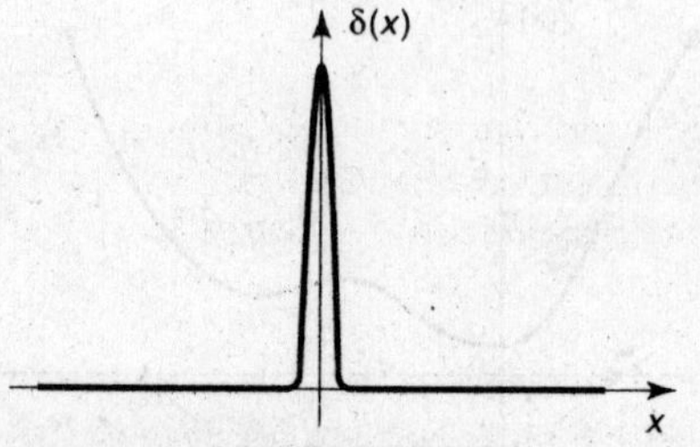

FIGURE 2.13: The Dirac delta function (Equation 2.111).

2.5.2 The Delta-Function Well

The **Dirac delta function** is an infinitely high, infinitesimally narrow spike at the origin, whose *area* is 1 (Figure 2.13):

$$\delta(x) \equiv \begin{Bmatrix} 0, & \text{if } x \neq 0 \\ \infty, & \text{if } x = 0 \end{Bmatrix}, \quad \text{with} \quad \int_{-\infty}^{+\infty} \delta(x)\, dx = 1. \qquad [2.111]$$

Technically, it isn't a function at all, since it is not finite at $x = 0$ (mathematicians call it a **generalized function**, or **distribution**).[35] Nevertheless, it is an extremely useful construct in theoretical physics. (For example, in electrodynamics the charge *density* of a point charge is a delta function.) Notice that $\delta(x-a)$ would be a spike of area 1 at the point a. If you multiply $\delta(x-a)$ by an *ordinary* function $f(x)$, it's the same as multiplying by $f(a)$,

$$f(x)\delta(x-a) = f(a)\delta(x-a), \qquad [2.112]$$

because the product is *zero* anyway except at the point a. In particular,

$$\int_{-\infty}^{+\infty} f(x)\delta(x-a)\, dx = f(a) \int_{-\infty}^{+\infty} \delta(x-a)\, dx = f(a). \qquad [2.113]$$

That's the most important property of the delta function: Under the integral sign it serves to "pick out" the value of $f(x)$ at the point a. (Of course, the integral need not go from $-\infty$ to $+\infty$; all that matters is that the domain of integration include the point a, so $a - \epsilon$ to $a + \epsilon$ would do, for any $\epsilon > 0$.)

Let's consider a potential of the form

$$V(x) = -\alpha\delta(x), \qquad [2.114]$$

note that *even linear combinations* of these solutions cannot be normalized. The positive energy solutions by themselves constitute a complete set.

[35]The delta function can be thought of as the *limit* of a *sequence* of functions, such as rectangles (or triangles) of ever-increasing height and ever-decreasing width.

where α is some positive constant.[36] This is an artificial potential, to be sure (so was the infinite square well), but it's delightfully simple to work with, and illuminates the basic theory with a minimum of analytical clutter. The Schrödinger equation for the delta-function well reads

$$-\frac{\hbar^2}{2m}\frac{d^2\psi}{dx^2} - \alpha\delta(x)\psi = E\psi; \qquad [2.115]$$

it yields both bound states ($E < 0$) and scattering states ($E > 0$).

We'll look first at the bound states. In the region $x < 0$, $V(x) = 0$, so

$$\frac{d^2\psi}{dx^2} = -\frac{2mE}{\hbar^2}\psi = \kappa^2\psi, \qquad [2.116]$$

where

$$\kappa \equiv \frac{\sqrt{-2mE}}{\hbar}. \qquad [2.117]$$

(E is negative, by assumption, so κ is real and positive.) The general solution to Equation 2.116 is

$$\psi(x) = Ae^{-\kappa x} + Be^{\kappa x}, \qquad [2.118]$$

but the first term blows up as $x \to -\infty$, so we must choose $A = 0$:

$$\psi(x) = Be^{\kappa x}, \quad (x < 0). \qquad [2.119]$$

In the region $x > 0$, $V(x)$ is again zero, and the general solution is of the form $F\exp(-\kappa x) + G\exp(\kappa x)$; this time it's the second term that blows up (as $x \to +\infty$), so

$$\psi(x) = Fe^{-\kappa x}, \quad (x > 0). \qquad [2.120]$$

It remains only to stitch these two functions together, using the appropriate boundary conditions at $x = 0$. I quoted earlier the standard boundary conditions for ψ:

$$\begin{cases} 1.\ \psi & \text{is always continuous;} \\ 2.\ d\psi/dx & \text{is continuous except at points where the potential is infinite.} \end{cases} \qquad [2.121]$$

In this case the first boundary condition tells us that $F = B$, so

$$\psi(x) = \begin{cases} Be^{\kappa x}, & (x \le 0), \\ Be^{-\kappa x}, & (x \ge 0); \end{cases} \qquad [2.122]$$

---

[36] The delta function itself carries units of 1/*length* (see Equation 2.111), so α has the dimensions *energy* × *length*.

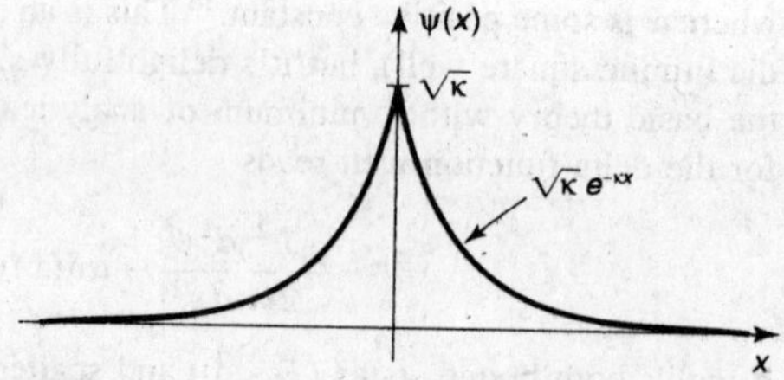

FIGURE 2.14: Bound state wave function for the delta-function potential (Equation 2.122).

$\psi(x)$ is plotted in Figure 2.14. The second boundary condition tells us nothing; this is (like the infinite square well) the exceptional case where V is infinite at the join, and it's clear from the graph that this function has a kink at $x = 0$. Moreover, up to this point the delta function has not come into the story at all. Evidently the delta function must determine the discontinuity in the derivative of ψ, at $x = 0$. I'll show you now how this works, and as a by-product we'll see why $d\psi/dx$ is ordinarily continuous.

The idea is to *integrate* the Schrödinger equation, from $-\epsilon$ to $+\epsilon$, and then take the limit as $\epsilon \to 0$:

$$-\frac{\hbar^2}{2m}\int_{-\epsilon}^{+\epsilon}\frac{d^2\psi}{dx^2}\,dx + \int_{-\epsilon}^{+\epsilon} V(x)\psi(x)\,dx = E\int_{-\epsilon}^{+\epsilon}\psi(x)\,dx. \qquad [2.123]$$

The first integral is nothing but $d\psi/dx$, evaluated at the two end points; the last integral is *zero*, in the limit $\epsilon \to 0$, since it's the area of a sliver with vanishing width and finite height. Thus

$$\Delta\left(\frac{d\psi}{dx}\right) \equiv \left.\frac{\partial\psi}{\partial x}\right|_{+\epsilon} - \left.\frac{\partial\psi}{\partial x}\right|_{-\epsilon} = \frac{2m}{\hbar^2}\lim_{\epsilon\to 0}\int_{-\epsilon}^{+\epsilon} V(x)\psi(x)\,dx. \qquad [2.124]$$

Typically, the limit on the right is again zero, and that's why $d\psi/dx$ is ordinarily continuous. But when $V(x)$ is *infinite* at the boundary, this argument fails. In particular, if $V(x) = -\alpha\delta(x)$, Equation 2.113 yields

$$\Delta\left(\frac{d\psi}{dx}\right) = -\frac{2m\alpha}{\hbar^2}\psi(0). \qquad [2.125]$$

For the case at hand (Equation 2.122),

$$\begin{cases} d\psi/dx = -B\kappa e^{-\kappa x}, & \text{for } (x > 0), \quad \text{so } d\psi/dx\big|_+ = -B\kappa, \\ d\psi/dx = +B\kappa e^{+\kappa x}, & \text{for } (x < 0), \quad \text{so } d\psi/dx\big|_- = +B\kappa, \end{cases}$$

and hence $\Delta(d\psi/dx) = -2B\kappa$. And $\psi(0) = B$. So Equation 2.125 says

$$\kappa = \frac{m\alpha}{\hbar^2}, \qquad [2.126]$$

and the allowed energy (Equation 2.117) is

$$E = -\frac{\hbar^2\kappa^2}{2m} = -\frac{m\alpha^2}{2\hbar^2}. \quad [2.127]$$

Finally, we normalize ψ:

$$\int_{-\infty}^{+\infty} |\psi(x)|^2\, dx = 2|B|^2 \int_0^{\infty} e^{-2\kappa x}\, dx = \frac{|B|^2}{\kappa} = 1,$$

so (choosing, for convenience, the positive real root):

$$B = \sqrt{\kappa} = \frac{\sqrt{m\alpha}}{\hbar}. \quad [2.128]$$

Evidently the delta-function well, regardless of its "strength" α, has *exactly one* bound state:

$$\boxed{\psi(x) = \frac{\sqrt{m\alpha}}{\hbar} e^{-m\alpha|x|/\hbar^2}; \quad E = -\frac{m\alpha^2}{2\hbar^2}.} \quad [2.129]$$

What about *scattering* states, with $E > 0$? For $x < 0$ the Schrödinger equation reads

$$\frac{d^2\psi}{dx^2} = -\frac{2mE}{\hbar^2}\psi = -k^2\psi,$$

where

$$k \equiv \frac{\sqrt{2mE}}{\hbar} \quad [2.130]$$

is real and positive. The general solution is

$$\psi(x) = Ae^{ikx} + Be^{-ikx}, \quad [2.131]$$

and this time we cannot rule out either term, since neither of them blows up. Similarly, for $x > 0$,

$$\psi(x) = Fe^{ikx} + Ge^{-ikx}. \quad [2.132]$$

The continuity of $\psi(x)$ at $x = 0$ requires that

$$F + G = A + B. \quad [2.133]$$

The derivatives are

$$\begin{cases} d\psi/dx = ik\left(Fe^{ikx} - Ge^{-ikx}\right), & \text{for } (x > 0), \quad \text{so } d\psi/dx\big|_+ = ik(F - G), \\ d\psi/dx = ik\left(Ae^{ikx} - Be^{-ikx}\right), & \text{for } (x < 0), \quad \text{so } d\psi/dx\big|_- = ik(A - B), \end{cases}$$

and hence $\Delta(d\psi/dx) = ik(F - G - A + B)$. Meanwhile, $\psi(0) = (A + B)$, so the second boundary condition (Equation 2.125) says

$$ik(F - G - A + B) = -\frac{2m\alpha}{\hbar^2}(A + B), \quad [2.134]$$

or, more compactly,

$$F - G = A(1 + 2i\beta) - B(1 - 2i\beta), \quad \text{where } \beta \equiv \frac{m\alpha}{\hbar^2 k}. \quad [2.135]$$

Having imposed both boundary conditions, we are left with two equations (Equations 2.133 and 2.135) in four unknowns (A, B, F, and G)—*five*, if you count k. Normalization won't help—this isn't a normalizable state. Perhaps we'd better pause, then, and examine the physical significance of these various constants. Recall that $\exp(ikx)$ gives rise (when coupled with the time-dependent factor $\exp(-iEt/\hbar)$) to a wave function propagating to the *right*, and $\exp(-ikx)$ leads to a wave propagating to the *left*. It follows that A (in Equation 2.131) is the amplitude of a wave coming in from the left, B is the amplitude of a wave returning to the left, F (Equation 2.132) is the amplitude of a wave traveling off to the right, and G is the amplitude of a wave coming in from the right (see Figure 2.15). In a typical scattering experiment particles are fired in from one direction—let's say, from the left. In that case the amplitude of the wave coming in from the *right* will be *zero*:

$$G = 0, \quad \text{(for scattering from the left);} \quad [2.136]$$

A is the amplitude of the **incident wave**, B is the amplitude of the **reflected wave**, and F is the amplitude of the **transmitted wave**. Solving Equations 2.133 and 2.135 for B and F, we find

$$B = \frac{i\beta}{1 - i\beta}A, \quad F = \frac{1}{1 - i\beta}A. \quad [2.137]$$

(If you want to study scattering from the *right*, set $A = 0$; then G is the incident amplitude, F is the reflected amplitude, and B is the transmitted amplitude.)

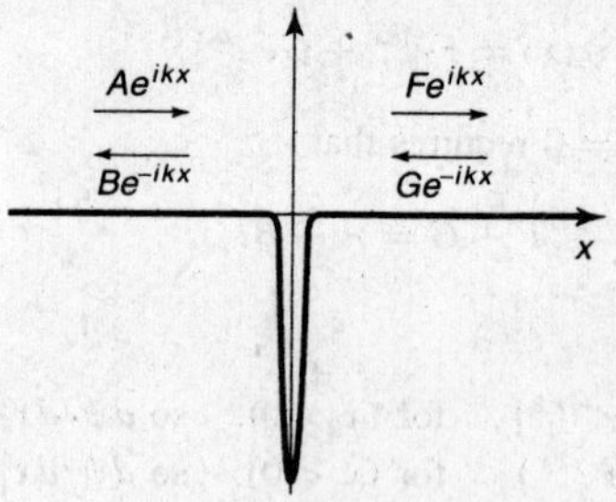

FIGURE 2.15: Scattering from a delta function well.

Now, the probability of finding the particle at a specified location is given by $|\Psi|^2$, so the *relative*[37] probability that an incident particle will be reflected back is

$$R \equiv \frac{|B|^2}{|A|^2} = \frac{\beta^2}{1+\beta^2}. \qquad [2.138]$$

R is called the **reflection coefficient**. (If you have a *beam* of particles, it tells you the *fraction* of the incoming number that will bounce back.) Meanwhile, the probability of transmission is given by the **transmission coefficient**

$$T \equiv \frac{|F|^2}{|A|^2} = \frac{1}{1+\beta^2}. \qquad [2.139]$$

Of course, the *sum* of these probabilities should be 1—and it *is*:

$$R + T = 1. \qquad [2.140]$$

Notice that R and T are functions of β, and hence (Equations 2.130 and 2.135) of E:

$$\boxed{R = \frac{1}{1 + (2\hbar^2 E/m\alpha^2)}, \quad T = \frac{1}{1 + (m\alpha^2/2\hbar^2 E)}.} \qquad [2.141]$$

The higher the energy, the greater the probability of transmission (which certainly seems reasonable).

This is all very tidy, but there is a sticky matter of principle that we cannot altogether ignore: These scattering wave functions are not normalizable, so they don't actually represent possible particle states. But we know what the resolution to this problem is: We must form normalizable linear combinations of the stationary states, just as we did for the free particle—true physical particles are represented by the resulting wave packets. Though straightforward in principle, this is a messy business in practice, and at this point it is best to turn the problem over to a computer.[38] Meanwhile, since it is impossible to create a normalizable free-particle wave function without involving a *range* of energies, R and T should be interpreted as the *approximate* reflection and transmission probabilities for particles in the *vicinity* of E.

Incidentally, it might strike you as peculiar that we were able to analyze a quintessentially time-dependent problem (particle comes in, scatters off a potential,

---

[37]This is not a normalizable wave function, so the *absolute* probability of finding the particle at a particular location is not well defined; nevertheless, the *ratio* of probabilities for the incident and reflected waves *is* meaningful. More on this in the next paragraph.

[38]Numerical studies of wave packets scattering off wells and barriers reveal extraordinarily rich structure. The classic analysis is A. Goldberg, H. M. Schey, and J. L. Schwartz, *Am. J. Phys.* **35**, 177 (1967); more recent work can be found on the Web.

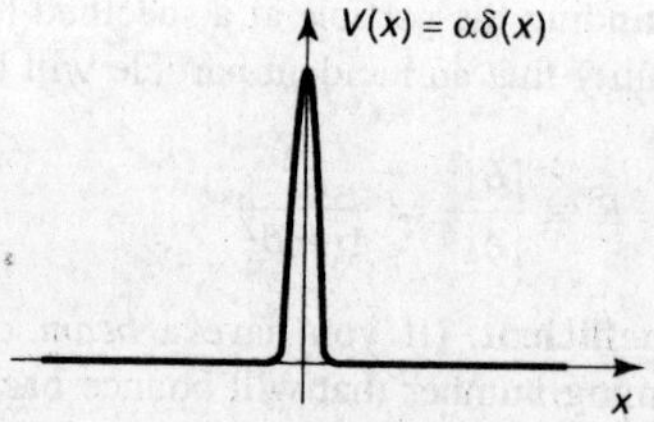

FIGURE 2.16: The delta-function barrier.

and flies off to infinity) using *stationary* states. After all, ψ (in Equations 2.131 and 2.132) is simply a complex, time-independent, sinusoidal function, extending (with constant amplitude) to infinity in both directions. And yet, by imposing appropriate boundary conditions on this function we were able to determine the probability that a particle (represented by a *localized* wave packet) would bounce off, or pass through, the potential. The mathematical miracle behind this is, I suppose, the fact that by taking linear combinations of states spread over all space, and with essentially trivial time dependence, we can *construct* wave functions that are concentrated about a (moving) point, with quite elaborate behavior in time (see Problem 2.43).

As long as we've got the relevant equations on the table, let's look briefly at the case of a delta-function *barrier* (Figure 2.16). Formally, all we have to do is change the sign of α. This kills the bound state, of course (Problem 2.2). On the other hand, the reflection and transmission coefficients, which depend only on α^2, are unchanged. Strange to say, the particle is just as likely to pass through the barrier as to cross over the well! *Classically*, of course, a particle cannot make it over an infinitely high barrier, regardless of its energy. In fact, classical scattering problems are pretty dull: If $E > V_{\text{max}}$, then $T = 1$ and $R = 0$—the particle certainly makes it over; if $E < V_{\text{max}}$ then $T = 0$ and $R = 1$—it rides up the hill until it runs out of steam, and then returns the same way it came. *Quantum* scattering problems are much richer: The particle has some nonzero probability of passing through the potential even if $E < V_{\text{max}}$. We call this phenomenon **tunneling**; it is the mechanism that makes possible much of modern electronics—not to mention spectacular advances in microscopy. Conversely, even if $E > V_{\text{max}}$ there is a possibility that the particle will bounce back—though I wouldn't advise driving off a cliff in the hope that quantum mechanics will save you (see Problem 2.35).

***Problem 2.23** Evaluate the following integrals:

(a) $\int_{-3}^{+1}(x^3 - 3x^2 + 2x - 1)\delta(x+2)\,dx$.

(b) $\int_0^{\infty}[\cos(3x) + 2]\delta(x - \pi)\,dx$.

(c) $\int_{-1}^{+1}\exp(|x| + 3)\delta(x - 2)\,dx$.

Problem 2.24 Delta functions live under integral signs, and two expressions ($D_1(x)$ and $D_2(x)$) involving delta functions are said to be equal if

$$\int_{-\infty}^{+\infty} f(x)D_1(x)\,dx = \int_{-\infty}^{+\infty} f(x)D_2(x)\,dx,$$

for every (ordinary) function $f(x)$.

(a) Show that

$$\delta(cx) = \frac{1}{|c|}\delta(x), \qquad [2.142]$$

where c is a real constant. (Be sure to check the case where c is negative.)

(b) Let $\theta(x)$ be the **step function**:

$$\theta(x) \equiv \begin{cases} 1, & \text{if } x > 0, \\ 0, & \text{if } x < 0. \end{cases} \qquad [2.143]$$

(In the rare case where it actually matters, we define $\theta(0)$ to be 1/2.) Show that $d\theta/dx = \delta(x)$.

∗∗**Problem 2.25** Check the uncertainty principle for the wave function in Equation 2.129. *Hint:* Calculating $\langle p^2\rangle$ is tricky, because the derivative of ψ has a step discontinuity at $x = 0$. Use the result in Problem 2.24(b). *Partial answer:* $\langle p^2\rangle = (m\alpha/\hbar)^2$.

∗**Problem 2.26** What is the Fourier transform of $\delta(x)$? Using Plancherel's theorem, show that

$$\delta(x) = \frac{1}{2\pi}\int_{-\infty}^{+\infty} e^{ikx}\,dk. \qquad [2.144]$$

Comment: This formula gives any respectable mathematician apoplexy. Although the integral is clearly infinite when $x = 0$, it doesn't converge (to zero or anything else) when $x \neq 0$, since the integrand oscillates forever. There are ways to patch it up (for instance, you can integrate from $-L$ to $+L$, and interpret Equation 2.144 to mean the *average* value of the finite integral, as $L \to \infty$). The source of the problem is that the delta function doesn't meet the requirement (square-integrability) for Plancherel's theorem (see footnote 33). In spite of this, Equation 2.144 can be extremely useful, if handled with care.

∗**Problem 2.27** Consider the *double* delta-function potential

$$V(x) = -\alpha[\delta(x+a) + \delta(x-a)],$$

where α and a are positive constants.

(a) Sketch this potential.

(b) How many bound states does it possess? Find the allowed energies, for $\alpha = \hbar^2/ma$ and for $\alpha = \hbar^2/4ma$, and sketch the wave functions.

****Problem 2.28** Find the transmission coefficient for the potential in Problem 2.27.

2.6 THE FINITE SQUARE WELL

As a last example, consider the *finite* square well potential

$$V(x) = \begin{cases} -V_0, & \text{for } -a < x < a, \\ 0, & \text{for } |x| > a, \end{cases} \qquad [2.145]$$

where V_0 is a (positive) constant (Figure 2.17). Like the delta-function well, this potential admits both bound states (with $E < 0$) and scattering states (with $E > 0$). We'll look first at the bound states.

In the region $x < -a$ the potential is zero, so the Schrödinger equation reads

$$-\frac{\hbar^2}{2m}\frac{d^2\psi}{dx^2} = E\psi, \quad \text{or} \quad \frac{d^2\psi}{dx^2} = \kappa^2\psi,$$

where

$$\kappa \equiv \frac{\sqrt{-2mE}}{\hbar} \qquad [2.146]$$

is real and positive. The general solution is $\psi(x) = A\exp(-\kappa x) + B\exp(\kappa x)$, but the first term blows up (as $x \to -\infty$), so the physically admissible solution (as before—see Equation 2.119) is

$$\psi(x) = Be^{\kappa x}, \quad \text{for } x < -a. \qquad [2.147]$$

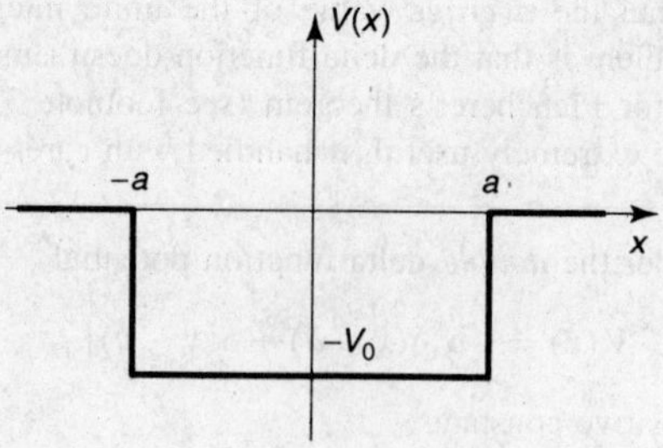

FIGURE 2.17: The finite square well (Equation 2.145).

In the region $-a < x < a$, $V(x) = -V_0$, and the Schrödinger equation reads

$$-\frac{\hbar^2}{2m}\frac{d^2\psi}{dx^2} - V_0\psi = E\psi, \quad \text{or} \quad \frac{d^2\psi}{dx^2} = -l^2\psi,$$

where

$$l \equiv \frac{\sqrt{2m(E+V_0)}}{\hbar}. \qquad [2.148]$$

Although E is negative, for bound states, it must be greater than $-V_0$, by the old theorem $E > V_{\min}$ (Problem 2.2); so l is also real and positive. The general solution is[39]

$$\psi(x) = C\sin(lx) + D\cos(lx), \quad \text{for } -a < x < a, \qquad [2.149]$$

where C and D are arbitrary constants. Finally, in the region $x > a$ the potential is again zero; the general solution is $\psi(x) = F\exp(-\kappa x) + G\exp(\kappa x)$, but the second term blows up (as $x \to \infty$), so we are left with

$$\psi(x) = Fe^{-\kappa x}, \quad \text{for } x > a. \qquad [2.150]$$

The next step is to impose boundary conditions: ψ and $d\psi/dx$ continuous at $-a$ and $+a$. But we can save a little time by noting that this potential is an even function, so we can assume with no loss of generality that the solutions are either even or odd (Problem 2.1(c)). The advantage of this is that we need only impose the boundary conditions on one side (say, at $+a$); the other side is then automatic, since $\psi(-x) = \pm\psi(x)$. I'll work out the even solutions; you get to do the odd ones in Problem 2.29. The cosine is even (and the sine is odd), so I'm looking for solutions of the form

$$\psi(x) = \begin{cases} Fe^{-\kappa x}, & \text{for } x > a, \\ D\cos(lx), & \text{for } 0 < x < a, \\ \psi(-x), & \text{for } x < 0. \end{cases} \qquad [2.151]$$

The continuity of $\psi(x)$, at $x = a$, says

$$Fe^{-\kappa a} = D\cos(la), \qquad [2.152]$$

and the continuity of $d\psi/dx$, says

$$-\kappa Fe^{-\kappa a} = -lD\sin(la). \qquad [2.153]$$

Dividing Equation 2.153 by Equation 2.152, we find that

$$\kappa = l\tan(la). \qquad [2.154]$$

---

[39] You can, if you like, write the general solution in exponential form ($C'e^{ilx} + D'e^{-ilx}$). This leads to the same final result, but since the potential is symmetric we know the solutions will be either even or odd, and the sine/cosine notation allows us to exploit this directly.

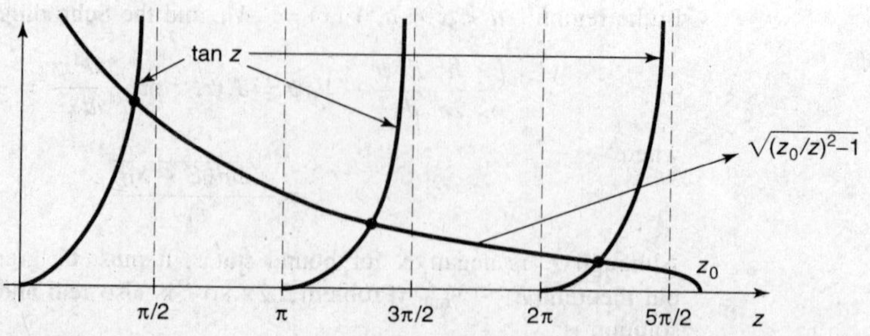

FIGURE 2.18: Graphical solution to Equation 2.156, for $z_0 = 8$ (*even* states).

This is a formula for the allowed energies, since κ and l are both functions of E. To solve for E, we first adopt some nicer notation: Let

$$z \equiv la, \quad \text{and} \quad z_0 \equiv \frac{a}{\hbar}\sqrt{2mV_0}. \tag{2.155}$$

According to Equations 2.146 and 2.148, $(\kappa^2 + l^2) = 2mV_0/\hbar^2$, so $\kappa a = \sqrt{z_0^2 - z^2}$, and Equation 2.154 reads

$$\tan z = \sqrt{(z_0/z)^2 - 1}. \tag{2.156}$$

This is a transcendental equation for z (and hence for E) as a function of z_0 (which is a measure of the "size" of the well). It can be solved numerically, using a computer, or graphically, by plotting $\tan z$ and $\sqrt{(z_0/z)^2 - 1}$ on the same grid, and looking for points of intersection (see Figure 2.18). Two limiting cases are of special interest:

1. Wide, deep well. If z_0 is very large, the intersections occur just slightly below $z_n = n\pi/2$, with n odd; it follows that

$$E_n + V_0 \cong \frac{n^2\pi^2\hbar^2}{2m(2a)^2}. \tag{2.157}$$

But $E + V_0$ is the energy *above the bottom of the well*, and on the right side we have precisely the infinite square well energies, for a well of width $2a$ (see Equation 2.27)—or rather, *half* of them, since this n is odd. (The other ones, of course, come from the *odd* wave functions, as you'll discover in Problem 2.29.) So the finite square well goes over to the infinite square well, as $V_0 \to \infty$; however, for any *finite* V_0 there are only a finite number of bound states.

2. Shallow, narrow well. As z_0 decreases, there are fewer and fewer bound states, until finally (for $z_0 < \pi/2$, where the lowest *odd* state disappears) only one remains. It is interesting to note, however, that there is always *one* bound state, no matter *how* "weak" the well becomes.

You're welcome to normalize ψ (Equation 2.151), if you're interested (Problem 2.30), but I'm going to move on now to the scattering states ($E > 0$). To the left, where $V(x) = 0$, we have

$$\psi(x) = Ae^{ikx} + Be^{-ikx}, \quad \text{for } (x < -a), \tag{2.158}$$

where (as usual)

$$k \equiv \frac{\sqrt{2mE}}{\hbar}. \tag{2.159}$$

Inside the well, where $V(x) = -V_0$,

$$\psi(x) = C\sin(lx) + D\cos(lx), \quad \text{for } (-a < x < a), \tag{2.160}$$

where, as before,

$$l \equiv \frac{\sqrt{2m(E + V_0)}}{\hbar}. \tag{2.161}$$

To the right, assuming there is no incoming wave in this region, we have

$$\psi(x) = Fe^{ikx}. \tag{2.162}$$

Here A is the incident amplitude, B is the reflected amplitude, and F is the transmitted amplitude.[40]

There are four boundary conditions: Continuity of $\psi(x)$ at $-a$ says

$$Ae^{-ika} + Be^{ika} = -C\sin(la) + D\cos(la), \tag{2.163}$$

continuity of $d\psi/dx$ at $-a$ gives

$$ik[Ae^{-ika} - Be^{ika}] = l[C\cos(la) + D\sin(la)] \tag{2.164}$$

continuity of $\psi(x)$ at $+a$ yields

$$C\sin(la) + D\cos(la) = Fe^{ika}, \tag{2.165}$$

and continuity of $d\psi/dx$ at $+a$ requires

$$l[C\cos(la) - D\sin(la)] = ikFe^{ika}. \tag{2.166}$$

[40] We *could* look for even and odd functions, as we did in the case of bound states, but the scattering problem is inherently asymmetric, since the waves come in from one side only, and the exponential notation (representing traveling waves) is more natural in this context.

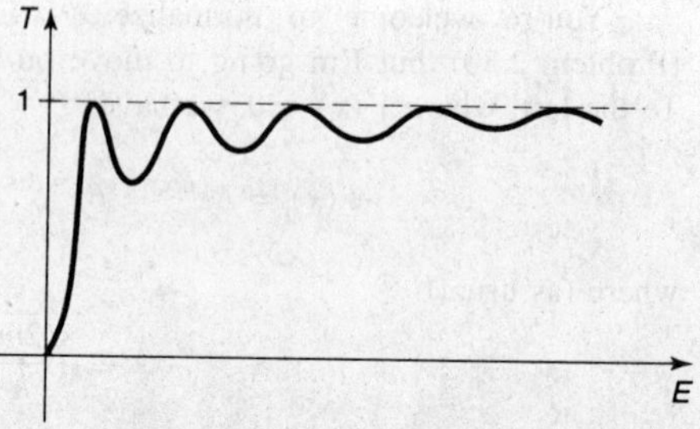

FIGURE 2.19: Transmission coefficient as a function of energy (Equation 2.169).

We can use two of these to eliminate C and D, and solve the remaining two for B and F (see Problem 2.32):

$$B = i\frac{\sin(2la)}{2kl}(l^2 - k^2)F, \tag{2.167}$$

$$F = \frac{e^{-2ika}A}{\cos(2la) - i\frac{(k^2+l^2)}{2kl}\sin(2la)}. \tag{2.168}$$

The transmission coefficient ($T = |F|^2/|A|^2$), expressed in terms of the original variables, is given by

$$T^{-1} = 1 + \frac{V_0^2}{4E(E+V_0)}\sin^2\left(\frac{2a}{\hbar}\sqrt{2m(E+V_0)}\right). \tag{2.169}$$

Notice that $T = 1$ (the well becomes "transparent") whenever the sine is zero, which is to say, when

$$\frac{2a}{\hbar}\sqrt{2m(E_n + V_0)} = n\pi, \tag{2.170}$$

where n is any integer. The energies for perfect transmission, then, are given by

$$E_n + V_0 = \frac{n^2\pi^2\hbar^2}{2m(2a)^2}, \tag{2.171}$$

which happen to be precisely the allowed energies for the *infinite* square well. T is plotted in Figure 2.19, as a function of energy.[41]

∗**Problem 2.29** Analyze the *odd* bound state wave functions for the finite square well. Derive the transcendental equation for the allowed energies, and solve it graphically. Examine the two limiting cases. Is there always an odd bound state?

---

[41] This remarkable phenomenon has been observed in the laboratory, in the form of the **Ramsauer-Townsend effect**. For an illuminating discussion see Richard W. Robinett, *Quantum Mechanics*, Oxford U.P., 1997, Section 12.4.1.

Problem 2.30 Normalize $\psi(x)$ in Equation 2.151, to determine the constants D and F.

Problem 2.31 The Dirac delta function can be thought of as the limiting case of a rectangle of area 1, as the height goes to infinity and the width goes to zero. Show that the delta-function well (Equation 2.114) is a "weak" potential (even though it is infinitely deep), in the sense that $z_0 \to 0$. Determine the bound state energy for the delta-function potential, by treating it as the limit of a finite square well. Check that your answer is consistent with Equation 2.129. Also show that Equation 2.169 reduces to Equation 2.141 in the appropriate limit.

Problem 2.32 Derive Equations 2.167 and 2.168. *Hint:* Use Equations 2.165 and 2.166 to solve for C and D in terms of F:

$$C = \left[\sin(la) + i\frac{k}{l}\cos(la)\right]e^{ika}F; \quad D = \left[\cos(la) - i\frac{k}{l}\sin(la)\right]e^{ika}F.$$

Plug these back into Equations 2.163 and 2.164. Obtain the transmission coefficient, and confirm Equation 2.169.

$**$**Problem 2.33** Determine the transmission coefficient for a rectangular *barrier* (same as Equation 2.145, only with $V(x) = +V_0 > 0$ in the region $-a < x < a$). Treat separately the three cases $E < V_0$, $E = V_0$, and $E > V_0$ (note that the wave function inside the barrier is different in the three cases). *Partial answer:* For $E < V_0$,[42]

$$T^{-1} = 1 + \frac{V_0^2}{4E(V_0 - E)}\sinh^2\left(\frac{2a}{\hbar}\sqrt{2m(V_0 - E)}\right).$$

$*$**Problem 2.34** Consider the "step" potential:

$$V(x) = \begin{cases} 0, & \text{if } x \le 0, \\ V_0, & \text{if } x > 0. \end{cases}$$

(a) Calculate the reflection coefficient, for the case $E < V_0$, and comment on the answer.

(b) Calculate the reflection coefficient for the case $E > V_0$.

(c) For a potential such as this, which does not go back to zero to the right of the barrier, the transmission coefficient is *not* simply $|F|^2/|A|^2$ (with A the

---

[42]This is a good example of tunneling—*classically* the particle would bounce back.

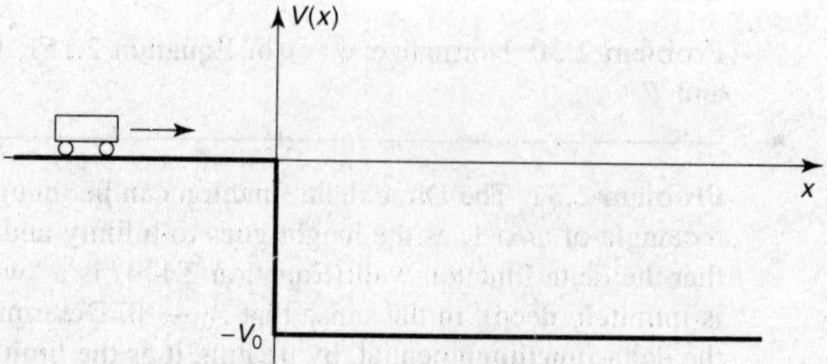

FIGURE 2.20: Scattering from a "cliff" (Problem 2.35).

incident amplitude and F the transmitted amplitude), because the transmitted wave travels at a different *speed*. Show that

$$T = \sqrt{\frac{E - V_0}{E}} \frac{|F|^2}{|A|^2}, \qquad [2.172]$$

for $E > V_0$. *Hint:* You can figure it out using Equation 2.98, or—more elegantly, but less informatively—from the probability current (Problem 2.19). What is T, for $E < V_0$?

(d) For $E > V_0$, calculate the transmission coefficient for the step potential, and check that $T + R = 1$.

Problem 2.35 A particle of mass m and kinetic energy $E > 0$ approaches an abrupt potential drop V_0 (Figure 2.20).

(a) What is the probability that it will "reflect" back, if $E = V_0/3$? *Hint:* This is just like Problem 2.34, except that the step now goes *down*, instead of up.

(b) I drew the figure so as to make you think of a car approaching a cliff, but obviously the probability of "bouncing back" from the edge of a cliff is *far* smaller than what you got in (a)—unless you're Bugs Bunny. Explain why this potential does *not* correctly represent a cliff. *Hint:* In Figure 2.20 the potential energy of the car drops *discontinuously* to $-V_0$, as it passes $x = 0$; would this be true for a falling car?

(c) When a free neutron enters a nucleus, it experiences a sudden drop in potential energy, from $V = 0$ outside to around -12 MeV (million electron volts) inside. Suppose a neutron, emitted with kinetic energy 4 MeV by a fission event, strikes such a nucleus. What is the probability it will be absorbed, thereby initiating another fission? *Hint:* You calculated the probability of *reflection* in part (a); use $T = 1 - R$ to get the probability of transmission through the surface.

FURTHER PROBLEMS FOR CHAPTER 2

Problem 2.36 Solve the time-independent Schrödinger equation with appropriate boundary conditions for the "centered" infinite square well: $V(x) = 0$ (for $-a < x < +a$), $V(x) = \infty$ (otherwise). Check that your allowed energies are consistent with mine (Equation 2.27), and confirm that your ψ's can be obtained from mine (Equation 2.28) by the substitution $x \to (x + a)/2$ (and appropriate renormalization). Sketch your first three solutions, and compare Figure 2.2. Note that the width of the well is now $2a$.

Problem 2.37 A particle in the infinite square well (Equation 2.19) has the initial wave function

$$\Psi(x, 0) = A \sin^3(\pi x/a) \quad (0 \le x \le a).$$

Determine A, find $\Psi(x, t)$, and calculate $\langle x \rangle$, as a function of time. What is the expectation value of the energy? *Hint:* $\sin^n \theta$ and $\cos^n \theta$ can be reduced, by repeated application of the trigonometric sum formulas, to linear combinations of $\sin(m\theta)$ and $\cos(m\theta)$, with $m = 0, 1, 2, \ldots, n$.

∗**Problem 2.38** A particle of mass m is in the ground state of the infinite square well (Equation 2.19). Suddenly the well expands to twice its original size—the right wall moving from a to $2a$—leaving the wave function (momentarily) undisturbed. The energy of the particle is now measured.

(a) What is the most probable result? What is the probability of getting that result?

(b) What is the *next* most probable result, and what is its probability?

(c) What is the *expectation value* of the energy? *Hint:* If you find yourself confronted with an infinite series, try another method.

Problem 2.39

(a) Show that the wave function of a particle in the infinite square well returns to its original form after a quantum **revival time** $T = 4ma^2/\pi\hbar$. That is: $\Psi(x, T) = \Psi(x, 0)$ for any state (*not* just a stationary state).

(b) What is the *classical* revival time, for a particle of energy E bouncing back and forth between the walls?

(c) For what energy are the two revival times equal?[43]

[43] The fact that the classical and quantum revival times bear no obvious relation to one another (and the quantum one doesn't even depend on the energy) is a curious paradox; see Daniel Styer, *Am. J. Phys.* **69**, 56 (2001).

Problem 2.40 A particle of mass m is in the potential

$$V(x) = \begin{cases} \infty & (x < 0), \\ -32\hbar^2/ma^2 & (0 \le x \le a), \\ 0 & (x > a). \end{cases}$$

(a) How many bound states are there?

(b) In the highest-energy bound state, what is the probability that the particle would be found *outside* the well ($x > a$)? *Answer:* 0.542, so even though it is "bound" by the well, it is more likely to be found outside than inside!

Problem 2.41 A particle of mass m in the harmonic oscillator potential (Equation 2.43) starts out in the state

$$\Psi(x, 0) = A\left(1 - 2\sqrt{\frac{m\omega}{\hbar}}x\right)^2 e^{-\frac{m\omega}{2\hbar}x^2}$$

for some constant A.

(a) What is the expectation value of the energy?

(b) At some later time T the wave function is

$$\Psi(x, T) = B\left(1 + 2\sqrt{\frac{m\omega}{\hbar}}x\right)^2 e^{-\frac{m\omega}{2\hbar}x^2},$$

for some constant B. What is the smallest possible value of T?

Problem 2.42 Find the allowed energies of the *half* harmonic oscillator

$$V(x) = \begin{cases} (1/2)m\omega^2x^2, & \text{for } x > 0, \\ \infty, & \text{for } x < 0. \end{cases}$$

(This represents, for example, a spring that can be stretched, but not compressed.) *Hint:* This requires some careful thought, but very little actual computation.

****Problem 2.43** In Problem 2.22 you analyzed the *stationary* gaussian free particle wave packet. Now solve the same problem for the *traveling* gaussian wave packet, starting with the initial wave function

$$\Psi(x, 0) = Ae^{-ax^2}e^{ilx},$$

where l is a real constant.

$**$**Problem 2.44** Solve the time-independent Schrödinger equation for a centered infinite square well with a delta-function barrier in the middle:

$$V(x) = \begin{cases} \alpha\delta(x), & \text{for } -a < x < +a, \\ \infty, & \text{for } |x| \geq a. \end{cases}$$

Treat the even and odd wave functions separately. Don't bother to normalize them. Find the allowed energies (graphically, if necessary). How do they compare with the corresponding energies in the absence of the delta function? Explain why the odd solutions are not affected by the delta function. Comment on the limiting cases $\alpha \to 0$ and $\alpha \to \infty$.

Problem 2.45 If two (or more) distinct[44] solutions to the (time-independent) Schrödinger equation have the same energy E, these states are said to be **degenerate**. For example, the free particle states are doubly degenerate—one solution representing motion to the right, and the other motion to the left. But we have never encountered *normalizable* degenerate solutions, and this is no accident. Prove the following theorem: *In one dimension*[45] *there are no degenerate bound states. Hint:* Suppose there are *two* solutions, ψ_1 and ψ_2, with the same energy E. Multiply the Schrödinger equation for ψ_1 by ψ_2, and the Schrödinger equation for ψ_2 by ψ_1, and subtract, to show that $(\psi_2 d\psi_1/dx - \psi_1 d\psi_2/dx)$ is a constant. Use the fact that for normalizable solutions $\psi \to 0$ at $\pm\infty$ to demonstrate that this constant is in fact zero. Conclude that ψ_2 is a multiple of ψ_1, and hence that the two solutions are not distinct.

Problem 2.46 Imagine a bead of mass m that slides frictionlessly around a circular wire ring of circumference L. (This is just like a free particle, except that $\psi(x + L) = \psi(x)$.) Find the stationary states (with appropriate normalization) and the corresponding allowed energies. Note that there are *two* independent solutions for each energy E_n—corresponding to clockwise and counter-clockwise circulation; call them $\psi_n^+(x)$ and $\psi_n^-(x)$. How do you account for this degeneracy, in view of the theorem in Problem 2.45 (why does the theorem fail, in this case)?

$**$**Problem 2.47** *Attention*: This is a *strictly qualitative* problem—no calculations allowed! Consider the "double square well" potential (Figure 2.21). Suppose the

[44] If two solutions differ only by a multiplicative constant (so that, once normalized, they differ only by a phase factor $e^{i\phi}$), they represent the same physical state, and in this sense they are *not* distinct solutions. Technically, by "distinct" I mean "linearly independent."

[45] In higher dimensions such degeneracy is very common, as we shall see in Chapter 4. Assume that the potential does not consist of isolated pieces separated by regions where $V = \infty$—two isolated infinite square wells, for instance, would give rise to degenerate bound states, for which the particle is either in the one or in the other.

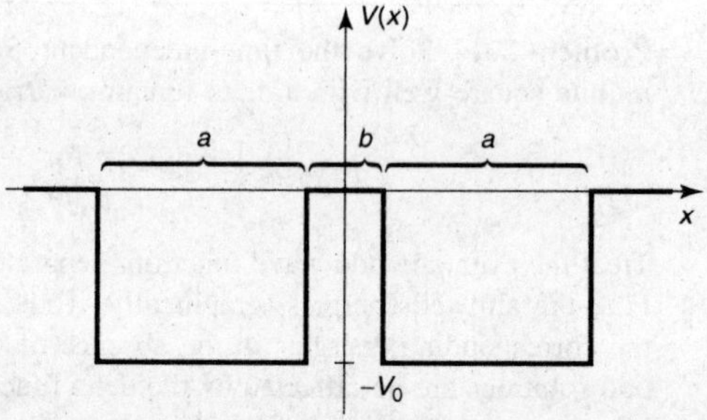

FIGURE 2.21: The double square well (Problem 2.47).

depth V_0 and the width a are fixed, and large enough so that several bound states occur.

(a) Sketch the ground state wave function ψ_1 and the first excited state ψ_2, (i) for the case $b = 0$, (ii) for $b \approx a$, and (iii) for $b \gg a$.

(b) Qualitatively, how do the corresponding energies (E_1 and E_2) vary, as b goes from 0 to ∞? Sketch $E_1(b)$ and $E_2(b)$ on the same graph.

(c) The double well is a very primitive one-dimensional model for the potential experienced by an electron in a diatomic molecule (the two wells represent the attractive force of the nuclei). If the nuclei are free to move, they will adopt the configuration of minimum energy. In view of your conclusions in (b), does the electron tend to draw the nuclei together, or push them apart? (Of course, there is also the internuclear repulsion to consider, but that's a separate problem.)

Problem 2.48 In Problem 2.7(d) you got the expectation value of the energy by summing the series in Equation 2.39, but I warned you (in footnote 15) not to try it the "old fashioned way," $\langle H \rangle = \int \Psi(x, 0)^* H \Psi(x, 0)\, dx$, because the discontinuous first derivative of $\Psi(x, 0)$ renders the second derivative problematic. Actually, you *could* have done it using integration by parts, but the Dirac delta function affords a much cleaner way to handle such anomalies.

(a) Calculate the first derivative of $\Psi(x, 0)$ (in Problem 2.7), and express the answer in terms of the step function, $\theta(x - a/2)$, defined in Equation 2.143. (Don't worry about the end points—just the interior region $0 < x < a$.)

(b) Exploit the result of Problem 2.24(b) to write the second derivative of $\Psi(x, 0)$ in terms of the delta function.

(c) Evaluate the integral $\int \Psi(x, 0)^* H \Psi(x, 0)\, dx$, and check that you get the same answer as before.

---

* * *Problem 2.49**

(a) Show that

$$\Psi(x,t)=\left(\frac{m\omega}{\pi\hbar}\right)^{1/4}\exp\left[-\frac{m\omega}{2\hbar}\left(x^2+\frac{a^2}{2}(1+e^{-2i\omega t})+\frac{i\hbar t}{m}-2axe^{-i\omega t}\right)\right]$$

satisfies the time-*dependent* Schrödinger equation for the harmonic oscillator potential (Equation 2.43). Here a is any real constant with the dimensions of length.[46]

(b) Find $|\Psi(x,t)|^2$, and describe the motion of the wave packet.

(c) Compute $\langle x\rangle$ and $\langle p\rangle$, and check that Ehrenfest's theorem (Equation 1.38) is satisfied.

* *Problem 2.50** Consider the *moving* delta-function well:

$$V(x,t)=-\alpha\delta(x-vt),$$

where v is the (constant) velocity of the well.

(a) Show that the time-dependent Schrödinger equation admits the exact solution

$$\Psi(x,t)=\frac{\sqrt{m\alpha}}{\hbar}e^{-m\alpha|x-vt|/\hbar^2}e^{-i[(E+(1/2)mv^2)t-mvx]/\hbar},$$

where $E=-m\alpha^2/2\hbar^2$ is the bound-state energy of the *stationary* delta function. *Hint:* Plug it in and *check* it! Use the result of Problem 2.24(b).

(b) Find the expectation value of the Hamiltonian in this state, and comment on the result.

* * *Problem 2.51** Consider the potential

$$V(x)=-\frac{\hbar^2a^2}{m}\operatorname{sech}^2(ax),$$

where a is a positive constant, and "sech" stands for the hyperbolic secant.

(a) Graph this potential.

(b) Check that this potential has the ground state

$$\psi_0(x)=A\operatorname{sech}(ax),$$

and find its energy. Normalize ψ_0, and sketch its graph.

[46]This rare example of an exact closed-form solution to the time-dependent Schrödinger equation was discovered by Schrödinger himself, in 1926.

(c) Show that the function

$$\psi_k(x) = A\left(\frac{ik - a\tanh(ax)}{ik + a}\right)e^{ikx},$$

(where $k \equiv \sqrt{2mE}/\hbar$, as usual) solves the Schrödinger equation for any (positive) energy E. Since $\tanh z \to -1$ as $z \to -\infty$,

$$\psi_k(x) \approx Ae^{ikx}, \quad \text{for large negative } x.$$

This represents, then, a wave coming in from the left with *no accompanying reflected wave* (i.e., no term $\exp(-ikx)$). What is the asymptotic form of $\psi_k(x)$ at large *positive* x? What are R and T, for this potential? *Comment:* This is a famous example of a **reflectionless potential**—every incident particle, regardless of its energy, passes right through.[47]

Problem 2.52 The scattering matrix. The theory of scattering generalizes in a pretty obvious way to arbitrary localized potentials (Figure 2.22). To the left (Region I), $V(x) = 0$, so

$$\psi(x) = Ae^{ikx} + Be^{-ikx}, \quad \text{where } k \equiv \frac{\sqrt{2mE}}{\hbar}. \qquad [2.173]$$

To the right (Region III), $V(x)$ is again zero, so

$$\psi(x) = Fe^{ikx} + Ge^{-ikx}. \qquad [2.174]$$

In between (Region II), of course, I can't tell you what ψ is until you specify the potential, but because the Schrödinger equation is a linear, second-order differential equation, the general solution has got to be of the form

$$\psi(x) = Cf(x) + Dg(x),$$

where $f(x)$ and $g(x)$ are two linearly independent particular solutions.[48] There will be four boundary conditions (two joining Regions I and II, and two joining

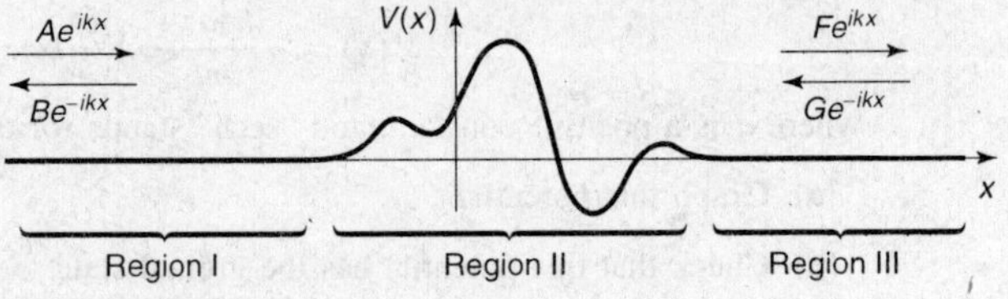

FIGURE 2.22: Scattering from an arbitrary localized potential ($V(x) = 0$ except in Region II); Problem 2.52.

---

[47] R. E. Crandall and B. R. Litt, *Annals of Physics*, **146**, 458 (1983).

[48] See any book on differential equations—for example, J. L. Van Iwaarden, *Ordinary Differential Equations with Numerical Techniques*, Harcourt Brace Jovanovich, San Diego, 1985, Chapter 3.

Regions II and III). Two of these can be used to eliminate C and D, and the other two can be "solved" for B and F in terms of A and G:

$$B = S_{11}A + S_{12}G, \quad F = S_{21}A + S_{22}G.$$

The four coefficients S_{ij}, which depend on k (and hence on E), constitute a 2×2 matrix $\mathbf{S}$, called the **scattering matrix** (or **S-matrix**, for short). The S-matrix tells you the outgoing amplitudes (B and F) in terms of the incoming amplitudes (A and G):

$$\begin{pmatrix} B \\ F \end{pmatrix} = \begin{pmatrix} S_{11} & S_{12} \\ S_{21} & S_{22} \end{pmatrix} \begin{pmatrix} A \\ G \end{pmatrix}. \qquad [2.175]$$

In the typical case of scattering from the left, $G = 0$, so the reflection and transmission coefficients are

$$R_l = \left.\frac{|B|^2}{|A|^2}\right|_{G=0} = |S_{11}|^2, \quad T_l = \left.\frac{|F|^2}{|A|^2}\right|_{G=0} = |S_{21}|^2. \qquad [2.176]$$

For scattering from the right, $A = 0$, and

$$R_r = \left.\frac{|F|^2}{|G|^2}\right|_{A=0} = |S_{22}|^2, \quad T_r = \left.\frac{|B|^2}{|G|^2}\right|_{A=0} = |S_{12}|^2. \qquad [2.177]$$

(a) Construct the S-matrix for scattering from a delta-function well (Equation 2.114).

(b) Construct the S-matrix for the finite square well (Equation 2.145). *Hint:* This requires no new work, if you carefully exploit the symmetry of the problem.

∗∗∗**Problem 2.53 The transfer matrix.** The S-matrix (Problem 2.52) tells you the *outgoing* amplitudes (B and F) in terms of the *incoming* amplitudes (A and G)—Equation 2.175. For some purposes it is more convenient to work with the **transfer matrix, M**, which gives you the amplitudes to the *right* of the potential (F and G) in terms of those to the *left* (A and B):

$$\begin{pmatrix} F \\ G \end{pmatrix} = \begin{pmatrix} M_{11} & M_{12} \\ M_{21} & M_{22} \end{pmatrix} \begin{pmatrix} A \\ B \end{pmatrix}. \qquad [2.178]$$

(a) Find the four elements of the M-matrix, in terms of the elements of the S-matrix, and vice versa. Express R_l, T_l, R_r, and T_r (Equations 2.176 and 2.177) in terms of elements of the M-matrix.

(b) Suppose you have a potential consisting of two isolated pieces (Figure 2.23). Show that the M-matrix for the combination is the *product* of the two M-matrices for each section separately:

$$\mathbf{M} = \mathbf{M}_2\mathbf{M}_1. \qquad [2.179]$$

(This obviously generalizes to any number of pieces, and accounts for the usefulness of the M-matrix.)

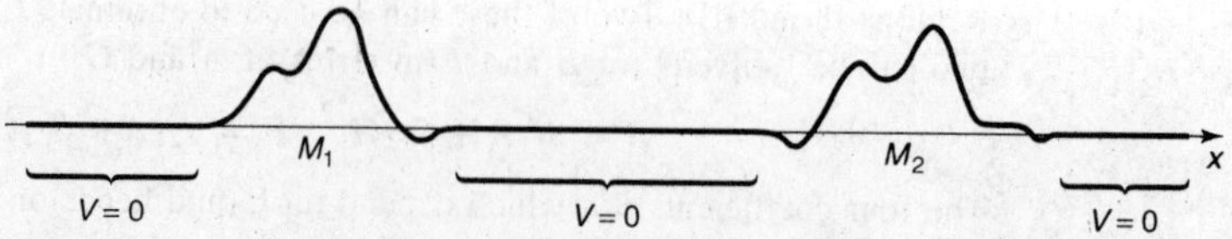

FIGURE 2.23: A potential consisting of two isolated pieces (Problem 2.53).

(c) Construct the M-matrix for scattering from a single delta-function potential at point a:

$$V(x) = -\alpha\delta(x-a).$$

(d) By the method of part (b), find the M-matrix for scattering from the double delta function

$$V(x) = -\alpha[\delta(x+a) + \delta(x-a)].$$

What is the transmission coefficient for this potential?

Problem 2.54 Find the ground state energy of the harmonic oscillator, to five significant digits, by the "wag-the-dog" method. That is, solve Equation 2.72 numerically, varying K until you get a wave function that goes to zero at large ξ. In Mathematica, appropriate input code would be

```
Plot[Evaluate[u[x]/.[NDSolve[{u''[x] -(x^2 - K)*u[x] == 0, u[0] == 1,
u'[0] == 0}, u[x], {x, 10^-8, 10}, MaxSteps -> 10000]], {x, a, b},
PlotRange -> {c, d}];
```

(Here (a, b) is the horizontal range of the graph, and (c, d) is the vertical range—start with $a = 0$, $b = 10$, $c = -10$, $d = 10$.) We know that the correct solution is $K = 1$, so you might start with a "guess" of $K = 0.9$. Notice what the "tail" of the wave function does. Now try $K = 1.1$, and note that the tail flips over. Somewhere in between those values lies the correct solution. Zero in on it by bracketing K tighter and tighter. As you do so, you may want to adjust a, b, c, and d, to zero in on the cross-over point.

Problem 2.55 Find the first three excited state energies (to five significant digits) for the harmonic oscillator, by wagging the dog (Problem 2.54). For the first (and third) excited state you will need to set $u[0] == 0$, $u'[0] == 1$.

Problem 2.56 Find the first four allowed energies (to five significant digits) for the infinite square well, by wagging the dog. *Hint:* Refer to Problem 2.54, making appropriate changes to the differential equation. This time the condition you are looking for is $u(1) = 0$.

CHAPTER 3

FORMALISM

3.1 HILBERT SPACE

In the last two chapters we have stumbled on a number of interesting properties of simple quantum systems. Some of these are "accidental" features of specific potentials (the even spacing of energy levels for the harmonic oscillator, for example), but others seem to be more general, and it would be nice to prove them once and for all (the uncertainty principle, for instance, and the orthogonality of stationary states). The purpose of this chapter is to recast the theory in a more powerful form, with that in mind. There is not much here that is genuinely *new*; the idea, rather, is to make coherent sense of what we have already discovered in particular cases.

Quantum theory is based on two constructs: *wave functions* and *operators*. The state of a system is represented by its wave function, observables are represented by operators. Mathematically, wave functions satisfy the defining conditions for abstract **vectors**, and operators act on them as **linear transformations**. So the natural language of quantum mechanics is **linear algebra**.[1]

But it is not, I suspect, a form of linear algebra with which you are immediately familiar. In an N-dimensional space it is simplest to represent a vector, $|\alpha\rangle$, by the N-tuple of its components, $\{a_n\}$, with respect to a specified orthonormal basis:

$$|\alpha\rangle \rightarrow \mathbf{a} = \begin{pmatrix} a_1 \\ a_2 \\ \vdots \\ a_N \end{pmatrix} \qquad [3.1]$$

[1] If you have never studied linear algebra, you should read the Appendix before continuing.

The **inner product**, $\langle\alpha|\beta\rangle$, of two vectors (generalizing the dot product in three dimensions) is the complex number,

$$\langle\alpha|\beta\rangle = a_1^* b_1 + a_2^* b_2 + \cdots + a_N^* b_N. \tag{3.2}$$

Linear transformations, T, are represented by **matrices** (with respect to the specified basis), which act on vectors (to produce new vectors) by the ordinary rules of matrix multiplication:

$$|\beta\rangle = T|\alpha\rangle \rightarrow \mathbf{b} = \mathbf{Ta} = \begin{pmatrix} t_{11} & t_{12} & \cdots & t_{1N} \\ t_{21} & t_{22} & \cdots & t_{2N} \\ \vdots & \vdots & & \vdots \\ t_{N1} & t_{N2} & \cdots & t_{NN} \end{pmatrix} \begin{pmatrix} a_1 \\ a_2 \\ \vdots \\ a_N \end{pmatrix}. \tag{3.3}$$

But the "vectors" we encounter in quantum mechanics are (for the most part) *functions*, and they live in *infinite*-dimensional spaces. For them the N-tuple/matrix notation is awkward, at best, and manipulations that are well-behaved in the finite-dimensional case can be problematic. (The underlying reason is that whereas the *finite* sum in Equation 3.2 always exists, an *infinite* sum—or an integral—may not converge, in which case the inner product does not exist, and any argument involving inner products is immediately suspect.) So even though most of the terminology and notation should be familiar, it pays to approach this subject with caution.

The collection of *all* functions of x constitutes a vector space, but for our purposes it is much too large. To represent a possible physical state, the wave function Ψ must be *normalized*:

$$\int |\Psi|^2\, dx = 1.$$

The set of all **square-integrable functions**, on a specified interval,[2]

$$f(x) \quad \text{such that} \quad \int_a^b |f(x)|^2\, dx < \infty, \tag{3.4}$$

constitutes a (much smaller) vector space (see Problem 3.1(a)). Mathematicians call it $L_2(a, b)$; physicists call it **Hilbert space.**[3] In quantum mechanics, then,

$$\boxed{\textbf{Wave functions live in Hilbert space.}} \tag{3.5}$$

[2]For us, the limits (a and b) will almost always be $\pm\infty$, but we might as well keep things more general for the moment.

[3]Technically, a Hilbert space is a **complete inner product space**, and the collection of square-integrable functions is only *one example* of a Hilbert space—indeed, every finite-dimensional vector space is trivially a Hilbert space. But since L_2 is the arena of quantum mechanics, it's what physicists generally *mean* when they say "Hilbert space." By the way, the word **complete** here means that any Cauchy sequence of functions in Hilbert space converges to a function that is also in the space: it has no "holes" in it, just as the set of all real numbers has no holes (by contrast, the space of all *polynomials*, for example, like the set of all *rational* numbers, certainly *does* have holes in it). The completeness of a *space* has nothing to do with the completeness (same word, unfortunately) of a *set of functions*, which is the property that any other function can be expressed as a linear combination of them.

We define the **inner product of two functions**, $f(x)$ and $g(x)$, as follows:

$$\langle f|g\rangle \equiv \int_a^b f(x)^* g(x)\,dx. \qquad [3.6]$$

If f and g are both square-integrable (that is, if they are both in Hilbert space), their inner product is guaranteed to exist (the integral in Equation 3.6 converges to a finite number).[4] This follows from the integral **Schwarz inequality**:[5]

$$\left|\int_a^b f(x)^* g(x)\,dx\right| \le \sqrt{\int_a^b |f(x)|^2\,dx \int_a^b |g(x)|^2\,dx}. \qquad [3.7]$$

You can check for yourself that Equation 3.6 satisfies all the conditions for an inner product (Problem 3.1(b)). Notice in particular that

$$\langle g|f\rangle = \langle f|g\rangle^*. \qquad [3.8]$$

Moreover, the inner product of $f(x)$ with *itself*,

$$\langle f|f\rangle = \int_a^b |f(x)|^2\,dx, \qquad [3.9]$$

is *real* and non-negative; it's *zero* only[6] when $f(x) = 0$.

A function is said to be **normalized** if its inner product with itself is 1; two functions are **orthogonal** if their inner product is 0; and a *set* of functions, $\{f_n\}$, is **orthonormal** if they are normalized and mutually orthogonal:

$$\langle f_m|f_n\rangle = \delta_{mn}. \qquad [3.10]$$

Finally, a set of functions is **complete** if any *other* function (in Hilbert space) can be expressed as a linear combination of them:

$$f(x) = \sum_{n=1}^{\infty} c_n f_n(x). \qquad [3.11]$$

[4]In Chapter 2 we were obliged on occasion to work with functions that were *not* normalizable. Such functions lie *outside* Hilbert space, and we are going to have to handle them with special care, as you will see shortly. For the moment, I shall assume that all the functions we encounter *are* in Hilbert space.

[5]For a proof, see F. Riesz and B. Sz.-Nagy, *Functional Analysis* (Unger, New York, 1955), Section 21. In a *finite* dimensional vector space the Schwarz inequality, $|\langle\alpha|\beta\rangle|^2 \le \langle\alpha|\alpha\rangle\langle\beta|\beta\rangle$, is easy to prove (see Problem A.5). But that proof *assumes* the existence of the inner products, which is precisely what we are trying to *establish* here.

[6]What about a function that is zero everywhere except at a few isolated points? The integral (Equation 3.9) would still vanish, even though the function itself does not. If this bothers you, you should have been a math major. In physics such pathological functions do not occur, but in any case, in Hilbert space two functions that have the same square integral are considered equivalent. Technically, vectors in Hilbert space represent **equivalence classes** of functions.

If the functions $\{f_n(x)\}$ are orthonormal, the coefficients are given by Fourier's trick:

$$c_n = \langle f_n | f \rangle, \tag{3.12}$$

as you can check for yourself. I anticipated this terminology, of course, back in Chapter 2. (The stationary states for the infinite square well (Equation 2.28) constitute a complete orthonormal set on the interval $(0, a)$; the stationary states for the harmonic oscillator (Equation 2.67 or 2.85) are a complete orthonormal set on the interval $(-\infty, \infty)$.)

Problem 3.1

(a) Show that the set of all square-integrable functions is a vector space (refer to Section A.1 for the definition). *Hint:* The main problem is to show that the sum of two square-integrable functions is itself square-integrable. Use Equation 3.7. Is the set of all *normalized* functions a vector space?

(b) Show that the integral in Equation 3.6 satisfies the conditions for an inner product (Section A.2).

$*$**Problem 3.2**

(a) For what range of ν is the function $f(x) = x^\nu$ in Hilbert space, on the interval $(0, 1)$? Assume ν is real, but not necessarily positive.

(b) For the specific case $\nu = 1/2$, is $f(x)$ in Hilbert space? What about $xf(x)$? How about $(d/dx)f(x)$?

3.2 OBSERVABLES

3.2.1 Hermitian Operators

The expectation value of an observable $Q(x, p)$ can be expressed very neatly in inner-product notation:[7]

$$\langle Q \rangle = \int \Psi^* \hat{Q} \Psi \, dx = \langle \Psi | \hat{Q} \Psi \rangle. \tag{3.13}$$

---

[7]Remember that $\hat{Q}$ is the operator constructed from Q by the replacement $p \to \hat{p} \equiv (\hbar/i)d/dx$. These operators are **linear**, in the sense that

$$\hat{Q}[af(x) + bg(x)] = a\hat{Q}f(x) + b\hat{Q}g(x),$$

for any functions f and g and any complex numbers a and b. They constitute *linear transformations* (Section A.3) on the space of all functions. However, they sometimes carry a function *inside* Hilbert

Now, the outcome of a measurement has got to be *real*, and so, *a fortiori*, is the *average* of many measurements:

$$\langle Q\rangle = \langle Q\rangle^*. \qquad [3.14]$$

But the complex conjugate of an inner product reverses the order (Equation 3.8), so

$$\langle\Psi|\hat{Q}\Psi\rangle = \langle\hat{Q}\Psi|\Psi\rangle, \qquad [3.15]$$

and this must hold true for any wave function Ψ. Thus operators representing *observables* have the very special property that

$$\langle f|\hat{Q}f\rangle = \langle\hat{Q}f|f\rangle \quad \text{for all } f(x). \qquad [3.16]$$

We call such operators **hermitian**.

Actually, most books require an ostensibly stronger condition:

$$\langle f|\hat{Q}g\rangle = \langle\hat{Q}f|g\rangle \quad \text{for all } f(x) \text{ and all } g(x). \qquad [3.17]$$

But it turns out, in spite of appearances, that this is perfectly equivalent to my definition (Equation 3.16), as you will prove in Problem 3.3. So use whichever you like. The essential point is that a hermitian operator can be applied either to the first member of an inner product or to the second, with the same result, and hermitian operators naturally arise in quantum mechanics because their expectation values are real:

Observables are represented by hermitian operators.	[3.18]

Well, let's *check* this. Is the momentum operator, for example, hermitian?

$$\langle f|\hat{p}g\rangle = \int_{-\infty}^{\infty} f^*\frac{\hbar}{i}\frac{dg}{dx}\,dx = \frac{\hbar}{i}f^*g\Big|_{-\infty}^{\infty} + \int_{-\infty}^{\infty}\left(\frac{\hbar}{i}\frac{df}{dx}\right)^* g\,dx = \langle\hat{p}f|g\rangle. \qquad [3.19]$$

I used integration by parts, of course, and threw away the boundary term for the usual reason: If $f(x)$ and $g(x)$ are square integrable, they must go to zero at $\pm\infty$.[8]

space into a function *outside* it (see Problem 3.2(b)), and in this case the domain of the operator may have to be restricted.

[8] Actually, this is not quite true. As I mention in Chapter 1, there exist pathological functions that are square-integrable but do *not* go to zero at infinity. However, such functions do not arise in physics, and if you are worried about it we will simply restrict the domain of our operators to exclude them. On *finite* intervals, though, you really *do* have to be more careful with the boundary terms, and an operator that is hermitian on $(-\infty, \infty)$ may *not* be hermitian on $(0, \infty)$ or $(-\pi, \pi)$. If you're wondering about the infinite square well, it's safest to think of those wave functions as residing on the infinite line—they just happen to be *zero* outside $(0, a)$.

Notice how the complex conjugation of i compensates for the minus sign picked up from integration by parts—the operator d/dx (without the i) is *not* hermitian, and it does not represent a possible observable.

$*$**Problem 3.3** Show that if $\langle h|\hat{Q}h\rangle = \langle \hat{Q}h|h\rangle$ for all functions h (in Hilbert space), then $\langle f|\hat{Q}g\rangle = \langle \hat{Q}f|g\rangle$ for all f and g (i.e., the two definitions of "hermitian"—Equations 3.16 and 3.17—are equivalent). *Hint:* First let $h = f + g$, and then let $h = f + ig$.

Problem 3.4

(a) Show that the *sum* of two hermitian operators is hermitian.

(b) Suppose $\hat{Q}$ is hermitian, and α is a complex number. Under what condition (on α) is $\alpha\hat{Q}$ hermitian?

(c) When is the *product* of two hermitian operators hermitian?

(d) Show that the position operator ($\hat{x} = x$) and the hamiltonian operator ($\hat{H} = -(\hbar^2/2m)d^2/dx^2 + V(x)$) are hermitian.

Problem 3.5 The **hermitian conjugate** (or **adjoint**) of an operator $\hat{Q}$ is the operator $\hat{Q}^\dagger$ such that

$$\langle f|\hat{Q}g\rangle = \langle \hat{Q}^\dagger f|g\rangle \quad \text{(for all } f \text{ and } g\text{)}. \tag{3.20}$$

(A hermitian operator, then, is equal to its hermitian conjugate: $\hat{Q} = \hat{Q}^\dagger$.)

(a) Find the hermitian conjugates of x, i, and d/dx.

(b) Construct the hermitian conjugate of the harmonic oscillator raising operator, a_+ (Equation 2.47).

(c) Show that $(\hat{Q}\hat{R})^\dagger = \hat{R}^\dagger\hat{Q}^\dagger$.

3.2.2 Determinate States

Ordinarily, when you measure an observable Q on an ensemble of identically prepared systems, all in the same state Ψ, you do *not* get the same result each time—this is the *indeterminacy* of quantum mechanics.[9] *Question:* Would it be possible to prepare a state such that *every* measurement of Q is certain to return the *same* value (call it q)? This would be, if you like, a **determinate state**, for the observable Q. (Actually, we already know one example: Stationary states are determinate states of the Hamiltonian; a measurement of the total energy, on a

---

[9] I'm talking about *competent* measurements, of course—it's always possible to make a *mistake*, and simply get the wrong answer, but that's not the fault of quantum mechanics.

particle in the stationary state Ψ_n, is certain to yield the corresponding "allowed" energy E_n.)

Well, the standard deviation of Q, in a determinate state, would be *zero*, which is to say,

$$\sigma^2 = \langle(\hat{Q} - \langle Q\rangle)^2\rangle = \langle\Psi|(\hat{Q} - q)^2\Psi\rangle = \langle(\hat{Q} - q)\Psi|(\hat{Q} - q)\Psi\rangle = 0. \quad [3.21]$$

(Of course, if every measurement gives q, their average is also q: $\langle Q\rangle = q$. I also used the fact that $\hat{Q}$, and hence also $\hat{Q} - q$, is a *hermitian* operator, to move one factor over to the first term in the inner product.) But the only function whose inner product with itself vanishes is 0, so

$$\hat{Q}\Psi = q\Psi. \quad [3.22]$$

This is the **eigenvalue equation** for the operator $\hat{Q}$; Ψ is an **eigenfunction** of $\hat{Q}$, and q is the corresponding **eigenvalue**. Thus

Determinate states are eigenfunctions of $\hat{Q}$. [3.23]

Measurement of Q on such a state is certain to yield the eigenvalue, q.

Note that the eigen*value* is a *number* (not an operator or a function). You can multiply any eigenfunction by a constant, and it is still an eigenfunction, with the same eigenvalue. Zero does not count as an eigenfunction (we exclude it by definition—otherwise *every* number would be an eigenvalue, since $\hat{Q}\,0 = q\,0 = 0$ for any operator $\hat{Q}$ and all q). But there's nothing wrong with zero as an eigen*value*. The collection of all the eigenvalues of an operator is called its **spectrum**. Sometimes two (or more) linearly independent eigenfunctions share the same eigenvalue; in that case the spectrum is said to be **degenerate**.

For example, determinate states of the total energy are eigenfunctions of the Hamiltonian:

$$\hat{H}\psi = E\psi, \quad [3.24]$$

which is precisely the time-independent Schrödinger equation. In this context we use the letter E for the eigenvalue, and the lower case ψ for the eigenfunction (tack on the factor $\exp(-iEt/\hbar)$ to make it Ψ, if you like; it's still an eigenfunction of H).

Example 3.1 Consider the operator

$$\hat{Q} \equiv i\frac{d}{d\phi}, \quad [3.25]$$

where ϕ is the usual polar coordinate in two dimensions. (This operator might arise in a physical context if we were studying the bead-on-a-ring; see Problem 2.46.) Is $\hat{Q}$ hermitian? Find its eigenfunctions and eigenvalues.

Solution: Here we are working with functions $f(\phi)$ on the *finite* interval $0 \le \phi \le 2\pi$, and stipulate that

$$f(\phi + 2\pi) = f(\phi), \tag{3.26}$$

since ϕ and $\phi + 2\pi$ describe the same physical point. Using integration by parts,

$$\langle f|\hat{Q}g\rangle = \int_0^{2\pi} f^* \left(i\frac{dg}{d\phi}\right) d\phi = if^*g\Big|_0^{2\pi} - \int_0^{2\pi} i\left(\frac{df^*}{d\phi}\right) g\, d\phi = \langle \hat{Q}f|g\rangle,$$

so $\hat{Q}$ *is* hermitian (this time the boundary term disappears by virtue of Equation 3.26).

The eigenvalue equation,

$$i\frac{d}{d\phi} f(\phi) = qf(\phi), \tag{3.27}$$

has the general solution

$$f(\phi) = Ae^{-iq\phi}. \tag{3.28}$$

Equation 3.26 restricts the possible values of the q:

$$e^{-iq2\pi} = 1 \quad \Rightarrow \quad q = 0, \pm 1, \pm 2, \ldots \tag{3.29}$$

The spectrum of this operator is the set of all integers, and it is nondegenerate.

Problem 3.6 Consider the operator $\hat{Q} = d^2/d\phi^2$, where (as in Example 3.1) ϕ is the azimuthal angle in polar coordinates, and the functions are subject to Equation 3.26. Is $\hat{Q}$ hermitian? Find its eigenfunctions and eigenvalues. What is the spectrum of $\hat{Q}$? Is the spectrum degenerate?

3.3 EIGENFUNCTIONS OF A HERMITIAN OPERATOR

Our attention is thus directed to the *eigenfunctions of hermitian operators* (physically: determinate states of observables). These fall into two categories: If the spectrum is **discrete** (i.e., the eigenvalues are separated from one another) then the eigenfunctions lie in Hilbert space and they constitute physically realizable states. If the spectrum is **continuous** (i.e., the eigenvalues fill out an entire range) then the eigenfunctions are not normalizable, and they do not represent possible wave functions (though *linear combinations* of them—involving necessarily a spread in eigenvalues—may be normalizable). Some operators have a discrete spectrum only (for example, the Hamiltonian for the harmonic oscillator), some have only a continuous spectrum (for example, the free particle Hamiltonian), and some have both a discrete part and a continuous part (for example, the Hamiltonian for a finite square well). The discrete case is easier to handle, because the relevant inner products are guaranteed to exist—in fact, it is very similar to the finite-dimensional theory (the eigenvectors of a hermitian *matrix*). I'll treat the discrete case first, and then the continuous one.

3.3.1 Discrete Spectra

Mathematically, the normalizable eigenfunctions of a hermitian operator have two important properties:

Theorem 1: Their eigen*values* are *real*.

Proof: Suppose

$$\hat{Q}f = qf,$$

(i.e., $f(x)$ is an eigenfunction of $\hat{Q}$, with eigenvalue q), and[10]

$$\langle f|\hat{Q}f\rangle = \langle\hat{Q}f|f\rangle$$

($\hat{Q}$ is hermitian). Then

$$q\langle f|f\rangle = q^*\langle f|f\rangle$$

(q is a *number*, so it comes outside the integral, and because the first function in the inner product is complex conjugated (Equation 3.6), so too is the q on the right). But $\langle f|f\rangle$ cannot be zero ($f(x) = 0$ is not a legal eigenfunction), so $q = q^*$, and hence q is real. QED

This is comforting: If you measure an observable on a particle in a determinate state, you will at least get a real number.

Theorem 2: Eigenfunctions belonging to distinct eigenvalues are *orthogonal*.

Proof: Suppose

$$\hat{Q}f = qf, \quad \text{and} \quad \hat{Q}g = q'g,$$

and $\hat{Q}$ is hermitian. Then $\langle f|\hat{Q}g\rangle = \langle\hat{Q}f|g\rangle$, so

$$q'\langle f|g\rangle = q^*\langle f|g\rangle$$

(again, the inner products exist because the eigenfunctions are in Hilbert space by assumption). But q is real (from Theorem 1), so if $q' \neq q$ it must be that $\langle f|g\rangle = 0$. QED

---

[10] It is here that we assume the eigenfunctions are in Hilbert space—otherwise the inner product might not exist at all.

That's why the stationary states of the infinite square well, for example, or the harmonic oscillator, are orthogonal—they are eigenfunctions of the Hamiltonian with distinct eigenvalues. But this property is not peculiar to them, or even to the Hamiltonian—the same holds for determinate states of *any* observable.

Unfortunately, Theorem 2 tells us nothing about degenerate states ($q' = q$). However, if two (or more) eigenfunctions share the same eigenvalue, any linear combination of them is itself an eigenfunction, with the same eigenvalue (Problem 3.7(a)), and we can use the **Gram-Schmidt orthogonalization procedure** (Problem A.4) to *construct* orthogonal eigenfunctions within each degenerate subspace. It is almost never necessary to do this explicitly (thank God!), but it can always be done in principle. So *even in the presence of degeneracy* the eigenfunctions can be *chosen* to be orthogonal, and in setting up the formalism of quantum mechanics we shall assume that this has already been done. That licenses the use of Fourier's trick, which depends on the orthonormality of the basis functions.

In a *finite*-dimensional vector space the eigenvectors of a hermitian matrix have a third fundamental property: They span the space (every vector can be expressed as a linear combination of them). Unfortunately, the proof does not generalize to infinite-dimensional spaces. But the property itself is essential to the internal consistency of quantum mechanics, so (following Dirac[11]) we will take it as an *axiom* (or, more precisely, as a restriction on the class of hermitian operators that can represent observables):

> **Axiom:** The eigenfunctions of an observable operator are *complete*: Any function (in Hilbert space) can be expressed as a linear combination of them.[12]

Problem 3.7

(a) Suppose that $f(x)$ and $g(x)$ are two eigenfunctions of an operator $\hat{Q}$, with the same eigenvalue q. Show that any linear combination of f and g is itself an eigenfunction of $\hat{Q}$, with eigenvalue q.

(b) Check that $f(x) = \exp(x)$ and $g(x) = \exp(-x)$ are eigenfunctions of the operator d^2/dx^2, with the same eigenvalue. Construct two linear combinations of f and g that are *orthogonal* eigenfunctions on the interval $(-1, 1)$.

---

[11] P. A. M. Dirac, *The Principles of Quantum Mechanics*, Oxford University Press, New York (1958).

[12] In some specific cases completeness is provable (we know that the stationary states of the infinite square well, for example, are complete, because of Dirichlet's theorem). It is a little awkward to call something an "axiom" that is *provable* in some cases, but I don't know a better way to handle it.

Problem 3.8

(a) Check that the eigenvalues of the hermitian operator in Example 3.1 are real. Show that the eigenfunctions (for distinct eigenvalues) are orthogonal.

(b) Do the same for the operator in Problem 3.6.

3.3.2 Continuous Spectra

If the spectrum of a hermitian operator is *continuous*, the eigenfunctions are not normalizable, and the proofs of Theorems 1 and 2 fail, because the inner products may not exist. Nevertheless, there is a sense in which the three essential properties (reality, orthogonality, and completeness) still hold. I think it's best to approach this subtle case through specific examples.

Example 3.2 Find the eigenfunctions and eigenvalues of the momentum operator.

Solution: Let $f_p(x)$ be the eigenfunction and p the eigenvalue:

$$\frac{\hbar}{i}\frac{d}{dx} f_p(x) = p f_p(x). \qquad [3.30]$$

The general solution is

$$f_p(x) = Ae^{ipx/\hbar}.$$

This is not square-integrable, for *any* (complex) value of p—the momentum operator has *no* eigenfunctions in Hilbert space. And yet, if we restrict ourselves to *real* eigenvalues, we do recover a kind of *ersatz* "orthonormality." Referring to Problems 2.24(a) and 2.26,

$$\int_{-\infty}^{\infty} f_{p'}^*(x)\, f_p(x)\, dx = |A|^2 \int_{-\infty}^{\infty} e^{i(p-p')x/\hbar}\, dx = |A|^2 2\pi\hbar\, \delta(p-p'). \qquad [3.31]$$

If we pick $A = 1/\sqrt{2\pi\hbar}$, so that

$$f_p(x) = \frac{1}{\sqrt{2\pi\hbar}} e^{ipx/\hbar}, \qquad [3.32]$$

then

$$\langle f_{p'} | f_p \rangle = \delta(p-p'), \qquad [3.33]$$

which is strikingly reminiscent of *true* orthonormality (Equation 3.10)—the indices are now continuous variables, and the Kronecker delta has become a Dirac delta, but otherwise it looks just the same. I'll call Equation 3.33 **Dirac orthonormality**.

Most important, the eigenfunctions are *complete*, with the sum (in Equation 3.11) replaced by an integral: Any (square-integrable) function $f(x)$ can be written in the form

$$f(x) = \int_{-\infty}^{\infty} c(p)\, f_p(x)\, dp = \frac{1}{\sqrt{2\pi\hbar}} \int_{-\infty}^{\infty} c(p) e^{ipx/\hbar}\, dp. \qquad [3.34]$$

The expansion coefficient (now a *function*, $c(p)$) is obtained, as always, by Fourier's trick:

$$\langle f_{p'} | f \rangle = \int_{-\infty}^{\infty} c(p) \langle f_{p'} | f_p \rangle\, dp = \int_{-\infty}^{\infty} c(p)\delta(p - p')\, dp = c(p'). \qquad [3.35]$$

Alternatively, you can get them from Plancherel's theorem (Equation 2.102), for the expansion (Equation 3.34) is nothing but a Fourier transform.

The eigenfunctions of momentum (Equation 3.32) are sinusoidal, with wavelength

$$\lambda = \frac{2\pi\hbar}{p}. \qquad [3.36]$$

This is the old de Broglie formula (Equation 1.39), which I promised to prove at the appropriate time. It turns out to be a little more subtle than de Broglie imagined, because we now know that there is actually *no such thing* as a particle with determinate momentum. But we could make a normalizable wave *packet* with a narrow range of momenta, and it is to such an object that the de Broglie relation applies.

What are we to make of Example 3.2? Although none of the eigenfunctions of $\hat{p}$ lives in Hilbert space, a certain family of them (those with real eigenvalues) reside in the nearby "suburbs," with a kind of quasi-normalizability. They do not represent possible physical states, but they are still very useful (as we have already seen, in our study of one-dimensional scattering).[13]

Example 3.3 Find the eigenfunctions and eigenvalues of the position operator.

Solution: Let $g_y(x)$ be the eigenfunction and y the eigenvalue:

$$x\, g_y(x) = y\, g_y(x). \qquad [3.37]$$

---

[13] What about the eigenfunctions with *non*real eigenvalues? These are not merely non-normalizable—they actually blow up at $\pm\infty$. Functions in what I called the "suburbs" of Hilbert space (the entire metropolitan area is sometimes called a "rigged Hilbert space"; see, for example, Leslie Ballentine's *Quantum Mechanics: A Modern Development*, World Scientific, 1998) have the property that although they have no (finite) inner product with *themselves*, they *do* admit inner products with all members of Hilbert space. This is *not* true for eigenfunctions of $\hat{p}$ with nonreal eigenvalues. In particular, I showed that the momentum operator is hermitian *for functions in Hilbert space*, but the argument depended on dropping the boundary term (in Equation 3.19). That term is still zero if g is an eigenfunction of $\hat{p}$ with a real eigenvalue (as long as f is in Hilbert space), but not if the eigenvalue has an imaginary part. In this sense *any* complex number is an eigenvalue of the operator $\hat{p}$, but only *real* numbers are eigenvalues of the *hermitian* operator $\hat{p}$—the others lie outside the space over which $\hat{p}$ is hermitian.

Here y is a fixed number (for any given eigenfunction), but x is a continuous variable. What function of x has the property that multiplying it by x is the same as multiplying it by the constant y? Obviously it's got to be *zero*, except at the one point $x = y$; in fact, it is nothing but the Dirac delta function:

$$g_y(x) = A\delta(x - y).$$

This time the eigenvalue *has* to be real; the eigenfunctions are not square-integrable, but again they admit *Dirac* orthonormality:

$$\int_{-\infty}^{\infty} g_{y'}^*(x)\, g_y(x)\, dx = |A|^2 \int_{-\infty}^{\infty} \delta(x - y')\delta(x - y)\, dx = |A|^2 \delta(y - y'). \quad [3.38]$$

If we pick $A = 1$, so

$$g_y(x) = \delta(x - y), \quad [3.39]$$

then

$$\langle g_{y'} | g_y \rangle = \delta(y - y'). \quad [3.40]$$

These eigenfunctions are also *complete*:

$$f(x) = \int_{-\infty}^{\infty} c(y)\, g_y(x)\, dy = \int_{-\infty}^{\infty} c(y)\delta(x - y)\, dy, \quad [3.41]$$

with

$$c(y) = f(y) \quad [3.42]$$

(trivial, in this case, but you can get it from Fourier's trick if you insist).

If the spectrum of a hermitian operator is *continuous* (so the eigenvalues are labeled by a continuous variable—p or y, in the examples; z, generically, in what follows), the eigenfunctions are not normalizable, they are not in Hilbert space and they do not represent possible physical states; nevertheless, the eigenfunctions with real eigenvalues are *Dirac* orthonormalizable and complete (with the sum now an integral). Luckily, this is all we really require.

Problem 3.9

(a) Cite a Hamiltonian from Chapter 2 (*other* than the harmonic oscillator) that has only a *discrete* spectrum.

(b) Cite a Hamiltonian from Chapter 2 (*other* than the free particle) that has only a *continuous* spectrum.

(c) Cite a Hamiltonian from Chapter 2 (*other* than the finite square well) that has both a discrete and a continuous part to its spectrum.

Problem 3.10 Is the ground state of the infinite square well an eigenfunction of momentum? If so, what is its momentum? If not, *why* not?

3.4 GENERALIZED STATISTICAL INTERPRETATION

In Chapter 1 I showed you how to calculate the probability that a particle would be found in a particular location, and how to determine the expectation value of any observable quantity. In Chapter 2 you learned how to find the possible outcomes of an energy measurement and their probabilities. I am now in a position to state the **generalized statistical interpretation**, which subsumes all of this and enables you to figure out the possible results of *any* measurement, and their probabilities. Together with the Schrödinger equation (which tells you how the wave function evolves in time) it is the foundation of quantum mechanics.

Generalized statistical interpretation: If you measure an observable $Q(x, p)$ on a particle in the state $\Psi(x, t)$, you are certain to get *one of the eigenvalues* of the hermitian operator $\hat{Q}(x, -i\hbar d/dx)$. If the spectrum of $\hat{Q}$ is discrete, the probability of getting the particular eigenvalue q_n associated with the orthonormalized eigenfunction $f_n(x)$ is

$$|c_n|^2, \quad \text{where} \quad c_n = \langle f_n|\Psi\rangle. \tag{3.43}$$

If the spectrum is continuous, with real eigenvalues $q(z)$ and associated Dirac-orthonormalized eigenfunctions $f_z(x)$, the probability of getting a result in the range dz is

$$|c(z)|^2\, dz \quad \text{where} \quad c(z) = \langle f_z|\Psi\rangle. \tag{3.44}$$

Upon measurement, the wave function "collapses" to the corresponding eigenstate.[14]

The statistical interpretation is radically different from anything we encounter in classical physics. A somewhat different perspective helps to make it plausible: The eigenfunctions of an observable operator are *complete*, so the wave function can be written as a linear combination of them:

$$\Psi(x, t) = \sum_n c_n f_n(x) \tag{3.45}$$

---

[14]In the case of continuous spectra the collapse is to a narrow *range* about the measured value, depending on the precision of the measuring device.

(For simplicity, I'll assume that the spectrum is discrete; it's easy to generalize this argument to the continuous case.) Because the eigenfunctions are *orthonormal*, the coefficients are given by Fourier's trick:[15]

$$c_n = \langle f_n | \Psi \rangle = \int f_n(x)^* \Psi(x, t)\, dx. \quad [3.46]$$

Qualitatively, c_n tells you "how much f_n is contained in Ψ," and given that a measurement has to return one of the eigenvalues of $\hat{Q}$, it seems reasonable that the probability of getting the particular eigenvalue q_n would be determined by the "amount of f_n" in Ψ. But because probabilities are determined by the absolute *square* of the wave function, the precise measure is actually $|c_n|^2$. That's the essential burden of the generalized statistical interpretation.[16]

Of course, the *total* probability (summed over all possible outcomes) has got to be *one*:

$$\sum_n |c_n|^2 = 1, \quad [3.47]$$

and sure enough, this follows from the normalization of the wave function:

$$1 = \langle \Psi | \Psi \rangle = \left\langle \left(\sum_{n'} c_{n'} f_{n'} \right) \middle| \left(\sum_n c_n f_n \right) \right\rangle = \sum_{n'} \sum_n c_{n'}^* c_n \langle f_{n'} | f_n \rangle$$

$$= \sum_{n'} \sum_n c_{n'}^* c_n \delta_{n'n} = \sum_n c_n^* c_n = \sum_n |c_n|^2. \quad [3.48]$$

Similarly, the expectation value of Q should be the sum over all possible outcomes of the eigenvalue times the probability of getting that eigenvalue:

$$\langle Q \rangle = \sum_n q_n |c_n|^2. \quad [3.49]$$

Indeed,

$$\langle Q \rangle = \langle \Psi | \hat{Q} \Psi \rangle = \left\langle \left(\sum_{n'} c_{n'} f_{n'} \right) \middle| \left(\hat{Q} \sum_n c_n f_n \right) \right\rangle, \quad [3.50]$$

---

[15] Notice that the time dependence—which is not at issue here—is carried by the coefficients; to make this explicit, we should really write $c_n(t)$.

[16] Again, I am scrupulously avoiding the all-too-common assertion "$|c_n|^2$ is the probability that the particle is in the state f_n." This is nonsense. The particle is in the state Ψ, *period*. Rather, $|c_n|^2$ is the probability that a *measurement* of Q would yield the value q_n. It is true that such a measurement will collapse the state to the eigenfunction f_n, so one could correctly say "$|c_n|^2$ is the probability that a particle which is *now* in the state Ψ *will be* in the state f_n subsequent to a measurement of Q" ... but that's a completely different assertion.

but $\hat{Q} f_n = q_n f_n$, so

$$\langle Q \rangle = \sum_{n'} \sum_{n} c_{n'}^* c_n q_n \langle f_{n'} | f_n \rangle = \sum_{n'} \sum_{n} c_{n'}^* c_n q_n \delta_{n'n} = \sum_{n} q_n |c_n|^2. \qquad [3.51]$$

So far, at least, everything looks consistent.

Can we reproduce, in this language, the original statistical interpretation for position measurements? Sure—it's real overkill, but worth checking. A measurement of x on a particle in state Ψ must return one of the eigenvalues of the position operator. Well, in Example 3.3 we found that every (real) number y is an eigenvalue of x, and the corresponding (Dirac-orthonormalized) eigenfunction is $g_y(x) = \delta(x - y)$. Evidently

$$c(y) = \langle g_y | \Psi \rangle = \int_{-\infty}^{\infty} \delta(x - y) \Psi(x, t)\, dx = \Psi(y, t), \qquad [3.52]$$

so the probability of getting a result in the range dy is $|\Psi(y, t)|^2\, dy$, which is precisely the original statistical interpretation.

What about momentum? In Example 3.2 we found that the eigenfunctions of the momentum operator are $f_p(x) = (1/\sqrt{2\pi\hbar}) \exp(ipx/\hbar)$, so

$$c(p) = \langle f_p | \Psi \rangle = \frac{1}{\sqrt{2\pi\hbar}} \int_{-\infty}^{\infty} e^{-ipx/\hbar} \Psi(x, t)\, dx. \qquad [3.53]$$

This is such an important quantity that we give it a special name and symbol: the **momentum space wave function**, $\Phi(p, t)$. It is essentially the *Fourier transform* of the (**position space**) wave function $\Psi(x, t)$—which, by Plancherel's theorem, is its *inverse* Fourier transform:

$$\Phi(p, t) = \frac{1}{\sqrt{2\pi\hbar}} \int_{-\infty}^{\infty} e^{-ipx/\hbar} \Psi(x, t)\, dx; \qquad [3.54]$$

$$\Psi(x, t) = \frac{1}{\sqrt{2\pi\hbar}} \int_{-\infty}^{\infty} e^{ipx/\hbar} \Phi(p, t)\, dp. \qquad [3.55]$$

According to the generalized statistical interpretation, the probability that a measurement of momentum would yield a result in the range dp is

$$|\Phi(p, t)|^2\, dp. \qquad [3.56]$$

Example 3.4 A particle of mass m is bound in the delta function well $V(x) = -\alpha\delta(x)$. What is the probability that a measurement of its momentum would yield a value greater than $p_0 = m\alpha/\hbar$?

Solution: The (position space) wave function is (Equation 2.129)

$$\Psi(x,t) = \frac{\sqrt{m\alpha}}{\hbar} e^{-m\alpha|x|/\hbar^2} e^{-iEt/\hbar}$$

(where $E = -m\alpha^2/2\hbar^2$). The momentum space wave function is therefore

$$\Phi(p,t) = \frac{1}{\sqrt{2\pi\hbar}} \frac{\sqrt{m\alpha}}{\hbar} e^{-iEt/\hbar} \int_{-\infty}^{\infty} e^{-ipx/\hbar} e^{-m\alpha|x|/\hbar^2} \, dx = \sqrt{\frac{2}{\pi}} \frac{p_0^{3/2} e^{-iEt/\hbar}}{p^2 + p_0^2}$$

(I looked up the integral). The probability, then, is

$$\frac{2}{\pi} p_0^3 \int_{p_0}^{\infty} \frac{1}{(p^2 + p_0^2)^2} \, dp = \frac{1}{\pi} \left[\frac{pp_0}{p^2 + p_0^2} + \tan^{-1}\left(\frac{p}{p_0}\right) \right]\Bigg|_{p_0}^{\infty}$$

$$= \frac{1}{4} - \frac{1}{2\pi} = 0.0908$$

(again, I looked up the integral).

Problem 3.11 Find the momentum-space wave function, $\Phi(p, t)$, for a particle in the ground state of the harmonic oscillator. What is the probability (to 2 significant digits) that a measurement of p on a particle in this state would yield a value outside the classical range (for the same energy)? *Hint:* Look in a math table under "Normal Distribution" or "Error Function" for the numerical part—or use Mathematica.

Problem 3.12 Show that

$$\langle x \rangle = \int \Phi^* \left(-\frac{\hbar}{i} \frac{\partial}{\partial p} \right) \Phi \, dp. \qquad [3.57]$$

Hint: Notice that $x \exp(ipx/\hbar) = -i\hbar(d/dp)\exp(ipx/\hbar)$.

In momentum space, then, the position operator is $i\hbar\partial/\partial p$. More generally,

$$\langle Q(x,p) \rangle = \begin{cases} \displaystyle\int \Psi^* \hat{Q}\left(x, \frac{\hbar}{i}\frac{\partial}{\partial x}\right) \Psi \, dx, & \text{in position space;} \\ \displaystyle\int \Phi^* \hat{Q}\left(-\frac{\hbar}{i}\frac{\partial}{\partial p}, p\right) \Phi \, dp, & \text{in momentum space.} \end{cases} \qquad [3.58]$$

In principle you can do all calculations in momentum space just as well (though not always as *easily*) as in position space.

3.5 THE UNCERTAINTY PRINCIPLE

I stated the uncertainty principle (in the form $\sigma_x \sigma_p \geq \hbar/2$), back in Section 1.6, and you have checked it several times, in the problems. But we have never actually *proved* it. In this section I will prove a more general version of the uncertainty principle, and explore some of its implications. The argument is beautiful, but rather abstract, so watch closely.

3.5.1 Proof of the Generalized Uncertainty Principle

For any observable A, we have (Equation 3.21):

$$\sigma_A^2 = \langle(\hat{A} - \langle A \rangle)\Psi | (\hat{A} - \langle A \rangle)\Psi\rangle = \langle f | f \rangle,$$

where $f \equiv (\hat{A} - \langle A \rangle)\Psi$. Likewise, for any *other* observable, B,

$$\sigma_B^2 = \langle g | g \rangle, \quad \text{where } g \equiv (\hat{B} - \langle B \rangle)\Psi.$$

Therefore (invoking the Schwarz inequality, Equation 3.7),

$$\sigma_A^2 \sigma_B^2 = \langle f | f \rangle \langle g | g \rangle \geq |\langle f | g \rangle|^2. \qquad [3.59]$$

Now, for any complex number z,

$$|z|^2 = [\mathrm{Re}(z)]^2 + [\mathrm{Im}(z)]^2 \geq [\mathrm{Im}(z)]^2 = \left[\frac{1}{2i}(z - z^*)\right]^2. \qquad [3.60]$$

Therefore, letting $z = \langle f | g \rangle$,

$$\sigma_A^2 \sigma_B^2 \geq \left(\frac{1}{2i}[\langle f | g \rangle - \langle g | f \rangle]\right)^2. \qquad [3.61]$$

But

$$\begin{aligned}
\langle f | g \rangle &= \langle(\hat{A} - \langle A \rangle)\Psi | (\hat{B} - \langle B \rangle)\Psi\rangle = \langle\Psi | (\hat{A} - \langle A \rangle)(\hat{B} - \langle B \rangle)\Psi\rangle \\
&= \langle\Psi | (\hat{A}\hat{B} - \hat{A}\langle B \rangle - \hat{B}\langle A \rangle + \langle A \rangle\langle B \rangle)\Psi\rangle \\
&= \langle\Psi | \hat{A}\hat{B}\Psi\rangle - \langle B \rangle\langle\Psi | \hat{A}\Psi\rangle - \langle A \rangle\langle\Psi | \hat{B}\Psi\rangle + \langle A \rangle\langle B \rangle\langle\Psi | \Psi\rangle \\
&= \langle\hat{A}\hat{B}\rangle - \langle B \rangle\langle A \rangle - \langle A \rangle\langle B \rangle + \langle A \rangle\langle B \rangle \\
&= \langle\hat{A}\hat{B}\rangle - \langle A \rangle\langle B \rangle.
\end{aligned}$$

Similarly,

$$\langle g | f \rangle = \langle\hat{B}\hat{A}\rangle - \langle A \rangle\langle B \rangle,$$

so

$$\langle f|g\rangle - \langle g|f\rangle = \langle \hat{A}\hat{B}\rangle - \langle \hat{B}\hat{A}\rangle = \langle [\hat{A}, \hat{B}]\rangle,$$

where

$$[\hat{A}, \hat{B}] \equiv \hat{A}\hat{B} - \hat{B}\hat{A}$$

is the commutator of the two operators (Equation 2.48). *Conclusion:*

$$\boxed{\sigma_A^2 \sigma_B^2 \geq \left(\frac{1}{2i}\langle [\hat{A}, \hat{B}]\rangle\right)^2.} \quad [3.62]$$

This is the (generalized) **uncertainty principle**. You might think the i makes it trivial—isn't the right side *negative*? No, for the commutator of two hermitian operators carries its own factor of i, and the two cancel out.[17]

As an example, suppose the first observable is position ($\hat{A} = x$), and the second is momentum ($\hat{B} = (\hbar/i)d/dx$). We worked out their commutator back in Chapter 2 (Equation 2.51):

$$[\hat{x}, \hat{p}] = i\hbar.$$

So

$$\sigma_x^2 \sigma_p^2 \geq \left(\frac{1}{2i} i\hbar\right)^2 = \left(\frac{\hbar}{2}\right)^2,$$

or, since standard deviations are by their nature positive,

$$\sigma_x \sigma_p \geq \frac{\hbar}{2}. \quad [3.63]$$

That's the original Heisenberg uncertainty principle, but we now see that it is just one application of a much more general theorem.

There is, in fact, an "uncertainty principle" for *every pair of observables whose operators do not commute*—we call them **incompatible observables**. Incompatible observables do not have shared eigenfunctions—at least, they cannot have a *complete set* of common eigenfunctions (see Problem 3.15). By contrast, *compatible* (commuting) observables *do* admit complete sets of simultaneous eigenfunctions.[18]

---

[17] More precisely, the commutator of two hermitian operators is itself *anti*-hermitian ($\hat{Q}^\dagger = -\hat{Q}$), and its expectation value is imaginary (Problem 3.26).

[18] This corresponds to the fact that noncommuting matrices cannot be simultaneously diagonalized (that is, they cannot both be brought to diagonal form by the same similarity transformation), whereas commuting hermitian matrices *can* be simultaneously diagonalized. See Section A.5.

For example, in the hydrogen atom (as we shall see in Chapter 4) the Hamiltonian, the magnitude of the angular momentum, and the z component of angular momentum are mutually compatible observables, and we will construct simultaneous eigenfunctions of all three, labeled by their respective eigenvalues. But there is *no* eigenfunction of position that is also an eigenfunction of momentum, because these operators are *in*compatible.

Note that the uncertainty principle is not an *extra* assumption in quantum theory, but rather a *consequence* of the statistical interpretation. You might wonder how it is enforced in the laboratory—*why* can't you determine (say) both the position and the momentum of a particle? You can certainly measure the position of the particle, but the act of measurement collapses the wave function to a narrow spike, which necessarily carries a broad range of wavelengths (hence momenta) in its Fourier decomposition. If you now measure the momentum, the state will collapse to a long sinusoidal wave, with (now) a well-defined wavelength—but the particle no longer has the position you got in the first measurement.[19] The problem, then, is that the second measurement renders the outcome of the first measurement obsolete. Only if the wave function were simultaneously an eigenstate of both observables would it be possible to make the second measurement without disturbing the state of the particle (the second collapse wouldn't change anything, in that case). But this is only possible, in general, if the two observables are compatible.

∗Problem 3.13

(a) Prove the following commutator identity:

$$[AB, C] = A[B, C] + [A, C]B. \qquad [3.64]$$

(b) Show that

$$[x^n, p] = i\hbar n x^{n-1}.$$

(c) Show more generally that

$$[f(x), p] = i\hbar \frac{df}{dx}, \qquad [3.65]$$

for any function $f(x)$.

---

[19]Niels Bohr was at pains to track down the *mechanism* by which the measurement of x (for instance) destroys the previously existing value of p. The crux of the matter is that in order to determine the position of a particle you have to poke it with something—shine light on it, say. But these photons impart to the particle a momentum you cannot control. You now know the position, but you no longer know the momentum. His famous debates with Einstein include many delightful examples, showing in detail how experimental constraints enforce the uncertainty principle. For an inspired account see Bohr's article in *Albert Einstein: Philosopher-Scientist*, edited by P. A. Schilpp, Tudor, New York (1949).

***Problem 3.14** Prove the famous "(your name) uncertainty principle," relating the uncertainty in position ($A = x$) to the uncertainty in energy ($B = p^2/2m + V$):

$$\sigma_x \sigma_H \geq \frac{\hbar}{2m} |\langle p \rangle|.$$

For stationary states this doesn't tell you much—why not?

Problem 3.15 Show that two noncommuting operators cannot have a complete set of common eigenfunctions. *Hint:* Show that if $\hat{P}$ and $\hat{Q}$ have a complete set of common eigenfunctions, then $[\hat{P}, \hat{Q}]f = 0$ for any function in Hilbert space.

3.5.2 The Minimum-Uncertainty Wave Packet

We have twice encountered wave functions that *hit* the position-momentum uncertainty limit ($\sigma_x \sigma_p = \hbar/2$): the ground state of the harmonic oscillator (Problem 2.11) and the Gaussian wave packet for the free particle (Problem 2.22). This raises an interesting question: What is the *most general* minimum-uncertainty wave packet? Looking back at the proof of the uncertainty principle, we note that there were two points at which *in*equalities came into the argument: Equation 3.59 and Equation 3.60. Suppose we require that each of these be an *equality*, and see what this tells us about Ψ.

The Schwarz inequality becomes an equality when one function is a multiple of the other: $g(x) = cf(x)$, for some complex number c (see Problem A.5). Meanwhile, in Equation 3.60 I threw away the real part of z; equality results if $\text{Re}(z) = 0$, which is to say, if $\text{Re}\langle f|g\rangle = \text{Re}(c\langle f|f\rangle) = 0$. Now, $\langle f|f\rangle$ is certainly real, so this means the constant c must be purely imaginary—let's call it ia. The necessary and sufficient condition for minimum uncertainty, then, is

$$g(x) = iaf(x), \quad \text{where } a \text{ is real.} \tag{3.66}$$

For the position-momentum uncertainty principle this criterion becomes:

$$\left(\frac{\hbar}{i}\frac{d}{dx} - \langle p \rangle\right)\Psi = ia(x - \langle x \rangle)\Psi, \tag{3.67}$$

which is a differential equation for Ψ as a function of x. Its general solution (Problem 3.16) is

$$\Psi(x) = Ae^{-a(x-\langle x \rangle)^2/2\hbar} e^{i\langle p \rangle x/\hbar}. \tag{3.68}$$

Evidently the minimum-uncertainty wave packet is a *gaussian*—and the two examples we encountered earlier *were* gaussians.[20]

[20]Note that it is only the dependence of Ψ on x that is at issue here—the "constants" A, a, $\langle x \rangle$, and $\langle p \rangle$ may all be functions of time, and for that matter Ψ may evolve away from the minimal form. All I'm asserting is that if, at some instant, the wave function is gaussian in x, then (at that instant) the uncertainty product is minimal.

Problem 3.16 Solve Equation 3.67 for $\Psi(x)$. Note that $\langle x\rangle$ and $\langle p\rangle$ are *constants*.

3.5.3 The Energy-Time Uncertainty Principle

The position-momentum uncertainty principle is often written in the form

$$\Delta x\,\Delta p \geq \frac{\hbar}{2}; \tag{3.69}$$

Δx (the "uncertainty" in x) is loose notation (and sloppy language) for the standard deviation of the results of repeated measurements on identically prepared systems.[21] Equation 3.69 is often paired with the **energy-time uncertainty principle**,

$$\Delta t\,\Delta E \geq \frac{\hbar}{2}. \tag{3.70}$$

Indeed, in the context of special relativity the energy-time form might be thought of as a *consequence* of the position-momentum version, because x and t (or rather, ct) go together in the position-time four-vector, while p and E (or rather, E/c) go together in the energy-momentum four-vector. So in a relativistic theory Equation 3.70 would be a necessary concomitant to Equation 3.69. But we're not doing relativistic quantum mechanics. The Schrödinger equation is explicitly nonrelativistic: It treats t and x on a very unequal footing (as a differential equation it is *first*-order in t, but *second*-order in x), and Equation 3.70 is emphatically *not* implied by Equation 3.69. My purpose now is to *derive* the energy-time uncertainty principle, and in the course of that derivation to persuade you that it is really an altogether different beast, whose superficial resemblance to the position-momentum uncertainty principle is actually quite misleading.

After all, position, momentum, and energy are all dynamical variables—measurable characteristics of the system, at any given time. But time itself is not a dynamical variable (not, at any rate, in a nonrelativistic theory): You don't go out and measure the "time" of a particle, as you might its position or its energy. Time is the *independent* variable, of which the dynamical quantities are *functions*. In particular, the Δt in the energy-time uncertainty principle is not the standard deviation of a collection of time measurements; roughly speaking (I'll make this more precise in a moment) it is the *time it takes the system to change substantially*.

[21] Many casual applications of the uncertainty principle are actually based (often inadvertently) on a completely different—and sometimes quite unjustified—measure of "uncertainty." Conversely, some perfectly rigorous arguments use other definitions of "uncertainty." See Jan Hilgevoord, *Am. J. Phys.* **70**, 983 (2002).

As a measure of how fast the system is changing, let us compute the time derivative of the expectation value of some observable, $Q(x, p, t)$:

$$\frac{d}{dt}\langle Q\rangle = \frac{d}{dt}\langle\Psi|\hat{Q}\Psi\rangle = \left\langle\frac{\partial\Psi}{\partial t}\middle|\hat{Q}\Psi\right\rangle + \left\langle\Psi\middle|\frac{\partial\hat{Q}}{\partial t}\Psi\right\rangle + \left\langle\Psi\middle|\hat{Q}\frac{\partial\Psi}{\partial t}\right\rangle.$$

Now, the Schrödinger equation says

$$i\hbar\frac{\partial\Psi}{\partial t} = \hat{H}\Psi$$

(where $H = p^2/2m + V$ is the Hamiltonian). So

$$\frac{d}{dt}\langle Q\rangle = -\frac{1}{i\hbar}\langle\hat{H}\Psi|\hat{Q}\Psi\rangle + \frac{1}{i\hbar}\langle\Psi|\hat{Q}\hat{H}\Psi\rangle + \left\langle\frac{\partial\hat{Q}}{\partial t}\right\rangle.$$

But $\hat{H}$ is hermitian, so $\langle\hat{H}\Psi|\hat{Q}\Psi\rangle = \langle\Psi|\hat{H}\hat{Q}\Psi\rangle$, and hence

$$\frac{d}{dt}\langle Q\rangle = \frac{i}{\hbar}\langle[\hat{H}, \hat{Q}]\rangle + \left\langle\frac{\partial\hat{Q}}{\partial t}\right\rangle. \qquad [3.71]$$

This is an interesting and useful result in its own right (see Problems 3.17 and 3.31). In the typical case where the operator does not depend explicitly on time,[22] it tells us that the rate of change of the expectation value is determined by the commutator of the operator with the Hamiltonian. In particular, if $\hat{Q}$ *commutes* with $\hat{H}$, then $\langle Q\rangle$ is constant, and in this sense Q is a *conserved* quantity.

Now, suppose we pick $A = H$ and $B = Q$, in the generalized uncertainty principle (Equation 3.62), and assume that Q does not depend explicitly on t:

$$\sigma_H^2\sigma_Q^2 \geq \left(\frac{1}{2i}\langle[\hat{H}, \hat{Q}]\rangle\right)^2 = \left(\frac{1}{2i}\frac{\hbar}{i}\frac{d\langle Q\rangle}{dt}\right)^2 = \left(\frac{\hbar}{2}\right)^2\left(\frac{d\langle Q\rangle}{dt}\right)^2.$$

Or, more simply,

$$\sigma_H\sigma_Q \geq \frac{\hbar}{2}\left|\frac{d\langle Q\rangle}{dt}\right|. \qquad [3.72]$$

Let's define $\Delta E \equiv \sigma_H$, and

$$\Delta t \equiv \frac{\sigma_Q}{|d\langle Q\rangle/dt|}. \qquad [3.73]$$

[22] Operators that depend explicitly on t are quite rare, so *almost always* $\partial\hat{Q}/\partial t = 0$. As an example of *explicit* time dependence, consider the potential energy of a harmonic oscillator whose spring constant is changing (perhaps the temperature is rising, so the spring becomes more flexible): $Q = (1/2)m[\omega(t)]^2x^2$.

Then

$$\Delta E\,\Delta t \geq \frac{\hbar}{2}, \qquad [3.74]$$

and that's the energy-time uncertainty principle. But notice what is meant by Δt, here: Since

$$\sigma_Q = \left|\frac{d\langle Q\rangle}{dt}\right| \Delta t,$$

Δt represents the *amount of time it takes the expectation value of Q to change by one standard deviation.*[23] In particular, Δt depends entirely on what observable (Q) you care to look at—the change might be rapid for one observable and slow for another. But if ΔE is small, then the rate of change of *all* observables must be very gradual; or, to put it the other way around, if *any* observable changes rapidly, the "uncertainty" in the energy must be large.

Example 3.5 In the extreme case of a stationary state, for which the energy is uniquely determined, all expectation values are constant in time ($\Delta E = 0 \Rightarrow \Delta t = \infty$)—as in fact we noticed some time ago (see Equation 2.9). To make something *happen* you must take a linear combination of at least two stationary states—say:

$$\Psi(x,t) = a\psi_1(x)e^{-iE_1t/\hbar} + b\psi_2(x)e^{-iE_2t/\hbar}.$$

If a, b, ψ_1, and ψ_2 are real,

$$|\Psi(x,t)|^2 = a^2(\psi_1(x))^2 + b^2(\psi_2(x))^2 + 2ab\psi_1(x)\psi_2(x)\cos\left(\frac{E_2 - E_1}{\hbar}t\right).$$

The period of oscillation is $\tau = 2\pi\hbar/(E_2 - E_1)$. Roughly speaking, $\Delta E = E_2 - E_1$ and $\Delta t = \tau$ (for the *exact* calculation see Problem 3.18), so

$$\Delta E\,\Delta t = 2\pi\hbar,$$

which is indeed $\geq \hbar/2$.

Example 3.6 How long does it take a free-particle wave packet to pass by a particular point (Figure 3.1)? Qualitatively (an exact version is explored in Problem 3.19), $\Delta t = \Delta x/v \doteq m\Delta x/p$, but $E = p^2/2m$, so $\Delta E = p\Delta p/m$. Therefore,

$$\Delta E\,\Delta t = \frac{p\Delta p}{m}\frac{m\Delta x}{p} = \Delta x\Delta p,$$

which is $\geq \hbar/2$ by the position-momentum uncertainty principle.

---

[23] This is sometimes called the "Mandelstam-Tamm" formulation of the energy-time uncertainty principle. For a review of alternative approaches see Paul Busch, *Found. Phys.* **20**, 1 (1990).

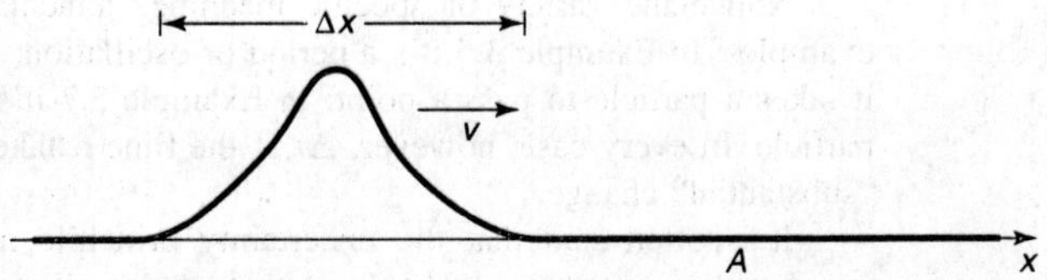

FIGURE 3.1: A free particle wave packet approaches the point *A* (Example 3.6).

Example 3.7 The Δ particle lasts about 10^{-23} seconds, before spontaneously disintegrating. If you make a histogram of all measurements of its mass, you get a kind of bell-shaped curve centered at 1232 MeV/c^2, with a width of about 120 MeV/c^2 (Figure 3.2). Why does the rest energy (mc^2) sometimes come out higher than 1232, and sometimes lower? Is this experimental error? No, for

$$\Delta E\,\Delta t = \left(\frac{120}{2}\text{ MeV}\right)(10^{-23}\text{ sec}) = 6\times 10^{-22}\text{ MeV sec},$$

whereas $\hbar/2 = 3\times 10^{-22}$ MeV sec. So the spread in m is about as small as the uncertainty principle allows—a particle with so short a lifetime just doesn't *have* a very well-defined mass.[24]

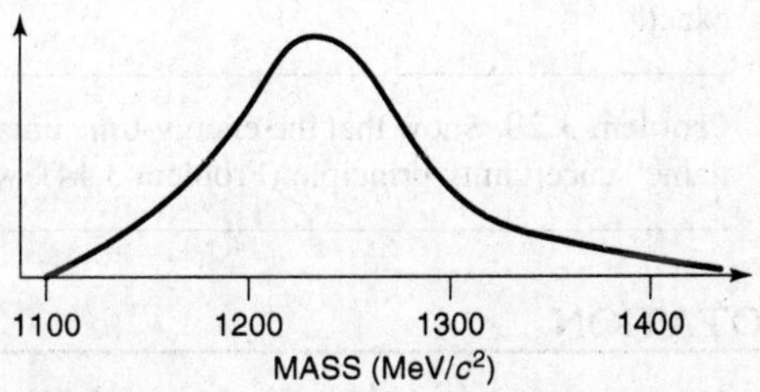

FIGURE 3.2: Histogram of measurements of the Δ mass (Example 3.7).

[24]Actually, Example 3.7 is a bit of a fraud. You can't measure 10^{-23} sec on a stop-watch, and in practice the lifetime of such a short-lived particle is *inferred* from the width of the mass plot, using the uncertainty principle as *input*. However, the point is valid, even if the logic is backwards. Moreover, if you assume the Δ is about the same size as a proton ($\sim 10^{-15}$ m), then 10^{-23} sec is roughly the time it takes light to cross the particle, and it's hard to imagine that the lifetime could be much *less* than that.

Notice the variety of specific meanings attaching to the term Δt in these examples: In Example 3.5 it's a period of oscillation; in Example 3.6 it's the time it takes a particle to pass a point; in Example 3.7 it's the lifetime of an unstable particle. In every case, however, Δt is the time it takes for the system to undergo "substantial" change.

It is often said that the uncertainty principle means energy is not strictly conserved in quantum mechanics—that you're allowed to "borrow" energy ΔE, as long as you "pay it back" in a time $\Delta t \approx \hbar/(2\Delta E)$; the greater the violation, the briefer the period over which it can occur. Now, there are many legitimate readings of the energy-time uncertainty principle, but this is not one of them. Nowhere does quantum mechanics license violation of energy conservation, and certainly no such authorization entered into the derivation of Equation 3.74. But the uncertainty principle is extraordinarily robust: It can be misused without leading to seriously incorrect results, and as a consequence physicists are in the habit of applying it rather carelessly.

***Problem 3.17** Apply Equation 3.71 to the following special cases: (a) $Q = 1$; (b) $Q = H$; (c) $Q = x$; (d) $Q = p$. In each case, comment on the result, with particular reference to Equations 1.27, 1.33, 1.38, and conservation of energy (comments following Equation 2.39).

Problem 3.18 Test the energy-time uncertainty principle for the wave function in Problem 2.5 and the observable x, by calculating σ_H, σ_x, and $d\langle x\rangle/dt$ exactly.

Problem 3.19 Test the energy-time uncertainty principle for the free particle wave packet in Problem 2.43 and the observable x, by calculating σ_H, σ_x, and $d\langle x\rangle/dt$ exactly.

Problem 3.20 Show that the energy-time uncertainty principle reduces to the "your name" uncertainty principle (Problem 3.14), when the observable in question is x.

3.6 DIRAC NOTATION

Imagine an ordinary vector $\mathbf{A}$ in two dimensions (Figure 3.3(a)). How would you describe this vector to someone? The most convenient way is to set up cartesian axes, x and y, and specify the components of $\mathbf{A}$: $A_x = \hat{\imath} \cdot \mathbf{A}$, $A_y = \hat{\jmath} \cdot \mathbf{A}$ (Figure 3.3(b)). Of course, your sister might have drawn a different set of axes, x' and y', and she would report different components: $A'_x = \hat{\imath}' \cdot \mathbf{A}$, $A'_y = \hat{\jmath}' \cdot \mathbf{A}$ (Figure 3.3(c)). But it's all the same *vector*—we're simply expressing it with respect to two different *bases* ($\{\hat{\imath}, \hat{\jmath}\}$ and $\{\hat{\imath}', \hat{\jmath}'\}$). The vector itself lives "out there in space," independent of anybody's (arbitrary) choice of coordinates.

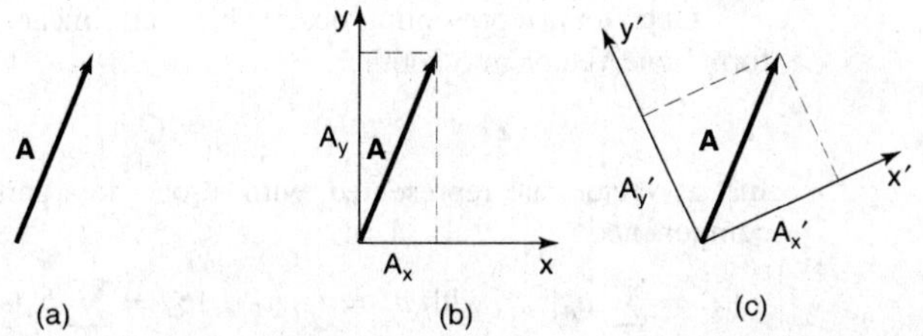

FIGURE 3.3: (a) Vector A. (b) Components of A with respect to xy axes. (c) Components of A with respect to $x'y'$ axes.

The same is true for the state of a system in quantum mechanics. It is represented by a *vector*, $|\mathcal{S}(t)\rangle$, that lives "out there in Hilbert space," but we can *express* it with respect to any number of different *bases*. The wave function $\Psi(x,t)$ is actually the coefficient in the expansion of $|\mathcal{S}\rangle$ in the basis of position eigenfunctions:

$$\Psi(x,t) = \langle x|\mathcal{S}(t)\rangle, \qquad [3.75]$$

(with $|x\rangle$ standing for the eigenfunction of $\hat{x}$ with eigenvalue x),[25] whereas the momentum space wavefunction $\Phi(p,t)$ is the expansion of $|\mathcal{S}\rangle$ in the basis of momentum eigenfunctions:

$$\Phi(p,t) = \langle p|\mathcal{S}(t)\rangle \qquad [3.76]$$

(with $|p\rangle$ standing for the eigenfunction of $\hat{p}$ with eigenvalue p).[26] Or we could expand $|\mathcal{S}\rangle$ in the basis of energy eigenfunctions (supposing for simplicity that the spectrum is discrete):

$$c_n(t) = \langle n|\mathcal{S}(t)\rangle \qquad [3.77]$$

(with $|n\rangle$ standing for the nth eigenfunction of $\hat{H}$)—Equation 3.46. But it's all the same state; the functions Ψ and Φ, and the collection of coefficients $\{c_n\}$, contain exactly the same information—they are simply three different ways of describing the same vector:

$$\Psi(x,t) = \int \Psi(y,t)\delta(x-y)\,dy = \int \Phi(p,t)\frac{1}{\sqrt{2\pi\hbar}}e^{ipx/\hbar}\,dp$$
$$= \sum c_n e^{-iE_nt/\hbar}\psi_n(x). \qquad [3.78]$$

[25] I don't want to call it g_x (Equation 3.39), because that is its form in the position basis, and the whole point here is to free ourselves from any particular basis. Indeed, when I first defined Hilbert space as the set of square-integrable functions—over x—that was already too restrictive, committing us to a specific representation (the position basis). I want now to think of it as an abstract vector space, whose members can be expressed with respect to any basis you like.

[26] In position space it would be $f_p(x)$ (Equation 3.32).

Operators (representing observables) are linear transformations—they "transform" one vector into another:

$$|\beta\rangle = \hat{Q}|\alpha\rangle. \tag{3.79}$$

Just as vectors are represented, with respect to a particular basis $\{|e_n\rangle\}$,[27] by their components,

$$|\alpha\rangle = \sum_n a_n |e_n\rangle, \text{ with } a_n = \langle e_n|\alpha\rangle; \;\; |\beta\rangle = \sum_n b_n |e_n\rangle, \text{ with } b_n = \langle e_n|\beta\rangle, \tag{3.80}$$

operators are represented (with respect to a particular basis) by their **matrix elements**[28]

$$\langle e_m|\hat{Q}|e_n\rangle \equiv Q_{mn}. \tag{3.81}$$

In this notation Equation 3.79 takes the form

$$\sum_n b_n |e_n\rangle = \sum_n a_n \hat{Q}|e_n\rangle, \tag{3.82}$$

or, taking the inner product with $|e_m\rangle$,

$$\sum_n b_n \langle e_m|e_n\rangle = \sum_n a_n \langle e_m|\hat{Q}|e_n\rangle, \tag{3.83}$$

and hence

$$b_m = \sum_n Q_{mn} a_n. \tag{3.84}$$

Thus the matrix elements tell you how the components transform.

Later on we will encounter systems that admit only a finite number (N) of linearly independent states. In that case $|\mathcal{S}(t)\rangle$ lives in an N-dimensional vector space; it can be represented as a column of (N) components (with respect to a given basis), and operators take the form of ordinary ($N \times N$) matrices. These are the simplest quantum systems—none of the subtleties associated with infinite-dimensional vector spaces arise. Easiest of all is the two-state system, which we explore in the following example.

Example 3.8 Imagine a system in which there are just *two* linearly independent states:[29]

$$|1\rangle = \begin{pmatrix} 1 \\ 0 \end{pmatrix} \quad \text{and} \quad |2\rangle = \begin{pmatrix} 0 \\ 1 \end{pmatrix}.$$

---

[27] I'll assume the basis is discrete; otherwise n becomes a continuous index and the sums are replaced by integrals.

[28] This terminology is inspired, obviously, by the finite-dimensional case, but the "matrix" will now typically have an infinite (maybe even uncountable) number of elements.

[29] Technically, the "equals" signs here mean "is represented by," but I don't think any confusion will arise if we adopt the customary informal notation.

The most general state is a normalized linear combination:

$$|\mathcal{S}\rangle = a|1\rangle + b|2\rangle = \begin{pmatrix} a \\ b \end{pmatrix}, \quad \text{with } |a|^2 + |b|^2 = 1.$$

The Hamiltonian can be expressed as a (hermitian) matrix; suppose it has the specific form

$$\mathsf{H} = \begin{pmatrix} h & g \\ g & h \end{pmatrix},$$

where g and h are real constants. If the system starts out (at $t = 0$) in state $|1\rangle$, what is its state at time t?

Solution: The (time-dependent) Schrödinger equation says

$$i\hbar \frac{d}{dt}|\mathcal{S}\rangle = H|\mathcal{S}\rangle. \qquad [3.85]$$

As always, we begin by solving the time-*in*dependent Schrödinger equation:

$$H|s\rangle = E|s\rangle; \qquad [3.86]$$

that is, we look for the eigenvectors and eigenvalues of H. The characteristic equation determines the eigenvalues:

$$\det\begin{pmatrix} h-E & g \\ g & h-E \end{pmatrix} = (h-E)^2 - g^2 = 0 \Rightarrow h - E = \mp g \Rightarrow E_{\pm} = h \pm g.$$

Evidently the allowed energies are $(h+g)$ and $(h-g)$. To determine the eigenvectors, we write

$$\begin{pmatrix} h & g \\ g & h \end{pmatrix}\begin{pmatrix} \alpha \\ \beta \end{pmatrix} = (h \pm g)\begin{pmatrix} \alpha \\ \beta \end{pmatrix} \Rightarrow h\alpha + g\beta = (h \pm g)\alpha \Rightarrow \beta = \pm\alpha,$$

so the normalized eigenvectors are

$$|s_{\pm}\rangle = \frac{1}{\sqrt{2}}\begin{pmatrix} 1 \\ \pm 1 \end{pmatrix}.$$

Next we expand the initial state as a linear combination of eigenvectors of the Hamiltonian:

$$|\mathcal{S}(0)\rangle = \begin{pmatrix} 1 \\ 0 \end{pmatrix} = \frac{1}{\sqrt{2}}\left(|s_+\rangle + |s_-\rangle\right).$$

Finally, we tack on the standard time-dependence $\exp(-iE_n t/\hbar)$:

$$|\mathcal{S}(t)\rangle = \frac{1}{\sqrt{2}}\left[e^{-i(h+g)t/\hbar}|s_+\rangle + e^{-i(h-g)t/\hbar}|s_-\rangle\right]$$

$$= \frac{1}{2} e^{-iht/\hbar} \left[e^{-igt/\hbar} \begin{pmatrix} 1 \\ 1 \end{pmatrix} + e^{igt/\hbar} \begin{pmatrix} 1 \\ -1 \end{pmatrix} \right]$$

$$= \frac{1}{2} e^{-iht/\hbar} \begin{pmatrix} e^{-igt/\hbar} + e^{igt/\hbar} \\ e^{-igt/\hbar} - e^{igt/\hbar} \end{pmatrix} = e^{-iht/\hbar} \begin{pmatrix} \cos(gt/\hbar) \\ -i \sin(gt/\hbar) \end{pmatrix}.$$

If you doubt this result, by all means *check* it: Does it satisfy the time-dependent Schrödinger equation? Does it match the initial state when $t = 0$?

This is a crude model for (among other things) **neutrino oscillations**. In that case $|1\rangle$ represents the electron neutrino, and $|2\rangle$ the muon neutrino; if the Hamiltonian has a nonvanishing off-diagonal term (g) then in the course of time the electron neutrino will turn into a muon neutrino (and back again).

Dirac proposed to chop the bracket notation for the inner product, $\langle\alpha|\beta\rangle$, into two pieces, which he called **bra**, $\langle\alpha|$, and **ket**, $|\beta\rangle$ (I don't know what happened to the c). The latter is a vector, but what exactly is the former? It's a *linear function* of vectors, in the sense that when it hits a vector (to its right) it yields a (complex) number—the inner product. (When an *operator* hits a vector, it delivers another vector; when a *bra* hits a vector, it delivers a number.) In a function space, the bra can be thought of as an instruction to integrate:

$$\langle f| = \int f^*[\cdots]\,dx,$$

with the ellipsis $[\cdots]$ waiting to be filled by whatever function the bra encounters in the ket to its right. In a finite-dimensional vector space, with the vectors expressed as columns,

$$|\alpha\rangle = \begin{pmatrix} a_1 \\ a_2 \\ \vdots \\ a_n \end{pmatrix}, \qquad [3.87]$$

the corresponding bra is a row vector:

$$\langle\alpha| = \left(a_1^* \, a_2^* \, \ldots \, a_n^*\right). \qquad [3.88]$$

The collection of all bras constitutes another vector space—the so-called **dual space**.

The license to treat bras as separate entities in their own right allows for some powerful and pretty notation (though I shall not exploit it in this book). For example, if $|\alpha\rangle$ is a normalized vector, the operator

$$\hat{P} \equiv |\alpha\rangle\langle\alpha| \qquad [3.89]$$

picks out the portion of any other vector that "lies along" $|\alpha\rangle$:

$$\hat{P}|\beta\rangle = \langle\alpha|\beta\rangle|\alpha\rangle;$$

we call it the **projection operator** onto the one-dimensional subspace spanned by $|\alpha\rangle$. If $\{|e_n\rangle\}$ is a discrete orthonormal basis,

$$\langle e_m|e_n\rangle = \delta_{mn}, \quad [3.90]$$

then

$$\sum_n |e_n\rangle\langle e_n| = 1 \quad [3.91]$$

(the identity operator). For if we let this operator act on any vector $|\alpha\rangle$, we recover the expansion of $|\alpha\rangle$ in the $\{|e_n\rangle\}$ basis:

$$\sum_n |e_n\rangle\langle e_n|\alpha\rangle = |\alpha\rangle. \quad [3.92]$$

Similarly, if $\{|e_z\rangle\}$ is a *Dirac* orthonormalized continuous basis,

$$\langle e_z|e_{z'}\rangle = \delta(z - z'), \quad [3.93]$$

then

$$\int |e_z\rangle\langle e_z|\, dz = 1. \quad [3.94]$$

Equations 3.91 and 3.94 are the tidiest ways to express completeness.

Problem 3.21 Show that projection operators are **idempotent**: $\hat{P}^2 = \hat{P}$. Determine the eigenvalues of $\hat{P}$, and characterize its eigenvectors.

Problem 3.22 Consider a three-dimensional vector space spanned by an orthonormal basis $|1\rangle$, $|2\rangle$, $|3\rangle$. Kets $|\alpha\rangle$ and $|\beta\rangle$ are given by

$$|\alpha\rangle = i|1\rangle - 2|2\rangle - i|3\rangle, \quad |\beta\rangle = i|1\rangle + 2|3\rangle.$$

(a) Construct $\langle\alpha|$ and $\langle\beta|$ (in terms of the dual basis $\langle 1|$, $\langle 2|$, $\langle 3|$).

(b) Find $\langle\alpha|\beta\rangle$ and $\langle\beta|\alpha\rangle$, and confirm that $\langle\beta|\alpha\rangle = \langle\alpha|\beta\rangle^*$.

(c) Find all nine matrix elements of the operator $\hat{A} \equiv |\alpha\rangle\langle\beta|$, in this basis, and construct the matrix **A**. Is it hermitian?

Problem 3.23 The Hamiltonian for a certain two-level system is

$$\hat{H} = E\,(|1\rangle\langle 1| - |2\rangle\langle 2| + |1\rangle\langle 2| + |2\rangle\langle 1|),$$

where $|1\rangle$, $|2\rangle$ is an orthonormal basis and E is a number with the dimensions of energy. Find its eigenvalues and (normalized) eigenvectors (as linear combinations of $|1\rangle$ and $|2\rangle$). What is the matrix $\mathbf{H}$ representing $\hat{H}$ with respect to this basis?

Problem 3.24 Let $\hat{Q}$ be an operator with a complete set of orthonormal eigenvectors:

$$\hat{Q}|e_n\rangle = q_n|e_n\rangle \quad (n = 1, 2, 3, \ldots).$$

Show that $\hat{Q}$ can be written in terms of its **spectral decomposition**:

$$\hat{Q} = \sum_n q_n |e_n\rangle\langle e_n|.$$

Hint: An operator is characterized by its action on all possible vectors, so what you must show is that

$$\hat{Q}|\alpha\rangle = \left\{\sum_n q_n |e_n\rangle\langle e_n|\right\}|\alpha\rangle,$$

for any vector $|\alpha\rangle$.

FURTHER PROBLEMS FOR CHAPTER 3

Problem 3.25 Legendre polynomials. Use the Gram-Schmidt procedure (Problem A.4) to orthonormalize the functions 1, x, x^2, and x^3, on the interval $-1 \leq x \leq 1$. You may recognize the results—they are (apart from the normalization)[30] **Legendre polynomials** (Table 4.1).

Problem 3.26 An **anti-hermitian** (or **skew-hermitian**) operator is equal to *minus* its hermitian conjugate:

$$\hat{Q}^\dagger = -\hat{Q}. \qquad [3.95]$$

(a) Show that the expectation value of an anti-hermitian operator is imaginary.

(b) Show that the commutator of two hermitian operators is anti-hermitian. How about the commutator of two *anti*-hermitian operators?

[30] Legendre didn't know what the best convention would be; he picked the overall factor so that all his functions would go to 1 at $x = 1$, and we're stuck with his unfortunate choice.

Problem 3.27 Sequential measurements. An operator $\hat{A}$, representing observable A, has two normalized eigenstates ψ_1 and ψ_2, with eigenvalues a_1 and a_2, respectively. Operator $\hat{B}$, representing observable B, has two normalized eigenstates ϕ_1 and ϕ_2, with eigenvalues b_1 and b_2. The eigenstates are related by

$$\psi_1 = (3\phi_1 + 4\phi_2)/5, \quad \psi_2 = (4\phi_1 - 3\phi_2)/5.$$

(a) Observable A is measured, and the value a_1 is obtained. What is the state of the system (immediately) after this measurement?

(b) If B is now measured, what are the possible results, and what are their probabilities?

(c) Right after the measurement of B, A is measured again. What is the probability of getting a_1? (Note that the answer would be quite different if I had told you the outcome of the B measurement.)

****Problem 3.28** Find the momentum-space wave function $\Phi_n(p, t)$ for the nth stationary state of the infinite square well. Graph $|\Phi_1(p, t)|^2$ and $|\Phi_2(p, t)|^2$, as functions of p (pay particular attention to the points $p = \pm n\pi\hbar/a$). Use $\Phi_n(p, t)$ to calculate the expectation value of p^2. Compare your answer to Problem 2.4.

Problem 3.29 Consider the wave function

$$\Psi(x, 0) = \begin{cases} \dfrac{1}{\sqrt{2n\lambda}} e^{i2\pi x/\lambda}, & -n\lambda < x < n\lambda, \\ 0, & \text{otherwise,} \end{cases}$$

where n is some positive integer. This function is purely sinusoidal (with wavelength λ) on the interval $-n\lambda < x < n\lambda$, but it still carries a *range* of momenta, because the oscillations do not continue out to infinity. Find the momentum space wave function $\Phi(p, 0)$. Sketch the graphs of $|\Psi(x, 0)|^2$ and $|\Phi(p, 0)|^2$, and determine their widths, w_x and w_p (the distance between zeros on either side of the main peak). Note what happens to each width as $n \to \infty$. Using w_x and w_p as estimates of Δx and Δp, check that the uncertainty principle is satisfied. *Warning:* If you try calculating σ_p, you're in for a rude surprise. Can you diagnose the problem?

Problem 3.30 Suppose

$$\Psi(x, 0) = \frac{A}{x^2 + a^2},$$

for constants A and a.

(a) Determine A, by normalizing $\Psi(x, 0)$.

(b) Find $\langle x \rangle$, $\langle x^2 \rangle$, and σ_x (at time $t = 0$).

(c) Find the momentum space wave function $\Phi(p, 0)$, and check that it is normalized.

(d) Use $\Phi(p, 0)$ to calculate $\langle p \rangle$, $\langle p^2 \rangle$, and σ_p (at time $t = 0$).

(e) Check the Heisenberg uncertainty principle for this state.

***Problem 3.31 Virial theorem.** Use Equation 3.71 to show that

$$\frac{d}{dt}\langle xp \rangle = 2\langle T \rangle - \left\langle x\frac{dV}{dx} \right\rangle, \quad [3.96]$$

where T is the kinetic energy ($H = T + V$). In a *stationary* state the left side is zero (why?) so

$$2\langle T \rangle = \left\langle x\frac{dV}{dx} \right\rangle. \quad [3.97]$$

This is called the **virial theorem**. Use it to prove that $\langle T \rangle = \langle V \rangle$ for stationary states of the harmonic oscillator, and check that this is consistent with the results you got in Problems 2.11 and 2.12.

Problem 3.32 In an interesting version of the energy-time uncertainty principle[31] $\Delta t = \tau/\pi$, where τ is the time it takes $\Psi(x, t)$ to evolve into a state orthogonal to $\Psi(x, 0)$. Test this out, using a wave function that is an equal admixture of two (orthonormal) stationary states of some (arbitrary) potential: $\Psi(x, 0) = (1/\sqrt{2})[\psi_1(x) + \psi_2(x)]$.

****Problem 3.33** Find the matrix elements $\langle n|x|n' \rangle$ and $\langle n|p|n' \rangle$ in the (orthonormal) basis of stationary states for the harmonic oscillator (Equation 2.67). You already calculated the "diagonal" elements ($n = n'$) in Problem 2.12; use the same technique for the general case. Construct the corresponding (infinite) matrices, X and P Show that $(1/2m)\mathsf{P}^2 + (m\omega^2/2)\mathsf{X}^2 = \mathsf{H}$ is *diagonal*, in this basis. Are its diagonal elements what you would expect? *Partial answer:*

$$\langle n|x|n' \rangle = \sqrt{\frac{\hbar}{2m\omega}}\left(\sqrt{n'}\delta_{n,n'-1} + \sqrt{n}\delta_{n',n-1}\right). \quad [3.98]$$

---

[31] See Lev Vaidman, *Am. J. Phys.* **60**, 182 (1992) for a proof.

Problem 3.34 A harmonic oscillator is in a state such that a measurement of the energy would yield either $(1/2)\hbar\omega$ or $(3/2)\hbar\omega$, with equal probability. What is the largest possible value of $\langle p \rangle$ in such a state? If it assumes this maximal value at time $t = 0$, what is $\Psi(x, t)$?

$***$**Problem 3.35 Coherent states of the harmonic oscillator.** Among the stationary states of the harmonic oscillator ($|n\rangle = \psi_n(x)$, Equation 2.67) only $n = 0$ hits the uncertainty limit ($\sigma_x \sigma_p = \hbar/2$); in general, $\sigma_x \sigma_p = (2n + 1)\hbar/2$, as you found in Problem 2.12. But certain *linear combinations* (known as **coherent states**) also minimize the uncertainty product. They are (as it turns out) *eigenfunctions of the lowering operator*:[32]

$$a_-|\alpha\rangle = \alpha|\alpha\rangle$$

(the eigenvalue α can be any complex number).

(a) Calculate $\langle x \rangle$, $\langle x^2 \rangle$, $\langle p \rangle$, $\langle p^2 \rangle$ in the state $|\alpha\rangle$. *Hint:* Use the technique in Example 2.5, and remember that a_+ is the hermitian conjugate of a_-. Do *not* assume α is real.

(b) Find σ_x and σ_p; show that $\sigma_x \sigma_p = \hbar/2$.

(c) Like any other wave function, a coherent state can be expanded in terms of energy eigenstates:

$$|\alpha\rangle = \sum_{n=0}^{\infty} c_n |n\rangle.$$

Show that the expansion coefficients are

$$c_n = \frac{\alpha^n}{\sqrt{n!}} c_0.$$

(d) Determine c_0 by normalizing $|\alpha\rangle$. *Answer:* $\exp(-|\alpha|^2/2)$.

(e) Now put in the time dependence:

$$|n\rangle \to e^{-iE_n t/\hbar}|n\rangle,$$

and show that $|\alpha(t)\rangle$ remains an eigenstate of a_-, but the eigen*value* evolves in time:

$$\alpha(t) = e^{-i\omega t}\alpha.$$

So a coherent state *stays* coherent, and continues to minimize the uncertainty product.

[32] There are no normalizable eigenfunctions of the *raising* operator.

(f) Is the ground state ($|n = 0\rangle$) itself a coherent state? If so, what is the eigenvalue?

Problem 3.36 Extended uncertainty principle.[33] The generalized uncertainty principle (Equation 3.62) states that

$$\sigma_A^2 \sigma_B^2 \geq \frac{1}{4}\langle C\rangle^2,$$

where $\hat{C} \equiv -i[\hat{A}, \hat{B}]$.

(a) Show that it can be strengthened to read

$$\sigma_A^2 \sigma_B^2 \geq \frac{1}{4}(\langle C\rangle^2 + \langle D\rangle^2), \qquad [3.99]$$

where $\hat{D} \equiv \hat{A}\hat{B} + \hat{B}\hat{A} - 2\langle A\rangle\langle B\rangle$. *Hint:* Keep the Re($z$) term in Equation 3.60.

(b) Check Equation 3.99 for the case $B = A$ (the standard uncertainty principle is trivial, in this case, since $\hat{C} = 0$; unfortunately, the extended uncertainty principle doesn't help much either).

Problem 3.37 The Hamiltonian for a certain three-level system is represented by the matrix

$$\mathbf{H} = \begin{pmatrix} a & 0 & b \\ 0 & c & 0 \\ b & 0 & a \end{pmatrix},$$

where a, b, and c are real numbers.

(a) If the system starts out in the state

$$|\mathcal{S}(0)\rangle = \begin{pmatrix} 0 \\ 1 \\ 0 \end{pmatrix},$$

what is $|\mathcal{S}(t)\rangle$?

(b) If the system starts out in the state

$$|\mathcal{S}(0)\rangle = \begin{pmatrix} 0 \\ 0 \\ 1 \end{pmatrix},$$

what is $|\mathcal{S}(t)\rangle$?

[33] For interesting commentary and references, see R. R. Puri, *Phys. Rev. A* **49**, 2178 (1994).

Problem 3.38 The Hamiltonian for a certain three-level system is represented by the matrix

$$\mathbf{H} = \hbar\omega \begin{pmatrix} 1 & 0 & 0 \\ 0 & 2 & 0 \\ 0 & 0 & 2 \end{pmatrix}$$

Two other observables, A and B, are represented by the matrices

$$\mathbf{A} = \lambda \begin{pmatrix} 0 & 1 & 0 \\ 1 & 0 & 0 \\ 0 & 0 & 2 \end{pmatrix}, \quad \mathbf{B} = \mu \begin{pmatrix} 2 & 0 & 0 \\ 0 & 0 & 1 \\ 0 & 1 & 0 \end{pmatrix},$$

where ω, λ, and μ are positive real numbers.

(a) Find the eigenvalues and (normalized) eigenvectors of $\mathbf{H}$, $\mathbf{A}$, and $\mathbf{B}$.

(b) Suppose the system starts out in the generic state

$$|\mathcal{S}(0)\rangle = \begin{pmatrix} c_1 \\ c_2 \\ c_3 \end{pmatrix},$$

with $|c_1|^2 + |c_2|^2 + |c_3|^2 = 1$. Find the expectation values (at $t = 0$) of H, A, and B.

(c) What is $|\mathcal{S}(t)\rangle$? If you measured the energy of this state (at time t), what values might you get, and what is the probability of each? Answer the same questions for A and for B.

$**$**Problem 3.39**

(a) For a function $f(x)$ that can be expanded in a Taylor series, show that

$$f(x + x_0) = e^{i\hat{p}x_0/\hbar} f(x)$$

(where x_0 is any constant distance). For this reason, $\hat{p}/\hbar$ is called the **generator of translations in space**. *Note:* The exponential of an *operator* is defined by the power series expansion: $e^{\hat{Q}} \equiv 1 + \hat{Q} + (1/2)\hat{Q}^2 + (1/3!)\hat{Q}^3 + \ldots$.

(b) If $\Psi(x, t)$ satisfies the (time-dependent) Schrödinger equation, show that

$$\Psi(x, t + t_0) = e^{-i\hat{H}t_0/\hbar} \Psi(x, t)$$

(where t_0 is any constant time); $-\hat{H}/\hbar$ is called the **generator of translations in time**.

(c) Show that the expectation value of a dynamical variable $Q(x, p, t)$, at time $t + t_0$, can be written[34]

$$\langle Q\rangle_{t+t_0} = \langle\Psi(x, t)|e^{i\hat{H}t_0/\hbar}\,\hat{Q}(\hat{x}, \hat{p}, t + t_0)e^{-i\hat{H}t_0/\hbar}|\Psi(x, t)\rangle.$$

Use this to recover Equation 3.71. *Hint:* Let $t_0 = dt$, and expand to first order in dt.

∗∗Problem 3.40

(a) Write down the time-dependent "Schrödinger equation" in momentum space, for a free particle, and solve it. *Answer:* $\exp(-ip^2t/2m\hbar)\,\Phi(p, 0)$.

(b) Find $\Phi(p, 0)$ for the traveling gaussian wave packet (Problem 2.43), and construct $\Phi(p, t)$ for this case. Also construct $|\Phi(p, t)|^2$, and note that it is independent of time.

(c) Calculate $\langle p\rangle$ and $\langle p^2\rangle$ by evaluating the appropriate integrals involving Φ, and compare your answers to Problem 2.43.

(d) Show that $\langle H\rangle = \langle p\rangle^2/2m + \langle H\rangle_0$ (where the subscript 0 denotes the *stationary* gaussian), and comment on this result.

---

[34] In particular, if we set $t = 0$, and drop the subscript on t_0,

$$\langle Q(t)\rangle = \langle\Psi(x, t)|\hat{Q}|\Psi(x, t)\rangle = \langle\Psi(x, 0)|\hat{U}^{-1}\hat{Q}\hat{U}|\Psi(x, 0)\rangle,$$

where $\hat{U} \equiv \exp(-i\hat{H}t/\hbar)$. This says that you can calculate expectation values of Q *either* by sandwiching $\hat{Q}$ between $\Psi(x, t)^*$ and $\Psi(x, t)$, as we have always done (letting the wave functions carry the time dependence), or *else* by sandwiching $\hat{U}^{-1}\hat{Q}\hat{U}$ between $\Psi(x, 0)^*$ and $\Psi(x, 0)$, letting the *operator* carry the time dependence. The former is called the **Schrödinger picture**, and the latter the **Heisenberg picture**.

CHAPTER 4

QUANTUM MECHANICS IN THREE DIMENSIONS

4.1 SCHRÖDINGER EQUATION IN SPHERICAL COORDINATES

The generalization to three dimensions is straightforward. Schrödinger's equation says

$$i\hbar\frac{\partial\Psi}{\partial t} = H\Psi; \tag{4.1}$$

the Hamiltonian operator[1] H is obtained from the classical energy

$$\frac{1}{2}mv^2 + V = \frac{1}{2m}(p_x^2 + p_y^2 + p_z^2) + V$$

by the standard prescription (applied now to y and z, as well as x):

$$p_x \rightarrow \frac{\hbar}{i}\frac{\partial}{\partial x}, \quad p_y \rightarrow \frac{\hbar}{i}\frac{\partial}{\partial y}, \quad p_z \rightarrow \frac{\hbar}{i}\frac{\partial}{\partial z}, \tag{4.2}$$

[1]Where confusion might otherwise occur I have been putting "hats" on operators, to distinguish them from the corresponding classical observables. I don't think there will be much occasion for ambiguity in this chapter, and the hats get to be cumbersome, so I am going to leave them off from now on.

or

$$\boxed{\mathbf{p} \to \frac{\hbar}{i}\nabla,} \qquad [4.3]$$

for short. Thus

$$\boxed{i\hbar\frac{\partial \Psi}{\partial t} = -\frac{\hbar^2}{2m}\nabla^2\Psi + V\Psi,} \qquad [4.4]$$

where

$$\nabla^2 \equiv \frac{\partial^2}{\partial x^2} + \frac{\partial^2}{\partial y^2} + \frac{\partial^2}{\partial z^2} \qquad [4.5]$$

is the **Laplacian**, in cartesian coordinates.

The potential energy V and the wave function Ψ are now functions of $\mathbf{r} = (x, y, z)$ and t. The probability of finding the particle in the infinitesimal volume $d^3\mathbf{r} = dx\,dy\,dz$ is $|\Psi(\mathbf{r}, t)|^2\,d^3\mathbf{r}$, and the normalization condition reads

$$\int |\Psi|^2\,d^3\mathbf{r} = 1, \qquad [4.6]$$

with the integral taken over all space. If the potential is independent of time, there will be a complete set of stationary states,

$$\Psi_n(\mathbf{r}, t) = \psi_n(\mathbf{r})e^{-iE_nt/\hbar}, \qquad [4.7]$$

where the spatial wave function ψ_n satisfies the time-*independent* Schrödinger equation:

$$\boxed{-\frac{\hbar^2}{2m}\nabla^2\psi + V\psi = E\psi.} \qquad [4.8]$$

The general solution to the (time-*dependent*) Schrödinger equation is

$$\Psi(\mathbf{r}, t) = \sum c_n\psi_n(\mathbf{r})e^{-iE_nt/\hbar}, \qquad [4.9]$$

with the constants c_n determined by the initial wave function, $\Psi(\mathbf{r}, 0)$, in the usual way. (If the potential admits continuum states, then the sum in Equation 4.9 becomes an integral.)

---

*Problem 4.1

(a) Work out all of the **canonical commutation relations** for components of the operators **r** and **p**: $[x, y]$, $[x, p_y]$, $[x, p_x]$, $[p_y, p_z]$, and so on. *Answer:*

$$[r_i, p_j] = -[p_i, r_j] = i\hbar\delta_{ij}, \quad [r_i, r_j] = [p_i, p_j] = 0, \tag{4.10}$$

where the indices stand for x, y, or z, and $r_x = x$, $r_y = y$, and $r_z = z$.

(b) Confirm Ehrenfest's theorem for 3-dimensions:

$$\frac{d}{dt}\langle \mathbf{r} \rangle = \frac{1}{m}\langle \mathbf{p} \rangle, \quad \text{and} \quad \frac{d}{dt}\langle \mathbf{p} \rangle = \langle -\nabla V \rangle. \tag{4.11}$$

(Each of these, of course, stands for *three* equations—one for each component.) *Hint:* First check that Equation 3.71 is valid in three dimensions.

(c) Formulate Heisenberg's uncertainty principle in three dimensions. *Answer:*

$$\sigma_x \sigma_{p_x} \geq \hbar/2, \quad \sigma_y \sigma_{p_y} \geq \hbar/2, \quad \sigma_z \sigma_{p_z} \geq \hbar/2, \tag{4.12}$$

but there is no restriction on, say, $\sigma_x \sigma_{p_y}$.

4.1.1 Separation of Variables

Typically, the potential is a function only of the distance from the origin. In that case it is natural to adopt **spherical coordinates**, (r, θ, ϕ) (see Figure 4.1). In spherical coordinates the Laplacian takes the form[2]

$$\nabla^2 = \frac{1}{r^2}\frac{\partial}{\partial r}\left(r^2 \frac{\partial}{\partial r}\right) + \frac{1}{r^2 \sin\theta}\frac{\partial}{\partial \theta}\left(\sin\theta \frac{\partial}{\partial \theta}\right) + \frac{1}{r^2 \sin^2\theta}\left(\frac{\partial^2}{\partial \phi^2}\right). \tag{4.13}$$

In spherical coordinates, then, the time-independent Schrödinger equation reads

$$-\frac{\hbar^2}{2m}\left[\frac{1}{r^2}\frac{\partial}{\partial r}\left(r^2 \frac{\partial \psi}{\partial r}\right) + \frac{1}{r^2 \sin\theta}\frac{\partial}{\partial \theta}\left(\sin\theta \frac{\partial \psi}{\partial \theta}\right) + \frac{1}{r^2 \sin^2\theta}\left(\frac{\partial^2 \psi}{\partial \phi^2}\right)\right] + V\psi = E\psi. \tag{4.14}$$

We begin by looking for solutions that are separable into products:

$$\psi(r, \theta, \phi) = R(r)Y(\theta, \phi). \tag{4.15}$$

---

[2]In principle, this can be obtained by change of variables from the cartesian expression (Equation 4.5). However, there are much more efficient ways of getting it; see, for instance, M. Boas, *Mathematical Methods in the Physical Sciences*, 2nd ed., (Wiley, New York, 1983), Chapter 10, Section 9.

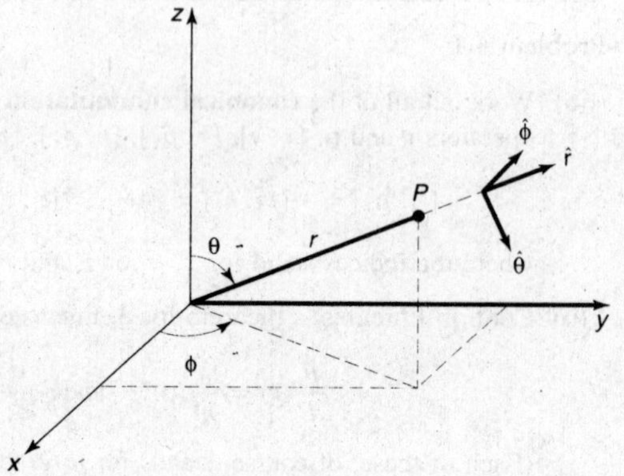

FIGURE 4.1: Spherical coordinates: radius r, polar angle θ, and azimuthal angle ϕ.

Putting this into Equation 4.14, we have

$$-\frac{\hbar^2}{2m}\left[\frac{Y}{r^2}\frac{d}{dr}\left(r^2\frac{dR}{dr}\right)+\frac{R}{r^2\sin\theta}\frac{\partial}{\partial\theta}\left(\sin\theta\frac{\partial Y}{\partial\theta}\right)+\frac{R}{r^2\sin^2\theta}\frac{\partial^2 Y}{\partial\phi^2}\right]$$
$$+VRY = ERY.$$

Dividing by RY and multiplying by $-2mr^2/\hbar^2$:

$$\left\{\frac{1}{R}\frac{d}{dr}\left(r^2\frac{dR}{dr}\right)-\frac{2mr^2}{\hbar^2}[V(r)-E]\right\}$$
$$+\frac{1}{Y}\left\{\frac{1}{\sin\theta}\frac{\partial}{\partial\theta}\left(\sin\theta\frac{\partial Y}{\partial\theta}\right)+\frac{1}{\sin^2\theta}\frac{\partial^2 Y}{\partial\phi^2}\right\}=0.$$

The term in the first curly bracket depends only on r, whereas the remainder depends only on θ and ϕ; accordingly, each must be a constant. For reasons that will appear in due course,[3] I will write this "separation constant" in the form $l(l+1)$:

$$\frac{1}{R}\frac{d}{dr}\left(r^2\frac{dR}{dr}\right)-\frac{2mr^2}{\hbar^2}[V(r)-E]=l(l+1); \qquad [4.16]$$

$$\frac{1}{Y}\left\{\frac{1}{\sin\theta}\frac{\partial}{\partial\theta}\left(\sin\theta\frac{\partial Y}{\partial\theta}\right)+\frac{1}{\sin^2\theta}\frac{\partial^2 Y}{\partial\phi^2}\right\}=-l(l+1). \qquad [4.17]$$

[3]Note that there is no loss of generality here—at this stage l could be any complex number. Later on we'll discover that l must in fact be an *integer*, and it is in anticipation of that result that I express the separation constant in a way that looks peculiar now.

***Problem 4.2** Use separation of variables in *cartesian* coordinates to solve the infinite *cubical* well (or "particle in a box"):

$$V(x, y, z) = \begin{cases} 0, & \text{if } x, y, z \text{ are all between 0 and } a; \\ \infty, & \text{otherwise.} \end{cases}$$

(a) Find the stationary states, and the corresponding energies.

(b) Call the distinct energies $E_1, E_2, E_3, \ldots$, in order of increasing energy. Find E_1, E_2, E_3, E_4, E_5, and E_6. Determine their degeneracies (that is, the number of different states that share the same energy). *Comment:* In *one* dimension degenerate bound states do not occur (see Problem 2.45), but in three dimensions they are very common.

(c) What is the degeneracy of E_{14}, and why is this case interesting?

4.1.2 The Angular Equation

Equation 4.17 determines the dependence of ψ on θ and ϕ; multiplying by $Y \sin^2\theta$, it becomes:

$$\sin\theta \frac{\partial}{\partial\theta}\left(\sin\theta \frac{\partial Y}{\partial\theta}\right) + \frac{\partial^2 Y}{\partial\phi^2} = -l(l+1)\sin^2\theta\, Y. \qquad [4.18]$$

You might recognize this equation—it occurs in the solution to Laplace's equation in classical electrodynamics. As always, we try separation of variables:

$$Y(\theta, \phi) = \Theta(\theta)\Phi(\phi). \qquad [4.19]$$

Plugging this in, and dividing by $\Theta\Phi$, we find:

$$\left\{\frac{1}{\Theta}\left[\sin\theta \frac{d}{d\theta}\left(\sin\theta \frac{d\Theta}{d\theta}\right)\right] + l(l+1)\sin^2\theta\right\} + \frac{1}{\Phi}\frac{d^2\Phi}{d\phi^2} = 0.$$

The first term is a function only of θ, and the second is a function only of ϕ, so each must be a constant. This time[4] I'll call the separation constant m^2:

$$\frac{1}{\Theta}\left[\sin\theta \frac{d}{d\theta}\left(\sin\theta \frac{d\Theta}{d\theta}\right)\right] + l(l+1)\sin^2\theta = m^2; \qquad [4.20]$$

$$\frac{1}{\Phi}\frac{d^2\Phi}{d\phi^2} = -m^2. \qquad [4.21]$$

[4] Again, there is no loss of generality here, since at this stage m could be any complex number; in a moment, though, we will discover that m must in fact be an *integer*. *Beware:* The letter m is now doing double duty, as *mass* and as a separation constant. There is no graceful way to avoid this confusion, since both uses are standard. Some authors now switch to M or μ for mass, but I hate to change notation in mid-stream, and I don't think confusion will arise, as long as you are aware of the problem.

The ϕ equation is easy:

$$\frac{d^2\Phi}{d\phi^2} = -m^2\Phi \Rightarrow \Phi(\phi) = e^{im\phi}. \quad [4.22]$$

[Actually, there are *two* solutions: $\exp(im\phi)$ and $\exp(-im\phi)$, but we'll cover the latter by allowing m to run negative. There could also be a constant factor in front, but we might as well absorb that into Θ. Incidentally, in electrodynamics we would write the azimuthal function (Φ) in terms of sines and cosines, instead of exponentials, because electric potentials must be *real*. In quantum mechanics there is no such constraint, and the exponentials are a lot easier to work with.] Now, when ϕ advances by 2π, we return to the same point in space (see Figure 4.1), so it is natural to require that[5]

$$\Phi(\phi + 2\pi) = \Phi(\phi). \quad [4.23]$$

In other words, $\exp[im(\phi + 2\pi)] = \exp(im\phi)$, or $\exp(2\pi im) = 1$. From this it follows that m must be an *integer*:

$$m = 0, \pm 1, \pm 2, \dots . \quad [4.24]$$

The θ equation,

$$\sin\theta \frac{d}{d\theta}\left(\sin\theta \frac{d\Theta}{d\theta}\right) + [l(l+1)\sin^2\theta - m^2]\Theta = 0, \quad [4.25]$$

is not so simple. The solution is

$$\Theta(\theta) = AP_l^m(\cos\theta), \quad [4.26]$$

where P_l^m is the **associated Legendre function**, defined by[6]

$$P_l^m(x) \equiv (1 - x^2)^{|m|/2}\left(\frac{d}{dx}\right)^{|m|} P_l(x), \quad [4.27]$$

and $P_l(x)$ is the lth **Legendre polynomial**, defined by the **Rodrigues formula**:

$$P_l(x) \equiv \frac{1}{2^l l!}\left(\frac{d}{dx}\right)^l (x^2 - 1)^l. \quad [4.28]$$

---

[5] This is more slippery than it looks. After all, the *probability* density ($|\Phi|^2$) is single-valued *regardless* of m. In Section 4.3 we'll obtain the condition on m by an entirely different—and more compelling—argument.

[6] Notice that $P_l^{-m} = P_l^m$. Some authors adopt a different sign convention for negative values of m; see Boas (footnote 2), p. 505.

TABLE 4.1: The first few Legendre polynomials, $P_l(x)$: (a) functional form, (b) graphs.

$P_0 = 1$
$P_1 = x$
$P_2 = \frac{1}{2}(3x^2 - 1)$
$P_3 = \frac{1}{2}(5x^3 - 3x)$
$P_4 = \frac{1}{8}(35x^4 - 30x^2 + 3)$
$P_5 = \frac{1}{8}(63x^5 - 70x^3 + 15x)$

(a)

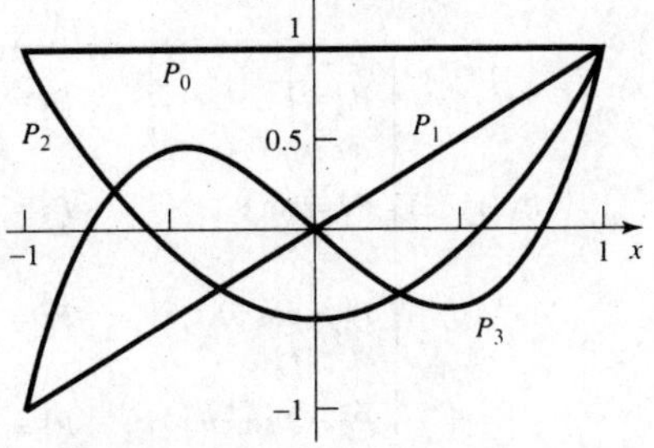

(b)

For example,

$$P_0(x) = 1, \quad P_1(x) = \frac{1}{2}\frac{d}{dx}(x^2 - 1) = x,$$

$$P_2(x) = \frac{1}{4 \cdot 2}\left(\frac{d}{dx}\right)^2 (x^2 - 1)^2 = \frac{1}{2}(3x^2 - 1),$$

and so on. The first few Legendre polynomials are listed in Table 4.1. As the name suggests, $P_l(x)$ is a polynomial (of degree l) in x, and is even or odd according to the parity of l. But $P_l^m(x)$ is not, in general, a polynomial—if m is odd it carries a factor of $\sqrt{1 - x^2}$:

$$P_2^0(x) = \frac{1}{2}(3x^2 - 1), \quad P_2^1(x) = (1 - x^2)^{1/2}\frac{d}{dx}\left[\frac{1}{2}(3x^2 - 1)\right] = 3x\sqrt{1 - x^2},$$

$$P_2^2(x) = (1 - x^2)\left(\frac{d}{dx}\right)^2\left[\frac{1}{2}(3x^2 - 1)\right] = 3(1 - x^2),$$

etc. (On the other hand, what *we* need is $P_l^m(\cos\theta)$, and $\sqrt{1 - \cos^2\theta} = \sin\theta$, so $P_l^m(\cos\theta)$ is always a polynomial in $\cos\theta$, multiplied—if m is odd—by $\sin\theta$. Some associated Legendre functions of $\cos\theta$ are listed in Table 4.2.)

Notice that l must be a nonnegative *integer*, for the Rodrigues formula to make any sense; moreover, if $|m| > l$, then Equation 4.27 says $P_l^m = 0$. For any given l, then, there are $(2l + 1)$ possible values of m:

$$l = 0,\ 1,\ 2, \ldots; \quad m = -l,\ -l + 1,\ \ldots,\ -1,\ 0,\ 1,\ \ldots,\ l - 1,\ l. \qquad [4.29]$$

But wait! Equation 4.25 is a second-order differential equation: It should have *two* linearly independent solutions, for *any old* values of l and m. Where are all the

TABLE 4.2: Some associated Legendre functions, $P_l^m(\cos\theta)$: (a) functional form, (b) graphs of $r = P_l^m(\cos\theta)$ (in these plots r tells you the magnitude of the function in the direction θ; each figure should be rotated about the z-axis).

$P_0^0 = 1$	$P_2^0 = \frac{1}{2}(3\cos^2\theta - 1)$
$P_1^1 = \sin\theta$	$P_3^3 = 15\sin\theta(1 - \cos^2\theta)$
$P_1^0 = \cos\theta$	$P_3^2 = 15\sin^2\theta\cos\theta$
$P_2^2 = 3\sin^2\theta$	$P_3^1 = \frac{3}{2}\sin\theta(5\cos^2\theta - 1)$
$P_2^1 = 3\sin\theta\cos\theta$	$P_3^0 = \frac{1}{2}(5\cos^3\theta - 3\cos\theta)$

(a)

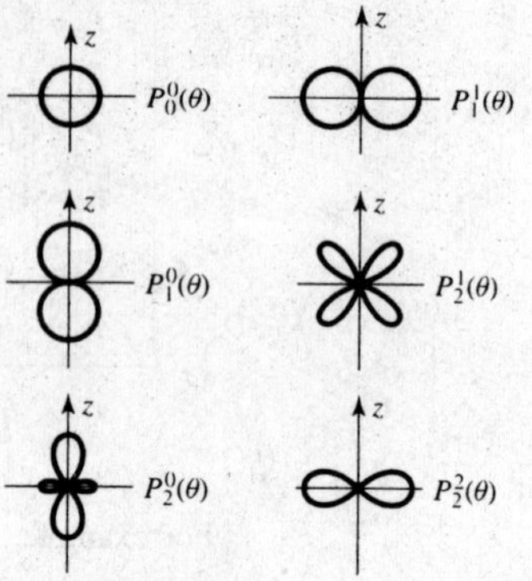

(b)

other solutions? *Answer:* They *exist*, of course, as mathematical solutions to the equation, but they are *physically* unacceptable, because they blow up at $\theta = 0$ and/or $\theta = \pi$ (see Problem 4.4).

Now, the volume element in spherical coordinates[7] is

$$d^3\mathbf{r} = r^2 \sin\theta \, dr \, d\theta \, d\phi, \qquad [4.30]$$

so the normalization condition (Equation 4.6) becomes

$$\int |\psi|^2 r^2 \sin\theta \, dr \, d\theta \, d\phi = \int |R|^2 r^2 \, dr \int |Y|^2 \sin\theta \, d\theta \, d\phi = 1.$$

It is convenient to normalize R and Y separately:

$$\int_0^\infty |R|^2 r^2 \, dr = 1 \quad \text{and} \quad \int_0^{2\pi} \int_0^{\pi} |Y|^2 \sin\theta \, d\theta \, d\phi = 1. \qquad [4.31]$$

---

[7]See, for instance, Boas (footnote 2), Chapter 5, Section 4.

TABLE 4.3: The first few spherical harmonics, $Y_l^m(\theta, \phi)$.

$Y_0^0 = \left(\frac{1}{4\pi}\right)^{1/2}$	$Y_2^{\pm 2} = \left(\frac{15}{32\pi}\right)^{1/2} \sin^2\theta e^{\pm 2i\phi}$
$Y_1^0 = \left(\frac{3}{4\pi}\right)^{1/2} \cos\theta$	$Y_3^0 = \left(\frac{7}{16\pi}\right)^{1/2} (5\cos^3\theta - 3\cos\theta)$
$Y_1^{\pm 1} = \mp\left(\frac{3}{8\pi}\right)^{1/2} \sin\theta e^{\pm i\phi}$	$Y_3^{\pm 1} = \mp\left(\frac{21}{64\pi}\right)^{1/2} \sin\theta\,(5\cos^2\theta - 1)e^{\pm i\phi}$
$Y_2^0 = \left(\frac{5}{16\pi}\right)^{1/2} (3\cos^2\theta - 1)$	$Y_3^{\pm 2} = \left(\frac{105}{32\pi}\right)^{1/2} \sin^2\theta\cos\theta e^{\pm 2i\phi}$
$Y_2^{\pm 1} = \mp\left(\frac{15}{8\pi}\right)^{1/2} \sin\theta\cos\theta e^{\pm i\phi}$	$Y_3^{\pm 3} = \mp\left(\frac{35}{64\pi}\right)^{1/2} \sin^3\theta e^{\pm 3i\phi}$

The normalized angular wave functions[8] are called **spherical harmonics**:

$$Y_l^m(\theta, \phi) = \epsilon\sqrt{\frac{(2l+1)}{4\pi}\frac{(l-|m|)!}{(l+|m|)!}}\, e^{im\phi} P_l^m(\cos\theta), \qquad [4.32]$$

where $\epsilon = (-1)^m$ for $m \geq 0$ and $\epsilon = 1$ for $m \leq 0$. As we shall prove later on, they are automatically orthogonal, so

$$\int_0^{2\pi}\int_0^{\pi} [Y_l^m(\theta, \phi)]^*[Y_{l'}^{m'}(\theta, \phi)] \sin\theta\, d\theta\, d\phi = \delta_{ll'}\delta_{mm'}, \qquad [4.33]$$

In Table 4.3 I have listed the first few spherical harmonics. For historical reasons, l is called the **azimuthal quantum number**, and m the **magnetic quantum number**.

***Problem 4.3** Use Equations 4.27, 4.28, and 4.32, to construct Y_0^0 and Y_2^1. Check that they are normalized and orthogonal.

Problem 4.4 Show that

$$\Theta(\theta) = A\ln[\tan(\theta/2)]$$

---

[8]The normalization factor is derived in Problem 4.54; ϵ (which is always 1 or -1) is chosen for consistency with the notation we will be using in the theory of angular momentum; it is reasonably standard, though some older books use other conventions. Notice that

$$Y_l^{-m} = (-1)^m\,(Y_l^m)^*$$

satisfies the θ equation (Equation 4.25), for $l = m = 0$. This is the unacceptable "second solution"—what's *wrong* with it?

***Problem 4.5** Use Equation 4.32 to construct $Y_l^l(\theta, \phi)$ and $Y_3^2(\theta, \phi)$. (You can take P_3^2 from Table 4.2, but you'll have to work out P_l^l from Equations 4.27 and 4.28.) Check that they satisfy the angular equation (Equation 4.18), for the appropriate values of l and m.

****Problem 4.6** Starting from the Rodrigues formula, derive the orthonormality condition for Legendre polynomials:

$$\int_{-1}^{1} P_l(x) P_{l'}(x)\, dx = \left(\frac{2}{2l+1}\right) \delta_{ll'}. \qquad [4.34]$$

Hint: Use integration by parts.

4.1.3 The Radial Equation

Notice that the angular part of the wave function, $Y(\theta, \phi)$, is the same for all spherically symmetric potentials; the actual *shape* of the potential, $V(r)$, affects only the *radial* part of the wave function, $R(r)$, which is determined by Equation 4.16:

$$\frac{d}{dr}\left(r^2 \frac{dR}{dr}\right) - \frac{2mr^2}{\hbar^2}[V(r) - E]R = l(l+1)R. \qquad [4.35]$$

This equation simplifies if we change variables: Let

$$u(r) \equiv rR(r), \qquad [4.36]$$

so that $R = u/r$, $dR/dr = [r(du/dr) - u]/r^2$, $(d/dr)[r^2(dR/dr)] = rd^2u/dr^2$, and hence

$$\boxed{-\frac{\hbar^2}{2m}\frac{d^2u}{dr^2} + \left[V + \frac{\hbar^2}{2m}\frac{l(l+1)}{r^2}\right]u = Eu.} \qquad [4.37]$$

This is called the **radial equation**;[9] it is *identical in form* to the one-dimensional Schrödinger equation (Equation 2.5), except that the **effective potential**,

$$V_{\text{eff}} = V + \frac{\hbar^2}{2m}\frac{l(l+1)}{r^2}, \qquad [4.38]$$

---

[9]Those m's are *masses*, of course—the separation constant m does not appear in the radial equation.

contains an extra piece, the so-called **centrifugal term,** $(\hbar^2/2m)[l(l+1)/r^2]$. It tends to throw the particle outward (away from the origin), just like the centrifugal (pseudo-)force in classical mechanics. Meanwhile, the normalization condition (Equation 4.31) becomes

$$\int_0^\infty |u|^2\, dr = 1. \tag{4.39}$$

That's as far as we can go until a specific potential $V(r)$ is provided.

Example 4.1 Consider the **infinite spherical well,**

$$V(r) = \begin{cases} 0, & \text{if } r < a; \\ \infty, & \text{if } r > a. \end{cases} \tag{4.40}$$

Find the wave functions and the allowed energies.

Solution: Outside the well, the wave function is zero; inside the well, the radial equation says

$$\frac{d^2u}{dr^2} = \left[\frac{l(l+1)}{r^2} - k^2\right] u, \tag{4.41}$$

where

$$k \equiv \frac{\sqrt{2mE}}{\hbar}, \tag{4.42}$$

as usual. Our problem is to solve this equation, subject to the boundary condition $u(a) = 0$. The case $l = 0$ is easy:

$$\frac{d^2u}{dr^2} = -k^2 u \;\Rightarrow\; u(r) = A\sin(kr) + B\cos(kr).$$

But remember, the actual radial wave function is $R(r) = u(r)/r$, and $[\cos(kr)]/r$ blows up as $r \to 0$. So[10] we must choose $B = 0$. The boundary condition then requires $\sin(ka) = 0$, and hence $ka = n\pi$, for some integer n. The allowed energies are evidently

$$E_{n0} = \frac{n^2\pi^2\hbar^2}{2ma^2}, \quad (n = 1, 2, 3, \ldots), \tag{4.43}$$

[10] Actually, all we require is that the wave function be *normalizable*, not that it be *finite*: $R(r) \sim 1/r$ at the origin *is* normalizable (because of the r^2 in Equation 4.31). For a more compelling proof that $B = 0$, see R. Shankar, *Principles of Quantum Mechanics* (Plenum, New York, 1980), p. 351.

the same as for the one-dimensional infinite square well (Equation 2.27). Normalizing $u(r)$ yields $A = \sqrt{2/a}$; tacking on the angular part (trivial, in this instance, since $Y_0^0(\theta, \phi) = 1/\sqrt{4\pi}$), we conclude that

$$\psi_{n00} = \frac{1}{\sqrt{2\pi a}} \frac{\sin(n\pi r/a)}{r}. \qquad [4.44]$$

[Notice that the stationary states are labeled by *three* **quantum numbers**, n, l, and m: $\psi_{nlm}(r, \theta, \phi)$. The *energy*, however, depends only on n and l: E_{nl}.]

The general solution to Equation 4.41 (for an *arbitrary* integer l) is not so familiar:

$$u(r) = Arj_l(kr) + Brn_l(kr), \qquad [4.45]$$

where $j_l(x)$ is the **spherical Bessel function** of order l, and $n_l(x)$ is the **spherical Neumann function** of order l. They are defined as follows:

$$j_l(x) \equiv (-x)^l \left(\frac{1}{x}\frac{d}{dx}\right)^l \frac{\sin x}{x}; \quad n_l(x) \equiv -(-x)^l \left(\frac{1}{x}\frac{d}{dx}\right)^l \frac{\cos x}{x}. \qquad [4.46]$$

For example,

$$j_0(x) = \frac{\sin x}{x}; \quad n_0(x) = -\frac{\cos x}{x};$$

$$j_1(x) = (-x)\frac{1}{x}\frac{d}{dx}\left(\frac{\sin x}{x}\right) = \frac{\sin x}{x^2} - \frac{\cos x}{x};$$

$$j_2(x) = (-x)^2\left(\frac{1}{x}\frac{d}{dx}\right)^2 \frac{\sin x}{x} = x^2\left(\frac{1}{x}\frac{d}{dx}\right)\frac{x\cos x - \sin x}{x^3}$$

$$= \frac{3\sin x - 3x\cos x - x^2 \sin x}{x^3};$$

and so on. The first few spherical Bessel and Neumann functions are listed in Table 4.4. For small x (where $\sin x \approx x - x^3/3! + x^5/5! - \cdots$ and $\cos x \approx 1 - x^2/2 + x^4/4! - \cdots$),

$$j_0(x) \approx 1; \quad n_0(x) \approx -\frac{1}{x}; \quad j_1(x) \approx \frac{x}{3}; \quad j_2(x) \approx \frac{x^2}{15};$$

etc. Notice that Bessel functions are *finite* at the origin, but *Neumann* functions *blow up* at the origin. Accordingly, we must have $B_l = 0$, and hence

$$R(r) = Aj_l(kr). \qquad [4.47]$$

TABLE 4.4: The first few spherical Bessel and Neumann functions, $j_n(x)$ and $n_l(x)$; asymptotic forms for small x.

$j_0 = \frac{\sin x}{x}$	$n_0 = -\frac{\cos x}{x}$
$j_1 = \frac{\sin x}{x^2} - \frac{\cos x}{x}$	$n_1 = -\frac{\cos x}{x^2} - \frac{\sin x}{x}$
$j_2 = \left(\frac{3}{x^3} - \frac{1}{x}\right)\sin x - \frac{3}{x^2}\cos x$	$n_2 = -\left(\frac{3}{x^3} - \frac{1}{x}\right)\cos x - \frac{3}{x^2}\sin x$
$j_l \rightarrow \frac{2^l l!}{(2l+1)!}x^l,$	$n_l \rightarrow -\frac{(2l)!}{2^l l!}\frac{1}{x^{l+1}},$ for $x << 1$.

There remains the boundary condition, $R(a) = 0$. Evidently k must be chosen such that

$$j_l(ka) = 0; \tag{4.48}$$

that is, (ka) is a zero of the lth-order spherical Bessel function. Now, the Bessel functions are oscillatory (see Figure 4.2); each one has an infinite number of zeros.

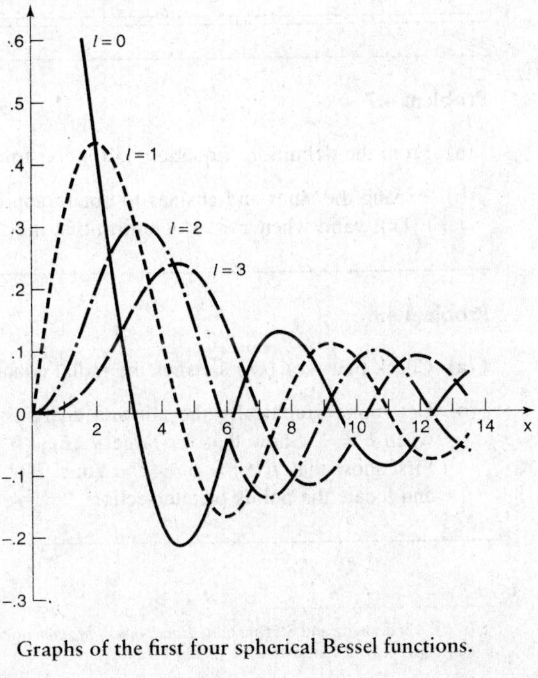

FIGURE 4.2: Graphs of the first four spherical Bessel functions.

But (unfortunately for us) they are not located at nice sensible points (such as n, or $n\pi$, or something); they have to be computed numerically.[11] At any rate, the boundary condition requires that

$$k = \frac{1}{a}\beta_{nl}, \quad [4.49]$$

where β_{nl} is the nth zero of the lth spherical Bessel function. The allowed energies, then, are given by

$$E_{nl} = \frac{\hbar^2}{2ma^2}\beta_{nl}^2, \quad [4.50]$$

and the wave functions are

$$\psi_{nlm}(r, \theta, \phi) = A_{nl}\, j_l(\beta_{nl}r/a)Y_l^m(\theta, \phi), \quad [4.51]$$

with the constant A_{nl} to be determined by normalization. Each energy level is $(2l + 1)$-fold degenerate, since there are $(2l + 1)$ different values of m for each value of l (see Equation 4.29).

Problem 4.7

(a) From the definition (Equation 4.46), construct $n_1(x)$ and $n_2(x)$.

(b) Expand the sines and cosines to obtain approximate formulas for $n_1(x)$ and $n_2(x)$, valid when $x \ll 1$. Confirm that they blow up at the origin.

Problem 4.8

(a) Check that $Arj_1(kr)$ satisfies the radial equation with $V(r) = 0$ and $l = 1$.

(b) Determine graphically the allowed energies for the infinite spherical well, when $l = 1$. Show that for large n, $E_{n1} \approx (\hbar^2\pi^2/2ma^2)(n + 1/2)^2$. *Hint:* First show that $j_1(x) = 0 \Rightarrow x = \tan x$. Plot x and $\tan x$ on the same graph, and locate the points of intersection.

---

[11] Abramowitz and Stegun, eds., *Handbook of Mathematical Functions*, (Dover, New York, 1965), Chapter 10, provides an extensive listing.

∗∗**Problem 4.9** A particle of mass m is placed in a *finite* spherical well:

$$V(r) = \begin{cases} -V_0, & \text{if } r \leq a; \\ 0, & \text{if } r > a. \end{cases}$$

Find the ground state, by solving the radial equation with $l = 0$. Show that there is no bound state if $V_0 a^2 < \pi^2 \hbar^2 / 8m$.

4.2 THE HYDROGEN ATOM

The hydrogen atom consists of a heavy, essentially motionless proton (we may as well put it at the origin), of charge e, together with a much lighter electron (charge $-e$) that orbits around it, bound by the mutual attraction of opposite charges (see Figure 4.3). From Coulomb's law, the potential energy (in SI units) is

$$V(r) = -\frac{e^2}{4\pi\epsilon_0} \frac{1}{r}, \tag{4.52}$$

and the radial equation (Equation 4.37) says

$$-\frac{\hbar^2}{2m} \frac{d^2 u}{dr^2} + \left[-\frac{e^2}{4\pi\epsilon_0} \frac{1}{r} + \frac{\hbar^2}{2m} \frac{l(l+1)}{r^2} \right] u = Eu. \tag{4.53}$$

Our problem is to solve this equation for $u(r)$, and determine the allowed energies, E. The hydrogen atom is such an important case that I'm not going to hand you the solutions this time—we'll work them out in detail, by the method we used in the analytical solution to the harmonic oscillator. (If any step in this process is unclear, you may wish to refer back to Section 2.3.2 for a more complete explanation.)

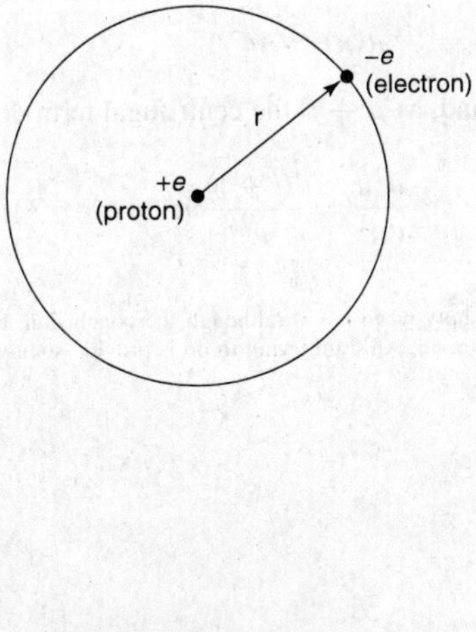

FIGURE 4.3: The hydrogen atom.

Incidentally, the Coulomb potential (Equation 4.52) admits *continuum* states (with $E > 0$), describing electron-proton scattering, as well as discrete *bound* states, representing the hydrogen atom, but we shall confine our attention to the latter.

4.2.1 The Radial Wave Function

Our first task is to tidy up the notation. Let

$$\kappa \equiv \frac{\sqrt{-2mE}}{\hbar}. \qquad [4.54]$$

(For bound states, E is negative, so κ is *real*.) Dividing Equation 4.53 by E, we have

$$\frac{1}{\kappa^2}\frac{d^2u}{dr^2} = \left[1 - \frac{me^2}{2\pi\epsilon_0\hbar^2\kappa}\frac{1}{(\kappa r)} + \frac{l(l+1)}{(\kappa r)^2}\right]u.$$

This suggests that we introduce

$$\rho \equiv \kappa r, \quad \text{and} \quad \rho_0 \equiv \frac{me^2}{2\pi\epsilon_0\hbar^2\kappa}, \qquad [4.55]$$

so that

$$\frac{d^2u}{d\rho^2} = \left[1 - \frac{\rho_0}{\rho} + \frac{l(l+1)}{\rho^2}\right]u. \qquad [4.56]$$

Next we examine the asymptotic form of the solutions. As $\rho \to \infty$, the constant term in the brackets dominates, so (approximately)

$$\frac{d^2u}{d\rho^2} = u.$$

The general solution is

$$u(\rho) = Ae^{-\rho} + Be^{\rho}, \qquad [4.57]$$

but e^{ρ} blows up (as $\rho \to \infty$), so $B = 0$. Evidently,

$$u(\rho) \sim Ae^{-\rho}, \qquad [4.58]$$

for large ρ. On the other hand, as $\rho \to 0$ the centrifugal term dominates;[12] approximately, then:

$$\frac{d^2u}{d\rho^2} = \frac{l(l+1)}{\rho^2}u.$$

[12] This argument does not apply when $l = 0$ (although the conclusion, Equation 4.59, is in fact valid for that case too). But never mind: All I am trying to do is provide some *motivation* for a change of variables (Equation 4.60).

The general solution (check it!) is

$$u(\rho) = C\rho^{l+1} + D\rho^{-l},$$

but ρ^{-l} blows up (as $\rho \to 0$), so $D = 0$. Thus

$$u(\rho) \sim C\rho^{l+1}, \tag{4.59}$$

for small ρ.

The next step is to peel off the asymptotic behavior, introducing the new function $v(\rho)$:

$$u(\rho) = \rho^{l+1}e^{-\rho}v(\rho), \tag{4.60}$$

in the hope that $v(\rho)$ will turn out to be simpler than $u(\rho)$. The first indications are not auspicious:

$$\frac{du}{d\rho} = \rho^l e^{-\rho}\left[(l+1-\rho)v + \rho\frac{dv}{d\rho}\right],$$

and

$$\frac{d^2u}{d\rho^2} = \rho^l e^{-\rho}\left\{\left[-2l-2+\rho+\frac{l(l+1)}{\rho}\right]v + 2(l+1-\rho)\frac{dv}{d\rho} + \rho\frac{d^2v}{d\rho^2}\right\}.$$

In terms of $v(\rho)$, then, the radial equation (Equation 4.56) reads

$$\rho\frac{d^2v}{d\rho^2} + 2(l+1-\rho)\frac{dv}{d\rho} + [\rho_0 - 2(l+1)]v = 0. \tag{4.61}$$

Finally, we assume the solution, $v(\rho)$, can be expressed as a power series in ρ:

$$v(\rho) = \sum_{j=0}^{\infty} c_j\rho^j. \tag{4.62}$$

Our problem is to determine the coefficients $(c_0, c_1, c_2, \ldots)$. Differentiating term by term:

$$\frac{dv}{d\rho} = \sum_{j=0}^{\infty} jc_j\rho^{j-1} = \sum_{j=0}^{\infty}(j+1)c_{j+1}\rho^j.$$

[In the second summation I have renamed the "dummy index": $j \to j+1$. If this troubles you, write out the first few terms explicitly, and *check* it. You may object that the sum should now begin at $j = -1$, but the factor $(j + 1)$ kills that term anyway, so we might as well start at zero.] Differentiating again,

$$\frac{d^2 v}{d\rho^2} = \sum_{j=0}^{\infty} j(j+1)c_{j+1}\rho^{j-1}.$$

Inserting these into Equation 4.61, we have

$$\sum_{j=0}^{\infty} j(j+1)c_{j+1}\rho^{j} + 2(l+1)\sum_{j=0}^{\infty}(j+1)c_{j+1}\rho^{j}$$

$$-2\sum_{j=0}^{\infty} jc_j\rho^j + [\rho_0 - 2(l+1)]\sum_{j=0}^{\infty} c_j\rho^j = 0.$$

Equating the coefficients of like powers yields

$$j(j+1)c_{j+1} + 2(l+1)(j+1)c_{j+1} - 2jc_j + [\rho_0 - 2(l+1)]c_j = 0,$$

or:

$$c_{j+1} = \left\{\frac{2(j+l+1) - \rho_0}{(j+1)(j+2l+2)}\right\} c_j. \qquad [4.63]$$

This recursion formula determines the coefficients, and hence the function $v(\rho)$: We start with c_0 (this becomes an overall constant, to be fixed eventually by normalization), and Equation 4.63 gives us c_1; putting this back in, we obtain c_2, and so on.[13]

Now let's see what the coefficients look like for large j (this corresponds to large ρ, where the higher powers dominate). In this regime the recursion formula says[14]

$$c_{j+1} \cong \frac{2j}{j(j+1)}c_j = \frac{2}{j+1}c_j.$$

---

[13]You might wonder why I didn't use the series method directly on $u(\rho)$—why factor out the asymptotic behavior before applying this procedure? Well, the reason for peeling off ρ^{l+1} is largely aesthetic: Without this, the sequence would begin with a long string of zeros (the first nonzero coefficient being c_{l+1}); by factoring out ρ^{l+1} we obtain a series that starts out with ρ^0. The $e^{-\rho}$ factor is more critical—if you *don't* pull that out, you get a three-term recursion formula, involving c_{j+2}, c_{j+1}, and c_j (*try* it!) and that is enormously more difficult to work with.

[14]You might ask why I don't drop the 1 in $j + 1$—after all, I am ignoring $2(l + 1) - \rho_0$ in the numerator, and $2l + 2$ in the denominator. In this approximation it would be fine to drop the 1 as well, but keeping it makes the argument a little cleaner. Try doing it without the 1, and you'll see what I mean.

Suppose for a moment that this were *exact*. Then

$$c_j = \frac{2^j}{j!} c_0, \tag{4.64}$$

so

$$v(\rho) = c_0 \sum_{j=0}^{\infty} \frac{2^j}{j!} \rho^j = c_0 e^{2\rho},$$

and hence

$$u(\rho) = c_0 \rho^{l+1} e^{\rho}, \tag{4.65}$$

which blows up at large ρ. The positive exponential is precisely the asymptotic behavior we *didn't* want, in Equation 4.57. (It's no accident that it reappears here; after all, it *does* represent the asymptotic form of *some* solutions to the radial equation—they just don't happen to be the ones we're interested in, because they aren't normalizable.) There is only one way out of this dilemma: *The series must terminate*. There must occur some maximal integer, $j_{\max}$, such that

$$c_{(j_{\max}+1)} = 0, \tag{4.66}$$

(and beyond which all coefficients vanish automatically). Evidently (Equation 4.63)

$$2(j_{\max} + l + 1) - \rho_0 = 0.$$

Defining

$$n \equiv j_{\max} + l + 1 \tag{4.67}$$

(the so-called **principal quantum number**), we have

$$\rho_0 = 2n. \tag{4.68}$$

But ρ_0 determines E (Equations 4.54 and 4.55):

$$E = -\frac{\hbar^2 \kappa^2}{2m} = -\frac{me^4}{8\pi^2 \epsilon_0^2 \hbar^2 \rho_0^2}, \tag{4.69}$$

so the allowed energies are

$$\boxed{E_n = -\left[\frac{m}{2\hbar^2}\left(\frac{e^2}{4\pi\epsilon_0}\right)^2\right]\frac{1}{n^2} = \frac{E_1}{n^2}, \quad n = 1, 2, 3, \ldots} \tag{4.70}$$

This is the famous **Bohr formula**—by any measure the most important result in all of quantum mechanics. Bohr obtained it in 1913 by a serendipitous mixture of inapplicable classical physics and premature quantum theory (the Schrödinger equation did not come until 1924).

Combining Equations 4.55 and 4.68, we find that

$$\kappa = \left(\frac{me^2}{4\pi\epsilon_0\hbar^2}\right)\frac{1}{n} = \frac{1}{an}, \tag{4.71}$$

where

$$\boxed{a \equiv \frac{4\pi\epsilon_0\hbar^2}{me^2} = 0.529 \times 10^{-10}\ \text{m}} \tag{4.72}$$

is the so-called **Bohr radius**.[15] It follows (again, from Equation 4.55) that

$$\rho = \frac{r}{an}. \tag{4.73}$$

The spatial wave functions for hydrogen are labeled by three quantum numbers (n, l, and m):

$$\psi_{nlm}(r, \theta, \phi) = R_{nl}(r)\, Y_l^m(\theta, \phi), \tag{4.74}$$

where (referring back to Equations 4.36 and 4.60)

$$R_{nl}(r) = \frac{1}{r}\rho^{l+1}e^{-\rho}v(\rho), \tag{4.75}$$

and $v(\rho)$ is a polynomial of degree $j_{\max} = n - l - 1$ in ρ, whose coefficients are determined (up to an overall normalization factor) by the recursion formula

$$c_{j+1} = \frac{2(j + l + 1 - n)}{(j + 1)(j + 2l + 2)}c_j. \tag{4.76}$$

The **ground state** (that is, the state of lowest energy) is the case $n = 1$; putting in the accepted values for the physical constants, we get:

$$\boxed{E_1 = -\left[\frac{m}{2\hbar^2}\left(\frac{e^2}{4\pi\epsilon_0}\right)^2\right] = -13.6\ \text{eV}.} \tag{4.77}$$

[15]It is traditional to write the Bohr radius with a subscript: a_0. But this is cumbersome and unnecessary, so I prefer to leave the subscript off.

Evidently the **binding energy** of hydrogen (the amount of energy you would have to impart to the electron in the ground state in order to ionize the atom) is 13.6 eV. Equation 4.67 forces $l = 0$, whence also $m = 0$ (see Equation 4.29), so

$$\psi_{100}(r, \theta, \phi) = R_{10}(r) Y_0^0(\theta, \phi). \qquad [4.78]$$

The recursion formula truncates after the first term (Equation 4.76 with $j = 0$ yields $c_1 = 0$), so $v(\rho)$ is a constant (c_0), and

$$R_{10}(r) = \frac{c_0}{a} e^{-r/a}. \qquad [4.79]$$

Normalizing it, in accordance with Equation 4.31:

$$\int_0^\infty |R_{10}|^2 r^2\, dr = \frac{|c_0|^2}{a^2} \int_0^\infty e^{-2r/a} r^2\, dr = |c_0|^2 \frac{a}{4} = 1,$$

so $c_0 = 2/\sqrt{a}$. Meanwhile, $Y_0^0 = 1/\sqrt{4\pi}$, and hence the ground state of hydrogen is

$$\boxed{\psi_{100}(r, \theta, \phi) = \frac{1}{\sqrt{\pi a^3}} e^{-r/a}.} \qquad [4.80]$$

If $n = 2$ the energy is

$$E_2 = \frac{-13.6 \text{ eV}}{4} = -3.4 \text{ eV}; \qquad [4.81]$$

this is the first excited state—or rather, *states*, since we can have either $l = 0$ (in which case $m = 0$) or $l = 1$ (with $m = -1$, 0, or $+1$); evidently four different states share this same energy. If $l = 0$, the recursion relation (Equation 4.76) gives

$$c_1 = -c_0 \text{ (using } j = 0), \quad \text{and} \quad c_2 = 0 \text{ (using } j = 1),$$

so $v(\rho) = c_0(1 - \rho)$, and therefore

$$R_{20}(r) = \frac{c_0}{2a}\left(1 - \frac{r}{2a}\right) e^{-r/2a}. \qquad [4.82]$$

[Notice that the expansion coefficients $\{c_j\}$ are completely different for different quantum numbers n and l.] If $l = 1$ the recursion formula terminates the series after a single term; $v(\rho)$ is a constant, and we find

$$R_{21}(r) = \frac{c_0}{4a^2} r e^{-r/2a}. \qquad [4.83]$$

(In each case the constant c_0 is to be determined by normalization—see Problem 4.11.)

For arbitrary n, the possible values of l (consistent with Equation 4.67) are

$$l = 0, 1, 2, \ldots, n-1, \tag{4.84}$$

and for each l there are $(2l+1)$ possible values of m (Equation 4.29), so the total degeneracy of the energy level E_n is

$$d(n) = \sum_{l=0}^{n-1}(2l+1) = n^2. \tag{4.85}$$

The polynomial $v(\rho)$ (defined by the recursion formula, Equation 4.76) is a function well known to applied mathematicians; apart from normalization, it can be written as

$$v(\rho) = L_{n-l-1}^{2l+1}(2\rho), \tag{4.86}$$

where

$$L_{q-p}^{p}(x) \equiv (-1)^p \left(\frac{d}{dx}\right)^p L_q(x) \tag{4.87}$$

is an **associated Laguerre polynomial**, and

$$L_q(x) \equiv e^x \left(\frac{d}{dx}\right)^q \left(e^{-x}x^q\right) \tag{4.88}$$

is the qth **Laguerre polynomial**.[16] (The first few Laguerre polynomials are listed in Table 4.5; some associated Laguerre polynomials are given in Table 4.6. The first few radial wave functions are listed in Table 4.7, and plotted in Figure 4.4.) The normalized hydrogen wave functions are[17]

$$\boxed{\psi_{nlm} = \sqrt{\left(\frac{2}{na}\right)^3 \frac{(n-l-1)!}{2n[(n+l)!]^3}}\, e^{-r/na}\left(\frac{2r}{na}\right)^l \left[L_{n-l-1}^{2l+1}(2r/na)\right] Y_l^m(\theta,\phi).} \tag{4.89}$$

They are not pretty, but don't complain—this is one of the very few realistic systems that can be solved at all, in exact closed form. Notice that whereas the wave functions depend on all three quantum numbers, the *energies* (Equation 4.70) are determined by n alone. This is a peculiarity of the Coulomb potential; in the

[16] As usual, there are rival normalization conventions in the literature; I have adopted the most nearly standard one.

[17] If you want to see how the normalization factor is calculated, study (for example), L. Schiff, *Quantum Mechanics*, 2nd ed., (McGraw-Hill, New York, 1968), page 93.

TABLE 4.5: The first few Laguerre polynomials, $L_q(x)$.

$L_0 = 1$
$L_1 = -x + 1$
$L_2 = x^2 - 4x + 2$
$L_3 = -x^3 + 9x^2 - 18x + 6$
$L_4 = x^4 - 16x^3 + 72x^2 - 96x + 24$
$L_5 = -x^5 + 25x^4 - 200x^3 + 600x^2 - 600x + 120$
$L_6 = x^6 - 36x^5 + 450x^4 - 2400x^3 + 5400x^2 - 4320x + 720$

TABLE 4.6: Some associated Laguerre polynomials, $L_{q-p}^p(x)$.

$L_0^0 = 1$	$L_0^2 = 2$
$L_1^0 = -x + 1$	$L_1^2 = -6x + 18$
$L_2^0 = x^2 - 4x + 2$	$L_2^2 = 12x^2 - 96x + 144$
$L_0^1 = 1$	$L_0^3 = 6$
$L_1^1 = -2x + 4$	$L_1^3 = -24x + 96$
$L_2^1 = 3x^2 - 18x + 18$	$L_2^3 = 60x^2 - 600x + 1200$

case of the spherical well, you may recall, the energies depend also on l (Equation 4.50). The wave functions are mutually orthogonal:

$$\int \psi_{nlm}^* \psi_{n'l'm'} r^2 \sin\theta \, dr \, d\theta \, d\phi = \delta_{nn'}\delta_{ll'}\delta_{mm'}. \qquad [4.90]$$

This follows from the orthogonality of the spherical harmonics (Equation 4.33) and (for $n \neq n'$) from the fact that they are eigenfunctions of H with distinct eigenvalues.

Visualizing the hydrogen wave functions is not easy. Chemists like to draw "density plots," in which the brightness of the cloud is proportional to $|\psi|^2$ (Figure 4.5). More quantitative (but perhaps harder to read) are surfaces of constant probability density (Figure 4.6).

∗**Problem 4.10** Work out the radial wave functions R_{30}, R_{31}, and R_{32}, using the recursion formula (Equation 4.76). Don't bother to normalize them.

TABLE 4.7: The first few radial wave functions for hydrogen, $R_{nl}(r)$.

$R_{10} = 2a^{-3/2}\exp(-r/a)$
$R_{20} = \frac{1}{\sqrt{2}}a^{-3/2}\left(1 - \frac{1}{2}\frac{r}{a}\right)\exp(-r/2a)$ $R_{21} = \frac{1}{\sqrt{24}}a^{-3/2}\frac{r}{a}\exp(-r/2a)$
$R_{30} = \frac{2}{\sqrt{27}}a^{-3/2}\left(1 - \frac{2}{3}\frac{r}{a} + \frac{2}{27}\left(\frac{r}{a}\right)^2\right)\exp(-r/3a)$ $R_{31} = \frac{8}{27\sqrt{6}}a^{-3/2}\left(1 - \frac{1}{6}\frac{r}{a}\right)\left(\frac{r}{a}\right)\exp(-r/3a)$ $R_{32} = \frac{4}{81\sqrt{30}}a^{-3/2}\left(\frac{r}{a}\right)^2\exp(-r/3a)$
$R_{40} = \frac{1}{4}a^{-3/2}\left(1 - \frac{3}{4}\frac{r}{a} + \frac{1}{8}\left(\frac{r}{a}\right)^2 - \frac{1}{192}\left(\frac{r}{a}\right)^3\right)\exp(-r/4a)$ $R_{41} = \frac{\sqrt{5}}{16\sqrt{3}}a^{-3/2}\left(1 - \frac{1}{4}\frac{r}{a} + \frac{1}{80}\left(\frac{r}{a}\right)^2\right)\frac{r}{a}\exp(-r/4a)$ $R_{42} = \frac{1}{64\sqrt{5}}a^{-3/2}\left(1 - \frac{1}{12}\frac{r}{a}\right)\left(\frac{r}{a}\right)^2\exp(-r/4a)$ $R_{43} = \frac{1}{768\sqrt{35}}a^{-3/2}\left(\frac{r}{a}\right)^3\exp(-r/4a)$

***Problem 4.11**

(a) Normalize R_{20} (Equation 4.82), and construct the function ψ_{200}.

(b) Normalize R_{21} (Equation 4.83), and construct ψ_{211}, ψ_{210}, and ψ_{21-1}.

***Problem 4.12**

(a) Using Equation 4.88, work out the first four Laguerre polynomials.

(b) Using Equations 4.86, 4.87, and 4.88, find $v(\rho)$, for the case $n = 5$, $l = 2$.

(c) Find $v(\rho)$ again (for the case $n = 5$, $l = 2$), but this time get it from the recursion formula (Equation 4.76).

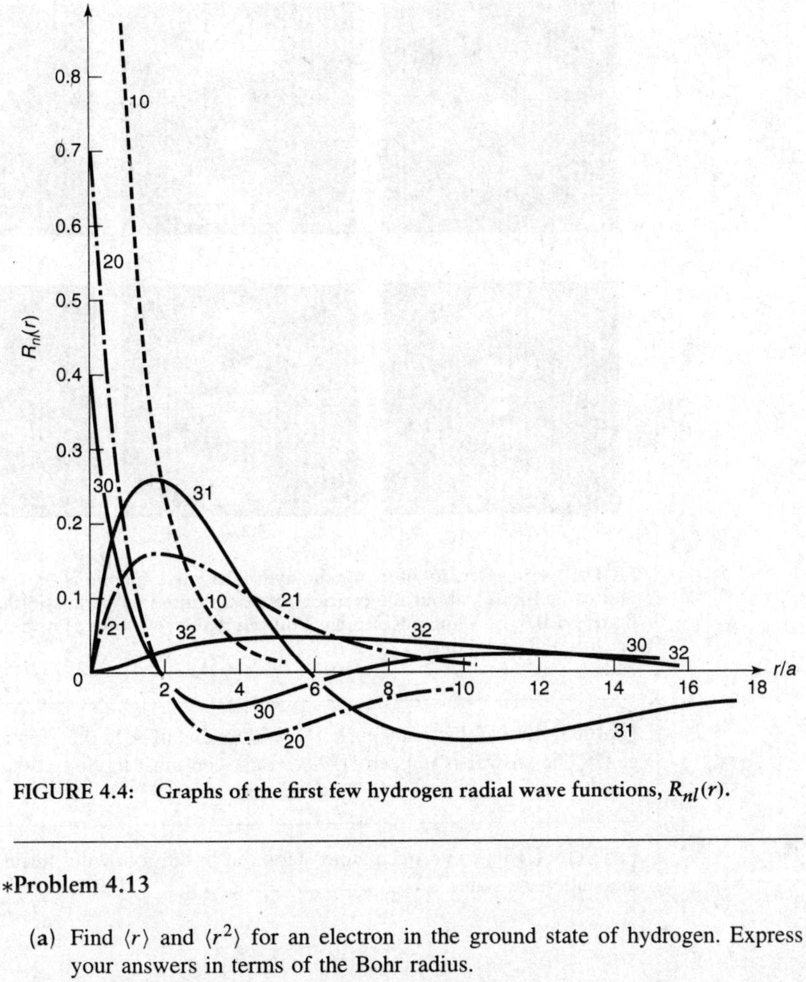

FIGURE 4.4: Graphs of the first few hydrogen radial wave functions, $R_{nl}(r)$.

∗Problem 4.13

(a) Find $\langle r \rangle$ and $\langle r^2 \rangle$ for an electron in the ground state of hydrogen. Express your answers in terms of the Bohr radius.

(b) Find $\langle x \rangle$ and $\langle x^2 \rangle$ for an electron in the ground state of hydrogen. *Hint:* This requires no new integration—note that $r^2 = x^2 + y^2 + z^2$, and exploit the symmetry of the ground state.

(c) Find $\langle x^2 \rangle$ in the state $n = 2$, $l = 1$, $m = 1$. *Warning:* This state is *not* symmetrical in x, y, z. Use $x = r \sin\theta \cos\phi$.

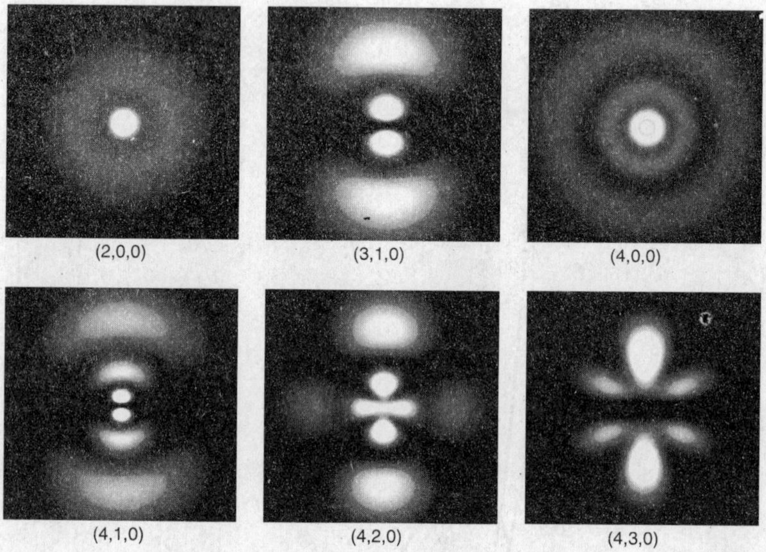

FIGURE 4.5: Density plots for the hydrogen wave functions (n, l, m). Imagine each plot to be rotated about the (vertical) z axis. Printed by permission using "Atom in a Box," v1.0.8, by Dauger Research. You can make your own plots by going to the Web site http://dauger.com.

Problem 4.14 What is the *most probable* value of r, in the ground state of hydrogen? (The answer is *not* zero!) *Hint:* First you must figure out the probability that the electron would be found between r and $r + dr$.

Problem 4.15 A hydrogen atom starts out in the following linear combination of the stationary states $n = 2, l = 1, m = 1$ and $n = 2, l = 1, m = -1$:

$$\Psi(\mathbf{r}, 0) = \frac{1}{\sqrt{2}}(\psi_{211} + \psi_{21-1}).$$

(a) Construct $\Psi(\mathbf{r}, t)$. Simplify it as much as you can.

(b) Find the expectation value of the potential energy, $\langle V \rangle$. (Does it depend on t?) Give both the formula and the actual number, in electron volts.

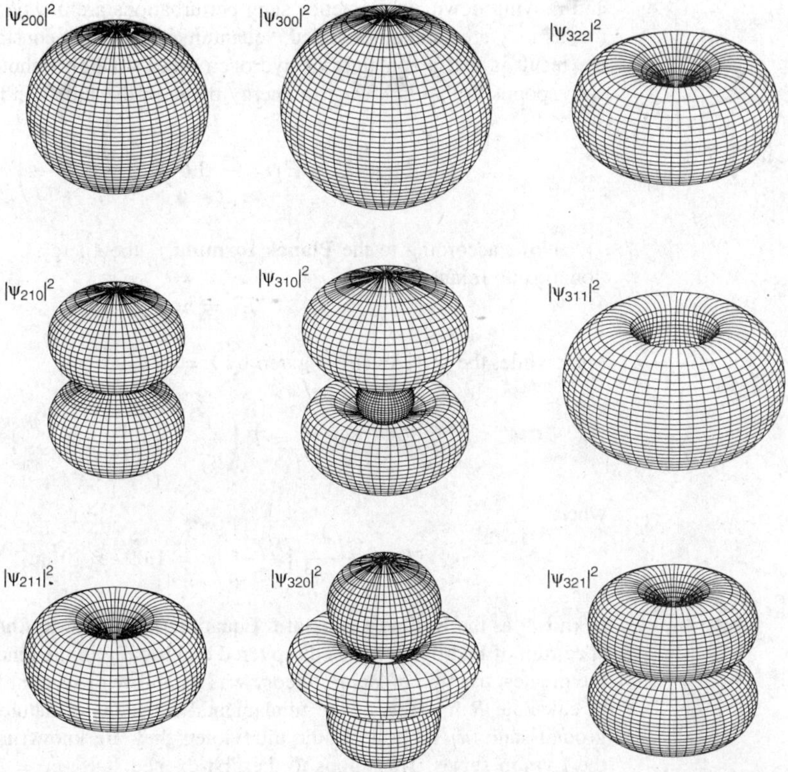

FIGURE 4.6: Surfaces of constant $|\psi|^2$ for the first few hydrogen wave functions. Reprinted by permission from Siegmund Brandt and Hans Dieter Dahmen, *The Picture Book of Quantum Mechanics*, 3rd ed., Springer, New York (2001).

4.2.2 The Spectrum of Hydrogen

In principle, if you put a hydrogen atom into some stationary state Ψ_{nlm}, it should stay there forever. However, if you *tickle* it slightly (by collision with another atom, say, or by shining light on it), the electron may undergo a **transition** to some other stationary state—either by *absorbing* energy, and moving up to a higher-energy state, or by *giving off* energy (typically in the form of electromagnetic radiation), and moving down.[18] In practice such perturbations are *always* present; transitions (or, as they are sometimes called, "quantum jumps") are constantly occurring, and the result is that a container of hydrogen gives off light (photons), whose energy corresponds to the *difference* in energy between the initial and final states:

$$E_\gamma = E_i - E_f = -13.6 \text{ eV}\left(\frac{1}{n_i^2} - \frac{1}{n_f^2}\right). \qquad [4.91]$$

Now, according to the **Planck formula**,[19] the energy of a photon is proportional to its frequency:

$$E_\gamma = h\nu. \qquad [4.92]$$

Meanwhile, the *wavelength* is given by $\lambda = c/\nu$, so

$$\frac{1}{\lambda} = R\left(\frac{1}{n_f^2} - \frac{1}{n_i^2}\right), \qquad [4.93]$$

where

$$R \equiv \frac{m}{4\pi c\hbar^3}\left(\frac{e^2}{4\pi\epsilon_0}\right)^2 = 1.097 \times 10^7 \text{ m}^{-1} \qquad [4.94]$$

is known as the **Rydberg constant**. Equation 4.93 is the **Rydberg formula** for the spectrum of hydrogen; it was discovered empirically in the nineteenth century, and the greatest triumph of Bohr's theory was its ability to account for this result—and to calculate R in terms of the fundamental constants of nature. Transitions to the ground state ($n_f = 1$) lie in the ultraviolet; they are known to spectroscopists as the **Lyman series**. Transitions to the first excited state ($n_f = 2$) fall in the visible region; they constitute the **Balmer series**. Transitions to $n_f = 3$ (the **Paschen series**) are in the infrared; and so on (see Figure 4.7). (At room temperature, most hydrogen atoms are in the ground state; to obtain the emission spectrum you must first populate the various excited states; typically this is done by passing an electric spark through the gas.)

∗**Problem 4.16** A **hydrogenic atom** consists of a single electron orbiting a nucleus with Z protons ($Z = 1$ would be hydrogen itself, $Z = 2$ is ionized helium, $Z = 3$

[18]By its nature, this involves a time-*de*pendent interaction, and the details will have to wait for Chapter 9; for our present purposes the actual mechanism involved is immaterial.

[19]The photon is a quantum of electromagnetic radiation; it's a relativistic object if there ever was one, and therefore outside the scope of nonrelativistic quantum mechanics. It will be useful in a few places to speak of photons, and to invoke the Planck formula for their energy, but please bear in mind that this is external to the theory we are developing.

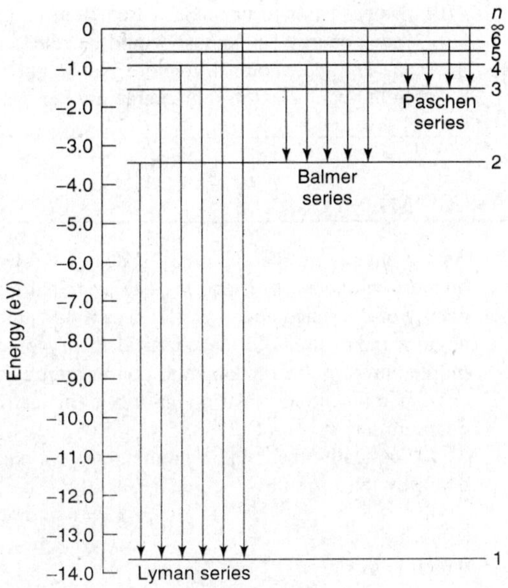

FIGURE 4.7: **Energy levels and transitions in the spectrum of hydrogen.**

is doubly ionized lithium, and so on). Determine the Bohr energies $E_n(Z)$, the binding energy $E_1(Z)$, the Bohr radius $a(Z)$, and the Rydberg constant $R(Z)$ for a hydrogenic atom. (Express your answers as appropriate multiples of the hydrogen values.) Where in the electromagnetic spectrum would the Lyman series fall, for $Z = 2$ and $Z = 3$? *Hint:* There's nothing much to *calculate* here—in the potential (Equation 4.52) $e^2 \rightarrow Ze^2$, so all you have to do is make the same substitution in all the final results.

Problem 4.17 Consider the earth-sun system as a gravitational analog to the hydrogen atom.

(a) What is the potential energy function (replacing Equation 4.52)? (Let m be the mass of the earth, and M the mass of the sun.)

(b) What is the "Bohr radius," a_g, for this system? Work out the actual number.

(c) Write down the gravitational "Bohr formula," and, by equating E_n to the classical energy of a planet in a circular orbit of radius r_o, show that $n = \sqrt{r_o/a_g}$. From this, estimate the quantum number n of the earth.

(d) Suppose the earth made a transition to the next lower level ($n - 1$). How much energy (in Joules) would be released? What would the wavelength of the emitted photon (or, more likely, graviton) be? (Express your answer in light years—is the remarkable answer[20] a coincidence?)

4.3 ANGULAR MOMENTUM

As we have seen, the stationary states of the hydrogen atom are labeled by three quantum numbers: n, l, and m. The principal quantum number (n) determines the energy of the state (Equation 4.70); as it turns out, l and m are related to the orbital angular momentum. In the classical theory of central forces, energy and angular momentum are the fundamental conserved quantities, and it is not surprising that angular momentum plays a significant (in fact, even *more* important) role in the quantum theory.

Classically, the angular momentum of a particle (with respect to the origin) is given by the formula

$$\mathbf{L} = \mathbf{r} \times \mathbf{p}, \tag{4.95}$$

which is to say,

$$L_x = yp_z - zp_y, \quad L_y = zp_x - xp_z, \quad L_z = xp_y - yp_x. \tag{4.96}$$

The corresponding quantum operators are obtained by the standard prescription $p_x \rightarrow -i\hbar\partial/\partial x$, $p_y \rightarrow -i\hbar\partial/\partial y$, $p_z \rightarrow -i\hbar\partial/\partial z$. In the following section we'll obtain the eigenvalues of the angular momentum operators by a purely algebraic technique reminiscent of the one we used in Chapter 2 to get the allowed energies of the harmonic oscillator; it is all based on the clever exploitation of commutation relations. After that we will turn to the more difficult problem of determining the eigenfunctions.

4.3.1 Eigenvalues

The operators L_x and L_y do not commute; in fact[21]

$$\begin{aligned}[L_x, L_y] &= [yp_z - zp_y, zp_x - xp_z] \\ &= [yp_z, zp_x] - [yp_z, xp_z] - [zp_y, zp_x] + [zp_y, xp_z].\end{aligned} \tag{4.97}$$

[20]Thanks to John Meyer for pointing this out.

[21]Note that all the operators we encounter in quantum mechanics (footnote 15, Chapter 1) are *distributive* with respect to addition: $A(B + C) = AB + AC$. In particular, $[A, B + C] = [A, B] + [A, C]$.

From the canonical commutation relations (Equation 4.10) we know that the only operators here that *fail* to commute are x with p_x, y with p_y, and z with p_z. So the two middle terms drop out, leaving

$$[L_x, L_y] = yp_x[p_z, z] + xp_y[z, p_z] = i\hbar(xp_y - yp_x) = i\hbar L_z. \qquad [4.98]$$

Of course, we could have started out with $[L_y, L_z]$ or $[L_z, L_x]$, but there is no need to calculate these separately—we can get them immediately by cyclic permutation of the indices ($x \to y$, $y \to z$, $z \to x$):

$$\boxed{[L_x, L_y] = i\hbar L_z; \quad [L_y, L_z] = i\hbar L_x; \quad [L_z, L_x] = i\hbar L_y.} \qquad [4.99]$$

These are the fundamental commutation relations for angular momentum; everything else follows from them.

Notice that L_x, L_y, and L_z are *incompatible* observables. According to the generalized uncertainty principle (Equation 3.62),

$$\sigma_{L_x}^2 \sigma_{L_y}^2 \geq \left(\frac{1}{2i}\langle i\hbar L_z\rangle\right)^2 = \frac{\hbar^2}{4}\langle L_z\rangle^2,$$

or

$$\sigma_{L_x}\sigma_{L_y} \geq \frac{\hbar}{2}|\langle L_z\rangle|. \qquad [4.100]$$

It would therefore be futile to look for states that are simultaneously eigenfunctions of L_x and L_y. On the other hand, the *square* of the *total* angular momentum,

$$L^2 \equiv L_x^2 + L_y^2 + L_z^2, \qquad [4.101]$$

does commute with L_x:

$$\begin{aligned}
[L^2, L_x] &= [L_x^2, L_x] + [L_y^2, L_x] + [L_z^2, L_x] \\
&= L_y[L_y, L_x] + [L_y, L_x]L_y + L_z[L_z, L_x] + [L_z, L_x]L_z \\
&= L_y(-i\hbar L_z) + (-i\hbar L_z)L_y + L_z(i\hbar L_y) + (i\hbar L_y)L_z \\
&= 0.
\end{aligned}$$

(I used Equation 3.64 to simplify the commutators; note also that *any* operator commutes with *itself*.) It follows, of course, that L^2 also commutes with L_y and L_z:

$$[L^2, L_x] = 0, \quad [L^2, L_y] = 0, \quad [L^2, L_z] = 0, \qquad [4.102]$$

or, more compactly,

$$[L^2, \mathbf{L}] = 0. \tag{4.103}$$

So L^2 *is* compatible with each component of **L**, and we *can* hope to find simultaneous eigenstates of L^2 and (say) L_z:

$$L^2 f = \lambda f \quad \text{and} \quad L_z f = \mu f. \tag{4.104}$$

We'll use a "ladder operator" technique, very similar to the one we applied to the harmonic oscillator back in Section 2.3.1. Let

$$\boxed{L_\pm \equiv L_x \pm i L_y.} \tag{4.105}$$

The commutator with L_z is

$$[L_z, L_\pm] = [L_z, L_x] \pm i[L_z, L_y] = i\hbar L_y \pm i(-i\hbar L_x) = \pm\hbar(L_x \pm i L_y),$$

so

$$[L_z, L_\pm] = \pm\hbar L_\pm. \tag{4.106}$$

And, of course,

$$[L^2, L_\pm] = 0. \tag{4.107}$$

I claim that if f is an eigenfunction of L^2 and L_z, so also is $L_\pm f$: Equation 4.107 says

$$L^2(L_\pm f) = L_\pm(L^2 f) = L_\pm(\lambda f) = \lambda(L_\pm f), \tag{4.108}$$

so $L_\pm f$ is an eigenfunction of L^2, with the same eigenvalue λ, and Equation 4.106 says

$$\begin{aligned} L_z(L_\pm f) &= (L_z L_\pm - L_\pm L_z)f + L_\pm L_z f = \pm\hbar L_\pm f + L_\pm(\mu f) \\ &= (\mu \pm \hbar)(L_\pm f), \end{aligned} \tag{4.109}$$

so $L_\pm f$ is an eigenfunction of L_z with the *new* eigenvalue $\mu \pm \hbar$. We call L_+ the "raising" operator, because it *increases* the eigenvalue of L_z by $\hbar$, and L_- the "lowering" operator, because it *lowers* the eigenvalue by $\hbar$.

For a given value of λ, then, we obtain a "ladder" of states, with each "rung" separated from its neighbors by one unit of $\hbar$ in the eigenvalue of L_z (see Figure 4.8). To ascend the ladder we apply the raising operator, and to descend, the lowering operator. But this process cannot go on forever: Eventually we're going

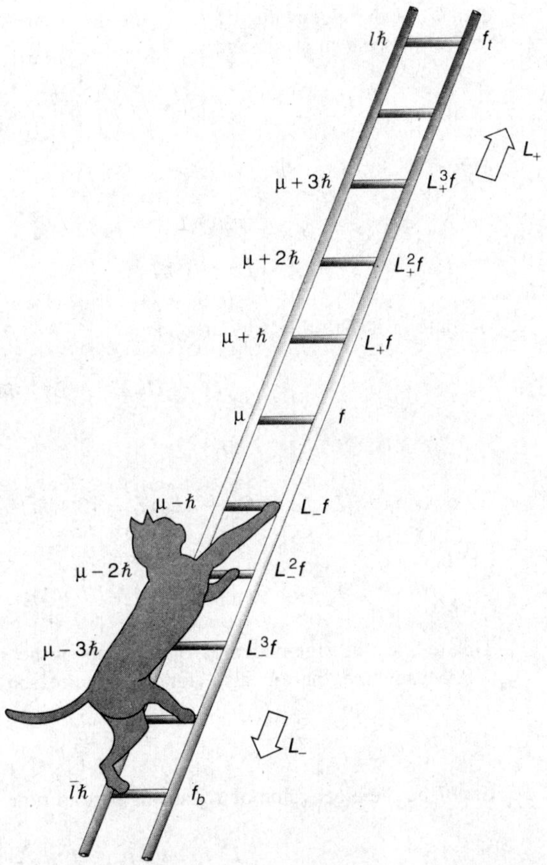

FIGURE 4.8: The "ladder" of angular momentum states.

to reach a state for which the z-component exceeds the *total*, and that cannot be.[22] There must exist a "top rung," f_t, such that[23]

$$L_+ f_t = 0. \tag{4.110}$$

[22]Formally, $\langle L^2 \rangle = \langle L_x^2 \rangle + \langle L_y^2 \rangle + \langle L_z^2 \rangle$, but $\langle L_x^2 \rangle = \langle f | L_x^2 f \rangle = \langle L_x f | L_x f \rangle \geq 0$ (and likewise for L_y), so $\lambda = \langle L_x^2 \rangle + \langle L_y^2 \rangle + \mu^2 \geq \mu^2$.

[23]Actually, all we can conclude is that $L_+ f_t$ is *not normalizable*—its norm could be *infinite*, instead of zero. Problem 4.18 explores this alternative.

Let $\hbar l$ be the eigenvalue of L_z at this top rung (the appropriateness of the letter "l" will appear in a moment):

$$L_z f_t = \hbar l f_t; \quad L^2 f_t = \lambda f_t. \qquad [4.111]$$

Now,

$$L_\pm L_\mp = (L_x \pm i L_y)(L_x \mp i L_y) = L_x^2 + L_y^2 \mp i(L_x L_y - L_y L_x)$$
$$= L^2 - L_z^2 \mp i(i\hbar L_z),$$

or, putting it the other way around,

$$L^2 = L_\pm L_\mp + L_z^2 \mp \hbar L_z. \qquad [4.112]$$

It follows that

$$L^2 f_t = (L_- L_+ + L_z^2 + \hbar L_z) f_t = (0 + \hbar^2 l^2 + \hbar^2 l) f_t = \hbar^2 l(l+1) f_t,$$

and hence

$$\lambda = \hbar^2 l(l+1). \qquad [4.113]$$

This tells us the eigenvalue of L^2 in terms of the *maximum* eigenvalue of L_z.

Meanwhile, there is also (for the same reason) a *bottom* rung, f_b, such that

$$L_- f_b = 0. \qquad [4.114]$$

Let $\hbar \bar{l}$ be the eigenvalue of L_z at this bottom rung:

$$L_z f_b = \hbar \bar{l} f_b; \quad L^2 f_b = \lambda f_b. \qquad [4.115]$$

Using Equation 4.112, we have

$$L^2 f_b = (L_+ L_- + L_z^2 - \hbar L_z) f_b = (0 + \hbar^2 \bar{l}^2 - \hbar^2 \bar{l}) f_b = \hbar^2 \bar{l}(\bar{l} - 1) f_b,$$

and therefore

$$\lambda = \hbar^2 \bar{l}(\bar{l} - 1). \qquad [4.116]$$

Comparing Equations 4.113 and 4.116, we see that $l(l+1) = \bar{l}(\bar{l} - 1)$, so either $\bar{l} = l + 1$ (which is absurd—the bottom rung would be higher than the top rung!) or else

$$\bar{l} = -l. \qquad [4.117]$$

Evidently the eigenvalues of L_z are $m\hbar$, where m (the appropriateness of this letter will also be clear in a moment) goes from $-l$ to $+l$ in N integer steps. In particular, it follows that $l = -l + N$, and hence $l = N/2$, so l must be *an integer or a half-integer*. The eigenfunctions are characterized by the numbers l and m:

$$L^2 f_l^m = \hbar^2 l(l+1) f_l^m; \quad L_z f_l^m = \hbar m f_l^m, \tag{4.118}$$

where

$$l = 0,\ 1/2,\ 1,\ 3/2,\ \ldots; \quad m = -l,\ -l+1,\ \ldots,\ l-1,\ l. \tag{4.119}$$

For a given value of l, there are $2l+1$ different values of m (i.e., $2l+1$ "rungs" on the "ladder").

Some people like to illustrate this result with the diagram in Figure 4.9 (drawn for the case $l = 2$). The arrows are supposed to represent possible angular momenta—in units of $\hbar$ they all have the same length $\sqrt{l(l+1)}$ (in this case $\sqrt{6} = 2.45$), and their z components are the allowed values of m ($-2, -1, 0, 1, 2$). Notice that the magnitude of the vectors (the radius of the sphere) is *greater* than the maximum z component! (In general, $\sqrt{l(l+1)} > l$, except for the "trivial" case $l = 0$.) Evidently you can't get the angular momentum to point perfectly along the z direction. At first, this sounds absurd. "Why can't I just *pick* my axes so that z points along the direction of the angular momentum vector?" Well, to do this you would have to know all three components simultaneously, and the

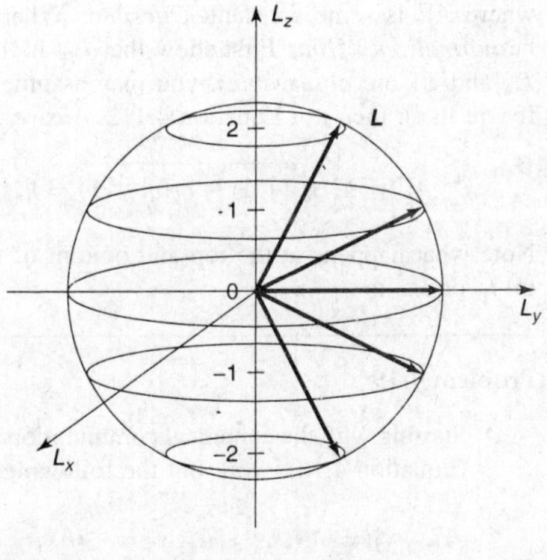

FIGURE 4.9: Angular momentum states (for $l = 2$).

uncertainty principle (Equation 4.100) says that's impossible. "Well, all right, but surely once in a while, by good fortune, I will just *happen* to aim my z-axis along the direction of **L**." No, no! You have missed the point. It's not merely that you don't *know* all three components of **L**; there simply *aren't* three components—a particle just cannot *have* a determinate angular momentum vector, any more than it can simultaneously have a determinate position and momentum. If L_z has a well-defined value, then L_x and L_y do *not*. It is misleading even to *draw* the vectors in Figure 4.9—at best they should be smeared out around the latitude lines, to indicate that L_x and L_y are indeterminate.

I hope you're impressed: By *purely algebraic means*, starting with the fundamental commutation relations for angular momentum (Equation 4.99), we have determined the eigenvalues of L^2 and L_z—without ever seeing the eigenfunctions themselves! We turn now to the problem of constructing the eigenfunctions, but I should warn you that this is a much messier business. Just so you know where we're headed, I'll begin with the punch line: $f_l^m = Y_l^m$—the eigenfunctions of L^2 and L_z are nothing but the old spherical harmonics, which we came upon by a quite different route in Section 4.1.2 (that's why I chose the letters l and m, of course). And I can now tell you why the spherical harmonics are orthogonal: They are eigenfunctions of hermitian operators (L^2 and L_z) belonging to distinct eigenvalues (Theorem 2, Section 3.3.1).

∗Problem 4.18 The raising and lowering operators change the value of m by one unit:

$$L_\pm f_l^m = (A_l^m) f_l^{m\pm 1}, \qquad [4.120]$$

where A_l^m is some constant. *Question:* What *is* A_l^m, if the eigenfunctions are to be *normalized*? *Hint:* First show that $L_\mp$ is the hermitian conjugate of $L_\pm$ (since L_x and L_y are *observables*, you may assume they are hermitian ... but *prove* it if you like); then use Equation 4.112. *Answer:*

$$A_l^m = \hbar\sqrt{l(l+1) - m(m\pm 1)} = \hbar\sqrt{(l\mp m)(l\pm m+1)}. \qquad [4.121]$$

Note what happens at the top and bottom of the ladder (i.e., when you apply L_+ to f_l^l or L_- to f_l^{-l}).

∗Problem 4.19

(a) Starting with the canonical commutation relations for position and momentum (Equation 4.10), work out the following commutators:

$$\begin{aligned} &[L_z, x] = i\hbar y, \quad &[L_z, y] = -i\hbar x, \quad &[L_z, z] = 0, \\ &[L_z, p_x] = i\hbar p_y, \quad &[L_z, p_y] = -i\hbar p_x, \quad &[L_z, p_z] = 0. \end{aligned} \qquad [4.122]$$

(b) Use these results to obtain $[L_z, L_x] = i\hbar L_y$ directly from Equation 4.96.

(c) Evaluate the commutators $[L_z, r^2]$ and $[L_z, p^2]$ (where, of course, $r^2 = x^2 + y^2 + z^2$ and $p^2 = p_x^2 + p_y^2 + p_z^2$).

(d) Show that the Hamiltonian $H = (p^2/2m) + V$ commutes with all three components of **L**, provided that V depends only on r. (Thus H, L^2, and L_z are mutually compatible observables.)

∗∗Problem 4.20

(a) Prove that for a particle in a potential $V(\mathbf{r})$ the rate of change of the expectation value of the orbital angular momentum **L** is equal to the expectation value of the torque:

$$\frac{d}{dt}\langle \mathbf{L} \rangle = \langle \mathbf{N} \rangle,$$

where

$$\mathbf{N} = \mathbf{r} \times (-\nabla V).$$

(This is the rotational analog to Ehrenfest's theorem.)

(b) Show that $d\langle \mathbf{L} \rangle / dt = 0$ for any spherically symmetric potential. (This is one form of the quantum statement of **conservation of angular momentum**.)

4.3.2 Eigenfunctions

First of all we need to rewrite L_x, L_y, and L_z in spherical coordinates. Now, $\mathbf{L} = (\hbar/i)(\mathbf{r} \times \nabla)$, and the gradient, in spherical coordinates, is:[24]

$$\nabla = \hat{r}\frac{\partial}{\partial r} + \hat{\theta}\frac{1}{r}\frac{\partial}{\partial \theta} + \hat{\phi}\frac{1}{r\sin\theta}\frac{\partial}{\partial \phi}; \qquad [4.123]$$

meanwhile, $\mathbf{r} = r\hat{r}$, so

$$\mathbf{L} = \frac{\hbar}{i}\left[r(\hat{r} \times \hat{r})\frac{\partial}{\partial r} + (\hat{r} \times \hat{\theta})\frac{\partial}{\partial \theta} + (\hat{r} \times \hat{\phi})\frac{1}{\sin\theta}\frac{\partial}{\partial \phi}\right].$$

[24] George Arfken and Hans-Jurgen Weber, *Mathematical Methods for Physicists*, 5th ed., Academic Press, Orlando (2000), Section 2.5.

But $(\hat{r} \times \hat{r}) = 0$, $(\hat{r} \times \hat{\theta}) = \hat{\phi}$, and $(\hat{r} \times \hat{\phi}) = -\hat{\theta}$ (see Figure 4.1), and hence

$$\mathbf{L} = \frac{\hbar}{i}\left(\hat{\phi}\frac{\partial}{\partial\theta} - \hat{\theta}\frac{1}{\sin\theta}\frac{\partial}{\partial\phi}\right). \quad [4.124]$$

The unit vectors $\hat{\theta}$ and $\hat{\phi}$ can be resolved into their cartesian components:

$$\hat{\theta} = (\cos\theta\cos\phi)\hat{\imath} + (\cos\theta\sin\phi)\hat{\jmath} - (\sin\theta)\hat{k}; \quad [4.125]$$

$$\hat{\phi} = -(\sin\phi)\hat{\imath} + (\cos\phi)\hat{\jmath}. \quad [4.126]$$

Thus

$$\mathbf{L} = \frac{\hbar}{i}\left[(-\sin\phi\,\hat{\imath} + \cos\phi\,\hat{\jmath})\frac{\partial}{\partial\theta} - (\cos\theta\cos\phi\,\hat{\imath} + \cos\theta\sin\phi\,\hat{\jmath} - \sin\theta\,\hat{k})\frac{1}{\sin\theta}\frac{\partial}{\partial\phi}\right].$$

Evidently

$$L_x = \frac{\hbar}{i}\left(-\sin\phi\frac{\partial}{\partial\theta} - \cos\phi\cot\theta\frac{\partial}{\partial\phi}\right), \quad [4.127]$$

$$L_y = \frac{\hbar}{i}\left(+\cos\phi\frac{\partial}{\partial\theta} - \sin\phi\cot\theta\frac{\partial}{\partial\phi}\right), \quad [4.128]$$

and

$$\boxed{L_z = \frac{\hbar}{i}\frac{\partial}{\partial\phi}.} \quad [4.129]$$

We shall also need the raising and lowering operators:

$$L_{\pm} = L_x \pm iL_y = \frac{\hbar}{i}\left[(-\sin\phi \pm i\cos\phi)\frac{\partial}{\partial\theta} - (\cos\phi \pm i\sin\phi)\cot\theta\frac{\partial}{\partial\phi}\right].$$

But $\cos\phi \pm i\sin\phi = e^{\pm i\phi}$, so

$$L_{\pm} = \pm\hbar e^{\pm i\phi}\left(\frac{\partial}{\partial\theta} \pm i\cot\theta\frac{\partial}{\partial\phi}\right) \quad [4.130]$$

In particular (Problem 4.21(a)):

$$L_+L_- = -\hbar^2\left(\frac{\partial^2}{\partial\theta^2} + \cot\theta\frac{\partial}{\partial\theta} + \cot^2\theta\frac{\partial^2}{\partial\phi^2} + i\frac{\partial}{\partial\phi}\right), \quad [4.131]$$

and hence (Problem 4.21(b)):

$$L^2 = -\hbar^2\left[\frac{1}{\sin\theta}\frac{\partial}{\partial\theta}\left(\sin\theta\frac{\partial}{\partial\theta}\right) + \frac{1}{\sin^2\theta}\frac{\partial^2}{\partial\phi^2}\right]. \tag{4.132}$$

We are now in a position to determine $f_l^m(\theta, \phi)$. It's an eigenfunction of L^2, with eigenvalue $\hbar^2 l(l+1)$:

$$L^2 f_l^m = -\hbar^2\left[\frac{1}{\sin\theta}\frac{\partial}{\partial\theta}\left(\sin\theta\frac{\partial}{\partial\theta}\right) + \frac{1}{\sin^2\theta}\frac{\partial^2}{\partial\phi^2}\right] f_l^m = \hbar^2 l(l+1) f_l^m.$$

But this is precisely the "angular equation" (Equation 4.18). And it's also an eigenfunction of L_z, with the eigenvalue $m\hbar$:

$$L_z f_l^m = \frac{\hbar}{i}\frac{\partial}{\partial\phi} f_l^m = \hbar m f_l^m,$$

but this is equivalent to the azimuthal equation (Equation 4.21). We have already solved this system of equations: The result (appropriately normalized) is the spherical harmonic, $Y_l^m(\theta, \phi)$. *Conclusion:* Spherical harmonics *are* eigenfunctions of L^2 and L_z. When we solved the Schrödinger equation by separation of variables, in Section 4.1, we were inadvertently constructing simultaneous eigenfunctions of the three commuting operators H, L^2, and L_z:

$$H\psi = E\psi, \quad L^2\psi = \hbar^2 l(l+1)\psi, \quad L_z\psi = \hbar m\psi. \tag{4.133}$$

Incidentally, we can use Equation 4.132 to rewrite the Schrödinger equation (Equation 4.14) more compactly:

$$\frac{1}{2mr^2}\left[-\hbar^2\frac{\partial}{\partial r}\left(r^2\frac{\partial}{\partial r}\right) + L^2\right]\psi + V\psi = E\psi.$$

There is a curious final twist to this story, for the *algebraic* theory of angular momentum permits l (and hence also m) to take on *half*-integer values (Equation 4.119), whereas separation of variables yielded eigenfunctions only for *integer* values (Equation 4.29). You might suppose that the half-integer solutions are spurious, but it turns out that they are of profound importance, as we shall see in the following sections.

∗Problem 4.21

(a) Derive Equation 4.131 from Equation 4.130. *Hint:* Use a test function; otherwise you're likely to drop some terms.

(b) Derive Equation 4.132 from Equations 4.129 and 4.131. *Hint:* Use Equation 4.112.

∗Problem 4.22

(a) What is $L_+ Y_l^l$? (No calculation allowed!)

(b) Use the result of (a), together with Equation 4.130 and the fact that $L_z Y_l^l = \hbar l Y_l^l$, to determine $Y_l^l(\theta, \phi)$, up to a normalization constant.

(c) Determine the normalization constant by direct integration. Compare your final answer to what you got in Problem 4.5.

Problem 4.23 In Problem 4.3 you showed that

$$Y_2^1(\theta, \phi) = -\sqrt{15/8\pi}\,\sin\theta\cos\theta e^{i\phi}.$$

Apply the raising operator to find $Y_2^2(\theta, \phi)$. Use Equation 4.121 to get the normalization.

Problem 4.24 Two particles of mass m are attached to the ends of a massless rigid rod of length a. The system is free to rotate in three dimensions about the center (but the center point itself is fixed).

(a) Show that the allowed energies of this **rigid rotor** are

$$E_n = \frac{\hbar^2 n(n+1)}{ma^2}, \qquad \text{for} \quad n = 0,\ 1,\ 2, \ldots$$

Hint: First express the (classical) energy in terms of the total angular momentum.

(b) What are the normalized eigenfunctions for this system? What is the degeneracy of the nth energy level?

4.4 SPIN

In *classical* mechanics, a rigid object admits two kinds of angular momentum: **orbital** ($\mathbf{L} = \mathbf{r} \times \mathbf{p}$), associated with the motion *of* the center of mass, and **spin** ($\mathbf{S} = I\boldsymbol{\omega}$), associated with motion *about* the center of mass. For example, the earth has orbital angular momentum attributable to its annual revolution around the sun, and spin angular momentum coming from its daily rotation about the north-south axis. In the classical context this distinction is largely a matter of convenience, for when you come right down to it, **S** is nothing but the sum total of the "orbital" angular momenta of all the rocks and dirt clods that go to make up the earth, as they circle around the axis. But an analogous thing happens in quantum mechanics, and here the distinction is absolutely fundamental. In addition to orbital angular momentum, associated (in the case of hydrogen) with the motion of the electron around the nucleus (and described by the spherical harmonics), the electron also carries *another* form of angular momentum, which has nothing to do with motion in space (and which is not, therefore, described by any function of the position variables r, θ, ϕ) but which is somewhat analogous to classical spin (and for which, therefore, we use the same word). It doesn't pay to press this analogy too far: The electron (as far as we know) is a structureless point particle, and its spin angular momentum cannot be decomposed into orbital angular momenta of constituent parts (see Problem 4.25).[25] Suffice it to say that elementary particles carry **intrinsic** angular momentum (**S**) in addition to their "extrinsic" angular momentum (**L**).

The *algebraic* theory of spin is a carbon copy of the theory of orbital angular momentum, beginning with the fundamental commutation relations:[26]

$$[S_x, S_y] = i\hbar S_z, \quad [S_y, S_z] = i\hbar S_x, \quad [S_z, S_x] = i\hbar S_y. \qquad [4.134]$$

It follows (as before) that the eigenvectors of S^2 and S_z satisfy[27]

$$S^2|s\,m\rangle = \hbar^2 s(s+1)|s\,m\rangle; \quad S_z|s\,m\rangle = \hbar m|s\,m\rangle; \qquad [4.135]$$

[25] For a contrary interpretation, see Hans C. Ohanian, "What is Spin?", *Am. J. Phys.* **54**, 500 (1986).

[26] We shall take these as *postulates* for the theory of spin; the analogous formulas for *orbital* angular momentum (Equation 4.99) were *derived* from the known form of the operators (Equation 4.96). In a more sophisticated treatment they can both be obtained from rotational invariance in three dimensions (see, for example, Leslie E. Ballentine, *Quantum Mechanics: A Modern Development*, World Scientific, Singapore (1998), Section 3.3). Indeed, these fundamental commutation relations apply to *all* forms of angular momentum, whether spin, orbital, or the combined angular momentum of a composite system, which could include some spin and some orbital.

[27] Because the eigenstates of spin are not *functions*, I will use the "ket" notation for them. (I could have done the same in Section 4.3, writing $|l\,m\rangle$ in place of Y_l^m, but in that context the function notation seems more natural.) By the way, I'm running out of letters, so I'll use m for the eigenvalue of S_z, just as I did for L_z (some authors write m_l and m_s at this stage, just to be absolutely clear).

and

$$S_{\pm}|s\,m\rangle = \hbar\sqrt{s(s+1) - m(m\pm 1)}\;|s\,(m\pm 1)\rangle, \qquad [4.136]$$

where $S_{\pm} \equiv S_x \pm iS_y$. But this time the eigenvectors are not spherical harmonics (they're not functions of θ and ϕ at *all*), and there is no *a priori* reason to exclude the half-integer values of s and m:

$$s = 0, \frac{1}{2}, 1, \frac{3}{2}, \ldots; \quad m = -s, -s+1, \ldots, s-1, s. \qquad [4.137]$$

It so happens that every elementary particle has a *specific and immutable* value of s, which we call **the spin** of that particular species: pi mesons have spin 0; electrons have spin 1/2; photons have spin 1; deltas have spin 3/2; gravitons have spin 2; and so on. By contrast, the *orbital* angular momentum quantum number l (for an electron in a hydrogen atom, say) can take on any (integer) value you please, and will change from one to another when the system is perturbed. But s is *fixed*, for any given particle, and this makes the theory of spin comparatively simple.[28]

Problem 4.25 If the electron were a classical solid sphere, with radius

$$r_c = \frac{e^2}{4\pi\epsilon_0 mc^2} \qquad [4.138]$$

(the so-called **classical electron radius**, obtained by assuming the electron's mass is attributable to energy stored in its electric field, via the Einstein formula $E = mc^2$), and its angular momentum is $(1/2)\hbar$, then how fast (in m/s) would a point on the "equator" be moving? Does this model make sense? (Actually, the radius of the electron is known experimentally to be much less than r_c, but this only makes matters worse.)

---

[28] Indeed, in a mathematical sense, spin 1/2 is the simplest possible nontrivial quantum system, for it admits just two basis states. In place of an infinite-dimensional Hilbert space, with all its subtleties and complications, we find ourselves working in an ordinary 2-dimensional vector space; in place of unfamiliar differential equations and fancy functions, we are confronted with 2 × 2 matrices and 2-component vectors. For this reason, some authors *begin* quantum mechanics with the study of spin. (An outstanding example is John S. Townsend, *A Modern Approach to Quantum Mechanics*, University Books, Sausalito, CA, 2000.) But the price of mathematical simplicity is conceptual abstraction, and I prefer not to do it that way.

4.4.1 Spin 1/2

By far the most important case is $s = 1/2$, for this is the spin of the particles that make up ordinary matter (protons, neutrons, and electrons), as well as all quarks and all leptons. Moreover, once you understand spin 1/2, it is a simple matter to work out the formalism for any higher spin. There are just *two* eigenstates: $|\frac{1}{2}\,\frac{1}{2}\rangle$, which we call **spin up** (informally, $\uparrow$), and $|\frac{1}{2}\,(-\frac{1}{2})\rangle$, which we call **spin down** ($\downarrow$). Using these as basis vectors, the general state of a spin-1/2 particle can be expressed as a two-element column matrix (or **spinor**):

$$\chi = \begin{pmatrix} a \\ b \end{pmatrix} = a\chi_+ + b\chi_-, \qquad [4.139]$$

with

$$\chi_+ = \begin{pmatrix} 1 \\ 0 \end{pmatrix} \qquad [4.140]$$

representing spin up, and

$$\chi_- = \begin{pmatrix} 0 \\ 1 \end{pmatrix} \qquad [4.141]$$

for spin down.

Meanwhile, the spin operators become 2×2 matrices, which we can work out by noting their effect on χ_+ and χ_-. Equation 4.135 says

$$\mathbf{S}^2\chi_+ = \frac{3}{4}\hbar^2\chi_+ \quad \text{and} \quad \mathbf{S}^2\chi_- = \frac{3}{4}\hbar^2\chi_-. \qquad [4.142]$$

If we write $\mathbf{S}^2$ as a matrix with (as yet) undetermined elements,

$$\mathbf{S}^2 = \begin{pmatrix} c & d \\ e & f \end{pmatrix},$$

then the first equation says

$$\begin{pmatrix} c & d \\ e & f \end{pmatrix}\begin{pmatrix} 1 \\ 0 \end{pmatrix} = \frac{3}{4}\hbar^2\begin{pmatrix} 1 \\ 0 \end{pmatrix}, \quad \text{or} \quad \begin{pmatrix} c \\ e \end{pmatrix} = \begin{pmatrix} \frac{3}{4}\hbar^2 \\ 0 \end{pmatrix},$$

so $c = (3/4)\hbar^2$ and $e = 0$. The second equation says

$$\begin{pmatrix} c & d \\ e & f \end{pmatrix}\begin{pmatrix} 0 \\ 1 \end{pmatrix} = \frac{3}{4}\hbar^2\begin{pmatrix} 0 \\ 1 \end{pmatrix}, \quad \text{or} \quad \begin{pmatrix} d \\ f \end{pmatrix} = \begin{pmatrix} 0 \\ \frac{3}{4}\hbar^2 \end{pmatrix},$$

so $d = 0$ and $f = (3/4)\hbar^2$. *Conclusion:*

$$\mathbf{S}^2 = \frac{3}{4}\hbar^2\begin{pmatrix} 1 & 0 \\ 0 & 1 \end{pmatrix}. \qquad [4.143]$$

Similarly,

$$\mathbf{S}_z\chi_+ = \frac{\hbar}{2}\chi_+, \quad \mathbf{S}_z\chi_- = -\frac{\hbar}{2}\chi_-, \qquad [4.144]$$

from which it follows that

$$\mathbf{S}_z = \frac{\hbar}{2}\begin{pmatrix} 1 & 0 \\ 0 & -1 \end{pmatrix}. \qquad [4.145]$$

Meanwhile, Equation 4.136 says

$$\mathbf{S}_+\chi_- = \hbar\chi_+, \quad \mathbf{S}_-\chi_+ = \hbar\chi_-, \quad \mathbf{S}_+\chi_+ = \mathbf{S}_-\chi_- = 0,$$

so

$$\mathbf{S}_+ = \hbar\begin{pmatrix} 0 & 1 \\ 0 & 0 \end{pmatrix}, \quad \mathbf{S}_- = \hbar\begin{pmatrix} 0 & 0 \\ 1 & 0 \end{pmatrix}. \qquad [4.146]$$

Now $S_\pm = S_x \pm iS_y$, so $S_x = (1/2)(S_+ + S_-)$ and $S_y = (1/2i)(S_+ - S_-)$, and hence

$$\mathbf{S}_x = \frac{\hbar}{2}\begin{pmatrix} 0 & 1 \\ 1 & 0 \end{pmatrix}, \quad \mathbf{S}_y = \frac{\hbar}{2}\begin{pmatrix} 0 & -i \\ i & 0 \end{pmatrix}. \qquad [4.147]$$

Since $\mathbf{S}_x$, $\mathbf{S}_y$, and $\mathbf{S}_z$ all carry a factor of $\hbar/2$, it is tidier to write $\mathbf{S} = (\hbar/2)\boldsymbol{\sigma}$, where

$$\boxed{\sigma_x \equiv \begin{pmatrix} 0 & 1 \\ 1 & 0 \end{pmatrix}, \quad \sigma_y \equiv \begin{pmatrix} 0 & -i \\ i & 0 \end{pmatrix}, \quad \sigma_z \equiv \begin{pmatrix} 1 & 0 \\ 0 & -1 \end{pmatrix}.} \qquad [4.148]$$

These are the famous **Pauli spin matrices**. Notice that $\mathbf{S}_x$, $\mathbf{S}_y$, $\mathbf{S}_z$, and $\mathbf{S}^2$ are all *hermitian* (as they *should* be, since they represent observables). On the other hand, $\mathbf{S}_+$ and $\mathbf{S}_-$ are *not* hermitian—evidently they are not observable.

The eigenspinors of $\mathbf{S}_z$ are (of course):

$$\chi_+ = \begin{pmatrix} 1 \\ 0 \end{pmatrix}, \left(\text{eigenvalue} + \frac{\hbar}{2}\right); \quad \chi_- = \begin{pmatrix} 0 \\ 1 \end{pmatrix}, \left(\text{eigenvalue} - \frac{\hbar}{2}\right). \qquad [4.149]$$

If you measure S_z on a particle in the general state χ (Equation 4.139), you could get $+\hbar/2$, with probability $|a|^2$, or $-\hbar/2$, with probability $|b|^2$. Since these are the *only* possibilities,

$$|a|^2 + |b|^2 = 1 \qquad [4.150]$$

(i.e., the spinor must be *normalized*).[29]

---

[29]People often say that $|a|^2$ is the "probability that the particle is in the spin-up state," but this is sloppy language; what they *mean* is that if you *measured* S_z, $|a|^2$ is the probability you'd get $\hbar/2$. See footnote 16 in Chapter 3.

But what if, instead, you chose to measure S_x? What are the possible results, and what are their respective probabilities? According to the generalized statistical interpretation, we need to know the eigenvalues and eigenspinors of $\mathbf{S}_x$. The characteristic equation is

$$\begin{vmatrix} -\lambda & \hbar/2 \\ \hbar/2 & -\lambda \end{vmatrix} = 0 \Rightarrow \lambda^2 = \left(\frac{\hbar}{2}\right)^2 \Rightarrow \lambda = \pm\frac{\hbar}{2}.$$

Not surprisingly, the possible values for S_x are the same as those for S_z. The eigenspinors are obtained in the usual way:

$$\frac{\hbar}{2}\begin{pmatrix} 0 & 1 \\ 1 & 0 \end{pmatrix}\begin{pmatrix} \alpha \\ \beta \end{pmatrix} = \pm\frac{\hbar}{2}\begin{pmatrix} \alpha \\ \beta \end{pmatrix} \Rightarrow \begin{pmatrix} \beta \\ \alpha \end{pmatrix} = \pm\begin{pmatrix} \alpha \\ \beta \end{pmatrix},$$

so $\beta = \pm\alpha$. Evidently the (normalized) eigenspinors of $\mathbf{S}_x$ are

$$\chi_+^{(x)} = \begin{pmatrix} \frac{1}{\sqrt{2}} \\ \frac{1}{\sqrt{2}} \end{pmatrix}, \left(\text{eigenvalue} + \frac{\hbar}{2}\right); \quad \chi_-^{(x)} = \begin{pmatrix} \frac{1}{\sqrt{2}} \\ \frac{-1}{\sqrt{2}} \end{pmatrix}, \left(\text{eigenvalue} - \frac{\hbar}{2}\right). \quad [4.151]$$

As the eigenvectors of a hermitian matrix, they span the space; the generic spinor χ (Equation 4.139) can be expressed as a linear combination of them:

$$\chi = \left(\frac{a+b}{\sqrt{2}}\right)\chi_+^{(x)} + \left(\frac{a-b}{\sqrt{2}}\right)\chi_-^{(x)}. \quad [4.152]$$

If you measure S_x, the probability of getting $+\hbar/2$ is $(1/2)|a+b|^2$, and the probability of getting $-\hbar/2$ is $(1/2)|a-b|^2$. (You should check for yourself that these probabilities add up to 1.)

Example 4.2 Suppose a spin-1/2 particle is in the state

$$\chi = \frac{1}{\sqrt{6}}\begin{pmatrix} 1+i \\ 2 \end{pmatrix}.$$

What are the probabilities of getting $+\hbar/2$ and $-\hbar/2$, if you measure S_z and S_x?

Solution: Here $a = (1+i)/\sqrt{6}$ and $b = 2/\sqrt{6}$, so for S_z the probability of getting $+\hbar/2$ is $|(1+i)/\sqrt{6}|^2 = 1/3$, and the probability of getting $-\hbar/2$ is $|2/\sqrt{6}|^2 = 2/3$. For S_x the probability of getting $+\hbar/2$ is $(1/2)|(3+i)/\sqrt{6}|^2 = 5/6$, and the probability of getting $-\hbar/2$ is $(1/2)|(-1+i)/\sqrt{6}|^2 = 1/6$. Incidentally, the *expectation* value of S_x is

$$\frac{5}{6}\left(+\frac{\hbar}{2}\right)+\frac{1}{6}\left(-\frac{\hbar}{2}\right)=\frac{\hbar}{3},$$

which we could also have obtained more directly:

$$\langle S_x\rangle = \chi^{\dagger}\mathbf{S}_x\chi = \begin{pmatrix}\frac{(1-i)}{\sqrt{6}} & \frac{2}{\sqrt{6}}\end{pmatrix}\begin{pmatrix}0 & \hbar/2\\ \hbar/2 & 0\end{pmatrix}\begin{pmatrix}(1+i)/\sqrt{6}\\ 2/\sqrt{6}\end{pmatrix}=\frac{\hbar}{3}.$$

I'd like now to walk you through an imaginary measurement scenario involving spin 1/2, because it serves to illustrate in very concrete terms some of the abstract ideas we discussed back in Chapter 1. Let's say we start out with a particle in the state χ_+. If someone asks, "What is the z-component of that particle's spin angular momentum?", we could answer unambiguously: $+\hbar/2$. For a measurement of S_z is certain to return that value. But if our interrogator asks instead, "What is the x-component of that particle's spin angular momentum?" we are obliged to equivocate: If you measure S_x, the chances are fifty-fifty of getting either $\hbar/2$ or $-\hbar/2$. If the questioner is a classical physicist, or a "realist" (in the sense of Section 1.2), he will regard this as an inadequate—not to say impertinent—response: "Are you telling me that you *don't know* the true state of that particle?" On the contrary; I know *precisely* what the state of the particle is: χ_+. "Well, then, how come you can't tell me what the x-component of its spin is?" Because it simply *does not have* a particular x-component of spin. Indeed, it *cannot*, for if both S_x and S_z were well-defined, the uncertainty principle would be violated.

At this point our challenger grabs the test-tube and *measures* the x-component of its spin; let's say he gets the value $+\hbar/2$. "Aha!" (he shouts in triumph), "You *lied!* This particle has a perfectly well-defined value of S_x: $\hbar/2$." Well, sure—it does *now*, but that doesn't prove it *had* that value, prior to your measurement. "You have obviously been reduced to splitting hairs. And anyway, what happened to your uncertainty principle? I now know both S_x *and* S_z." I'm sorry, but you do *not:* In the course of your measurement, you altered the particle's state; it is now in the state $\chi_+^{(x)}$, and whereas you know the value of S_x, you no longer know the value of S_z. "But I was extremely careful not to disturb the particle when I measured S_x." Very well, if you don't believe me, *check it out:* Measure S_z, and see what you get. (Of course, he *may* get $+\hbar/2$, which will be embarrassing to my case—but if we repeat this whole scenario over and over, half the time he will get $-\hbar/2$.)

To the layman, the philosopher, or the classical physicist, a statement of the form "this particle doesn't have a well-defined position" (or momentum, or x-component of spin angular momentum, or whatever) sounds vague, incompetent, or (worst of all) profound. It is none of these. But its precise meaning is, I think, almost impossible to convey to anyone who has not studied quantum mechanics in some depth. If you find your own comprehension slipping, from time to time (if you *don't*, you probably haven't understood the problem), come back to the spin-1/2 system: It is the simplest and cleanest context for thinking through the conceptual paradoxes of quantum mechanics.

Problem 4.26

(a) Check that the spin matrices (Equations 4.145 and 4.147) obey the fundamental commutation relations for angular momentum, Equation 4.134.

(b) Show that the Pauli spin matrices (Equation 4.148) satisfy the product rule

$$\sigma_j \sigma_k = \delta_{jk} + i \sum_l \epsilon_{jkl} \sigma_l, \tag{4.153}$$

where the indices stand for x, y, or z, and ϵ_{jkl} is the **Levi-Civita** symbol: $+1$ if $jkl = 123$, 231, or 312; -1 if $jkl = 132$, 213, or 321; 0 otherwise.

∗Problem 4.27 An electron is in the spin state

$$\chi = A \begin{pmatrix} 3i \\ 4 \end{pmatrix}.$$

(a) Determine the normalization constant A.

(b) Find the expectation values of S_x, S_y, and S_z.

(c) Find the "uncertainties" σ_{S_x}, σ_{S_y}, and σ_{S_z}. (*Note:* These sigmas are standard deviations, not Pauli matrices!)

(d) Confirm that your results are consistent with all three uncertainty principles (Equation 4.100 and its cyclic permutations—only with S in place of L, of course).

∗Problem 4.28 For the most general normalized spinor χ (Equation 4.139), compute $\langle S_x \rangle$, $\langle S_y \rangle$, $\langle S_z \rangle$, $\langle S_x^2 \rangle$, $\langle S_y^2 \rangle$, and $\langle S_z^2 \rangle$. Check that $\langle S_x^2 \rangle + \langle S_y^2 \rangle + \langle S_z^2 \rangle = \langle S^2 \rangle$.

∗Problem 4.29

(a) Find the eigenvalues and eigenspinors of $\mathbf{S}_y$.

(b) If you measured S_y on a particle in the general state χ (Equation 4.139), what values might you get, and what is the probability of each? Check that the probabilities add up to 1. *Note: a* and *b* need not be real!

(c) If you measured S_y^2, what values might you get, and with what probabilities?

∗∗Problem 4.30 Construct the matrix $\mathbf{S}_r$ representing the component of spin angular momentum along an arbitrary direction $\hat{r}$. Use spherical coordinates, for which

$$\hat{r} = \sin\theta\cos\phi\,\hat{\imath} + \sin\theta\sin\phi\,\hat{\jmath} + \cos\theta\,\hat{k}. \quad [4.154]$$

Find the eigenvalues and (normalized) eigenspinors of $\mathbf{S}_r$. *Answer:*

$$\chi_+^{(r)} = \begin{pmatrix} \cos(\theta/2) \\ e^{i\phi}\sin(\theta/2) \end{pmatrix}; \quad \chi_-^{(r)} = \begin{pmatrix} e^{-i\phi}\sin(\theta/2) \\ -\cos(\theta/2) \end{pmatrix}. \quad [4.155]$$

Note: You're always free to multiply by an arbitrary phase factor—say, $e^{i\phi}$—so your answer may not *look* exactly the same as mine.

Problem 4.31 Construct the spin matrices ($\mathbf{S}_x$, $\mathbf{S}_y$, and $\mathbf{S}_z$) for a particle of spin 1. *Hint:* How many eigenstates of S_z are there? Determine the action of S_z, S_+, and S_- on each of these states. Follow the procedure used in the text for spin 1/2.

4.4.2 Electron in a Magnetic Field

A spinning charged particle constitutes a magnetic dipole. Its **magnetic dipole moment,** $\boldsymbol{\mu}$, is proportional to its spin angular momentum, **S**:

$$\boldsymbol{\mu} = \gamma\mathbf{S}; \quad [4.156]$$

the proportionality constant, γ, is called the **gyromagnetic ratio.**[30] When a magnetic dipole is placed in a magnetic field **B**, it experiences a torque, $\boldsymbol{\mu} \times \mathbf{B}$, which

---

[30]See, for example, D. Griffiths, *Introduction to Electrodynamics*, 3rd ed. (Prentice Hall, Upper Saddle River, NJ, 1999), page 252. Classically, the gyromagnetic ratio of an object whose charge and mass are identically distributed is $q/2m$, where q is the charge and m is the mass. For reasons that are fully explained only in relativistic quantum theory, the gyromagnetic ratio of the electron is (almost) exactly *twice* the classical value: $\gamma = -e/m$.

tends to line it up parallel to the field (just like a compass needle). The energy associated with this torque is[31]

$$H = -\boldsymbol{\mu} \cdot \mathbf{B}, \qquad [4.157]$$

so the Hamiltonian of a spinning charged particle, at rest[32] in a magnetic field $\mathbf{B}$, is

$$H = -\gamma \mathbf{B} \cdot \mathbf{S}. \qquad [4.158]$$

Example 4.3 **Larmor precession:** Imagine a particle of spin 1/2 at rest in a uniform magnetic field, which points in the z-direction:

$$\mathbf{B} = B_0 \hat{k}. \qquad [4.159]$$

The Hamiltonian (Equation 4.158), in matrix form, is

$$\mathsf{H} = -\gamma B_0 \mathsf{S}_z = -\frac{\gamma B_0 \hbar}{2} \begin{pmatrix} 1 & 0 \\ 0 & -1 \end{pmatrix}. \qquad [4.160]$$

The eigenstates of H are the same as those of S_z:

$$\begin{cases} \chi_+, & \text{with energy } E_+ = -(\gamma B_0 \hbar)/2, \\ \chi_-, & \text{with energy } E_- = +(\gamma B_0 \hbar)/2. \end{cases} \qquad [4.161]$$

Evidently the energy is lowest when the dipole moment is parallel to the field—just as it would be classically.

Since the Hamiltonian is time-independent, the general solution to the time-dependent Schrödinger equation,

$$i\hbar \frac{\partial \chi}{\partial t} = \mathsf{H}\chi, \qquad [4.162]$$

can be expressed in terms of the stationary states:

$$\chi(t) = a\chi_+ e^{-iE_+ t/\hbar} + b\chi_- e^{-iE_- t/\hbar} = \begin{pmatrix} a e^{i\gamma B_0 t/2} \\ b e^{-i\gamma B_0 t/2} \end{pmatrix}.$$

---

[31]Griffiths (footnote 30), page 281.

[32]If the particle is allowed to *move*, there will also be kinetic energy to consider; moreover, it will be subject to the Lorentz force ($q\mathbf{v} \times \mathbf{B}$), which is not derivable from a potential energy function, and hence does not fit the Schrödinger equation as we have formulated it so far. I'll show you later on how to handle this (Problem 4.59), but for the moment let's just assume that the particle is free to *rotate*, but otherwise stationary.

The constants a and b are determined by the initial conditions:

$$\chi(0) = \begin{pmatrix} a \\ b \end{pmatrix},$$

(of course, $|a|^2 + |b|^2 = 1$). With no essential loss of generality[33] I'll write $a = \cos(\alpha/2)$ and $b = \sin(\alpha/2)$, where α is a fixed angle whose physical significance will appear in a moment. Thus

$$\chi(t) = \begin{pmatrix} \cos(\alpha/2)e^{i\gamma B_0 t/2} \\ \sin(\alpha/2)e^{-i\gamma B_0 t/2} \end{pmatrix}. \quad [4.163]$$

To get a feel for what is happening here, let's calculate the expectation value of **S**, as a function of time:

$$\begin{aligned} \langle S_x \rangle = \chi(t)^\dagger \mathbf{S}_x \chi(t) &= \left(\cos(\alpha/2)e^{-i\gamma B_0 t/2} \quad \sin(\alpha/2)e^{i\gamma B_0 t/2}\right) \\ &\times \frac{\hbar}{2}\begin{pmatrix} 0 & 1 \\ 1 & 0 \end{pmatrix}\begin{pmatrix} \cos(\alpha/2)e^{i\gamma B_0 t/2} \\ \sin(\alpha/2)e^{-i\gamma B_0 t/2} \end{pmatrix} \\ &= \frac{\hbar}{2}\sin\alpha\cos(\gamma B_0 t). \end{aligned} \quad [4.164]$$

Similarly,

$$\langle S_y \rangle = \chi(t)^\dagger \mathbf{S}_y \chi(t) = -\frac{\hbar}{2}\sin\alpha\sin(\gamma B_0 t), \quad [4.165]$$

and

$$\langle S_z \rangle = \chi(t)^\dagger \mathbf{S}_z \chi(t) = \frac{\hbar}{2}\cos\alpha. \quad [4.166]$$

Evidently $\langle \mathbf{S} \rangle$ is tilted at a constant angle α to the z-axis, and precesses about the field at the **Larmor frequency**

$$\omega = \gamma B_0, \quad [4.167]$$

just as it would classically[34] (see Figure 4.10). No *surprise* here—Ehrenfest's theorem (in the form derived in Problem 4.20) guarantees that $\langle \mathbf{S} \rangle$ evolves according to the classical laws. But it's nice to see how this works out in a specific context.

---

[33]This does assume that a and b are *real*; you can work out the general case if you like, but all it does is add a constant to t.

[34]See, for instance, *The Feynman Lectures on Physics* (Addison-Wesley, Reading, 1964), Volume II, Section 34-3. Of course, in the classical case it is the angular momentum vector itself, not just its expectation value, that precesses around the magnetic field.

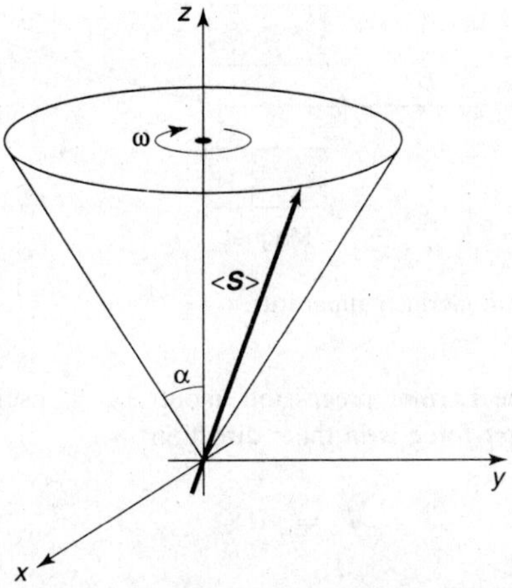

FIGURE 4.10: Precession of $\langle \mathbf{S} \rangle$ in a uniform magnetic field.

Example 4.4 The Stern-Gerlach experiment: In an *inhomogeneous* magnetic field, there is not only a *torque*, but also a *force*, on a magnetic dipole:[35]

$$\mathbf{F} = \nabla(\boldsymbol{\mu} \cdot \mathbf{B}). \quad [4.168]$$

This force can be used to separate out particles with a particular spin orientation, as follows. Imagine a beam of relatively heavy neutral atoms,[36] traveling in the y direction, which passes through a region of inhomogeneous magnetic field (Figure 4.11)—say,

$$\mathbf{B}(x, y, z) = -\alpha x \hat{\imath} + (B_0 + \alpha z)\hat{k}, \quad [4.169]$$

where B_0 is a strong uniform field and the constant α describes a small deviation from homogeneity. (Actually, what we'd *like* is just the z component, but unfortunately that's impossible—it would violate the electromagnetic law $\nabla \cdot \mathbf{B} = 0$; like it or not, an x component comes along for the ride.) The force on these atoms is

$$\mathbf{F} = \gamma\alpha(-S_x\hat{\imath} + S_z\hat{k}).$$

---

[35]Griffiths (footnote 30), page 258. Note that $\mathbf{F}$ is the negative gradient of the energy (Equation 4.157).

[36]We make them neutral so as to avoid the large-scale deflection that would otherwise result from the Lorentz force, and heavy so we can construct localized wave packets and treat the motion in terms of classical particle trajectories. In practice, the Stern-Gerlach experiment doesn't work, for example, with a beam of free electrons.

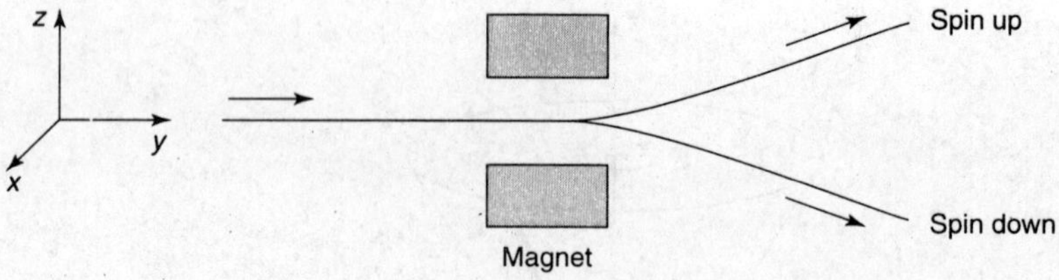

FIGURE 4.11: The Stern-Gerlach apparatus.

But because of the Larmor precession about B_0, S_x oscillates rapidly, and *averages* to zero; the *net* force is in the z direction:

$$F_z = \gamma \alpha S_z, \quad [4.170]$$

and the beam is deflected up or down, in proportion to the z component of the spin angular momentum. *Classically* we'd expect a *smear* (because S_z would not be quantized), but in fact the beam splits into $2s + 1$ separate streams, beautifully demonstrating the quantization of angular momentum. (If you use silver atoms, for example, all the inner electrons are paired, in such a way that their spin and orbital angular momenta cancel. The net spin is simply that of the outermost—unpaired—electron, so in this case $s = 1/2$, and the beam splits in two.)

Now, that argument was purely *classical*, up to the very final step; "force" has no place in a proper quantum calculation, and you might therefore prefer the following approach to the same problem.[37] We examine the process from the perspective of a reference frame that moves along with the beam. In this frame the Hamiltonian starts out zero, turns on for a time T (as the particle passes through the magnet), and then turns off again:

$$H(t) = \begin{cases} 0, & \text{for } t < 0, \\ -\gamma(B_0 + \alpha z)S_z, & \text{for } 0 \leq t \leq T, \\ 0, & \text{for } t > T. \end{cases} \quad [4.171]$$

(I ignore the pesky x component of **B**, which—for reasons indicated above—is irrelevant to the problem.) Suppose the atom has spin 1/2, and starts out in the state

$$\chi(t) = a\chi_+ + b\chi_-, \quad \text{for } t \leq 0.$$

[37]This argument follows L. Ballentine (footnote 26) Section 9.1.

While the Hamiltonian acts, $\chi(t)$ evolves in the usual way:

$$\chi(t) = a\chi_+ e^{-iE_+t/\hbar} + b\chi_- e^{-iE_-t/\hbar}, \quad \text{for } 0 \leq t \leq T,$$

where (from Equation 4.158)

$$E_\pm = \mp\gamma(B_0 + \alpha z)\frac{\hbar}{2}, \tag{4.172}$$

and hence it emerges in the state

$$\chi(t) = \left(ae^{i\gamma T B_0/2}\chi_+\right)e^{i(\alpha\gamma T/2)z} + \left(be^{-i\gamma T B_0/2}\chi_-\right)e^{-i(\alpha\gamma T/2)z}, \tag{4.173}$$

(for $t \geq T$). The two terms now carry *momentum* in the z direction (see Equation 3.32); the spin-up component has momentum

$$p_z = \frac{\alpha\gamma T\hbar}{2}, \tag{4.174}$$

and it moves in the plus-z direction; the spin-down component has the opposite momentum, and it moves in the minus-z direction. Thus the beam splits in two, as before. (Note that Equation 4.174 is consistent with the earlier result (Equation 4.170), for in this case $S_z = \hbar/2$, and $p_z = F_z T$.)

The Stern-Gerlach experiment has played an important role in the philosophy of quantum mechanics, where it serves both as the prototype for the preparation of a quantum state and as an illuminating model for a certain kind of quantum measurement. We tend casually to assume that the *initial* state of a system is *known* (the Schrödinger equation tells us how it subsequently evolves)—but it is natural to wonder how you get a system into a particular state in the first place. Well, if you want to prepare a beam of atoms in a given spin configuration, you pass an unpolarized beam through a Stern-Gerlach magnet, and select the outgoing stream you are interested in (closing off the others with suitable baffles and shutters). Conversely, if you want to *measure* the z component of an atom's spin, you send it through a Stern-Gerlach apparatus, and record which bin it lands in. I do not claim that this is always the most *practical* way to do the job, but it is *conceptually* very clean, and hence a useful context in which to explore the problems of state preparation and measurement.

Problem 4.32 In Example 4.3:

(a) If you measured the component of spin angular momentum along the x direction, at time t, what is the probability that you would get $+\hbar/2$?

(b) Same question, but for the y component.

(c) Same, for the z component.

****Problem 4.33** An electron is at rest in an oscillating magnetic field

$$\mathbf{B} = B_0 \cos(\omega t)\hat{k},$$

where B_0 and ω are constants.

(a) Construct the Hamiltonian matrix for this system.

(b) The electron starts out (at $t = 0$) in the spin-up state with respect to the x-axis (that is: $\chi(0) = \chi_+^{(x)}$). Determine $\chi(t)$ at any subsequent time. *Beware:* This is a time-*dependent* Hamiltonian, so you cannot get $\chi(t)$ in the usual way from stationary states. Fortunately, in this case you can solve the time-dependent Schrödinger equation (Equation 4.162) directly.

(c) Find the probability of getting $-\hbar/2$, if you measure S_x. *Answer:*

$$\sin^2\left(\frac{\gamma B_0}{2\omega}\sin(\omega t)\right).$$

(d) What is the minimum field (B_0) required to force a complete flip in S_x?

4.4.3 Addition of Angular Momenta

Suppose now that we have *two* spin-1/2 particles—for example, the electron and the proton in the ground state[38] of hydrogen. Each can have spin up or spin down, so there are four possibilities in all:[39]

$$\uparrow\uparrow,\ \uparrow\downarrow,\ \downarrow\uparrow,\ \downarrow\downarrow, \qquad [4.175]$$

where the first arrow refers to the electron and the second to the proton. *Question:* What is the *total* angular momentum of the atom? Let

$$\mathbf{S} \equiv \mathbf{S}^{(1)} + \mathbf{S}^{(2)}. \qquad [4.176]$$

---

[38] I put them in the ground state so there won't be any *orbital* angular momentum to worry about.

[39] More precisely, each particle is in a *linear combination* of spin up and spin down, and the composite system is in a *linear combination* of the four states listed.

Each of these four composite states is an eigenstate of S_z—the z components simply *add:*

$$S_z\chi_1\chi_2 = (S_z^{(1)} + S_z^{(2)})\chi_1\chi_2 = (S_z^{(1)}\chi_1)\chi_2 + \chi_1(S_z^{(2)}\chi_2)$$
$$= (\hbar m_1\chi_1)\chi_2 + \chi_1(\hbar m_2\chi_2) = \hbar(m_1 + m_2)\chi_1\chi_2,$$

(note that $\mathbf{S}^{(1)}$ acts only on χ_1, and $\mathbf{S}^{(2)}$ acts only on χ_2; this notation may not be elegant, but it does the job). So m (the quantum number for the composite system) is just $m_1 + m_2$:

$$\uparrow\uparrow:\ m = 1;$$
$$\uparrow\downarrow:\ m = 0;$$
$$\downarrow\uparrow:\ m = 0;$$
$$\downarrow\downarrow:\ m = -1.$$

At first glance, this doesn't look right: m is supposed to advance in integer steps, from $-s$ to $+s$, so it appears that $s = 1$—but there is an "extra" state with $m = 0$. One way to untangle this problem is to apply the lowering operator, $S_- = S_-^{(1)} + S_-^{(2)}$ to the state $\uparrow\uparrow$, using Equation 4.146:

$$S_-(\uparrow\uparrow) = (S_-^{(1)}\uparrow)\uparrow + \uparrow(S_-^{(2)}\uparrow)$$
$$= (\hbar\downarrow)\uparrow + \uparrow(\hbar\downarrow) = \hbar(\downarrow\uparrow + \uparrow\downarrow).$$

Evidently the three states with $s = 1$ are (in the notation $|s\,m\rangle$):

$$\left.\begin{cases} |1\,1\rangle &= \uparrow\uparrow \\ |1\,0\rangle &= \frac{1}{\sqrt{2}}(\uparrow\downarrow + \downarrow\uparrow) \\ |1\,{-1}\rangle &= \downarrow\downarrow \end{cases}\right\} \quad s = 1 \text{ (triplet).} \qquad [4.177]$$

(As a check, try applying the lowering operator to $|1\,0\rangle$; what *should* you get? See Problem 4.34(a).) This is called the **triplet** combination, for the obvious reason. Meanwhile, the orthogonal state with $m = 0$ carries $s = 0$:

$$\left\{|0\,0\rangle = \frac{1}{\sqrt{2}}(\uparrow\downarrow - \downarrow\uparrow)\right\} \quad s = 0 \text{ (singlet).} \qquad [4.178]$$

(If you apply the raising or lowering operator to *this* state, you'll get *zero*. See Problem 4.34(b).)

I claim, then, that the combination of two spin-1/2 particles can carry a total spin of 1 or 0, depending on whether they occupy the triplet or the singlet configuration. To *confirm* this, I need to prove that the triplet states are eigenvectors of S^2 with eigenvalue $2\hbar^2$, and the singlet is an eigenvector of S^2 with eigenvalue 0. Now,

$$S^2 = (\mathbf{S}^{(1)} + \mathbf{S}^{(2)}) \cdot (\mathbf{S}^{(1)} + \mathbf{S}^{(2)}) = (S^{(1)})^2 + (S^{(2)})^2 + 2\mathbf{S}^{(1)} \cdot \mathbf{S}^{(2)}. \qquad [4.179]$$

Using Equations 4.145 and 4.147, we have

$$\begin{aligned}\mathbf{S}^{(1)} \cdot \mathbf{S}^{(2)}(\uparrow\downarrow) &= (S_x^{(1)}\uparrow)(S_x^{(2)}\downarrow) + (S_y^{(1)}\uparrow)(S_y^{(2)}\downarrow) + (S_z^{(1)}\uparrow)(S_z^{(2)}\downarrow)\\ &= \left(\frac{\hbar}{2}\downarrow\right)\left(\frac{\hbar}{2}\uparrow\right) + \left(\frac{i\hbar}{2}\downarrow\right)\left(\frac{-i\hbar}{2}\uparrow\right) + \left(\frac{\hbar}{2}\uparrow\right)\left(\frac{-\hbar}{2}\downarrow\right)\\ &= \frac{\hbar^2}{4}(2\downarrow\uparrow - \uparrow\downarrow).\end{aligned}$$

Similarly,

$$\mathbf{S}^{(1)} \cdot \mathbf{S}^{(2)}(\downarrow\uparrow) = \frac{\hbar^2}{4}(2\uparrow\downarrow - \downarrow\uparrow).$$

It follows that

$$\mathbf{S}^{(1)} \cdot \mathbf{S}^{(2)}|1\,0\rangle = \frac{\hbar^2}{4}\frac{1}{\sqrt{2}}(2\downarrow\uparrow - \uparrow\downarrow + 2\uparrow\downarrow - \downarrow\uparrow) = \frac{\hbar^2}{4}|1\,0\rangle, \qquad [4.180]$$

and

$$\mathbf{S}^{(1)} \cdot \mathbf{S}^{(2)}|0\,0\rangle = \frac{\hbar^2}{4}\frac{1}{\sqrt{2}}(2\downarrow\uparrow - \uparrow\downarrow - 2\uparrow\downarrow + \downarrow\uparrow) = -\frac{3\hbar^2}{4}|0\,0\rangle. \qquad [4.181]$$

Returning to Equation 4.179 (and using Equation 4.142), we conclude that

$$S^2|1\,0\rangle = \left(\frac{3\hbar^2}{4} + \frac{3\hbar^2}{4} + 2\frac{\hbar^2}{4}\right)|1\,0\rangle = 2\hbar^2|1\,0\rangle, \qquad [4.182]$$

so $|1\,0\rangle$ is indeed an eigenstate of S^2 with eigenvalue $2\hbar^2$; and

$$S^2|0\,0\rangle = \left(\frac{3\hbar^2}{4} + \frac{3\hbar^2}{4} - 2\frac{3\hbar^2}{4}\right)|0\,0\rangle = 0, \qquad [4.183]$$

so $|0\,0\rangle$ is an eigenstate of S^2 with eigenvalue 0. (I will leave it for you to confirm that $|1\,1\rangle$ and $|1\,-1\rangle$ are eigenstates of S^2, with the appropriate eigenvalue—see Problem 4.34(c).)

What we have just done (combining spin 1/2 with spin 1/2 to get spin 1 and spin 0) is the simplest example of a larger problem: If you combine spin s_1 with spin s_2, what total spins s can you get?[40] The answer[41] is that you get every spin from $(s_1 + s_2)$ down to $(s_1 - s_2)$—or $(s_2 - s_1)$, if $s_2 > s_1$—in integer steps:

$$\boxed{s = (s_1 + s_2),\ (s_1 + s_2 - 1),\ (s_1 + s_2 - 2),\ \ldots,\ |s_1 - s_2|.} \qquad [4.184]$$

(Roughly speaking, the highest total spin occurs when the individual spins are aligned parallel to one another, and the lowest occurs when they are antiparallel.) For example, if you package together a particle of spin 3/2 with a particle of spin 2, you could get a total spin of 7/2, 5/2, 3/2, or 1/2, depending on the configuration. Another example: If a hydrogen atom is in the state ψ_{nlm}, the net angular momentum of the electron (spin plus orbital) is $l + 1/2$ or $l - 1/2$; if you now throw in spin of the *proton*, the atom's *total* angular momentum quantum number is $l + 1$, l, or $l - 1$ (and l can be achieved in two distinct ways, depending on whether the electron alone is in the $l + 1/2$ configuration or the $l - 1/2$ configuration).

The combined state $|s\,m\rangle$ with total spin s and z-component m will be some linear combination of the composite states $|s_1\,m_1\rangle|s_2\,m_2\rangle$:

$$|s\,m\rangle = \sum_{m_1+m_2=m} C^{s_1 s_2 s}_{m_1 m_2 m} |s_1\,m_1\rangle|s_2\,m_2\rangle \qquad [4.185]$$

(because the z components *add*, the only composite states that contribute are those for which $m_1 + m_2 = m$). Equations 4.177 and 4.178 are special cases of this general form, with $s_1 = s_2 = 1/2$ (I used the informal notation $\uparrow = |\frac{1}{2}\,\frac{1}{2}\rangle$, $\downarrow = |\frac{1}{2}\,(-\frac{1}{2})\rangle$). The constants $C^{s_1 s_2 s}_{m_1 m_2 m}$ are called **Clebsch-Gordan coefficients**. A few of the simplest cases are listed in Table 4.8.[42] For example, the shaded column of the 2×1 table tells us that

$$|3\,0\rangle = \frac{1}{\sqrt{5}}|2\,1\rangle|1\,-1\rangle + \sqrt{\frac{3}{5}}|2\,0\rangle|1\,0\rangle + \frac{1}{\sqrt{5}}|2\,-1\rangle|1\,1\rangle.$$

In particular, if two particles (of spin 2 and spin 1) are at rest in a box, and the *total* spin is 3, and its z component is 0, then a measurement of $S_z^{(1)}$ could return the value $\hbar$ (with probability 1/5), or 0 (with probability 3/5), or $-\hbar$ (with probability

---

[40] I say *spins*, for simplicity, but either one (or both) could just as well be *orbital* angular momentum (for which, however, we would use the letter l).

[41] For a proof you must look in a more advanced text; see, for instance, Claude Cohen-Tannoudji, Bernard Diu, and Franck Laloë, *Quantum Mechanics*, (Wiley, New York, 1977), Vol. 2, Chapter X.

[42] The general formula is derived in Arno Bohm, *Quantum Mechanics: Foundations and Applications*, 2nd ed., (Springer, 1986), p. 172.

TABLE 4.8: Clebsch-Gordan coefficients. (A square root sign is understood for every entry; the minus sign, if present, goes *outside* the radical.)

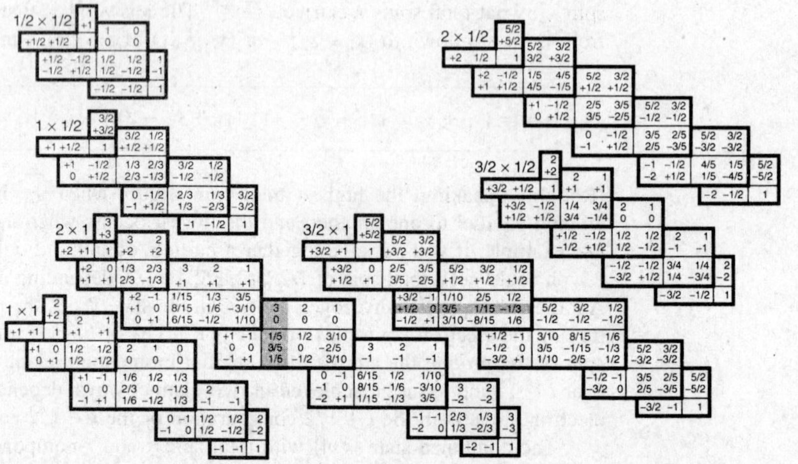

1/5). Notice that the probabilities add up to 1 (the sum of the squares of any column on the Clebsch-Gordan table is 1).

These tables also work the other way around:

$$|s_1\, m_1\rangle|s_2\, m_2\rangle = \sum_s C^{s_1 s_2 s}_{m_1 m_2 m}\,|s\, m\rangle. \qquad [4.186]$$

For example, the shaded *row* in the $3/2 \times 1$ table tells us that

$$|\tfrac{3}{2}\,\tfrac{1}{2}\rangle|1\,0\rangle = \sqrt{\tfrac{3}{5}}\,|\tfrac{5}{2}\,\tfrac{1}{2}\rangle + \sqrt{\tfrac{1}{15}}\,|\tfrac{3}{2}\,\tfrac{1}{2}\rangle - \sqrt{\tfrac{1}{3}}\,|\tfrac{1}{2}\,\tfrac{1}{2}\rangle.$$

If you put particles of spin 3/2 and spin 1 in the box, and you know that the first has $m_1 = 1/2$ and the second has $m_2 = 0$ (so m is necessarily 1/2), and you measure the *total* spin, s, you could get 5/2 (with probability 3/5), or 3/2 (with probability 1/15), or 1/2 (with probability 1/3). Again, the sum of the probabilities is 1 (the sum of the squares of each *row* on the Clebsch-Gordan table is 1).

If you think this is starting to sound like mystical numerology, I don't blame you. We will not be using the Clebsch-Gordan tables much in the rest of the book, but I wanted you to know where they fit into the scheme of things, in case you encounter them later on. In a mathematical sense this is all applied **group theory**—what we are talking about is the decomposition of the direct product of two irreducible representations of the rotation group into a direct sum of irreducible representations (you can quote that, to impress your friends).

∗Problem 4.34

(a) Apply S_- to $|1\,0\rangle$ (Equation 4.177), and confirm that you get $\sqrt{2}\hbar|1\,-1\rangle$.

(b) Apply $S_\pm$ to $|0\,0\rangle$ (Equation 4.178), and confirm that you get zero.

(c) Show that $|1\,1\rangle$ and $|1\,-1\rangle$ (Equation 4.177) are eigenstates of S^2, with the appropriate eigenvalue.

Problem 4.35 **Quarks** carry spin 1/2. Three quarks bind together to make a **baryon** (such as the proton or neutron); two quarks (or more precisely a quark and an antiquark) bind together to make a **meson** (such as the pion or the kaon). Assume the quarks are in the ground state (so the *orbital* angular momentum is zero).

(a) What spins are possible for baryons?

(b) What spins are possible for mesons?

Problem 4.36

(a) A particle of spin 1 and a particle of spin 2 are at rest in a configuration such that the total spin is 3, and its z component is $\hbar$. If you measured the z component of the angular momentum of the spin-2 particle, what values might you get, and what is the probability of each one?

(b) An electron with spin down is in the state ψ_{510} of the hydrogen atom. If you could measure the total angular momentum squared of the electron alone (*not* including the proton spin), what values might you get, and what is the probability of each?

Problem 4.37 Determine the commutator of S^2 with $S_z^{(1)}$ (where $\mathbf{S} \equiv \mathbf{S}^{(1)} + \mathbf{S}^{(2)}$). Generalize your result to show that

$$[S^2, \mathbf{S}^{(1)}] = 2i\hbar(\mathbf{S}^{(1)} \times \mathbf{S}^{(2)}). \qquad [4.187]$$

Comment: Because $S_z^{(1)}$ does not commute with S^2, we cannot hope to find states that are simultaneous eigenvectors of both. In order to form eigenstates of S^2 we need *linear combinations* of eigenstates of $S_z^{(1)}$. This is precisely what the Clebsch-Gordan coefficients (in Equation 4.185) do for us. On the other hand, it follows by obvious inference from Equation 4.187 that the *sum* $\mathbf{S}^{(1)}+\mathbf{S}^{(2)}$ *does* commute with S^2, which is a special case of something we already knew (see Equation 4.103).

FURTHER PROBLEMS FOR CHAPTER 4

∗**Problem 4.38** Consider the **three-dimensional harmonic oscillator**, for which the potential is

$$V(r) = \frac{1}{2}m\omega^2 r^2. \qquad [4.188]$$

(a) Show that separation of variables in cartesian coordinates turns this into three one-dimensional oscillators, and exploit your knowledge of the latter to determine the allowed energies. *Answer:*

$$E_n = (n + 3/2)\hbar\omega. \qquad [4.189]$$

(b) Determine the degeneracy $d(n)$ of E_n.

∗∗∗**Problem 4.39** Because the three-dimensional harmonic oscillator potential (Equation 4.188) is spherically symmetric, the Schrödinger equation can be handled by separation of variables in *spherical* coordinates, as well as cartesian coordinates. Use the power series method to solve the radial equation. Find the recursion formula for the coefficients, and determine the allowed energies. Check your answer against Equation 4.189.

∗∗**Problem 4.40**

(a) Prove the **three-dimensional virial theorem**:

$$2\langle T\rangle = \langle \mathbf{r}\cdot\nabla V\rangle \qquad [4.190]$$

(for stationary states). *Hint:* Refer to Problem 3.31.

(b) Apply the virial theorem to the case of hydrogen, and show that

$$\langle T\rangle = -E_n; \quad \langle V\rangle = 2E_n. \qquad [4.191]$$

(c) Apply the virial theorem to the three-dimensional harmonic oscillator (Problem 4.38), and show that in this case

$$\langle T \rangle = \langle V \rangle = E_n/2. \qquad [4.192]$$

$***$**Problem 4.41** [Attempt this problem only if you are familiar with vector calculus.] Define the (three-dimensional) **probability current** by generalization of Problem 1.14:

$$\mathbf{J} \equiv \frac{i\hbar}{2m}\left(\Psi\,\nabla\Psi^* - \Psi^*\,\nabla\Psi\right). \qquad [4.193]$$

(a) Show that $\mathbf{J}$ satisfies the **continuity equation**

$$\nabla \cdot \mathbf{J} = -\frac{\partial}{\partial t}|\Psi|^2, \qquad [4.194]$$

which expresses local **conservation of probability**. It follows (from the divergence theorem) that

$$\int_{\mathcal{S}} \mathbf{J} \cdot d\mathbf{a} = -\frac{d}{dt}\int_{\mathcal{V}} |\Psi|^2\, d^3\mathbf{r}, \qquad [4.195]$$

where $\mathcal{V}$ is a (fixed) volume and $\mathcal{S}$ is its boundary surface. In words: The flow of probability out through the surface is equal to the decrease in probability of finding the particle in the volume.

(b) Find $\mathbf{J}$ for hydrogen in the state $n = 2$, $l = 1$, $m = 1$. *Answer:*

$$\frac{\hbar}{64\pi m a^5} r e^{-r/a} \sin\theta\,\hat{\phi}.$$

(c) If we interpret $m\mathbf{J}$ as the flow of *mass*, the angular momentum is

$$\mathbf{L} = m\int (\mathbf{r} \times \mathbf{J})\, d^3\mathbf{r}.$$

Use this to calculate L_z for the state ψ_{211}, and comment on the result.

$***$**Problem 4.42** The (time independent) **momentum space wave function** in three dimensions is defined by the natural generalization of Equation 3.54:

$$\phi(\mathbf{p}) \equiv \frac{1}{(2\pi\hbar)^{3/2}}\int e^{-i(\mathbf{p}\cdot\mathbf{r})/\hbar}\,\psi(\mathbf{r})\, d^3\mathbf{r}. \qquad [4.196]$$

(a) Find the momentum space wave function for the ground state of hydrogen (Equation 4.80). *Hint:* Use spherical coordinates, setting the polar axis along the direction of **p**. Do the θ integral first. *Answer:*

$$\phi(\mathbf{p}) = \frac{1}{\pi}\left(\frac{2a}{\hbar}\right)^{3/2} \frac{1}{[1+(ap/\hbar)^2]^2}. \qquad [4.197]$$

(b) Check that $\phi(\mathbf{p})$ is normalized.

(c) Use $\phi(\mathbf{p})$ to calculate $\langle p^2 \rangle$, in the ground state of hydrogen.

(d) What is the expectation value of the kinetic energy in this state? Express your answer as a multiple of E_1, and check that it is consistent with the virial theorem (Equation 4.191).

Problem 4.43

(a) Construct the spatial wave function (ψ) for hydrogen in the state $n = 3$, $l = 2$, $m = 1$. Express your answer as a function of r, θ, ϕ, and a (the Bohr radius) *only*—no other variables (ρ, z, etc.) or functions (Y, v, etc.), or constants (A, c_0, etc.), or derivatives, allowed (π is okay, and e, and 2, etc.).

(b) Check that this wave function is properly normalized, by carrying out the appropriate integrals over r, θ, and ϕ.

(c) Find the expectation value of r^s in this state. For what range of s (positive and negative) is the result finite?

Problem 4.44

(a) Construct the wave function for hydrogen in the state $n = 4$, $l = 3$, $m = 3$. Express your answer as a function of the spherical coordinates r, θ, and ϕ.

(b) Find the expectation value of r in this state. (As always, look up any nontrivial integrals.)

(c) If you could somehow measure the observable $L_x^2 + L_y^2$ on an atom in this state, what value (or values) could you get, and what is the probability of each?

Problem 4.45 What is the probability that an electron in the ground state of hydrogen will be found *inside the nucleus*?

(a) First calculate the *exact* answer, assuming the wave function (Equation 4.80) is correct all the way down to $r = 0$. Let b be the radius of the nucleus.

(b) Expand your result as a power series in the small number $\epsilon \equiv 2b/a$, and show that the lowest-order term is the cubic: $P \approx (4/3)(b/a)^3$. This should be a suitable approximation, provided that $b \ll a$ (which it *is*).

(c) Alternatively, we might assume that $\psi(r)$ is essentially constant over the (tiny) volume of the nucleus, so that $P \approx (4/3)\pi b^3|\psi(0)|^2$. Check that you get the same answer this way.

(d) Use $b \approx 10^{-15}$ m and $a \approx 0.5 \times 10^{-10}$ m to get a numerical estimate for P. Roughly speaking, this represents the "fraction of its time that the electron spends inside the nucleus."

Problem 4.46

(a) Use the recursion formula (Equation 4.76) to confirm that when $l = n - 1$ the radial wave function takes the form

$$R_{n(n-1)} = N_n r^{n-1} e^{-r/na},$$

and determine the normalization constant N_n by direct integration.

(b) Calculate $\langle r \rangle$ and $\langle r^2 \rangle$ for states of the form $\psi_{n(n-1)m}$.

(c) Show that the "uncertainty" in r (σ_r) is $\langle r \rangle/\sqrt{2n+1}$ for such states. Note that the fractional spread in r decreases, with increasing n (in this sense the system "begins to look classical," with identifiable circular "orbits," for large n). Sketch the radial wave functions for several values of n, to illustrate this point.

Problem 4.47 Coincident spectral lines.[43] According to the Rydberg formula (Equation 4.93) the wavelength of a line in the hydrogen spectrum is determined by the principal quantum numbers of the initial and final states. Find two distinct pairs $\{n_i, n_f\}$ that yield the *same* λ. For example, {6851, 6409} and {15283, 11687} will do it, but you're not allowed to use those!

Problem 4.48 Consider the observables $A = x^2$ and $B = L_z$.

(a) Construct the uncertainty principle for $\sigma_A \sigma_B$.

(b) Evaluate σ_B in the hydrogen state ψ_{nlm}.

(c) What can you conclude about $\langle xy \rangle$ in this state?

---

[43] Nicholas Wheeler, "Coincident Spectral Lines" (unpublished Reed College report, 2001).

Problem 4.49 An electron is in the spin state

$$\chi = A\begin{pmatrix} 1-2i \\ 2 \end{pmatrix}.$$

(a) Determine the constant A by normalizing χ.

(b) If you measured S_z on this electron, what values could you get, and what is the probability of each? What is the expectation value of S_z?

(c) If you measured S_x on this electron, what values could you get, and what is the probability of each? What is the expectation value of S_x?

(d) If you measured S_y on this electron, what values could you get, and what is the probability of each? What is the expectation value of S_y?

*****Problem 4.50** Suppose two spin-1/2 particles are known to be in the singlet configuration (Equation 4.178). Let $S_a^{(1)}$ be the component of the spin angular momentum of particle number 1 in the direction defined by the unit vector $\hat{a}$. Similarly, let $S_b^{(2)}$ be the component of 2's angular momentum in the direction $\hat{b}$. Show that

$$\langle S_a^{(1)} S_b^{(2)} \rangle = -\frac{\hbar^2}{4}\cos\theta, \qquad [4.198]$$

where θ is the angle between $\hat{a}$ and $\hat{b}$.

*****Problem 4.51**

(a) Work out the Clebsch-Gordan coefficients for the case $s_1 = 1/2$, $s_2 =$ anything. *Hint:* You're looking for the coefficients A and B in

$$|s\,m\rangle = A|\tfrac{1}{2}\,\tfrac{1}{2}\rangle|s_2\,(m-\tfrac{1}{2})\rangle + B|\tfrac{1}{2}\,(-\tfrac{1}{2})\rangle|s_2\,(m+\tfrac{1}{2})\rangle,$$

such that $|s\,m\rangle$ is an eigenstate of S^2. Use the method of Equations 4.179 through 4.182. If you can't figure out what $S_x^{(2)}$ (for instance) does to $|s_2\,m_2\rangle$, refer back to Equation 4.136 and the line before Equation 4.147. *Answer:*

$$A = \sqrt{\frac{s_2 \pm m + 1/2}{2s_2+1}}; \quad B = \pm\sqrt{\frac{s_2 \mp m + 1/2}{2s_2+1}},$$

where the signs are determined by $s = s_2 \pm 1/2$.

(b) Check this general result against three or four entries in Table 4.8.

Problem 4.52 Find the matrix representing S_x for a particle of spin 3/2 (using, as always, the basis of eigenstates of S_z). Solve the characteristic equation to determine the eigenvalues of S_x.

$***$**Problem 4.53** Work out the spin matrices for arbitrary spin s, generalizing spin 1/2 (Equations 4.145 and 4.147), spin 1 (Problem 4.31), and spin 3/2 (Problem 4.52). *Answer:*

$$\mathsf{S}_z = \hbar \begin{pmatrix} s & 0 & 0 & \cdots & 0 \\ 0 & s-1 & 0 & \cdots & 0 \\ 0 & 0 & s-2 & \cdots & 0 \\ \vdots & \vdots & \vdots & \cdots & \vdots \\ 0 & 0 & 0 & \cdots & -s \end{pmatrix}$$

$$\mathsf{S}_x = \frac{\hbar}{2} \begin{pmatrix} 0 & b_s & 0 & 0 & \cdots & 0 & 0 \\ b_s & 0 & b_{s-1} & 0 & \cdots & 0 & 0 \\ 0 & b_{s-1} & 0 & b_{s-2} & \cdots & 0 & 0 \\ 0 & 0 & b_{s-2} & 0 & \cdots & 0 & 0 \\ \vdots & \vdots & \vdots & \vdots & \ddots & \vdots & \vdots \\ 0 & 0 & 0 & 0 & \cdots & 0 & b_{-s+1} \\ 0 & 0 & 0 & 0 & \cdots & b_{-s+1} & 0 \end{pmatrix}$$

$$\mathsf{S}_y = \frac{\hbar}{2} \begin{pmatrix} 0 & -ib_s & 0 & 0 & \cdots & 0 & 0 \\ ib_s & 0 & -ib_{s-1} & 0 & \cdots & 0 & 0 \\ 0 & ib_{s-1} & 0 & -ib_{s-2} & \cdots & 0 & 0 \\ 0 & 0 & ib_{s-2} & 0 & \cdots & 0 & 0 \\ \vdots & \vdots & \vdots & \vdots & \cdots & \vdots & \vdots \\ 0 & 0 & 0 & 0 & \cdots & 0 & -ib_{-s+1} \\ 0 & 0 & 0 & 0 & \cdots & ib_{-s+1} & 0 \end{pmatrix}$$

where

$$b_j \equiv \sqrt{(s+j)(s+1-j)}.$$

$***$**Problem 4.54** Work out the normalization factor for the spherical harmonics, as follows. From Section 4.1.2 we know that

$$Y_l^m = B_l^m e^{im\phi} P_l^m(\cos\theta);$$

the problem is to determine the factor B_l^m (which I *quoted*, but did not derive, in Equation 4.32). Use Equations 4.120, 4.121, and 4.130 to obtain a recursion relation giving B_l^{m+1} in terms of B_l^m. Solve it by induction on m to get B_l^m up to an overall constant, $C(l)$. Finally, use the result of Problem 4.22 to fix the constant. You may find the following formula for the derivative of an associated Legendre function useful:

$$(1-x^2)\frac{dP_l^m}{dx} = \sqrt{1-x^2}\,P_l^{m+1} - mxP_l^m. \qquad [4.199]$$

Problem 4.55 The electron in a hydrogen atom occupies the combined spin and position state

$$R_{21}\left(\sqrt{1/3}\,Y_1^0\chi_+ + \sqrt{2/3}\,Y_1^1\chi_-\right).$$

(a) If you measured the orbital angular momentum squared (L^2), what values might you get, and what is the probability of each?

(b) Same for the z component of orbital angular momentum (L_z).

(c) Same for the spin angular momentum squared (S^2).

(d) Same for the z component of spin angular momentum (S_z).

Let $\mathbf{J} \equiv \mathbf{L} + \mathbf{S}$ be the *total* angular momentum.

(e) If you measured J^2, what values might you get, and what is the probability of each?

(f) Same for J_z.

(g) If you measured the *position* of the particle, what is the probability density for finding it at r, θ, ϕ?

(h) If you measured both the z component of the spin *and* the distance from the origin (note that these are compatible observables), what is the probability density for finding the particle with spin up and at radius r?

*****Problem 4.56**

(a) For a function $f(\phi)$ that can be expanded in a Taylor series, show that

$$f(\phi + \varphi) = e^{iL_z\varphi/\hbar} f(\phi)$$

(where φ is an arbitrary angle). For this reason, $L_z/\hbar$ is called the **generator of rotations** about the z-axis. *Hint:* Use Equation 4.129, and refer to Problem 3.39.

More generally, $\mathbf{L}\cdot\hat{n}/\hbar$ is the generator of rotations about the direction $\hat{n}$, in the sense that $\exp(i\mathbf{L}\cdot\hat{n}\varphi/\hbar)$ effects a rotation through angle φ (in the right-hand sense) about the axis $\hat{n}$. In the case of *spin*, the generator of rotations is $\mathbf{S}\cdot\hat{n}/\hbar$. In particular, for spin 1/2

$$\chi' = e^{i(\boldsymbol{\sigma}\cdot\hat{n})\varphi/2}\chi \qquad [4.200]$$

tells us how *spinors* rotate.

(b) Construct the (2×2) matrix representing rotation by 180° about the x-axis, and show that it converts "spin up" (χ_+) into "spin down" (χ_-), as you would expect.

(c) Construct the matrix representing rotation by 90° about the y-axis, and check what it does to χ_+.

(d) Construct the matrix representing rotation by 360° about the z-axis. If the answer is not quite what you expected, discuss its implications.

(e) Show that

$$e^{i(\boldsymbol{\sigma}\cdot\hat{n})\varphi/2} = \cos(\varphi/2) + i(\hat{n}\cdot\boldsymbol{\sigma})\sin(\varphi/2). \qquad [4.201]$$

****Problem 4.57** The fundamental commutation relations for angular momentum (Equation 4.99) allow for half-integer (as well as integer) eigenvalues. But for *orbital* angular momentum only the integer values occur. There must be some *extra* constraint in the specific form $\mathbf{L} = \mathbf{r}\times\mathbf{p}$ that excludes half-integer values.[44] Let a be some convenient constant with the dimensions of length (the Bohr radius, say, if we're talking about hydrogen), and define the operators

$$q_1 \equiv \frac{1}{\sqrt{2}}\left[x + (a^2/\hbar)p_y\right]; \quad p_1 \equiv \frac{1}{\sqrt{2}}\left[p_x - (\hbar/a^2)y\right];$$

$$q_2 \equiv \frac{1}{\sqrt{2}}\left[x - (a^2/\hbar)p_y\right]; \quad p_2 \equiv \frac{1}{\sqrt{2}}\left[p_x + (\hbar/a^2)y\right].$$

(a) Verify that $[q_1, q_2] = [p_1, p_2] = 0$; $[q_1, p_1] = [q_2, p_2] = i\hbar$. Thus the q's and the p's satisfy the canonical commutation relations for position and momentum, and those of index 1 are compatible with those of index 2.

(b) Show that

$$L_z = \frac{\hbar}{2a^2}(q_1^2 - q_2^2) + \frac{a^2}{2\hbar}(p_1^2 - p_2^2).$$

---

[44] This problem is based on an argument in Ballentine (footnote 26), page 127.

(c) Check that $L_z = H_1 - H_2$, where each H is the Hamiltonian for a harmonic oscillator with mass $m = \hbar/a^2$ and frequency $\omega = 1$.

(d) We know that the eigenvalues of the harmonic oscillator Hamiltonian are $(n + 1/2)\hbar\omega$, where $n = 0, 1, 2, \ldots$ (In the algebraic theory of Section 2.3.1 this follows from the form of the Hamiltonian and the canonical commutation relations). Use this to conclude that the eigenvalues of L_z must be integers.

Problem 4.58 Deduce the condition for minimum uncertainty in S_x and S_y (that is, *equality* in the expression $\sigma_{S_x}\sigma_{S_y} \geq (\hbar/2)|\langle S_z\rangle|$), for a particle of spin 1/2 in the generic state (Equation 4.139). *Answer:* With no loss of generality we can pick a to be real; then the condition for minimum uncertainty is that b is either pure real or else pure imaginary.

*****Problem 4.59** In classical electrodynamics the force on a particle of charge q moving with velocity $\mathbf{v}$ through electric and magnetic fields $\mathbf{E}$ and $\mathbf{B}$ is given by the **Lorentz force law**:

$$\mathbf{F} = q(\mathbf{E} + \mathbf{v} \times \mathbf{B}). \qquad [4.202]$$

This force cannot be expressed as the gradient of a scalar potential energy function, and therefore the Schrödinger equation in its original form (Equation 1.1) cannot accommodate it. But in the more sophisticated form

$$i\hbar\frac{\partial\Psi}{\partial t} = H\Psi \qquad [4.203]$$

there is no problem; the classical Hamiltonian[45] is

$$H = \frac{1}{2m}(\mathbf{p} - q\mathbf{A})^2 + q\varphi, \qquad [4.204]$$

where $\mathbf{A}$ is the vector potential ($\mathbf{B} = \nabla \times \mathbf{A}$) and φ is the scalar potential ($\mathbf{E} = -\nabla\varphi - \partial\mathbf{A}/\partial t$), so the Schrödinger equation (making the canonical substitution $\mathbf{p} \to (\hbar/i)\nabla$) becomes

$$\boxed{i\hbar\frac{\partial\Psi}{\partial t} = \left[\frac{1}{2m}\left(\frac{\hbar}{i}\nabla - q\mathbf{A}\right)^2 + q\varphi\right]\Psi.} \qquad [4.205]$$

---

[45] See, for example, H. Goldstein, C. P. Poole, and J. L. Safko, *Classical Mechanics*, 3rd ed., (Prentice Hall, Upper Saddle River, NJ, 2002), page 342.

(a) Show that

$$\frac{d\langle \mathbf{r}\rangle}{dt} = \frac{1}{m}\langle(\mathbf{p} - q\mathbf{A})\rangle. \qquad [4.206]$$

(b) As always (see Equation 1.32) we identify $d\langle \mathbf{r}\rangle/dt$ with $\langle \mathbf{v}\rangle$. Show that

$$m\frac{d\langle \mathbf{v}\rangle}{dt} = q\langle \mathbf{E}\rangle + \frac{q}{2m}\langle(\mathbf{p}\times\mathbf{B} - \mathbf{B}\times\mathbf{p})\rangle - \frac{q^2}{m}\langle(\mathbf{A}\times\mathbf{B})\rangle. \qquad [4.207]$$

(c) In particular, if the fields **E** and **B** are *uniform* over the volume of the wave packet, show that

$$m\frac{d\langle \mathbf{v}\rangle}{dt} = q(\mathbf{E} + \langle \mathbf{v}\rangle \times \mathbf{B}), \qquad [4.208]$$

so the *expectation value* of $\langle \mathbf{v}\rangle$ moves according to the Lorentz force law, as we would expect from Ehrenfest's theorem.

****Problem 4.60** [Refer to Problem 4.59 for background.] Suppose

$$\mathbf{A} = \frac{B_0}{2}(x\hat{j} - y\hat{i}), \quad \text{and} \quad \varphi = Kz^2,$$

where B_0 and K are constants.

(a) Find the fields **E** and **B**.

(b) Find the stationary states and the allowed energies, for a particle of mass m and charge q, in these fields. *Answer:*

$$E(n_1, n_2) = (n_1 + \tfrac{1}{2})\hbar\omega_1 + (n_2 + \tfrac{1}{2})\hbar\omega_2, \quad (n_1, n_2 = 0, 1, 2, \ldots), \qquad [4.209]$$

where $\omega_1 \equiv qB_0/m$ and $\omega_2 \equiv \sqrt{2qK/m}$. *Comment:* If $K = 0$ this is the quantum analog to **cyclotron motion**; ω_1 is the classical cyclotron frequency, and it's a free particle in the z direction. The allowed energies, $(n_1 + \frac{1}{2})\hbar\omega_1$, are called **Landau Levels**.[46]

****Problem 4.61** [Refer to Problem 4.59 for background.] In classical electrodynamics the potentials **A** and φ are not uniquely determined;[47] the *physical* quantities are the *fields*, **E** and **B**.

[46] For further discussion see Ballentine (footnote 26) Section 11.3.

[47] See, for example, Griffiths (footnote 30) Section 10.1.2.

(a) Show that the potentials

$$\varphi' \equiv \varphi - \frac{\partial \Lambda}{\partial t}, \quad \mathbf{A}' \equiv \mathbf{A} + \nabla\Lambda \qquad [4.210]$$

(where Λ is an arbitrary real function of position and time) yield the same fields as φ and **A**. Equation 4.210 is called a **gauge transformation**, and the theory is said to be **gauge invariant**.

(b) In quantum mechanics the potentials play a more direct role, and it is of interest to know whether the theory remains gauge invariant. Show that

$$\Psi' \equiv e^{iq\Lambda/\hbar}\Psi \qquad [4.211]$$

satisfies the Schrödinger equation (4.205) with the gauge-transformed potentials φ' and $\mathbf{A}'$. Since Ψ' differs from Ψ only by a *phase factor*, it represents the same physical state,[48] and the theory *is* gauge invariant (see Section 10.2.3 for further discussion).

---

[48]That is to say, $\langle \mathbf{r} \rangle$, $d\langle \mathbf{r} \rangle/dt$, etc. are unchanged. Because Λ depends on position, $\langle \mathbf{p} \rangle$ (with $\mathbf{p}$ represented by the operator $(\hbar/i)\nabla$) *does* change, but as we found in Equation 4.206, $\mathbf{p}$ does not represent the mechanical momentum ($m\mathbf{v}$) in this context (in lagrangian mechanics it is so-called **canonical momentum**).

CHAPTER 5

IDENTICAL PARTICLES

5.1 TWO-PARTICLE SYSTEMS

For a *single* particle, $\Psi(\mathbf{r}, t)$ is a function of the spatial coordinates, $\mathbf{r}$, and the time, t (we'll ignore spin, for the moment). The state of a *two*-particle system is a function of the coordinates of particle one ($\mathbf{r}_1$), the coordinates of particle two ($\mathbf{r}_2$), and the time:

$$\Psi(\mathbf{r}_1, \mathbf{r}_2, t). \tag{5.1}$$

Its time evolution is determined (as always) by the Schrödinger equation:

$$i\hbar\frac{\partial\Psi}{\partial t} = H\Psi, \tag{5.2}$$

where H is the Hamiltonian for the whole system:

$$H = -\frac{\hbar^2}{2m_1}\nabla_1^2 - \frac{\hbar^2}{2m_2}\nabla_2^2 + V(\mathbf{r}_1, \mathbf{r}_2, t) \tag{5.3}$$

(the subscript on ∇ indicates differentiation with respect to the coordinates of particle 1 or particle 2, as the case may be). The statistical interpretation carries over in the obvious way:

$$|\Psi(\mathbf{r}_1, \mathbf{r}_2, t)|^2\, d^3\mathbf{r}_1\, d^3\mathbf{r}_2 \tag{5.4}$$

is the probability of finding particle 1 in the volume $d^3\mathbf{r}_1$ *and* particle 2 in the volume $d^3\mathbf{r}_2$; evidently Ψ must be normalized in such a way that

$$\int |\Psi(\mathbf{r}_1, \mathbf{r}_2, t)|^2\, d^3\mathbf{r}_1\, d^3\mathbf{r}_2 = 1. \tag{5.5}$$

For time-independent potentials, we obtain a complete set of solutions by separation of variables:

$$\Psi(\mathbf{r}_1, \mathbf{r}_2, t) = \psi(\mathbf{r}_1, \mathbf{r}_2)e^{-iEt/\hbar}, \tag{5.6}$$

where the spatial wave function (ψ) satisfies the time-independent Schrödinger equation:

$$-\frac{\hbar^2}{2m_1}\nabla_1^2\psi - \frac{\hbar^2}{2m_2}\nabla_2^2\psi + V\psi = E\psi, \tag{5.7}$$

and E is the total energy of the system.

$**$**Problem 5.1** Typically, the interaction potential depends only on the vector $\mathbf{r} \equiv \mathbf{r}_1 - \mathbf{r}_2$ between the two particles. In that case the Schrödinger equation separates, if we change variables from $\mathbf{r}_1$, $\mathbf{r}_2$ to $\mathbf{r}$ and $\mathbf{R} \equiv (m_1\mathbf{r}_1 + m_2\mathbf{r}_2)/(m_1 + m_2)$ (the center of mass).

(a) Show that $\mathbf{r}_1 = \mathbf{R} + (\mu/m_1)\mathbf{r}$, $\mathbf{r}_2 = \mathbf{R} - (\mu/m_2)\mathbf{r}$, and $\nabla_1 = (\mu/m_2)\nabla_R + \nabla_r$, $\nabla_2 = (\mu/m_1)\nabla_R - \nabla_r$, where

$$\mu \equiv \frac{m_1 m_2}{m_1 + m_2} \tag{5.8}$$

is the **reduced mass** of the system.

(b) Show that the (time-independent) Schrödinger equation becomes

$$-\frac{\hbar^2}{2(m_1 + m_2)}\nabla_R^2\psi - \frac{\hbar^2}{2\mu}\nabla_r^2\psi + V(\mathbf{r})\psi = E\psi.$$

(c) Separate the variables, letting $\psi(\mathbf{R}, \mathbf{r}) = \psi_R(\mathbf{R})\psi_r(\mathbf{r})$. Note that ψ_R satisfies the one-particle Schrödinger equation, with the *total* mass $(m_1 + m_2)$ in place of m, potential zero, and energy E_R, while ψ_r satisfies the one-particle Schrödinger equation with the *reduced* mass in place of m, potential $V(\mathbf{r})$, and energy E_r. The total energy is the sum: $E = E_R + E_r$. What this tells us is that the center of mass moves like a free particle, and the *relative* motion (that is, the motion of particle 2 with respect to particle 1) is the same as if we had a *single* particle with the *reduced* mass, subject to the potential V. Exactly the same decomposition occurs in *classical* mechanics;[1] it reduces the two-body problem to an equivalent one-body problem.

---

[1] See, for example, Jerry B. Marion and Stephen T. Thornton, *Classical Dynamics of Particles and Systems*, 4th ed., Saunders, Fort Worth, TX (1995), Section 8.2.

Problem 5.2 In view of Problem 5.1, we can correct for the motion of the nucleus in hydrogen by simply replacing the electron mass with the reduced mass.

(a) Find (to two significant digits) the percent error in the binding energy of hydrogen (Equation 4.77) introduced by our use of m instead of μ.

(b) Find the separation in wavelength between the red Balmer lines ($n = 3 \rightarrow n = 2$) for hydrogen and deuterium.

(c) Find the binding energy of **positronium** (in which the proton is replaced by a positron—positrons have the same mass as electrons, but opposite charge).

(d) Suppose you wanted to confirm the existence of **muonic hydrogen**, in which the electron is replaced by a muon (same charge, but 206.77 times heavier). Where (i.e., at what wavelength) would you look for the "Lyman-α" line ($n = 2 \rightarrow n = 1$)?

Problem 5.3 Chlorine has two naturally occurring isotopes, Cl^{35} and Cl^{37}. Show that the vibrational spectrum of HCl should consist of closely spaced doublets, with a splitting given by $\Delta\nu = 7.51 \times 10^{-4}\nu$, where ν is the frequency of the emitted photon. *Hint:* Think of it as a harmonic oscillator, with $\omega = \sqrt{k/\mu}$, where μ is the reduced mass (Equation 5.8) and k is presumably the same for both isotopes.

5.1.1 Bosons and Fermions

Suppose particle 1 is in the (one-particle) state $\psi_a(\mathbf{r})$, and particle 2 is in the state $\psi_b(\mathbf{r})$. (Remember: I'm ignoring spin, for the moment.) In that case $\psi(\mathbf{r}_1, \mathbf{r}_2)$ is a simple *product*:[2]

$$\psi(\mathbf{r}_1, \mathbf{r}_2) = \psi_a(\mathbf{r}_1)\psi_b(\mathbf{r}_2). \qquad [5.9]$$

Of course, this assumes that we can tell the particles apart—otherwise it wouldn't make any sense to claim that number 1 is in state ψ_a and number 2 is in state ψ_b; all we could say is that *one* of them is in the state ψ_a and the other is in state ψ_b, but we wouldn't know which is which. If we were talking *classical* mechanics this would be a silly objection: You can *always* tell the particles apart, in principle—just

[2]It is emphatically *not* true that every two-particle wave function is a product of two one-particle wave functions. There exist so-called **entangled states** that *cannot* be decomposed this way. However: If particle 1 is in state a and particle 2 is in state b, *then* the two-particle state is a product. I know what you're thinking: "How could particle 1 *not* be in *some* state, and particle 2 in some other state?" The classic example is the singlet spin configuration (Equation 4.178)—I can't tell you the state of particle 1 by itself, because it is "entangled" (Schrödinger's lovely word) with the state of particle 2. If 2 is measured, and found to be spin *up*, then 1 is spin *down*, but if 2 is spin *down*, then 1 is spin *up*.

paint one of them red and the other one blue, or stamp identification numbers on them, or hire private detectives to follow them around. But in quantum mechanics the situation is fundamentally different: You can't paint an electron red, or pin a label on it, and a detective's observations will inevitably and unpredictably alter its state, raising doubts as to whether the two had perhaps switched places. The fact is, all electrons are *utterly identical*, in a way that no two classical objects can ever be. It's not just that *we* don't happen to know which electron is which; *God* doesn't know which is which, because there is *no such thing* as "this" electron, or "that" electron; all we can legitimately speak about is "an" electron.

Quantum mechanics neatly accommodates the existence of particles that are *indistinguishable in principle*: We simply construct a wave function that is *non-committal* as to which particle is in which state. There are actually *two* ways to do it:

$$\psi_{\pm}(\mathbf{r}_1, \mathbf{r}_2) = A[\psi_a(\mathbf{r}_1)\psi_b(\mathbf{r}_2) \pm \psi_b(\mathbf{r}_1)\psi_a(\mathbf{r}_2)]. \qquad [5.10]$$

Thus the theory admits two kinds of identical particles: **bosons**, for which we use the plus sign, and **fermions**, for which we use the minus sign. Photons and mesons are bosons; protons and electrons are fermions. It so happens that

$$\left\{\begin{array}{l}\text{all particles with } \textit{integer} \text{ spin are bosons, and}\\ \text{all particles with } \textit{half integer} \text{ spin are fermions.}\end{array}\right. \qquad [5.11]$$

This connection between **spin and statistics** (as we shall see, bosons and fermions have quite different statistical properties) can be *proved* in *relativistic* quantum mechanics; in the nonrelativistic theory it is taken as an axiom.[3]

It follows, in particular, that *two identical fermions* (for example, two electrons) *cannot occupy the same state*. For if $\psi_a = \psi_b$, then

$$\psi_{-}(\mathbf{r}_1, \mathbf{r}_2) = A[\psi_a(\mathbf{r}_1)\psi_a(\mathbf{r}_2) - \psi_a(\mathbf{r}_1)\psi_a(\mathbf{r}_2)] = 0,$$

and we are left with no wave function at all.[4] This is the famous **Pauli exclusion principle**. It is not (as you may have been led to believe) a weird ad hoc assumption applying only to electrons, but rather a consequence of the rules for constructing two-particle wave functions, applying to *all* identical fermions.

I assumed, for the sake of argument, that one particle was in the state ψ_a and the other in state ψ_b, but there is a more general (and more sophisticated)

---

[3]It seems bizarre that *relativity* should have anything to do with it, and there has been a lot of discussion recently as to whether it might be possible to prove the spin-statistics connection in other (simpler) ways. See, for example, Robert C. Hilborn, *Am. J. Phys.* **63**, 298 (1995); Ian Duck and E. C. G. Sudarshan, *Pauli and the Spin-Statistics Theorem*, World Scientific, Singapore (1997).

[4]I'm still leaving out the spin, don't forget—if this bothers you (after all, a spinless fermion is a contradiction in terms), assume they're in the *same* spin state. I'll incorporate spin explicitly in a moment.

way to formulate the problem. Let us define the **exchange operator**, P, which interchanges the two particles:

$$Pf(\mathbf{r}_1, \mathbf{r}_2) = f(\mathbf{r}_2, \mathbf{r}_1). \qquad [5.12]$$

Clearly, $P^2 = 1$, and it follows (prove it for yourself) that the eigenvalues of P are ± 1. Now, if the two particles are identical, the Hamiltonian must treat them the same: $m_1 = m_2$ and $V(\mathbf{r}_1, \mathbf{r}_2) = V(\mathbf{r}_2, \mathbf{r}_1)$. It follows that P and H are compatible observables,

$$[P, H] = 0, \qquad [5.13]$$

and hence we can find a complete set of functions that are simultaneous eigenstates of both. That is to say, we can find solutions to the Schrödinger equation that are either symmetric (eigenvalue $+1$) or antisymmetric (eigenvalue -1) under exchange:

$$\boxed{\psi(\mathbf{r}_1, \mathbf{r}_2) = \pm\psi(\mathbf{r}_2, \mathbf{r}_1).} \qquad [5.14]$$

Moreover, if a system starts out in such a state, it will remain in such a state. The *new* law (I'll call it the **symmetrization requirement**) is that for identical particles the wave function is not merely *allowed*, but *required* to satisfy Equation 5.14, with the plus sign for bosons, and the minus sign for fermions.[5] This is the *general* statement, of which Equation 5.10 is a special case.

Example 5.1 Suppose we have two noninteracting—they pass right through one another ... never mind how you would set this up in practice!—particles, both of mass m, in the infinite square well (Section 2.2). The one-particle states are

$$\psi_n(x) = \sqrt{\frac{2}{a}}\sin\left(\frac{n\pi}{a}x\right), \quad E_n = n^2 K$$

(where $K \equiv \pi^2\hbar^2/2ma^2$, for convenience). If the particles are *distinguishable*, with #1 in state n_1 and #2 in state n_2, the composite wave function is a simple product:

$$\psi_{n_1 n_2}(x_1, x_2) = \psi_{n_1}(x_1)\psi_{n_2}(x_2), \quad E_{n_1 n_2} = (n_1^2 + n_2^2)K.$$

[5]It is sometimes suggested that the symmetrization requirement (Equation 5.14) is *forced* by the fact that P and H commute. This is false: It is perfectly possible to imagine a system of two *distinguishable* particles (say, an electron and a positron) for which the Hamiltonian is symmetric, and yet there is no requirement that the wave function be symmetric (or antisymmetric). But *identical* particles *have* to occupy symmetric or antisymmetric states, and this is a completely *new fundamental law*—on a par, logically, with Schrödinger's equation and the statistical interpretation. Of course, there didn't *have* to be any such things as identical particles; it could have been that every single particle in nature was distinguishable from every other one. Quantum mechanics allows for the *possibility* of identical particles, and nature (being lazy) seized the opportunity. (But I'm not complaining—this makes matters enormously simpler!)

For example, the ground state is

$$\psi_{11} = \frac{2}{a}\sin(\pi x_1/a)\sin(\pi x_2/a), \quad E_{11} = 2K;$$

the first excited state is doubly degenerate:

$$\psi_{12} = \frac{2}{a}\sin(\pi x_1/a)\sin(2\pi x_2/a), \quad E_{12} = 5K,$$

$$\psi_{21} = \frac{2}{a}\sin(2\pi x_1/a)\sin(\pi x_2/a), \quad E_{21} = 5K;$$

and so on. If the two particles are identical *bosons*, the ground state is unchanged, but the first excited state is *nondegenerate*:

$$\frac{\sqrt{2}}{a}\left[\sin(\pi x_1/a)\sin(2\pi x_2/a) + \sin(2\pi x_1/a)\sin(\pi x_2/a)\right]$$

(still with energy $5K$). And if the particles are identical *fermions*, there is *no* state with energy $2K$; the ground state is

$$\frac{\sqrt{2}}{a}\left[\sin(\pi x_1/a)\sin(2\pi x_2/a) - \sin(2\pi x_1/a)\sin(\pi x_2/a)\right],$$

and its energy is $5K$.

∗Problem 5.4

(a) If ψ_a and ψ_b are orthogonal, and both normalized, what is the constant A in Equation 5.10?

(b) If $\psi_a = \psi_b$ (and it is normalized), what is A? (This case, of course, occurs only for bosons.)

Problem 5.5

(a) Write down the Hamiltonian for two noninteracting identical particles in the infinite square well. Verify that the fermion ground state given in Example 5.1 is an eigenfunction of H, with the appropriate eigenvalue.

(b) Find the next two excited states (beyond the ones in Example 5.1)—wave functions and energies—for each of the three cases (distinguishable, identical bosons, identical fermions).

5.1.2 Exchange Forces

To give you some sense of what the symmetrization requirement actually *does*, I'm going to work out a simple one-dimensional example. Suppose one particle is in state $\psi_a(x)$, and the other is in state $\psi_b(x)$, and these two states are orthogonal and normalized. If the two particles are distinguishable, and number 1 is the one in state ψ_a, then the combined wave function is

$$\psi(x_1, x_2) = \psi_a(x_1)\psi_b(x_2); \qquad [5.15]$$

if they are identical bosons, the composite wave function is (see Problem 5.4 for the normalization)

$$\psi_+(x_1, x_2) = \frac{1}{\sqrt{2}}[\psi_a(x_1)\psi_b(x_2) + \psi_b(x_1)\psi_a(x_2)]; \qquad [5.16]$$

and if they are identical fermions, it is

$$\psi_-(x_1, x_2) = \frac{1}{\sqrt{2}}[\psi_a(x_1)\psi_b(x_2) - \psi_b(x_1)\psi_a(x_2)]. \qquad [5.17]$$

Let's calculate the expectation value of the square of the separation distance between the two particles,

$$\langle(x_1 - x_2)^2\rangle = \langle x_1^2\rangle + \langle x_2^2\rangle - 2\langle x_1 x_2\rangle. \qquad [5.18]$$

Case 1: Distinguishable particles. For the wavefunction in Equation 5.15,

$$\langle x_1^2\rangle = \int x_1^2|\psi_a(x_1)|^2\, dx_1 \int |\psi_b(x_2)|^2\, dx_2 = \langle x^2\rangle_a$$

(the expectation value of x^2 in the one-particle state ψ_a),

$$\langle x_2^2\rangle = \int |\psi_a(x_1)|^2\, dx_1 \int x_2^2|\psi_b(x_2)|^2\, dx_2 = \langle x^2\rangle_b,$$

and

$$\langle x_1 x_2\rangle = \int x_1|\psi_a(x_1)|^2\, dx_1 \int x_2|\psi_b(x_2)|^2\, dx_2 = \langle x\rangle_a\langle x\rangle_b.$$

In this case, then,

$$\langle(x_1 - x_2)^2\rangle_d = \langle x^2\rangle_a + \langle x^2\rangle_b - 2\langle x\rangle_a\langle x\rangle_b. \qquad [5.19]$$

(Incidentally, the answer would, of course, be the same if particle 1 had been in state ψ_b, and particle 2 in state ψ_a.)

Case 2: Identical particles. For the wave functions in Equations 5.16 and 5.17,

$$\begin{aligned}\langle x_1^2\rangle &= \frac{1}{2}\Big[\int x_1^2|\psi_a(x_1)|^2\,dx_1 \int |\psi_b(x_2)|^2\,dx_2 \\ &+ \int x_1^2|\psi_b(x_1)|^2\,dx_1 \int |\psi_a(x_2)|^2\,dx_2 \\ &\pm \int x_1^2\psi_a(x_1)^*\psi_b(x_1)\,dx_1 \int \psi_b(x_2)^*\psi_a(x_2)\,dx_2 \\ &\pm \int x_1^2\psi_b(x_1)^*\psi_a(x_1)\,dx_1 \int \psi_a(x_2)^*\psi_b(x_2)\,dx_2\Big] \\ &= \frac{1}{2}\Big[\langle x^2\rangle_a + \langle x^2\rangle_b \pm 0 \pm 0\Big] = \frac{1}{2}\Big(\langle x^2\rangle_a + \langle x^2\rangle_b\Big).\end{aligned}$$

Similarly,

$$\langle x_2^2\rangle = \frac{1}{2}\Big(\langle x^2\rangle_b + \langle x^2\rangle_a\Big).$$

(Naturally, $\langle x_2^2\rangle = \langle x_1^2\rangle$, since you can't tell them apart.) But

$$\begin{aligned}\langle x_1x_2\rangle &= \frac{1}{2}\Big[\int x_1|\psi_a(x_1)|^2\,dx_1 \int x_2|\psi_b(x_2)|^2\,dx_2 \\ &+ \int x_1|\psi_b(x_1)|^2\,dx_1 \int x_2|\psi_a(x_2)|^2\,dx_2 \\ &\pm \int x_1\psi_a(x_1)^*\psi_b(x_1)\,dx_1 \int x_2\psi_b(x_2)^*\psi_a(x_2)\,dx_2 \\ &\pm \int x_1\psi_b(x_1)^*\psi_a(x_1)\,dx_1 \int x_2\psi_a(x_2)^*\psi_b(x_2)\,dx_2\Big] \\ &= \frac{1}{2}\big(\langle x\rangle_a\langle x\rangle_b + \langle x\rangle_b\langle x\rangle_a \pm \langle x\rangle_{ab}\langle x\rangle_{ba} \pm \langle x\rangle_{ba}\langle x\rangle_{ab}\big) \\ &= \langle x\rangle_a\langle x\rangle_b \pm |\langle x\rangle_{ab}|^2,\end{aligned}$$

where

$$\langle x\rangle_{ab} \equiv \int x\psi_a(x)^*\psi_b(x)\,dx. \qquad [5.20]$$

Evidently

$$\langle(x_1-x_2)^2\rangle_\pm = \langle x^2\rangle_a + \langle x^2\rangle_b - 2\langle x\rangle_a\langle x\rangle_b \mp 2|\langle x\rangle_{ab}|^2. \qquad [5.21]$$

Comparing Equations 5.19 and 5.21, we see that the difference resides in the final term:

$$\langle(\Delta x)^2\rangle_\pm = \langle(\Delta x)^2\rangle_d \mp 2|\langle x\rangle_{ab}|^2. \qquad [5.22]$$

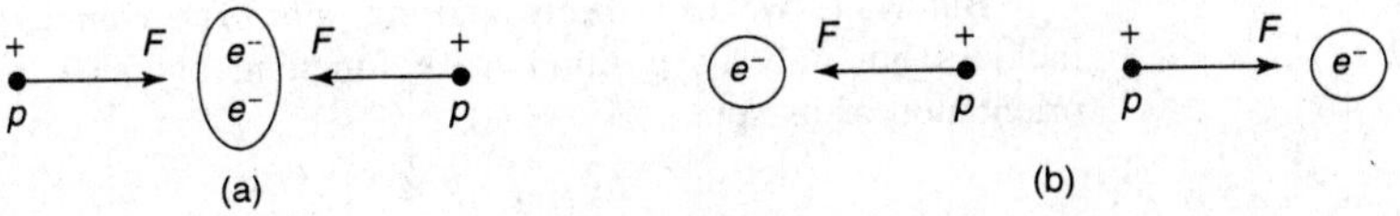

FIGURE 5.1: Schematic picture of the covalent bond: (a) Symmetric configuration produces attractive force. (b) Antisymmetric configuration produces repulsive force.

Identical bosons (the upper signs) tend to be somewhat closer together, and identical fermions (the lower signs) somewhat farther apart, than distinguishable particles in the same two states. Notice that $\langle x \rangle_{ab}$ *vanishes* unless the two wave functions actually *overlap* [if $\psi_a(x)$ is zero wherever $\psi_b(x)$ is *non*zero, the integral in Equation 5.20 is zero]. So if ψ_a represents an electron in an atom in Chicago, and ψ_b represents an electron in an atom in Seattle, it's not going to make any difference whether you antisymmetrize the wave function or not. As a *practical* matter, therefore, it's okay to pretend that electrons with nonoverlapping wave functions are distinguishable. (Indeed, this is the only thing that allows physicists and chemists to proceed at *all*, for in *principle* every electron in the universe is linked to every other one, via the antisymmetrization of their wave functions, and if this really *mattered*, you wouldn't be able to talk about any *one* unless you were prepared to deal with them *all*!)

The *interesting* case is when there *is* some overlap of the wave functions. The system behaves as though there were a "force of attraction" between identical bosons, pulling them closer together, and a "force of repulsion" between identical fermions, pushing them apart (remember that we are for the moment ignoring spin). We call it an **exchange force**, although it's not really a force at all—no physical agency is pushing on the particles; rather, it is a purely *geometrical* consequence of the symmetrization requirement. It is also a strictly quantum mechanical phenomenon, with no classical counterpart. Nevertheless, it has profound consequences. Consider, for example, the hydrogen molecule (H_2). Roughly speaking, the ground state consists of one electron in the atomic ground state (Equation 4.80) centered on nucleus 1, and one electron in the atomic ground state centered at nucleus 2. If electrons were *bosons*, the symmetrization requirement (or, if you like, the "exchange force") would tend to concentrate the electrons toward the middle, between the two protons (Figure 5.1(a)), and the resulting accumulation of negative charge would attract the protons inward, accounting for the **covalent bond**.[6] Unfortunately, electrons *aren't* bosons, they're fermions, and this means that the concentration of negative charge should actually be shifted to the wings (Figure 5.1(b)), tearing the molecule apart!

---

[6] A covalent bond occurs when shared electrons congregate between the nuclei, pulling the atoms together. It need not involve *two* electrons—in Section 7.3 we'll encounter a covalent bond with just *one* electron.

But wait! We have been ignoring *spin*. The *complete* state of the electron includes not only its position wave function, but also a spinor, describing the orientation of its spin:[7]

$$\psi(\mathbf{r})\chi(\mathbf{s}). \qquad [5.23]$$

When we put together the two-electron state, it is the *whole works*, not just the spatial part, that has to be antisymmetric with respect to exchange. Now, a glance back at the composite spin states (Equations 4.177 and 4.178) reveals that the singlet combination is antisymmetric (and hence would have to be joined with a *symmetric* spatial function), whereas the three triplet states are all symmetric (and would require an *antisymmetric* spatial function). Evidently, then, the singlet state should lead to *bonding*, and the triplet to *anti*bonding. Sure enough, the chemists tell us that covalent bonding requires the two electrons to occupy the singlet state, with total spin zero.[8]

$*$**Problem 5.6** Imagine two noninteracting particles, each of mass m, in the infinite square well. If one is in the state ψ_n (Equation 2.28), and the other in state ψ_l $(l \neq n)$, calculate $\langle (x_1 - x_2)^2 \rangle$, assuming (a) they are distinguishable particles, (b) they are identical bosons, and (c) they are identical fermions.

Problem 5.7 Suppose you had *three* particles, one in state $\psi_a(x)$, one in state $\psi_b(x)$, and one in state $\psi_c(x)$. Assuming ψ_a, ψ_b, and ψ_c are orthonormal, construct the three-particle states (analogous to Equations 5.15, 5.16, and 5.17) representing (a) distinguishable particles, (b) identical bosons, and (c) identical fermions. Keep in mind that (b) must be completely symmetric, under interchange of *any* pair of particles, and (c) must be completely *anti*symmetric, in the same sense. *Comment:* There's a cute trick for constructing completely antisymmetric wave functions: Form the **Slater determinant**, whose first row is $\psi_a(x_1)$, $\psi_b(x_1)$, $\psi_c(x_1)$, etc., whose second row is $\psi_a(x_2)$, $\psi_b(x_2)$, $\psi_c(x_2)$, etc., and so on (this device works for any number of particles).

5.2 ATOMS

A neutral atom, of atomic number Z, consists of a heavy nucleus, with electric charge Ze, surrounded by Z electrons (mass m and charge $-e$). The Hamiltonian

---

[7]In the absence of coupling between spin and position, we are free to assume that the state is *separable* in its spin and spatial coordinates. This just says that the probability of getting spin up is independent of the *location* of the particle. In the *presence* of coupling, the general state would take the form of a linear combination: $\psi_+(\mathbf{r})\chi_+ + \psi_-(\mathbf{r})\chi_-$, as in Problem 4.55.

[8]In casual language, it is often said that the electrons are "oppositely aligned" (one with spin up, and the other with spin down). This is something of an oversimplification, since the same could be said of the $m = 0$ triplet state. The precise statement is that they are in the singlet configuration.

for this system is[9]

$$H = \sum_{j=1}^{Z}\left\{-\frac{\hbar^2}{2m}\nabla_j^2 - \left(\frac{1}{4\pi\epsilon_0}\right)\frac{Ze^2}{r_j}\right\} + \frac{1}{2}\left(\frac{1}{4\pi\epsilon_0}\right)\sum_{j\neq k}^{Z}\frac{e^2}{|\mathbf{r}_j - \mathbf{r}_k|}. \qquad [5.24]$$

The term in curly brackets represents the kinetic plus potential energy of the jth electron, in the electric field of the nucleus; the second sum (which runs over all values of j and k except $j = k$) is the potential energy associated with the mutual repulsion of the electrons (the factor of 1/2 in front corrects for the fact that the summation counts each pair twice). The problem is to solve Schrödinger's equation,

$$H\psi = E\psi, \qquad [5.25]$$

for the wave function $\psi(\mathbf{r}_1, \mathbf{r}_2, \ldots, \mathbf{r}_Z)$. Because electrons are identical fermions, however, not all solutions are acceptable: only those for which the complete state (position and spin),

$$\psi(\mathbf{r}_1, \mathbf{r}_2, \ldots, \mathbf{r}_Z)\chi(\mathbf{s}_1, \mathbf{s}_2, \ldots, \mathbf{s}_Z), \qquad [5.26]$$

is antisymmetric with respect to interchange of any two electrons. In particular, no two electrons can occupy the *same* state.

Unfortunately, the Schrödinger equation with Hamiltonian in Equation 5.24 cannot be solved exactly (at any rate, it *hasn't* been), except for the very simplest case, $Z = 1$ (hydrogen). In practice, one must resort to elaborate approximation methods. Some of these we shall explore in Part II; for now I plan only to sketch some qualitative features of the solutions, obtained by neglecting the electron repulsion term altogether. In Section 5.2.1 we'll study the ground state and excited states of helium, and in Section 5.2.2 we'll examine the ground states of higher atoms.

Problem 5.8 Suppose you could find a solution ($\psi(\mathbf{r}_1, \mathbf{r}_2, \ldots, \mathbf{r}_Z)$) to the Schrödinger equation (Equation 5.25), for the Hamiltonian in Equation 5.24. Describe how you would construct from it a completely symmetric function and a completely antisymmetric function, which also satisfy the Schrödinger equation, with the same energy.

---

[9] I'm assuming the nucleus is *stationary*. The trick of accounting for nuclear motion by using the reduced mass (Problem 5.1) works only for the *two*-body problem; fortunately, the nucleus is so much more massive than the electrons that the correction is extremely small even in the case of hydrogen (see Problem 5.2(a)), and it is smaller still for the heavier atoms. There are more interesting effects, due to magnetic interactions associated with electron spin, relativistic corrections, and the finite size of the nucleus. We'll look into these in later chapters, but all of them are minute corrections to the "purely coulombic" atom described by Equation 5.24.

5.2.1 Helium

After hydrogen, the simplest atom is helium ($Z = 2$). The Hamiltonian,

$$H = \left\{ -\frac{\hbar^2}{2m}\nabla_1^2 - \frac{1}{4\pi\epsilon_0}\frac{2e^2}{r_1} \right\} + \left\{ -\frac{\hbar^2}{2m}\nabla_2^2 - \frac{1}{4\pi\epsilon_0}\frac{2e^2}{r_2} \right\} + \frac{1}{4\pi\epsilon_0}\frac{e^2}{|\mathbf{r}_1 - \mathbf{r}_2|}, \quad [5.27]$$

consists of two *hydrogenic* Hamiltonians (with nuclear charge $2e$), one for electron 1 and one for electron 2, together with a final term describing the repulsion of the two electrons. It is this last term that causes all the trouble. If we simply *ignore* it, the Schrödinger equation separates, and the solutions can be written as products of *hydrogen* wave functions:

$$\psi(\mathbf{r}_1, \mathbf{r}_2) = \psi_{nlm}(\mathbf{r}_1)\psi_{n'l'm'}(\mathbf{r}_2), \quad [5.28]$$

only with half the Bohr radius (Equation 4.72), and four times the Bohr energies (Equation 4.70)—if you don't see why, refer back to Problem 4.16. The total energy would be

$$E = 4(E_n + E_{n'}), \quad [5.29]$$

where $E_n = -13.6/n^2$ eV. In particular, the ground state would be

$$\psi_0(\mathbf{r}_1, \mathbf{r}_2) = \psi_{100}(\mathbf{r}_1)\psi_{100}(\mathbf{r}_2) = \frac{8}{\pi a^3}e^{-2(r_1+r_2)/a}, \quad [5.30]$$

(see Equation 4.80), and its energy would be

$$E_0 = 8(-13.6 \text{ eV}) = -109 \text{ eV}. \quad [5.31]$$

Because ψ_0 is a symmetric function, the spin state has to be *antisymmetric*, so the ground state of helium should be a *singlet* configuration, with the spins "oppositely aligned." The *actual* ground state of helium is indeed a singlet, but the experimentally determined energy is -78.975 eV, so the agreement is not very good. But this is hardly surprising: We ignored electron repulsion, which is certainly *not* a small contribution. It is clearly *positive* (see Equation 5.27), which is comforting—evidently it brings the total energy up from -109 to -79 eV (see Problem 5.11).

The excited states of helium consist of one electron in the hydrogenic ground state, and the other in an excited state:

$$\psi_{nlm}\psi_{100}. \quad [5.32]$$

[If you try to put *both* electrons in excited states, one immediately drops to the ground state, releasing enough energy to knock the other one into the continuum ($E > 0$), leaving you with a helium *ion* (He^+) and a free electron. This is an interesting system in its own right—see Problem 5.9—but it is not our present concern.] We can construct both symmetric and antisymmetric combinations, in the usual way (Equation 5.10); the former go with the *antisymmetric* spin configuration (the singlet), and they are called **parahelium**, while the latter require a *symmetric* spin configuration (the triplet), and they are known as **orthohelium**. The ground state is necessarily parahelium; the excited states come in both forms. Because the symmetric spatial state brings the electrons closer together (as we discovered in Section 5.1.2), we expect a higher interaction energy in parahelium, and indeed, it is experimentally confirmed that the parahelium states have somewhat higher energy than their orthohelium counterparts (see Figure 5.2).

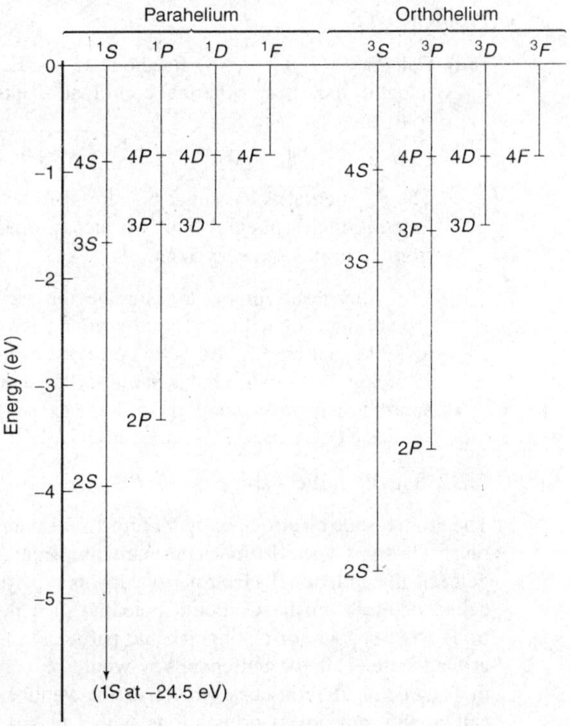

FIGURE 5.2: Energy level diagram for helium (the notation is explained in Section 5.2.2). Note that parahelium energies are uniformly higher than their orthohelium counterparts. The numerical values on the vertical scale are relative to the ground state of ionized helium (He^+): $4 \times (-13.6)$ eV $= -54.4$ eV; to get the *total* energy of the state, subtract 54.4 eV.

Problem 5.9

(a) Suppose you put both electrons in a helium atom into the $n = 2$ state; what would the energy of the emitted electron be?

(b) Describe (quantitatively) the spectrum of the helium ion, He^+.

Problem 5.10 Discuss (qualitatively) the energy level scheme for helium if (a) electrons were identical bosons, and (b) if electrons were distinguishable particles (but with the same mass and charge). Pretend these "electrons" still have spin 1/2, so the spin configurations are the singlet and the triplet.

$**$**Problem 5.11**

(a) Calculate $\langle(1/|\mathbf{r}_1 - \mathbf{r}_2|)\rangle$ for the state ψ_0 (Equation 5.30). *Hint:* Do the $d^3\mathbf{r}_2$ integral first, using spherical coordinates, and setting the polar axis along $\mathbf{r}_1$, so that

$$|\mathbf{r}_1 - \mathbf{r}_2| = \sqrt{r_1^2 + r_2^2 - 2r_1r_2\cos\theta_2}.$$

The θ_2 integral is easy, but be careful to take the *positive root*. You'll have to break the r_2 integral into two pieces, one ranging from 0 to r_1, the other from r_1 to ∞. *Answer:* $5/4a$.

(b) Use your result in (a) to estimate the electron interaction energy in the ground state of helium. Express your answer in electron volts, and add it to E_0 (Equation 5.31) to get a corrected estimate of the ground state energy. Compare the experimental value. (Of course, we're still working with an approximate wave function, so don't expect *perfect* agreement.)

5.2.2 The Periodic Table

The ground state electron configurations for heavier atoms can be pieced together in much the same way. To first approximation (ignoring their mutual repulsion altogether), the individual electrons occupy one-particle hydrogenic states (n, l, m), called **orbitals**, in the Coulomb potential of a nucleus with charge Ze. If electrons were bosons (or distinguishable particles) they would all shake down to the ground state $(1, 0, 0)$, and chemistry would be very dull indeed. But electrons are in fact identical fermions, subject to the Pauli exclusion principle, so only *two* can occupy any given orbital (one with spin up, and one with spin down—or, more precisely, in the singlet configuration). There are n^2 hydrogenic wave functions (all with the same energy E_n) for a given value of n, so the $n = 1$ **shell** has room for 2 electrons, the $n = 2$ shell holds 8, $n = 3$ takes 18, and in general the nth shell can accommodate $2n^2$ electrons. Qualitatively, the horizontal rows on the **Periodic Table** correspond to filling out each shell (if this were the whole story, they would have lengths 2, 8, 18, 32, 50, etc., instead of 2, 8, 8, 18, 18, etc.; we'll see in a moment how the electron-electron repulsion throws the counting off).

With helium, the $n = 1$ shell is filled, so the next atom, lithium ($Z = 3$), has to put one electron into the $n = 2$ shell. Now, for $n = 2$ we can have $l = 0$ or $l = 1$; which of these will the third electron choose? In the absence of electron-electron interactions, they both have the same energy (the Bohr energies depend on n, remember, but not on l). But the effect of electron repulsion is to favor the lowest value of l, for the following reason. Angular momentum tends to throw the electron outward, and the farther out it gets, the more effectively the inner electrons **screen** the nucleus (roughly speaking, the innermost electron "sees" the full nuclear charge Ze, but the outermost electron sees an effective charge hardly greater than e). Within a given shell, therefore, the state with lowest energy (which is to say, the most tightly bound electron) is $l = 0$, and the energy increases with increasing l. Thus the third electron in lithium occupies the orbital $(2, 0, 0)$. The next atom (beryllium, with $Z = 4$) also fits into this state (only with "opposite spin"), but boron ($Z = 5$) has to make use of $l = 1$.

Continuing in this way, we reach neon ($Z = 10$), at which point the $n = 2$ shell is filled, and we advance to the next row of the periodic table and begin to populate the $n = 3$ shell. First there are two atoms (sodium and magnesium) with $l = 0$, and then there are six with $l = 1$ (aluminum through argon). Following argon there "should" be 10 atoms with $n = 3$ and $l = 2$; however, by this time the screening effect is so strong that it overlaps the next shell, so potassium ($Z = 19$) and calcium ($Z = 20$) choose $n = 4$, $l = 0$, in preference to $n = 3$, $l = 2$. After that we drop back to pick up the $n = 3$, $l = 2$ stragglers (scandium through zinc), followed by $n = 4$, $l = 1$ (gallium through krypton), at which point we again make a premature jump to the next row ($n = 5$), and wait until later to slip in the $l = 2$ and $l = 3$ orbitals from the $n = 4$ shell. For details of this intricate counterpoint I refer you to any book on atomic physics.[10]

I would be delinquent if I failed to mention the archaic nomenclature for atomic states, because all chemists and most physicists use it (and the people who make up the Graduate Record Exam *love* this kind of thing). For reasons known best to nineteenth century spectroscopists, $l = 0$ is called s (for "sharp"), $l = 1$ is p (for "principal"), $l = 2$ is d ("diffuse"), and $l = 3$ is f ("fundamental"); after that I guess they ran out of imagination, because it now continues alphabetically (g, h, i, but skip j—just to be utterly perverse, k, l, etc.).[11] The state of a particular electron is represented by the pair nl, with n (the number) giving the shell, and l (the letter)

---

[10]See, for example, U. Fano and L. Fano, *Basic Physics of Atoms and Molecules*, Wiley, New York (1959), Chapter 18, or the classic by G. Herzberg, *Atomic Spectra and Atomic Structure*, Dover, New York (1944).

[11]The shells themselves are assigned equally arbitrary nicknames, starting (don't ask me why) with K: The K shell is $n = 1$, the L shell is $n = 2$, M is $n = 3$, and so on (at least they're in alphabetical order).

specifying the orbital angular momentum; the magnetic quantum number m is not listed, but an exponent is used to indicate the number of electrons that occupy the state in question. Thus the configuration

$$(1s)^2(2s)^2(2p)^2 \qquad [5.33]$$

tells us that there are two electrons in the orbital (1, 0, 0), two in the orbital (2, 0, 0), and two in some combination of the orbitals (2, 1, 1), (2, 1, 0), and (2, 1, −1). This happens to be the ground state of carbon.

In that example there are two electrons with orbital angular momentum quantum number 1, so the *total* orbital angular momentum quantum number, L (capital L, instead of l, to indicate that this pertains to the *total*, not to any one particle) could be 2, 1, or 0. Meanwhile, the two $(1s)$ electrons are locked together in the singlet state, with total spin zero, and so are the two $(2s)$ electrons, but the two $(2p)$ electrons could be in the singlet configuration or the triplet configuration. So the *total* spin quantum number S (capital, again, because it's the *total*) could be 1 or 0. Evidently the *grand* total (orbital plus spin), J, could be 3, 2, 1, or 0. There exist rituals, known as **Hund's Rules** (see Problem 5.13) for figuring out what these totals will be, for a particular atom. The result is recorded as the following hieroglyphic:

$$^{2S+1}L_J, \qquad [5.34]$$

(where S and J are the numbers, and L the letter—capitalized, this time, because we're talking about the *totals*). The ground state of carbon happens to be $^3P_0$: the total spin is 1 (hence the 3), the total orbital angular momentum is 1 (hence the P), and the *grand* total angular momentum is zero (hence the 0). In Table 5.1 the individual configurations and the total angular momenta (in the notation of Equation 5.34) are listed, for the first four rows of the Periodic Table.[12]

---

*Problem 5.12

(a) Figure out the electron configurations (in the notation of Equation 5.33) for the first two rows of the Periodic Table (up to neon), and check your results against Table 5.1.

(b) Figure out the corresponding total angular momenta, in the notation of Equation 5.34, for the first four elements. List all the *possibilities* for boron, carbon, and nitrogen.

---

[12] After krypton—element 36—the situation gets more complicated (fine structure starts to play a significant role in the ordering of the states) so it is not for want of space that the table terminates there.

TABLE 5.1: Ground state electron configurations for the first four rows of the Periodic Table.

Z	Element	Configuration	
1	H	$(1s)$	$^2S_{1/2}$
2	He	$(1s)^2$	$^1S_0$
3	Li	$(\text{He})(2s)$	$^2S_{1/2}$
4	Be	$(\text{He})(2s)^2$	$^1S_0$
5	B	$(\text{He})(2s)^2(2p)$	$^2P_{1/2}$
6	C	$(\text{He})(2s)^2(2p)^2$	$^3P_0$
7	N	$(\text{He})(2s)^2(2p)^3$	$^4S_{3/2}$
8	O	$(\text{He})(2s)^2(2p)^4$	$^3P_2$
9	F	$(\text{He})(2s)^2(2p)^5$	$^2P_{3/2}$
10	Ne	$(\text{He})(2s)^2(2p)^6$	$^1S_0$
11	Na	$(\text{Ne})(3s)$	$^2S_{1/2}$
12	Mg	$(\text{Ne})(3s)^2$	$^1S_0$
13	Al	$(\text{Ne})(3s)^2(3p)$	$^2P_{1/2}$
14	Si	$(\text{Ne})(3s)^2(3p)^2$	$^3P_0$
15	P	$(\text{Ne})(3s)^2(3p)^3$	$^4S_{3/2}$
16	S	$(\text{Ne})(3s)^2(3p)^4$	$^3P_2$
17	Cl	$(\text{Ne})(3s)^2(3p)^5$	$^2P_{3/2}$
18	Ar	$(\text{Ne})(3s)^2(3p)^6$	$^1S_0$
19	K	$(\text{Ar})(4s)$	$^2S_{1/2}$
20	Ca	$(\text{Ar})(4s)^2$	$^1S_0$
21	Sc	$(\text{Ar})(4s)^2(3d)$	$^2D_{3/2}$
22	Ti	$(\text{Ar})(4s)^2(3d)^2$	$^3F_2$
23	V	$(\text{Ar})(4s)^2(3d)^3$	$^4F_{3/2}$
24	Cr	$(\text{Ar})(4s)(3d)^5$	$^7S_3$
25	Mn	$(\text{Ar})(4s)^2(3d)^5$	$^6S_{5/2}$
26	Fe	$(\text{Ar})(4s)^2(3d)^6$	$^5D_4$
27	Co	$(\text{Ar})(4s)^2(3d)^7$	$^4F_{9/2}$
28	Ni	$(\text{Ar})(4s)^2(3d)^8$	$^3F_4$
29	Cu	$(\text{Ar})(4s)(3d)^{10}$	$^2S_{1/2}$
30	Zn	$(\text{Ar})(4s)^2(3d)^{10}$	$^1S_0$
31	Ga	$(\text{Ar})(4s)^2(3d)^{10}(4p)$	$^2P_{1/2}$
32	Ge	$(\text{Ar})(4s)^2(3d)^{10}(4p)^2$	$^3P_0$
33	As	$(\text{Ar})(4s)^2(3d)^{10}(4p)^3$	$^4S_{3/2}$
34	Se	$(\text{Ar})(4s)^2(3d)^{10}(4p)^4$	$^3P_2$
35	Br	$(\text{Ar})(4s)^2(3d)^{10}(4p)^5$	$^2P_{3/2}$
36	Kr	$(\text{Ar})(4s)^2(3d)^{10}(4p)^6$	$^1S_0$

****Problem 5.13**

(a) **Hund's first rule** says that, all other things being equal, the state with the highest total spin (S) will have the lowest energy. What would this predict in the case of the excited states of helium?

(b) **Hund's second rule** says that, for a given spin, the state with the highest total orbital angular momentum (L), consistent with overall antisymmetrization, will have the lowest energy. Why doesn't carbon have $L = 2$? *Hint:* Note that the "top of the ladder" ($M_L = L$) is symmetric.

(c) **Hund's third rule** says that if a subshell (n, l) is no more than half filled, then the lowest energy level has $J = |L - S|$; if it is more than half filled, then $J = L + S$ has the lowest energy. Use this to resolve the boron ambiguity in Problem 5.12(b).

(d) Use Hund's rules, together with the fact that a symmetric spin state must go with an antisymmetric position state (and vice versa) to resolve the carbon and nitrogen ambiguities in Problem 5.12(b). *Hint:* Always go to the "top of the ladder" to figure out the symmetry of a state.

Problem 5.14 The ground state of dysprosium (element 66, in the 6th row of the Periodic Table) is listed as $^5I_8$. What are the total spin, total orbital, and grand total angular momentum quantum numbers? Suggest a likely electron configuration for dysprosium.

5.3 SOLIDS

In the solid state, a few of the loosely bound outermost **valence** electrons in each atom become detached, and roam around throughout the material, no longer subject only to the Coulomb field of a specific "parent" nucleus, but rather to the combined potential of the entire crystal lattice. In this section we will examine two extremely primitive models: first, the electron gas theory of Sommerfeld, which ignores *all* forces (except the confining boundaries), treating the wandering electrons as free particles in a box (the three-dimensional analog to an infinite square well); and second, Bloch's theory, which introduces a periodic potential representing the electrical attraction of the regularly spaced, positively charged, nuclei (but still ignores electron-electron repulsion). These models are no more than the first halting steps toward a quantum theory of solids, but already they reveal the critical role of the Pauli exclusion principle in accounting for "solidity," and provide illuminating insight into the remarkable electrical properties of conductors, semi-conductors, and insulators.

5.3.1 The Free Electron Gas

Suppose the object in question is a rectangular solid, with dimensions l_x, l_y, l_z, and imagine that an electron inside experiences no forces at all, except at the impenetrable walls:

$$V(x, y, z) = \begin{cases} 0, & \text{if } 0 < x < l_x, \quad 0 < y < l_y, \text{ and } 0 < z < l_z; \\ \infty, & \text{otherwise.} \end{cases} \qquad [5.35]$$

The Schrödinger equation,

$$-\frac{\hbar^2}{2m}\nabla^2\psi = E\psi,$$

separates, in cartesian coordinates: $\psi(x, y, z) = X(x)Y(y)Z(z)$, with

$$-\frac{\hbar^2}{2m}\frac{d^2X}{dx^2} = E_x X; \quad -\frac{\hbar^2}{2m}\frac{d^2Y}{dy^2} = E_y Y; \quad -\frac{\hbar^2}{2m}\frac{d^2Z}{dz^2} = E_z Z,$$

and $E = E_x + E_y + E_z$. Letting

$$k_x \equiv \frac{\sqrt{2mE_x}}{\hbar}, \quad k_y \equiv \frac{\sqrt{2mE_y}}{\hbar}, \quad k_z \equiv \frac{\sqrt{2mE_z}}{\hbar},$$

we obtain the general solutions

$$X(x) = A_x \sin(k_x x) + B_x \cos(k_x x), \quad Y(y) = A_y \sin(k_y y) + B_y \cos(k_y y),$$

$$Z(z) = A_z \sin(k_z z) + B_z \cos(k_z z).$$

The boundary conditions require that $X(0) = Y(0) = Z(0) = 0$, so $B_x = B_y = B_z = 0$, and $X(l_x) = Y(l_y) = Z(l_z) = 0$, so that

$$k_x l_x = n_x \pi, \quad k_y l_y = n_y \pi, \quad k_z l_z = n_z \pi, \qquad [5.36]$$

where each n is a positive integer:

$$n_x = 1, 2, 3, \ldots, \quad n_y = 1, 2, 3, \ldots, \quad n_z = 1, 2, 3, \ldots. \qquad [5.37]$$

The (normalized) wave functions are

$$\psi_{n_x n_y n_z} = \sqrt{\frac{8}{l_x l_y l_z}} \sin\left(\frac{n_x \pi}{l_x}x\right) \sin\left(\frac{n_y \pi}{l_y}y\right) \sin\left(\frac{n_z \pi}{l_z}z\right). \qquad [5.38]$$

and the allowed energies are

$$E_{n_x n_y n_z} = \frac{\hbar^2 \pi^2}{2m}\left(\frac{n_x^2}{l_x^2} + \frac{n_y^2}{l_y^2} + \frac{n_z^2}{l_z^2}\right) = \frac{\hbar^2 k^2}{2m}, \qquad [5.39]$$

where k is the magnitude of the **wave vector**, $\mathbf{k} \equiv (k_x, k_y, k_z)$.

If you imagine a three-dimensional space, with axes k_x, k_y, k_z, and planes drawn in at $k_x = (\pi/l_x), (2\pi/l_x), (3\pi/l_x), \ldots$, at $k_y = (\pi/l_y), (2\pi/l_y), (3\pi/l_y), \ldots$, and at $k_z = (\pi/l_z), (2\pi/l_z), (3\pi/l_z), \ldots$, each intersection point represents

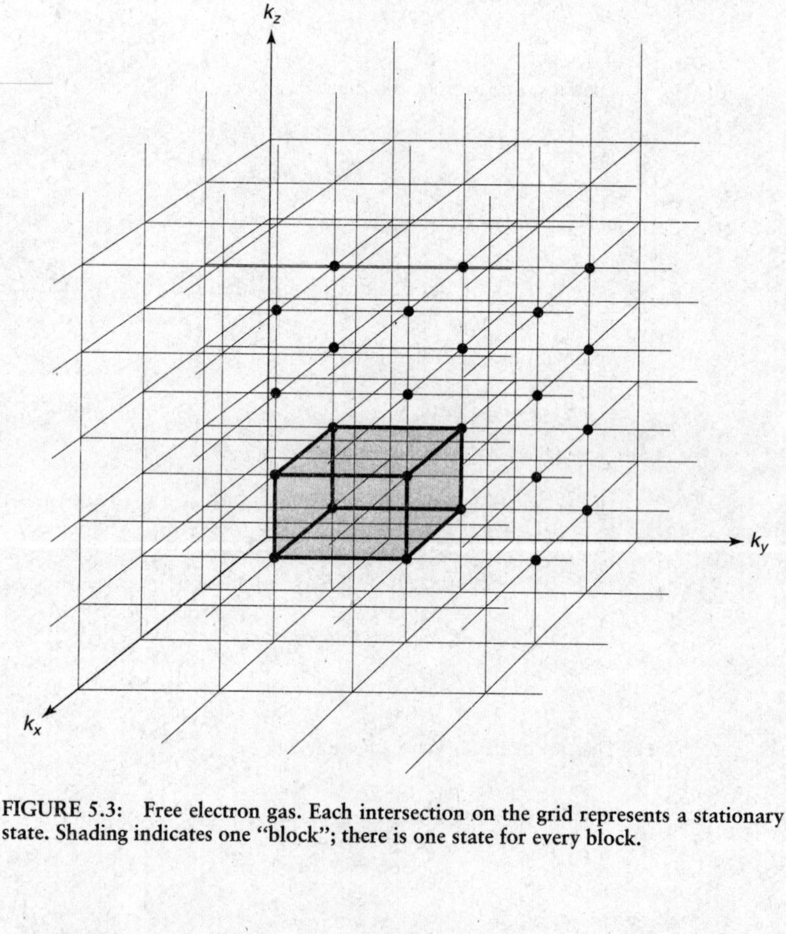

FIGURE 5.3: Free electron gas. Each intersection on the grid represents a stationary state. Shading indicates one "block"; there is one state for every block.

a distinct (one-particle) stationary state (Figure 5.3). Each block in this grid, and hence also each state, occupies a volume

$$\frac{\pi^3}{l_x l_y l_z} = \frac{\pi^3}{V} \qquad [5.40]$$

of "k-space," where $V \equiv l_x l_y l_z$ is the volume of the object itself. Suppose our sample contains N atoms, and each atom contributes q free electrons. (In practice, N will be enormous—on the order of Avogadro's number, for an object of macroscopic size—whereas q is a small number—1 or 2, typically.) If electrons were bosons (or distinguishable particles), they would all settle down to the ground state, ψ_{111}.[13] But electrons are in fact identical *fermions*, subject to the Pauli exclusion principle, so only two of them can occupy any given state. They will fill up one octant of a *sphere* in k-space,[14] whose radius, k_F, is determined by the fact that each pair of electrons requires a volume π^3/V (Equation 5.40):

$$\frac{1}{8}\left(\frac{4}{3}\pi k_F^3\right) = \frac{Nq}{2}\left(\frac{\pi^3}{V}\right).$$

Thus

$$k_F = (3\rho\pi^2)^{1/3}, \qquad [5.41]$$

where

$$\rho \equiv \frac{Nq}{V} \qquad [5.42]$$

is the *free electron density* (the number of free electrons per unit volume).

The boundary separating occupied and unoccupied states, in k-space, is called the **Fermi surface** (hence the subscript F). The corresponding energy is called the **Fermi energy**, E_F; for a free electron gas,

$$E_F = \frac{\hbar^2}{2m}(3\rho\pi^2)^{2/3}. \qquad [5.43]$$

The *total* energy of the electron gas can be calculated as follows: A shell of thickness dk (Figure 5.4) contains a volume

$$\frac{1}{8}(4\pi k^2)\,dk,$$

---

[13]I'm assuming there is no appreciable thermal excitation, or other disturbance, to lift the solid out of its collective ground state. If you like, I'm talking about a "cold" solid, though (as you will show in Problem 5.16(c)), typical solids are still "cold," in this sense, far above room temperature.

[14]Because N is such a huge number, we need not worry about the distinction between the actual jagged edge of the grid and the smooth spherical surface that approximates it.

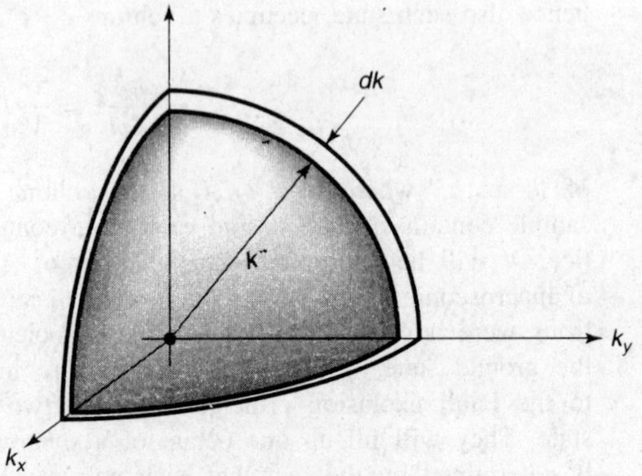

FIGURE 5.4: One octant of a spherical shell in *k*-space.

so the number of electron states in the shell is

$$\frac{2[(1/2)\pi k^2\,dk]}{(\pi^3/V)} = \frac{V}{\pi^2}k^2\,dk.$$

Each of these states carries an energy $\hbar^2k^2/2m$ (Equation 5.39), so the energy of the shell is

$$dE = \frac{\hbar^2k^2}{2m}\frac{V}{\pi^2}k^2\,dk, \qquad [5.44]$$

and hence the total energy is

$$E_{\text{tot}} = \frac{\hbar^2V}{2\pi^2m}\int_0^{k_F}k^4\,dk = \frac{\hbar^2k_F^5V}{10\pi^2m} = \frac{\hbar^2(3\pi^2Nq)^{5/3}}{10\pi^2m}V^{-2/3}. \qquad [5.45]$$

This quantum mechanical energy plays a role rather analogous to the internal *thermal* energy (U) of an ordinary gas. In particular, it exerts a *pressure* on the walls, for if the box expands by an amount dV, the total energy decreases:

$$dE_{\text{tot}} = -\frac{2}{3}\frac{\hbar^2(3\pi^2Nq)^{5/3}}{10\pi^2m}V^{-5/3}\,dV = -\frac{2}{3}E_{\text{tot}}\frac{dV}{V},$$

and this shows up as work done on the outside ($dW = P\,dV$) by the quantum pressure P. Evidently

$$P = \frac{2}{3}\frac{E_{\text{tot}}}{V} = \frac{2}{3}\frac{\hbar^2k_F^5}{10\pi^2m} = \frac{(3\pi^2)^{2/3}\hbar^2}{5m}\rho^{5/3}. \qquad [5.46]$$

Here, then, is a partial answer to the question of why a cold solid object doesn't simply *collapse*: There is a stabilizing internal pressure, having nothing to do with electron-electron repulsion (which we have ignored) or thermal motion (which we have excluded), but is strictly quantum mechanical, and derives ultimately from the antisymmetrization requirement for the wave functions of identical fermions. It is sometimes called **degeneracy pressure**, though "exclusion pressure" might be a better term.[15]

Problem 5.15 Find the average energy per free electron (E_{tot}/Nq), as a fraction of the Fermi energy. *Answer:* $(3/5)E_F$.

Problem 5.16 The density of copper is 8.96 gm/cm^3, and its atomic weight is 63.5 gm/mole.

(a) Calculate the Fermi energy for copper (Equation 5.43). Assume $q = 1$, and give your answer in electron volts.

(b) What is the corresponding electron velocity? *Hint:* Set $E_F = (1/2)mv^2$. Is it safe to assume the electrons in copper are nonrelativistic?

(c) At what temperature would the characteristic thermal energy ($k_B T$, where k_B is the Boltzmann constant and T is the Kelvin temperature) equal the Fermi energy, for copper? *Comment:* This is called the **Fermi temperature**. As long as the *actual* temperature is substantially below the Fermi temperature, the material can be regarded as "cold," with most of the electrons in the lowest accessible state. Since the melting point of copper is 1356 K, solid copper is *always* cold.

(d) Calculate the degeneracy pressure (Equation 5.46) of copper, in the electron gas model.

Problem 5.17 The **bulk modulus** of a substance is the ratio of a small decrease in pressure to the resulting fractional increase in volume:

$$B = -V\frac{dP}{dV}.$$

Show that $B = (5/3)P$, in the free electron gas model, and use your result in Problem 5.16(d) to estimate the bulk modulus of copper. *Comment:* The observed value is 13.4×10^{10} N/m^2, but don't expect perfect agreement—after all, we're

[15] We *derived* Equations 5.41, 5.43, 5.45, and 5.46 for the special case of an infinite rectangular well, but they hold for containers of any shape, as long as the number of particles is extremely large.

neglecting all electron-nucleus and electron-electron forces! Actually, it is rather surprising that this calculation comes as close as it *does*.

5.3.2 Band Structure

We're now going to improve on the free electron model by including the forces exerted on the electrons by the regularly spaced, positively charged, essentially stationary nuclei. The qualitative behavior of solids is dictated to a remarkable degree by the mere fact that this potential is *periodic*—its actual *shape* is relevant only to the finer details. To show you how it goes, I'm going to develop the simplest possible model: a one-dimensional **Dirac comb**, consisting of evenly spaced delta function spikes (Figure 5.5).[16] But first I need to introduce a powerful theorem that vastly simplifies the analysis of periodic potentials.

A periodic potential is one that repeats itself after some fixed distance a:

$$V(x+a) = V(x). \tag{5.47}$$

Bloch's theorem tells us that for such a potential the solutions to the Schrödinger equation,

$$-\frac{\hbar^2}{2m}\frac{d^2\psi}{dx^2} + V(x)\psi = E\psi, \tag{5.48}$$

can be taken to satisfy the condition

$$\psi(x+a) = e^{iKa}\psi(x), \tag{5.49}$$

for some constant K (by "constant" I mean that it is independent of x; it may well depend on E).

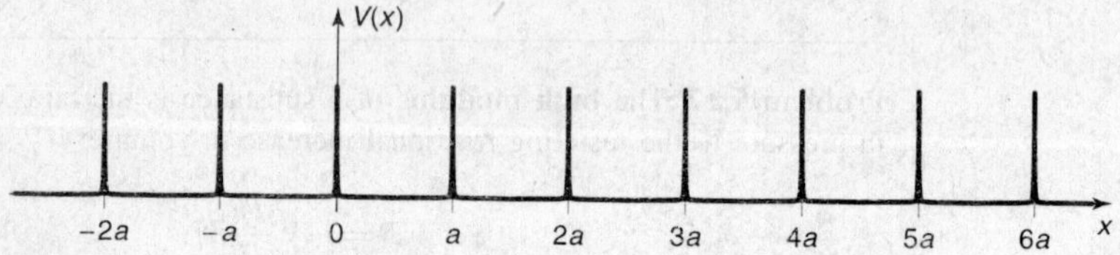

FIGURE 5.5: **The Dirac comb, Equation 5.57.**

[16]It would be more natural to let the delta functions go *down*, so as to represent the attractive force of the nuclei. But then there would be negative energy solutions as well as positive energy solutions, and that makes the calculations more cumbersome (see Problem 5.20). Since all we're trying to do here is explore the consequences of periodicity, it is simpler to adopt this less plausible shape; if it comforts you, think of the nuclei as residing at $\pm a/2, \pm 3a/2, \pm 5a/2, \ldots$.

Proof: Let D be the "displacement" operator:

$$Df(x) = f(x+a). \qquad [5.50]$$

For a periodic potential (Equation 5.47), D commutes with the Hamiltonian:

$$[D, H] = 0, \qquad [5.51]$$

and hence we are free to choose eigenfunctions of H that are simultaneously eigenfunctions of D: $D\psi = \lambda\psi$, or

$$\psi(x+a) = \lambda\psi(x). \qquad [5.52]$$

Now, λ is certainly not *zero* (if it *were*, then—since Equation 5.52 holds for *all* x—we would immediately obtain $\psi(x) = 0$, which is not a permissible eigenfunction); like any nonzero complex number, it can be expressed as an exponential:

$$\lambda = e^{iKa}, \qquad [5.53]$$

for some constant K. QED

At this stage Equation 5.53 is just a strange way to write the eigenvalue λ, but in a moment we will discover that K is in fact *real*, so that although $\psi(x)$ itself is not periodic, $|\psi(x)|^2$ *is*:

$$|\psi(x+a)|^2 = |\psi(x)|^2, \qquad [5.54]$$

as one would certainly expect.[17]

Of course, no *real* solid goes on forever, and the edges are going to spoil the periodicity of $V(x)$, and render Bloch's theorem inapplicable. However, for any macroscopic crystal, containing something on the order of Avogadro's number of atoms, it is hardly imaginable that edge effects can significantly influence the behavior of electrons deep inside. This suggests the following device to salvage Bloch's theorem: We wrap the x-axis around in a circle, and connect it onto its tail, after a large number $N \approx 10^{23}$ of periods; formally, we impose the boundary condition

$$\psi(x+Na) = \psi(x). \qquad [5.55]$$

It follows (from Equation 5.49) that

$$e^{iNKa}\psi(x) = \psi(x),$$

[17] Indeed, you might be tempted to reverse the argument, *starting* with Equation 5.54, as a way of proving Bloch's theorem. It doesn't work, for Equation 5.54 alone would allow the phase factor in Equation 5.49 to be a *function of x*.

so $e^{iNKa} = 1$, or $NKa = 2\pi n$, and hence

$$K = \frac{2\pi n}{Na}, \quad (n = 0, \pm 1, \pm 2, \ldots). \qquad [5.56]$$

In particular, for this arrangement K is necessarily real. The virtue of Bloch's theorem is that we need only solve the Schrödinger Equation *within a single cell* (say, on the interval $0 \leq x < a$); recursive application of Equation 5.49 generates the solution everywhere else.

Now, suppose the potential consists of a long string of delta-function spikes (the Dirac comb):

$$V(x) = \alpha \sum_{j=0}^{N-1} \delta(x - ja). \qquad [5.57]$$

(In Figure 5.5 you must imagine that the x-axis has been "wrapped around," so the Nth spike actually appears at $x = -a$.) No one would pretend that this is a *realistic* model, but remember, it is only the effect of *periodicity* that concerns us here; the classic study[18] used a repeating *rectangular* pattern, and many authors still prefer that one.[19] In the region $0 < x < a$ the potential is zero, so

$$-\frac{\hbar^2}{2m}\frac{d^2\psi}{dx^2} = E\psi,$$

or

$$\frac{d^2\psi}{dx^2} = -k^2\psi,$$

where

$$k \equiv \frac{\sqrt{2mE}}{\hbar}, \qquad [5.58]$$

as usual.

The general solution is

$$\psi(x) = A\sin(kx) + B\cos(kx), \quad (0 < x < a). \qquad [5.59]$$

According to Bloch's theorem, the wave function in the cell immediately to the *left* of the origin is

$$\psi(x) = e^{-iKa}[A\sin k(x + a) + B\cos k(x + a)], \quad (-a < x < 0). \qquad [5.60]$$

---

[18] R. de L. Kronig and W. G. Penney, *Proc. R. Soc. Lond.*, ser. A, **130**, 499 (1930).

[19] See, for instance, D. Park, *Introduction to the Quantum Theory*, 3rd ed., McGraw-Hill, New York (1992).

At $x = 0$, ψ must be continuous, so

$$B = e^{-iKa}[A\sin(ka) + B\cos(ka)]; \qquad [5.61]$$

its derivative suffers a discontinuity proportional to the strength of the delta function (Equation 2.125, with the sign of α switched, since these are spikes instead of wells):

$$kA - e^{-iKa}k[A\cos(ka) - B\sin(ka)] = \frac{2m\alpha}{\hbar^2}B. \qquad [5.62]$$

Solving Equation 5.61 for $A\sin(ka)$ yields

$$A\sin(ka) = [e^{iKa} - \cos(ka)]B. \qquad [5.63]$$

Substituting this into Equation 5.62, and cancelling kB, we find

$$[e^{iKa} - \cos(ka)][1 - e^{-iKa}\cos(ka)] + e^{-iKa}\sin^2(ka) = \frac{2m\alpha}{\hbar^2 k}\sin(ka),$$

which simplifies to

$$\cos(Ka) = \cos(ka) + \frac{m\alpha}{\hbar^2 k}\sin(ka). \qquad [5.64]$$

This is the fundamental result, from which all else follows. For the Kronig-Penney potential (see footnote 18), the formula is more complicated, but it shares the qualitative features we are about to explore.

Equation 5.64 determines the possible values of k, and hence the allowed energies. To simplify the notation, let

$$z \equiv ka, \quad \text{and} \quad \beta \equiv \frac{m\alpha a}{\hbar^2}, \qquad [5.65]$$

so the right side of Equation 5.64 can be written as

$$f(z) \equiv \cos(z) + \beta\frac{\sin(z)}{z}. \qquad [5.66]$$

The constant β is a dimensionless measure of the "strength" of the delta function. In Figure 5.6 I have plotted $f(z)$, for the case $\beta = 10$. The important thing to notice is that $f(z)$ strays outside the range $(-1, +1)$, and in such regions there is no hope of solving Equation 5.64, since $|\cos(Ka)|$, of course, cannot be greater than 1. These **gaps** represent *forbidden* energies; they are separated by **bands** of *allowed* energies. Within a given band, virtually any energy is allowed, since according to Equation 5.56 $Ka = 2\pi n/N$, where N is a huge number, and n can be any integer. You might imagine drawing N horizontal lines on Figure 5.6, at values of $\cos(2\pi n/N)$ ranging from +1 ($n = 0$) down to -1 ($n = N/2$), and back almost to +1 ($n = N - 1$)—at this point the Bloch factor e^{iKa} recycles, so no

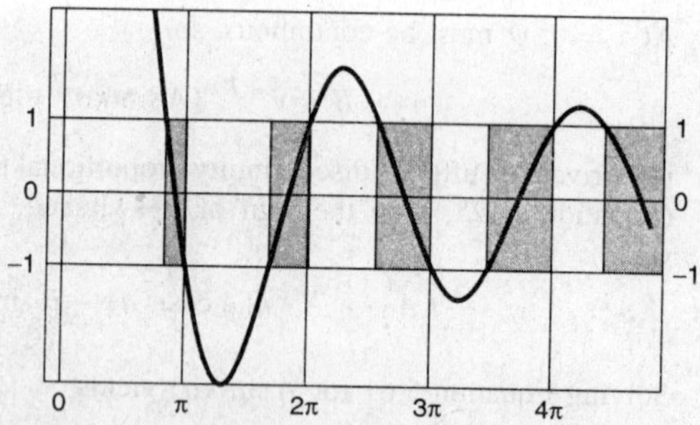

FIGURE 5.6: Graph of $f(z)$ (Equation 5.66) for $\beta = 10$, showing allowed bands (shaded) separated by forbidden gaps (where $|f(z)| > 1$).

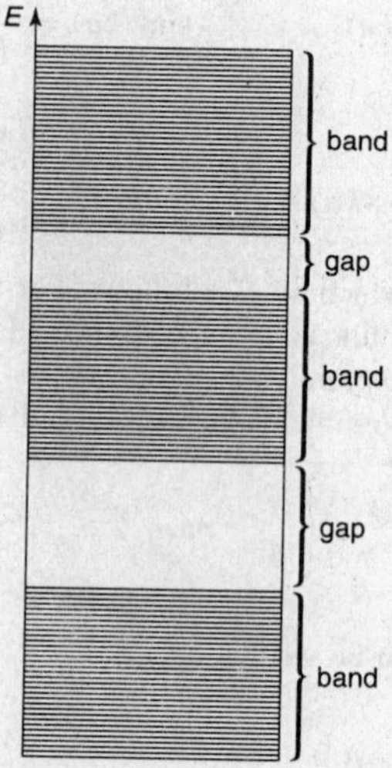

FIGURE 5.7: The allowed energies for a periodic potential form essentially continuous bands.

new solutions are generated by further increasing n. The intersection of each of these lines with $f(z)$ yields an allowed energy. Evidently there are N states in each band, so closely spaced that for most purposes we can regard them as forming a continuum (Figure 5.7).

So far, we've only put *one* electron in our potential. In practice there will be Nq of them, where q is again the number of "free" electrons per atom. Because of the Pauli exclusion principle, only two electrons can occupy a given spatial state, so if $q = 1$, they will half fill the first band, if $q = 2$ they will completely fill the first band, if $q = 3$ they half fill the second band, and so on—in the ground state. (In three dimensions, and with more realistic potentials, the band structure may be more complicated, but the *existence* of allowed bands, separated by forbidden gaps, persists—band structure is the *signature* of a periodic potential.)

Now, if a band is entirely filled, it takes a relatively large energy to excite an electron, since it has to jump across the forbidden zone. Such materials will be electrical **insulators**. On the other hand, if a band is only *partly* filled, it takes very little energy to excite an electron, and such materials are typically **conductors**. If you **dope** an insulator with a few atoms of larger or smaller q, this puts some "extra" electrons into the next higher band, or creates some **holes** in the previously filled one, allowing in either case for weak electric currents to flow; such materials are called **semiconductors**. In the free electron model *all* solids should be excellent conductors, since there are no large gaps in the spectrum of allowed energies. It takes the band theory to account for the extraordinary range of electrical conductivities exhibited by the solids in nature.

Problem 5.18

(a) Using Equations 5.59 and 5.63, show that the wave function for a particle in the periodic delta function potential can be written in the form

$$\psi(x) = C[\sin(kx) + e^{-iKa}\sin k(a-x)], \quad (0 \leq x \leq a).$$

(Don't bother to determine the normalization constant C.)

(b) There is an exception: At the top of a band, where z is an integer multiple of π (Figure 5.6), (a) yields $\psi(x) = 0$. Find the correct wave function for this case. Note what happens to ψ at each delta function.

Problem 5.19 Find the energy at the bottom of the first allowed band, for the case $\beta = 10$, correct to three significant digits. For the sake of argument, assume $\alpha/a = 1$ eV.

$**$**Problem 5.20** Suppose we use delta function *wells*, instead of *spikes* (i.e., switch the sign of α in Equation 5.57). Analyze this case, constructing the analogs to Figures 5.6 and 5.7. This requires no new calculation for the positive energy solutions (just make the appropriate modification in Equation 5.66), but you *do* need to work out the negative energy solutions—and be sure to include them on your graph (which will now extend to negative z). How many states are there in the first allowed band?

Problem 5.21 Show that *most* of the energies determined by Equation 5.64 are doubly degenerate. What are the exceptional cases? *Hint:* Try it for $N = 1, 2, 3, 4, \ldots$, to see how it goes. What are the possible values of $\cos(Ka)$ in each case?

5.4 QUANTUM STATISTICAL MECHANICS

At absolute zero, a physical system occupies its lowest allowed energy configuration. As we turn up the temperature, random thermal activity will begin to populate the excited states, and this raises the following question: If we have a large number N of particles, in thermal equilibrium at temperature T, what is the probability that a particle, selected at random, would be found to have the specific energy, E_j? Note that the "probability" in question has nothing to do with quantum indeterminacy—exactly the same question arises in *classical* statistical mechanics. The reason we must be content with a *probabilistic* answer is that we are typically dealing with enormous numbers of particles, and we could not possibly expect to keep track of each one separately, whether or *not* the underlying mechanics is deterministic.

The **fundamental assumption of statistical mechanics** is that in **thermal equilibrium** every distinct state with the same *total* energy, E, is equally probable. Random thermal motions constantly shift energy from one particle to another, and from one form (rotational, kinetic, vibrational, etc.) to another, but (absent external influences) the *total* is fixed by conservation of energy. The assumption (and it's a *deep* one, worth thinking about) is that this continual redistribution of energy does not favor any particular state. The **temperature**, T, is simply a measure of the total energy of a system in thermal equilibrium. The only new twist introduced by quantum mechanics has to do with *how we count the distinct states* (it's actually *easier* than in the classical theory, because the states are generally discrete), and this depends critically on whether the particles involved are distinguishable, identical bosons, or identical fermions. The arguments are relatively straightforward, but the *arithmetic* gets pretty dense, so I'm going to begin with an absurdly simple example, so you'll have a clear sense of what is at issue when we come to the general case.

5.4.1 An Example

Suppose we have just *three* noninteracting particles (all of mass m) in the one-dimensional infinite square well (Section 2.2). The total energy is

$$E = E_A + E_B + E_C = \frac{\pi^2\hbar^2}{2ma^2}(n_A^2 + n_B^2 + n_C^2) \qquad [5.67]$$

(see Equation 2.27), where n_A, n_B, and n_C are positive integers. Now suppose, for the sake of argument, that $E = 363(\pi^2\hbar^2/2ma^2)$, which is to say,

$$n_A^2 + n_B^2 + n_C^2 = 363. \qquad [5.68]$$

There are, as it happens, 13 combinations of three positive integers, the sum of whose squares is 363: All three could be 11, two could be 13 and one 5 (which occurs in three permutations), one could be 19 and two 1 (again, three permutations), or one could be 17, one 7, and one 5 (six permutations). Thus (n_A, n_B, n_C) is one of the following:

$$(11, 11, 11),$$
$$(13, 13, 5),\ (13, 5, 13),\ (5, 13, 13),$$
$$(1, 1, 19),\ (1, 19, 1),\ (19, 1, 1),$$
$$(5, 7, 17),\ (5, 17, 7),\ (7, 5, 17),\ (7, 17, 5),\ (17, 5, 7),\ (17, 7, 5).$$

If the particles are *distinguishable*, each of these represents a distinct quantum state, and the fundamental assumption of statistical mechanics says that in thermal equilibrium[20] they are all equally likely. But I'm not interested in knowing *which* particle is in *which* (one-particle) state, only the total *number* of particles in each state—the **occupation number**, N_n, for the state ψ_n. The collection of all occupation numbers for a given 3-particle state we will call the **configuration**. If all three are in ψ_{11}, the configuration is

$$(0, 0, 0, 0, 0, 0, 0, 0, 0, 0, 3, 0, 0, 0, 0, 0, 0, 0, 0, \ldots), \qquad [5.69]$$

(i.e., $N_{11} = 3$, all others zero). If two are in ψ_{13} and one is in ψ_5, the configuration is

$$(0, 0, 0, 0, 1, 0, 0, 0, 0, 0, 0, 0, 2, 0, 0, 0, 0, \ldots), \qquad [5.70]$$

(i.e., $N_5 = 1$, $N_{13} = 2$, all others zero). If two are in ψ_1 and one is in ψ_{19}, the configuration is

$$(2, 0, 0, 0, 0, 0, 0, 0, 0, 0, 0, 0, 0, 0, 0, 0, 0, 0, 1, 0, \ldots), \qquad [5.71]$$

(i.e., $N_1 = 2$, $N_{19} = 1$, all others zero). And if there is one particle in ψ_5, one in ψ_7, and one in ψ_{17}, the configuration is

$$(0, 0, 0, 0, 1, 0, 1, 0, 0, 0, 0, 0, 0, 0, 0, 0, 1, 0, 0, \ldots), \qquad [5.72]$$

(i.e., $N_5 = N_7 = N_{17} = 1$, all others zero.) Of these, the last is the *most probable* configuration, because it can be achieved in six different ways, whereas the middle two occur three ways, and the first only one.

---

[20] How the particles maintain thermal equilibrium, if they really don't interact at all, is a problem I'd rather not worry about—maybe God reaches in periodically and stirs things up (being careful not to add or remove any energy). In real life, of course, the continual redistribution of energy is caused precisely by interactions between the particles, so if you don't approve of divine intervention let there be extremely weak interactions—sufficient to thermalize the system (at least, over long time periods), but too small to alter the stationary states and the allowed energies appreciably.

Returning now to my original question, if we select one of these three particles at random, what is the probability (P_n) of getting a specific (allowed) energy E_n? The only way you can get E_1 is if it's in the third configuration (Equation 5.71); the chances of the system being in that configuration are 3 in 13, and in that configuration the probability of getting E_1 is 2/3, so $P_1 = (3/13) \times (2/3) = 2/13$. You could get E_5 either from configuration 2 (Equation 5.70)—chances 3 in 13—with probability 1/3, or from configuration 4 (Equation 5.72)—chances 6 in 13—with probability 1/3, so $P_5 = (3/13) \times (1/3) + (6/13) \times (1/3) = 3/13$. You can only get E_7 from configuration 4: $P_7 = (6/13) \times (1/3) = 2/13$. Likewise, E_{11} comes only from the first configuration (Equation 5.69)—chances 1 in 13—with probability 1: $P_{11} = (1/13)$. Similarly, $P_{13} = (3/13) \times (2/3) = 2/13$, $P_{17} = (6/13) \times (1/3) = 2/13$, and $P_{19} = (3/13) \times (1/3) = 1/13$. As a check, note that

$$P_1 + P_5 + P_7 + P_{11} + P_{13} + P_{17} + P_{19} = \frac{2}{13} + \frac{3}{13} + \frac{2}{13} + \frac{1}{13} + \frac{2}{13} + \frac{2}{13} + \frac{1}{13} = 1.$$

That's when the particles are distinguishable. If in fact they are *identical fermions*, the antisymmetrization requirement (leaving aside spin, for simplicity—or assuming they are all in the *same* spin state, if you prefer) excludes the first three configurations (which assign two—or, worse still, three—particles to the same state), and there is just *one* state in the fourth configuration (see Problem 5.22(a)). For identical fermions, then, $P_5 = P_7 = P_{17} = 1/3$ (and again the sum of the probabilities is 1). On the other hand, if they are *identical bosons* the symmetrization requirement allows for *one* state in each configuration (see Problem 5.22(b)), so $P_1 = (1/4) \times (2/3) = 1/6$, $P_5 = (1/4) \times (1/3) + (1/4) \times (1/3) = 1/6$, $P_7 = (1/4) \times (1/3) = 1/12$, $P_{11} = (1/4) \times (1) = 1/4$, $P_{13} = (1/4) \times (2/3) = 1/6$, $P_{17} = (1/4) \times (1/3) = 1/12$, and $P_{19} = (1/4) \times (1/3) = 1/12$. As always, the sum is 1.

The purpose of this example was to show you how the counting of states depends on the nature of the particles. In one respect it was actually *more complicated* than the realistic situation, in which N is a huge number. For as N grows, the most probable configuration (in this example, $N_5 = N_7 = N_{17} = 1$, for the case of distinguishable particles) becomes *overwhelmingly* more likely than its competitors, so that, for statistical purposes, we can afford to ignore the others altogether:[21] *The distribution of individual particle energies, at equilibrium, is simply their distribution in the most probable configuration.* (If this were true for $N = 3$—which, obviously, it is *not*—we would conclude that $P_5 = P_7 = P_{17} = 1/3$ for the case of distinguishable particles.) I'll return to this point in Section 5.4.3, but first we need to generalize the counting procedure itself.

---

[21] This is an astonishing and counterintuitive fact about the statistics of large numbers. For a good discussion see Ralph Baierlein, *Thermal Physics*, Cambridge U.P. (1999), Section 2.1.

∗**Problem 5.22**

(a) Construct the completely antisymmetric wave function $\psi(x_A, x_B, x_C)$ for three identical fermions, one in the state ψ_5, one in the state ψ_7, and one in the state ψ_{17}.

(b) Construct the completely symmetric wave function $\psi(x_A, x_B, x_C)$ for three identical bosons, (i) if all three are in state ψ_{11}, (ii) if two are in state ψ_1 and one is in state ψ_{19}, and (iii) if one is in the state ψ_5, one in the state ψ_7, and one in the state ψ_{17}.

∗**Problem 5.23** Suppose you had three (noninteracting) particles, in thermal equilibrium, in a one-dimensional harmonic oscillator potential, with a total energy $E = (9/2)\hbar\omega$.

(a) If they are distinguishable particles (but all with the same mass), what are the possible occupation-number configurations, and how many distinct (three-particle) states are there for each one? What is the most probable configuration? If you picked a particle at random and measured its energy, what values might you get, and what is the probability of each one? What is the most probable energy?

(b) Do the same for the case of identical fermions (ignoring spin, as we did in Section 5.4.1).

(c) Do the same for the case of identical bosons (ignoring spin).

5.4.2 The General Case

Now consider an arbitrary potential, for which the one-particle energies are E_1, E_2, $E_3, \ldots$, with degeneracies $d_1, d_2, d_3, \ldots$ (i.e., there are d_n distinct one-particle states with energy E_n). Suppose we put N particles (all with the same mass) into this potential; we are interested in the configuration $(N_1, N_2, N_3, \ldots)$, for which there are N_1 particles with energy E_1, N_2 particles with energy E_2, and so on. *Question:* How many different ways can this be achieved (or, more precisely, how many distinct states correspond to this particular configuration)? The answer, $Q(N_1, N_2, N_3, \ldots)$, depends on whether the particles are distinguishable, identical fermions, or identical bosons, so we'll treat the three cases separately.[22]

First, assume the particles are *distinguishable*. How many ways are there to select (from the N available candidates) the N_1 to be placed in the first "bin"?

[22]The presentation here follows closely that of Amnon Yariv, *An Introduction to Theory and Applications of Quantum Mechanics*, Wiley, New York (1982).

Answer: the **binomial coefficient**, "N choose N_1,"

$$\binom{N}{N_1} \equiv \frac{N!}{N_1!(N-N_1)!}. \qquad [5.73]$$

For there are N ways to pick the first particle, leaving $(N-1)$ for the second, and so on:

$$N(N-1)(N-2)\ldots(N-N_1+1) = \frac{N!}{(N-N_1)!}.$$

However, this counts separately the $N_1!$ different *permutations* of the N_1 particles, whereas we don't care whether number 37 was picked on the first draw, or on the 29th draw; so we divide by $N_1!$, confirming Equation 5.73. Now, how many different ways can those N_1 particles be arranged *within* the first bin? Well, there are d_1 states in the bin, so each particle has d_1 choices; evidently there are $(d_1)^{N_1}$ possibilities in all. Thus the number of ways to put N_1 particles, selected from a total population of N, into a bin containing d_1 distinct options, is

$$\frac{N!d_1^{N_1}}{N_1!(N-N_1)!}.$$

The same goes for bin 2, of course, except that there are now only $(N-N_1)$ particles left to work with:

$$\frac{(N-N_1)!d_2^{N_2}}{N_2!(N-N_1-N_2)!};$$

and so on. It follows that

$$Q(N_1, N_2, N_3, \ldots)$$
$$= \frac{N!d_1^{N_1}}{N_1!(N-N_1)!}\frac{(N-N_1)!d_2^{N_2}}{N_2!(N-N_1-N_2)!}\frac{(N-N_1-N_2)!d_3^{N_3}}{N_3!(N-N_1-N_2-N_3)!}\cdots$$
$$= N!\frac{d_1^{N_1}d_2^{N_2}d_3^{N_3}\ldots}{N_1!N_2!N_3!\ldots} = N!\prod_{n=1}^{\infty}\frac{d_n^{N_n}}{N_n!}. \qquad [5.74]$$

(You should pause right now and *check* this result, for the example in Section 5.4.1—see Problem 5.24.)

The problem is a lot easier for *identical fermions*. Because they are indistinguishable, it doesn't matter *which* particles are in *which* states—the antisymmetrization requirement means that there is just *one* N-particle state in which a specific set of one-particle states are occupied. Moreover, only one particle can occupy any given state. There are

$$\binom{d_n}{N_n}$$

ways to choose the N_n occupied states in the nth bin,[23] so

$$Q(N_1, N_2, N_3, \ldots) = \prod_{n=1}^{\infty} \frac{d_n!}{N_n!(d_n - N_n)!}. \qquad [5.75]$$

(Check it for the example in Section 5.4.1—see Problem 5.24.)

The calculation is hardest for the case of *identical bosons*. Again, the symmetrization requirement means that there is just one N-particle state in which a specific set of one-particle states are occupied, but this time there is no restriction on the number of particles that can share the same one-particle state. For the nth bin, the question becomes: How many different ways can we assign N_n identical particles to d_n different slots? There are many tricks to solve this combinatorial problem; an especially clever method is as follows: Let dots represent particles and crosses represent partitions, so that, for example, if $d_n = 5$ and $N_n = 7$,

$$\bullet\ \bullet\ \times\ \bullet\ \times\ \bullet\ \bullet\ \bullet\ \times\ \bullet\ \times$$

would indicate that there are two particles in the first state, one in the second, three in the third, one in the fourth, and none in the fifth. Note that there are N_n dots, and $(d_n - 1)$ crosses (partitioning the dots into d_n groups). If the individual dots and crosses were *labeled*, there would be $(N_n + d_n - 1)!$ different ways to arrange them. But for our purposes the dots are all equivalent—permuting them ($N_n!$ ways) does not change the state. Likewise, the crosses are all equivalent—permuting them ($(d_n - 1)!$ ways) changes nothing. So there are in fact

$$\frac{(N_n + d_n - 1)!}{N_n!(d_n - 1)!} = \binom{N_n + d_n - 1}{N_n} \qquad [5.76]$$

distinct ways of assigning the N_n particles to the d_n one-particle states in the nth bin, and we conclude that

$$Q(N_1, N_2, N_3, \ldots) = \prod_{n=1}^{\infty} \frac{(N_n + d_n - 1)!}{N_n!(d_n - 1)!}. \qquad [5.77]$$

(Check it for the Example in Section 5.4.1—see Problem 5.24.)

***Problem 5.24** Check Equations 5.74, 5.75, and 5.77, for the example in Section 5.4.1.

****Problem 5.25** Obtain Equation 5.76 by induction. The combinatorial question is this: How many different ways can you put N identical balls into d baskets (never

[23]This should be zero, of course, if $N_n > d_n$, and it *is*, provided we consider the factorial of a negative integer to be infinite.

mind the subscript n for this problem). You could stick all N of them into the third basket, or all but one in the second basket and one in the fifth, or two in the first and three in the third and all the rest in the seventh, etc. Work it out explicitly for the cases $N = 1$, $N = 2$, $N = 3$, and $N = 4$; by that stage you should be able to deduce the general formula.

5.4.3 The Most Probable Configuration

In thermal equilibrium, every state with a given total energy E and a given particle number N is equally likely. So the *most probable configuration* $(N_1, N_2, N_3, \ldots)$ is the one that can be achieved in the largest number of different ways—it is that particular configuration for which $Q(N_1, N_2, N_3, \ldots)$ is a maximum, subject to the constraints

$$\sum_{n=1}^{\infty} N_n = N, \qquad [5.78]$$

and

$$\sum_{n=1}^{\infty} N_n E_n = E. \qquad [5.79]$$

The problem of maximizing a function $F(x_1, x_2, x_3, \ldots)$ of several variables, subject to the constraints $f_1(x_1, x_2, x_3, \ldots) = 0$, $f_2(x_1, x_2, x_3, \ldots) = 0$, etc., is most conveniently handled by the method of **Lagrange multipliers**.[24] We introduce the new function

$$G(x_1, x_2, x_3, \ldots, \lambda_1, \lambda_2, \ldots) \equiv F + \lambda_1 f_1 + \lambda_2 f_2 + \cdots, \qquad [5.80]$$

and set *all* its derivatives equal to zero:

$$\frac{\partial G}{\partial x_n} = 0; \quad \frac{\partial G}{\partial \lambda_n} = 0. \qquad [5.81]$$

In our case it's a little easier to work with the *logarithm* of Q, instead of Q itself—this turns the *products* into *sums*. Since the logarithm is a monotonic function of its argument, the maxima of Q and $\ln(Q)$ occur at the same point. So we let

$$G \equiv \ln(Q) + \alpha\left[N - \sum_{n=1}^{\infty} N_n\right] + \beta\left[E - \sum_{n=1}^{\infty} N_n E_n\right], \qquad [5.82]$$

---

[24] See, for example, Mary Boas, *Mathematical Methods in the Physical Sciences*, 2nd ed., Wiley, New York (1983), Chapter 4, Section 9.

where α and β are the Lagrange multipliers. Setting the derivatives with respect to α and β equal to zero merely reproduces the constraints (Equations 5.78 and 5.79); it remains, then, to set the derivative with respect to N_n equal to zero.

If the particles are *distinguishable*, then Q is given by Equation 5.74, and we have

$$G = \ln(N!) + \sum_{n=1}^{\infty} [N_n \ln(d_n) - \ln(N_n!)] + \alpha\left[N - \sum_{n=1}^{\infty} N_n\right] + \beta\left[E - \sum_{n=1}^{\infty} N_n E_n\right]. \quad [5.83]$$

Assuming the relevant occupation numbers (N_n) are large, we can invoke **Stirling's approximation**:[25]

$$\ln(z!) \approx z\ln(z) - z \quad \text{for } z \gg 1, \quad [5.84]$$

to write

$$G \approx \sum_{n=1}^{\infty} [N_n \ln(d_n) - N_n \ln(N_n) + N_n - \alpha N_n - \beta E_n N_n] + \ln(N!) + \alpha N + \beta E. \quad [5.85]$$

It follows that

$$\frac{\partial G}{\partial N_n} = \ln(d_n) - \ln(N_n) - \alpha - \beta E_n. \quad [5.86]$$

Setting this equal to zero, and solving for N_n, we conclude that the *most probable occupation numbers*, for distinguishable particles, are

$$N_n = d_n e^{-(\alpha+\beta E_n)}. \quad [5.87]$$

If the particles are *identical fermions*, then Q is given by Equation 5.75, and we have

$$G = \sum_{n=1}^{\infty} \{\ln(d_n!) - \ln(N_n!) - \ln[(d_n - N_n)!]\} + \alpha\left[N - \sum_{n=1}^{\infty} N_n\right] + \beta\left[E - \sum_{n=1}^{\infty} N_n E_n\right]. \quad [5.88]$$

---

[25] Stirling's approximation can be improved by including more terms in the **Stirling series**, but the first two will suffice for our purposes. See George Arfken and Hans-Jurgen Weber, *Mathematical Methods for Physicists*, 5th ed., Academic Press, Orlando (2000), Section 10.3. If the relevant occupation numbers are *not* large—as in Section 5.4.1—then statistical mechanics simply doesn't apply. The whole point is to deal with such enormous numbers that statistical inference is a reliable predictor. Of course, there will always be one-particle states of extremely high energy that are not populated at *all*; fortunately, Stirling's approximation holds also for $z = 0$. I use the word "relevant" to exclude any stray states right at the margin, for which N_n is neither huge nor zero.

This time we must assume not only that N_n is large, but also that $d_n \gg N_n$,[26] so that Stirling's approximation applies to both terms. In that case

$$G \approx \sum_{n=1}^{\infty}\Big[\ln(d_n!) - N_n\ln(N_n) + N_n - (d_n - N_n)\ln(d_n - N_n) + (d_n - N_n) - \alpha N_n - \beta E_n N_n\Big] + \alpha N + \beta E, \tag{5.89}$$

so

$$\frac{\partial G}{\partial N_n} = -\ln(N_n) + \ln(d_n - N_n) - \alpha - \beta E_n. \tag{5.90}$$

Setting this equal to zero, and solving for N_n, we find the *most probable occupation numbers* for identical fermions:

$$N_n = \frac{d_n}{e^{(\alpha+\beta E_n)} + 1}. \tag{5.91}$$

Finally, if the particles are *identical bosons*, then Q is given by Equation 5.77, and we have

$$G = \sum_{n=1}^{\infty}\{\ln[(N_n + d_n - 1)!] - \ln(N_n!) - \ln[(d_n - 1)!]\} + \alpha\left[N - \sum_{n=1}^{\infty}N_n\right] + \beta\left[E - \sum_{n=1}^{\infty}N_nE_n\right]. \tag{5.92}$$

Assuming (as always) that $N_n \gg 1$, and using Stirling's approximation:

$$G \approx \sum_{n=1}^{\infty}\{(N_n + d_n - 1)\ln(N_n + d_n - 1) - (N_n + d_n - 1) - N_n\ln(N_n) + N_n - \ln[(d_n - 1)!] - \alpha N_n - \beta E_nN_n\} + \alpha N + \beta E, \tag{5.93}$$

so

$$\frac{\partial G}{\partial N_n} = \ln(N_n + d_n - 1) - \ln(N_n) - \alpha - \beta E_n. \tag{5.94}$$

---

[26]In *one* dimension the energies are nondegenerate (see Problem 2.45), but in three dimensions d_n typically increases rapidly with increasing n (for example, in the case of hydrogen, $d_n = n^2$). So it is not unreasonable to assume that for *most* of the occupied states $d_n \gg 1$. On the other hand, d_n is certainly *not* much greater than N_n at absolute zero, where all states up to the Fermi level are filled, and hence $d_n = N_n$. Here again we are rescued by the fact that Stirling's formula holds also for $z = 0$.

Setting this equal to zero, and solving for N_n, we find the *most probable occupation numbers* for identical bosons:

$$N_n = \frac{d_n - 1}{e^{(\alpha+\beta E_n)} - 1}. \quad [5.95]$$

(For consistency with the approximation already invoked in the case of fermions, we should really drop the 1 in the numerator, and I shall do so from now on.)

Problem 5.26 Use the method of Lagrange multipliers to find the rectangle of largest area, with sides parallel to the axes, that can be inscribed in the ellipse $(x/a)^2 + (y/b)^2 = 1$. What is that maximum area?

Problem 5.27

(a) Find the percent error in Stirling's approximation for $z = 10$.

(b) What is the smallest integer z such that the error is less than 1%?

5.4.4 Physical Significance of α and β

The parameters α and β came into the story as Lagrange multipliers, associated with the total number of particles and the total energy, respectively. Mathematically, they are determined by substituting the occupation numbers (Equations 5.87, 5.91, and 5.95) back into the constraints (Equations 5.78 and 5.79). To carry out the summation, however, we need to know the allowed energies (E_n), and their degeneracies (d_n), for the potential in question. As an example, I'll work out the case of an **ideal gas**—a large number of noninteracting particles, all with the same mass, in the three dimensional infinite square well. This will motivate the physical interpretation of α and β.

In Section 5.3.1 we found the allowed energies (Equation 5.39):

$$E_k = \frac{\hbar^2}{2m}k^2, \quad [5.96]$$

where

$$\mathbf{k} = \left(\frac{\pi n_x}{l_x}, \frac{\pi n_y}{l_y}, \frac{\pi n_z}{l_z}\right).$$

As before, we convert the sum into an integral, treating $\mathbf{k}$ as a continuous variable, with one state (or, for spin s, $2s+1$ states) per volume π^3/V of k-space. Taking as our "bins" the spherical shells in the first octant (see Figure 5.4), the "degeneracy" (that is, the number of states in the bin) is

$$d_k = \frac{1}{8}\frac{4\pi k^2\,dk}{(\pi^3/V)} = \frac{V}{2\pi^2}k^2\,dk. \qquad [5.97]$$

For distinguishable particles (Equation 5.87), the first constraint (Equation 5.78) becomes

$$N = \frac{V}{2\pi^2}e^{-\alpha}\int_0^\infty e^{-\beta\hbar^2k^2/2m}k^2\,dk = Ve^{-\alpha}\left(\frac{m}{2\pi\beta\hbar^2}\right)^{3/2},$$

so

$$e^{-\alpha} = \frac{N}{V}\left(\frac{2\pi\beta\hbar^2}{m}\right)^{3/2}. \qquad [5.98]$$

The second constraint (Equation 5.79) says

$$E = \frac{V}{2\pi^2}e^{-\alpha}\frac{\hbar^2}{2m}\int_0^\infty e^{-\beta\hbar^2k^2/2m}k^4\,dk = \frac{3V}{2\beta}e^{-\alpha}\left(\frac{m}{2\pi\beta\hbar^2}\right)^{3/2},$$

or, putting in Equation 5.98 for $e^{-\alpha}$:

$$E = \frac{3N}{2\beta}. \qquad [5.99]$$

(If you include the spin factor, $2s+1$, in Equation 5.97, it cancels out at this point, so Equation 5.99 is correct for all spins.)

This result (Equation 5.99) is reminiscent of the classical formula for the average kinetic energy of an atom at temperature T:[27]

$$\frac{E}{N} = \frac{3}{2}k_BT, \qquad [5.100]$$

where k_B is the Boltzmann constant. This suggests that β is related to the *temperature*:

$$\beta = \frac{1}{k_BT}. \qquad [5.101]$$

To prove that this holds in general, and not simply for distinguishable particles in the three-dimensional infinite square well, we would have to demonstrate that different substances in thermal equilibrium with one another have the same value of β. The argument is sketched in many books,[28] but I shall not reproduce it here—I will simply adopt Equation 5.101 as the *definition* of T.

---

[27] See, for example, David Halliday, Robert Resnick, and Jearl Walker, *Fundamentals of Physics*, 5th ed., Wiley, New York (1997), Section 20-5.

[28] See, for example, Yariv (footnote 22), Section 15.4.

It is customary to replace α (which, as is clear from the special case of Equation 5.98, is a function of T) by the so-called **chemical potential**,

$$\mu(T) \equiv -\alpha k_B T, \tag{5.102}$$

and rewrite Equations 5.87, 5.91, and 5.95 as formulas for the *most probable number of particles in a particular (one-particle) state with energy* ϵ (to go from the number of particles with a given energy to the number of particles in a particular *state* with that energy, we simply divide by the degeneracy of the state):

$$n(\epsilon) = \begin{cases} e^{-(\epsilon-\mu)/k_B T} & \text{MAXWELL-BOLTZMANN} \\ \dfrac{1}{e^{(\epsilon-\mu)/k_B T}+1} & \text{FERMI-DIRAC} \\ \dfrac{1}{e^{(\epsilon-\mu)/k_B T}-1} & \text{BOSE-EINSTEIN} \end{cases} \tag{5.103}$$

The **Maxwell-Boltzmann distribution** is the classical result, for *distinguishable* particles; the **Fermi-Dirac distribution** applies to *identical fermions*, and the **Bose-Einstein distribution** is for *identical bosons*.

The Fermi-Dirac distribution has a particularly simple behavior as $T \to 0$:

$$e^{(\epsilon-\mu)/k_B T} \to \begin{cases} 0, & \text{if } \epsilon < \mu(0), \\ \infty, & \text{if } \epsilon > \mu(0), \end{cases}$$

so

$$n(\epsilon) \to \begin{cases} 1, & \text{if } \epsilon < \mu(0), \\ 0, & \text{if } \epsilon > \mu(0). \end{cases} \tag{5.104}$$

All states are filled, up to an energy $\mu(0)$, and none are occupied for energies above this (Figure 5.8). Evidently the chemical potential at absolute zero is precisely the Fermi energy:

$$\mu(0) = E_F. \tag{5.105}$$

As the temperature rises, the Fermi-Dirac distribution "softens" the cutoff, as indicated by the rounded curve in Figure 5.8.

Returning now to the special case of an ideal gas, for distinguishable particles we found that the total energy at temperature T is (Equation 5.99)

$$E = \frac{3}{2} N k_B T, \tag{5.106}$$

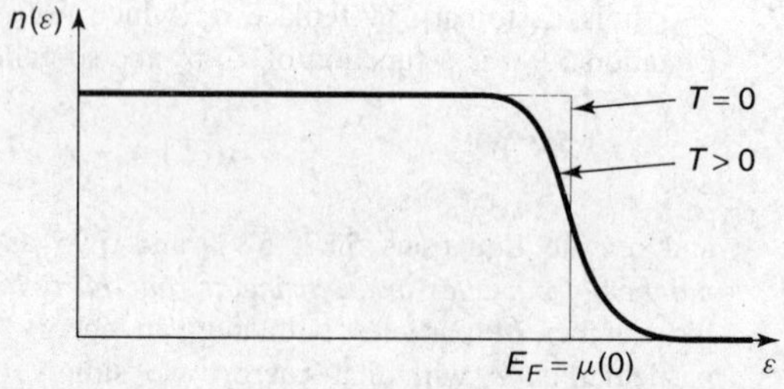

FIGURE 5.8: Fermi-Dirac distribution for $T = 0$ and for T somewhat above zero.

while (from Equation 5.98) the chemical potential is

$$\mu(T) = k_B T \left[\ln\left(\frac{N}{V}\right) + \frac{3}{2} \ln\left(\frac{2\pi\hbar^2}{mk_B T}\right) \right]. \qquad [5.107]$$

I would *like* to work out the corresponding formulas for an ideal gas of identical fermions and bosons, using Equations 5.91 and 5.95 in place of Equation 5.87. The first constraint (Equation 5.78) becomes

$$N = \frac{V}{2\pi^2} \int_0^\infty \frac{k^2}{e^{[(\hbar^2 k^2/2m)-\mu]/k_B T} \pm 1}\, dk, \qquad [5.108]$$

(with the plus sign for fermions and minus for bosons), and the second constraint (Equation 5.79) reads

$$E = \frac{V}{2\pi^2} \frac{\hbar^2}{2m} \int_0^\infty \frac{k^4}{e^{[(\hbar^2 k^2/2m)-\mu]/k_B T} \pm 1}\, dk. \qquad [5.109]$$

The first of these determines $\mu(T)$, and the second determines $E(T)$ (from the latter we obtain, for instance, the heat capacity: $C = \partial E/\partial T$). Unfortunately, these integrals cannot be evaluated in terms of elementary functions, and I shall leave it for you to explore the matter further (see Problems 5.28 and 5.29).

Problem 5.28 Evaluate the integrals (Equations 5.108 and 5.109) for the case of identical fermions at absolute zero. Compare your results with Equations 5.43 and 5.45. (Note that for electrons there is an extra factor of 2 in Equations 5.108 and 5.109, to account for the spin degeneracy.)

∗∗∗Problem 5.29

(a) Show that for bosons the chemical potential must always be less than the minimum allowed energy. *Hint:* $n(\epsilon)$ cannot be negative.

(b) In particular, for the ideal bose gas, $\mu(T) < 0$ for all T. Show that in this case $\mu(T)$ monotonically increases as T decreases, assuming N and V are held constant. *Hint:* Study Equation 5.108, with the minus sign.

(c) A crisis (called **Bose condensation**) occurs when (as we lower T) $\mu(T)$ hits zero. Evaluate the integral, for $\mu = 0$, and obtain the formula for the critical temperature T_c at which this happens. Below the critical temperature, the particles crowd into the ground state, and the calculational device of replacing the discrete sum (Equation 5.78) by a continuous integral (Equation 5.108) loses its validity.[29] *Hint:*

$$\int_0^\infty \frac{x^{s-1}}{e^x - 1}\,dx = \Gamma(s)\zeta(s), \qquad [5.110]$$

where Γ is Euler's **gamma function** and ζ is the **Riemann zeta function**. Look up the appropriate numerical values.

(d) Find the critical temperature for ^{4}He. Its density, at this temperature, is 0.15 gm/cm^3. *Comment:* The experimental value of the critical temperature in ^{4}He is 2.17 K. The remarkable properties of ^{4}He in the neighborhood of T_c are discussed in the reference cited in footnote 29.

5.4.5 The Blackbody Spectrum

Photons (quanta of the electromagnetic field) are identical bosons with spin 1, but they are very special, because they are *massless* particles, and hence intrinsically relativistic. We can include them here, if you are prepared to accept four assertions that do not belong to nonrelativistic quantum mechanics:

1. The energy of a photon is related to its frequency by the Planck formula, $E = h\nu = \hbar\omega$.
2. The wave number k is related to the frequency by $k = 2\pi/\lambda = \omega/c$, where c is the speed of light.
3. Only two spin states occur (the quantum number m can be $+1$ or -1, but not 0).

[29] See F. Mandl, *Statistical Physics*, Wiley, London (1971), Section 11.5.

4. The *number* of photons is not a conserved quantity; when the temperature rises, the number of photons (per unit volume) increases.

In view of item 4, the first constraint (Equation 5.78) does not apply. We can take account of this by simply setting $\alpha \to 0$, in Equation 5.82 and everything that follows. Thus the most probable occupation number, for photons, is (Equation 5.95):

$$N_\omega = \frac{d_k}{e^{\hbar\omega/k_B T} - 1}. \qquad [5.111]$$

For free photons in a box of volume V, d_k is given by Equation 5.97,[30] multiplied by 2 for spin (item 3), and expressed in terms of ω instead of k (item 2):

$$d_k = \frac{V}{\pi^2 c^3}\omega^2\, d\omega. \qquad [5.112]$$

So the energy density, $N_\omega \hbar\omega / V$, in the frequency range $d\omega$, is $\rho(\omega)\, d\omega$, where

$$\boxed{\rho(\omega) = \frac{\hbar\omega^3}{\pi^2 c^3 \left(e^{\hbar\omega/k_B T} - 1\right)}.} \qquad [5.113]$$

This is Planck's famous formula for the **blackbody spectrum**, giving the energy per unit volume, per unit frequency, for an electromagnetic field in equilibrium at temperature T. It is plotted, for three different temperatures, in Figure 5.9.

Problem 5.30

(a) Use Equation 5.113 to determine the energy density in the *wavelength* range $d\lambda$. *Hint;* Set $\rho(\omega)\, d\omega = \overline{\rho}(\lambda)\, d\lambda$, and solve for $\overline{\rho}(\lambda)$.

(b) Derive the **Wien displacement law** for the wavelength at which the blackbody energy density is a maximum:

$$\lambda_{\text{max}} = \frac{2.90 \times 10^{-3}\ \text{mK}}{T}. \qquad [5.114]$$

Hint: You'll need to solve the transcendental equation $(5 - x) = 5e^{-x}$, using a calculator or a computer; get the numerical answer accurate to three significant digits.

---

[30] In truth, we have no business using this formula, which came from the (nonrelativistic) Schrödinger equation; fortunately, the degeneracy is exactly the same for the relativistic case. See Problem 5.36.

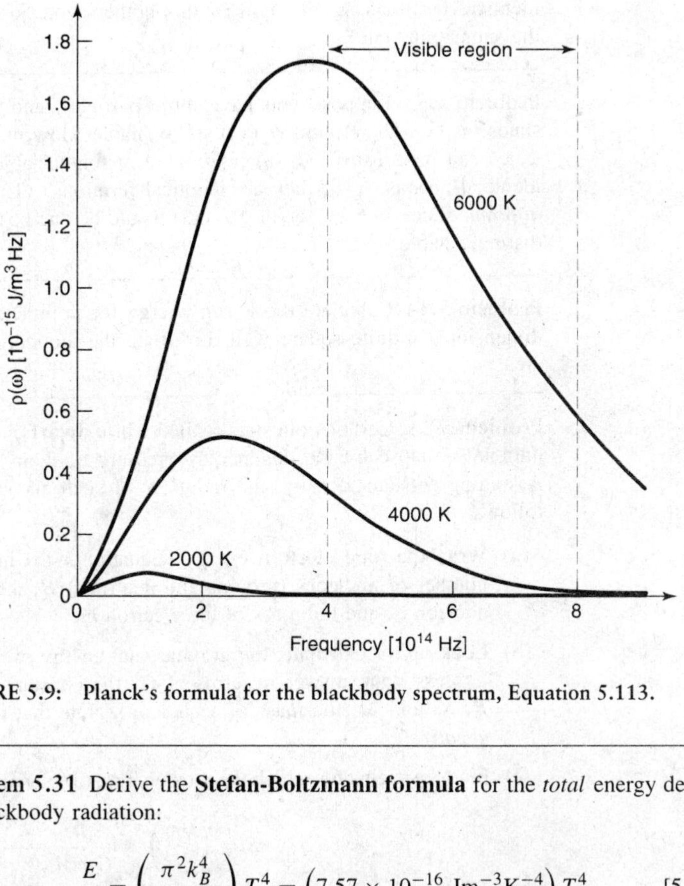

FIGURE 5.9: **Planck's formula for the blackbody spectrum, Equation 5.113.**

Problem 5.31 Derive the **Stefan-Boltzmann formula** for the *total* energy density in blackbody radiation:

$$\frac{E}{V} = \left(\frac{\pi^2 k_B^4}{15\hbar^3 c^3}\right) T^4 = \left(7.57 \times 10^{-16}\ \mathrm{Jm^{-3}K^{-4}}\right) T^4. \qquad [5.115]$$

Hint: Use Equation 5.110 to evaluate the integral. Note that $\zeta(4) = \pi^4/90$.

FURTHER PROBLEMS FOR CHAPTER 5

Problem 5.32 Imagine two noninteracting particles, each of mass m, in the one-dimensional harmonic oscillator potential (Equation 2.43). If one is in the ground state, and the other is in the first excited state, calculate $\langle(x_1 - x_2)^2\rangle$, assuming (a) they are distinguishable particles, (b) they are identical bosons, and (c) they are identical fermions. Ignore spin (if this bothers you, just assume they are both in the same spin state).

Problem 5.33 Suppose you have three particles, and three distinct one-particle states ($\psi_a(x)$, $\psi_b(x)$, and $\psi_c(x)$) are available. How many different three-particle states can be constructed, (a) if they are distinguishable particles, (b) if they are identical bosons, (c) if they are identical fermions? (The particles need not be in *different* states—$\psi_a(x_1)\psi_a(x_2)\psi_a(x_3)$ would be one possibility, if the particles are distinguishable.)

Problem 5.34 Calculate the Fermi energy for noninteracting electrons in a *two*-dimensional infinite square well. Let σ be the number of free electrons per unit area.

*****Problem 5.35** Certain cold stars (called **white dwarfs**) are stabilized against gravitational collapse by the degeneracy pressure of their electrons (Equation 5.46). Assuming constant density, the radius R of such an object can be calculated as follows:

(a) Write the total electron energy (Equation 5.45) in terms of the radius, the number of nucleons (protons and neutrons) N, the number of electrons per nucleon q, and the mass of the electron m.

(b) Look up, or calculate, the gravitational energy of a uniformly dense sphere. Express your answer in terms of G (the constant of universal gravitation), R, N, and M (the mass of a nucleon). Note that the gravitational energy is *negative*.

(c) Find the radius for which the total energy, (a) plus (b), is a minimum. *Answer:*

$$R = \left(\frac{9\pi}{4}\right)^{2/3} \frac{\hbar^2 q^{5/3}}{GmM^2N^{1/3}}.$$

(Note that the radius *decreases* as the total mass *increases*!) Put in the actual numbers, for everything except N, using $q = 1/2$ (actually, q decreases a bit as the atomic number increases, but this is close enough for our purposes). *Answer:* $R = 7.6 \times 10^{25} N^{-1/3}$.

(d) Determine the radius, in kilometers, of a white dwarf with the mass of the sun.

(e) Determine the Fermi energy, in electron volts, for the white dwarf in (d), and compare it with the rest energy of an electron. Note that this system is getting dangerously relativistic (see Problem 5.36).

****Problem 5.36** We can extend the theory of a free electron gas (Section 5.3.1) to the relativistic domain by replacing the classical kinetic energy, $E = p^2/2m$, with the relativistic formula, $E = \sqrt{p^2c^2 + m^2c^4} - mc^2$. Momentum is related to the wave vector in the usual way: $\mathbf{p} = \hbar\mathbf{k}$. In particular, in the *extreme* relativistic limit, $E \approx pc = \hbar ck$.

(a) Replace $\hbar^2k^2/2m$ in Equation 5.44 by the ultra-relativistic expression, $\hbar ck$, and calculate E_{tot} in this regime.

(b) Repeat parts (a) and (b) of Problem 5.35 for the ultra-relativistic electron gas. Notice that in this case there is *no* stable minimum, regardless of R; if the total energy is positive, degeneracy forces exceed gravitational forces, and the star will expand, whereas if the total is negative, gravitational forces win out, and the star will collapse. Find the critical number of nucleons, N_c, such that gravitational collapse occurs for $N > N_c$. This is called the **Chandrasekhar limit**. *Answer:* 2.4×10^{57}. What is the corresponding stellar mass (give your answer as a multiple of the sun's mass). Stars heavier than this will not form white dwarfs, but collapse further, becoming (if conditions are right) **neutron stars**.

(c) At extremely high density, **inverse beta decay**, $e^- + p^+ \rightarrow n + \nu$, converts virtually all of the protons and electrons into neutrons (liberating neutrinos, which carry off energy, in the process). Eventually *neutron* degeneracy pressure stabilizes the collapse, just as *electron* degeneracy does for the white dwarf (see Problem 5.35). Calculate the radius of a neutron star with the mass of the sun. Also calculate the (neutron) Fermi energy, and compare it to the rest energy of a neutron. Is it reasonable to treat a neutron star nonrelativistically?

****Problem 5.37**

(a) Find the chemical potential and the total energy for distinguishable particles in the three dimensional harmonic oscillator potential (Problem 4.38). *Hint:* The sums in Equations 5.78 and 5.79 can be evaluated exactly, in this case—no need to use an integral approximation, as we did for the infinite square well. Note that by differentiating the **geometric series**,

$$\frac{1}{1-x} = \sum_{n=0}^{\infty} x^n, \qquad [5.116]$$

you can get

$$\frac{d}{dx}\left(\frac{x}{1-x}\right) = \sum_{n=1}^{\infty} (n+1)x^n$$

and similar results for higher derivatives. *Answer:*

$$E = \frac{3}{2} N\hbar\omega \left(\frac{1 + e^{-\hbar\omega/k_B T}}{1 - e^{-\hbar\omega/k_B T}} \right). \quad [5.117]$$

(b) Discuss the limiting case $k_B T \ll \hbar\omega$.

(c) Discuss the classical limit, $k_B T \gg \hbar\omega$, in the light of the **equipartition theorem**.[31] How many **degrees of freedom** does a particle in the three dimensional harmonic oscillator possess?

---

[31] See, for example, Halliday and Resnick (footnote 27), Section 20-9.

PART II APPLICATIONS

CHAPTER 6

TIME-INDEPENDENT PERTURBATION THEORY

6.1 NONDEGENERATE PERTURBATION THEORY

6.1.1 General Formulation

Suppose we have solved the (time-independent) Schrödinger equation for some potential (say, the one-dimensional infinite square well):

$$H^0\psi_n^0 = E_n^0\psi_n^0, \qquad [6.1]$$

obtaining a complete set of orthonormal eigenfunctions, ψ_n^0,

$$\langle\psi_n^0|\psi_m^0\rangle = \delta_{nm}, \qquad [6.2]$$

and the corresponding eigenvalues E_n^0. Now we perturb the potential slightly (say, by putting a little bump in the bottom of the well—Figure 6.1). We'd *like* to find the new eigenfunctions and eigenvalues:

$$H\psi_n = E_n\psi_n, \qquad [6.3]$$

but unless we are very lucky, we're not going to be able to solve the Schrödinger equation exactly, for this more complicated potential. **Perturbation theory** is a systematic procedure for obtaining *approximate* solutions to the perturbed problem, by building on the known exact solutions to the *unperturbed* case.

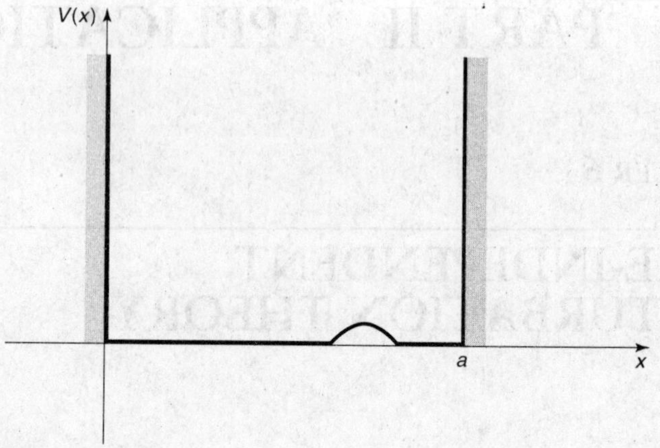

FIGURE 6.1: Infinite square well with small perturbation.

To begin with we write the new Hamiltonian as the sum of two terms:

$$H = H^0 + \lambda H', \qquad [6.4]$$

where H' is the perturbation (the superscript 0 always identifies the *un*perturbed quantity). For the moment we'll take λ to be a small number; later we'll crank it up to 1, and H will be the true Hamiltonian. Next we write ψ_n and E_n as power series in λ:

$$\psi_n = \psi_n^0 + \lambda\psi_n^1 + \lambda^2\psi_n^2 + \cdots ; \qquad [6.5]$$

$$E_n = E_n^0 + \lambda E_n^1 + \lambda^2 E_n^2 + \cdots . \qquad [6.6]$$

Here E_n^1 is the **first-order correction** to the nth eigenvalue, and ψ_n^1 is the first-order correction to the nth eigenfunction; E_n^2 and ψ_n^2 are the **second-order corrections**, and so on. Plugging Equations 6.5 and 6.6 into Equation 6.3, we have:

$$(H^0 + \lambda H')[\psi_n^0 + \lambda\psi_n^1 + \lambda^2\psi_n^2 + \cdots]$$
$$= (E_n^0 + \lambda E_n^1 + \lambda^2 E_n^2 + \cdots)[\psi_n^0 + \lambda\psi_n^1 + \lambda^2\psi_n^2 + \cdots],$$

or (collecting like powers of λ):

$$H^0\psi_n^0 + \lambda(H^0\psi_n^1 + H'\psi_n^0) + \lambda^2(H^0\psi_n^2 + H'\psi_n^1) + \cdots$$
$$= E_n^0\psi_n^0 + \lambda(E_n^0\psi_n^1 + E_n^1\psi_n^0) + \lambda^2(E_n^0\psi_n^2 + E_n^1\psi_n^1 + E_n^2\psi_n^0) + \cdots .$$

To lowest order[1] (λ^0) this yields $H^0\psi_n^0 = E_n^0\psi_n^0$, which is nothing new (Equation 6.1). To first order (λ^1),

$$H^0\psi_n^1 + H'\psi_n^0 = E_n^0\psi_n^1 + E_n^1\psi_n^0. \tag{6.7}$$

To second order (λ^2),

$$H^0\psi_n^2 + H'\psi_n^1 = E_n^0\psi_n^2 + E_n^1\psi_n^1 + E_n^2\psi_n^0, \tag{6.8}$$

and so on. (I'm done with λ, now—it was just a device to keep track of the different orders—so crank it up to 1.)

6.1.2 First-Order Theory

Taking the inner product of Equation 6.7 with ψ_n^0 (that is, multiplying by $(\psi_n^0)^*$ and integrating),

$$\langle\psi_n^0|H^0\psi_n^1\rangle + \langle\psi_n^0|H'\psi_n^0\rangle = E_n^0\langle\psi_n^0|\psi_n^1\rangle + E_n^1\langle\psi_n^0|\psi_n^0\rangle.$$

But H^0 is hermitian, so

$$\langle\psi_n^0|H^0\psi_n^1\rangle = \langle H^0\psi_n^0|\psi_n^1\rangle = \langle E_n^0\psi_n^0|\psi_n^1\rangle = E_n^0\langle\psi_n^0|\psi_n^1\rangle,$$

and this cancels the first term on the right. Moreover, $\langle\psi_n^0|\psi_n^0\rangle = 1$, so[2]

$$\boxed{E_n^1 = \langle\psi_n^0|H'|\psi_n^0\rangle.} \tag{6.9}$$

This is the fundamental result of first-order perturbation theory; as a *practical* matter, it may well be the most important equation in quantum mechanics. It says that the first-order correction to the energy is the *expectation value* of the perturbation, in the *unperturbed* state.

Example 6.1 The unperturbed wave functions for the infinite square well are (Equation 2.28)

$$\psi_n^0(x) = \sqrt{\frac{2}{a}}\sin\left(\frac{n\pi}{a}x\right).$$

[1] As always (Chapter 2, footnote 25) the uniqueness of power series expansions guarantees that the coefficients of like powers are equal.

[2] In this context it doesn't matter whether we write $\langle\psi_n^0|H'\psi_n^0\rangle$ or $\langle\psi_n^0|H'|\psi_n^0\rangle$ (with the extra vertical bar), because we are using the wave function itself to "label" the state. But the latter notation is preferable, because it frees us from this specific convention.

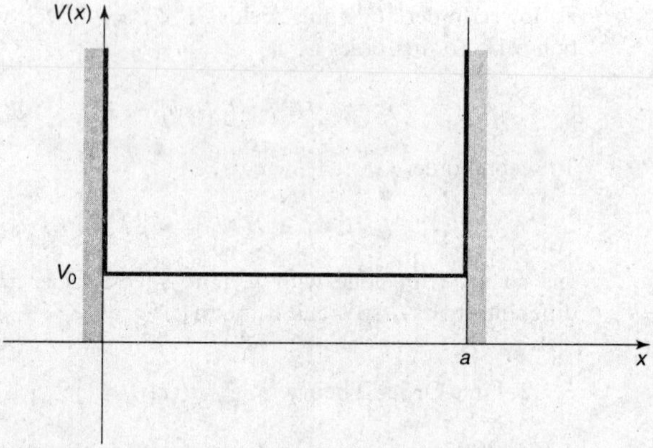

FIGURE 6.2: Constant perturbation over the whole well.

Suppose we perturb the system by simply raising the "floor" of the well a constant amount V_0 (Figure 6.2). Find the first-order correction to the energies.

Solution: In this case $H' = V_0$, and the first-order correction to the energy of the nth state is

$$E_n^1 = \langle\psi_n^0|V_0|\psi_n^0\rangle = V_0\langle\psi_n^0|\psi_n^0\rangle = V_0.$$

The corrected energy levels, then, are $E_n \cong E_n^0 + V_0$; they are simply lifted by the amount V_0. Of *course*! The only surprising thing is that in this case the first-order theory yields the *exact* answer. Evidently for a *constant* perturbation all the higher corrections vanish.[3] On the other hand, if the perturbation extends only half-way across the well (Figure 6.3), then

$$E_n^1 = \frac{2V_0}{a}\int_0^{a/2} \sin^2\left(\frac{n\pi}{a}x\right) dx = \frac{V_0}{2}.$$

In this case every energy level is lifted by $V_0/2$. That's not the *exact* result, presumably, but it does seem reasonable, as a first-order approximation.

Equation 6.9 is the first-order correction to the *energy*; to find the first-order correction to the *wave function* we first rewrite Equation 6.7:

$$(H^0 - E_n^0)\psi_n^1 = -(H' - E_n^1)\psi_n^0. \qquad [6.10]$$

[3]Incidentally, nothing here depends on the specific nature of the infinite square well—the same holds for *any* potential, when the perturbation is constant.

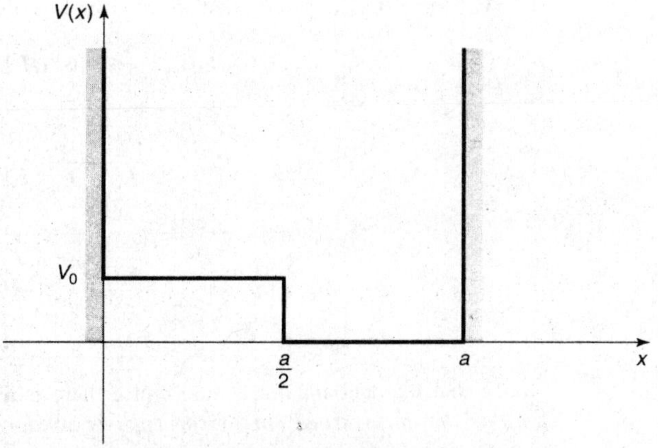

FIGURE 6.3: Constant perturbation over half the well.

The right side is a known function, so this amounts to an inhomogeneous differential equation for ψ_n^1. Now, the unperturbed wave functions constitute a complete set, so ψ_n^1 (like any other function) can be expressed as a linear combination of them:

$$\psi_n^1 = \sum_{m \neq n} c_m^{(n)} \psi_m^0. \qquad [6.11]$$

There is no need to include $m = n$ in the sum, for if ψ_n^1 satisfies Equation 6.10, so too does $(\psi_n^1 + \alpha\psi_n^0)$, for any constant α, and we can use this freedom to subtract off the ψ_n^0 term.[4] If we could determine the coefficients $c_m^{(n)}$, we'd be done.

Well, putting Equation 6.11 into Equation 6.10, and using the fact that the ψ_m^0 satisfies the unperturbed Schrödinger equation (Equation 6.1), we have

$$\sum_{m \neq n} (E_m^0 - E_n^0) c_m^{(n)} \psi_m^0 = -(H' - E_n^1)\psi_n^0.$$

Taking the inner product with ψ_l^0,

$$\sum_{m \neq n} (E_m^0 - E_n^0) c_m^{(n)} \langle \psi_l^0 | \psi_m^0 \rangle = -\langle \psi_l^0 | H' | \psi_n^0 \rangle + E_n^1 \langle \psi_l^0 | \psi_n^0 \rangle.$$

[4]Alternatively, a glance at Equation 6.5 reveals that any ψ_n^0 component in ψ_n^1 might as well be pulled out and combined with the first term. In fact, the choice $c_n^{(n)} = 0$ ensures that ψ_n—with 1 as the coefficient of ψ_n^0 in Equation 6.5—is *normalized* (to first order in λ): $\langle \psi_n | \psi_n \rangle = \langle \psi_n^0 | \psi_n^0 \rangle + \lambda(\langle \psi_n^1 | \psi_n^0 \rangle + \langle \psi_n^0 | \psi_n^1 \rangle) + \lambda^2(\cdots) + \cdots$, but the orthonormality of the unperturbed states means that the first term is 1 and $\langle \psi_n^1 | \psi_n^0 \rangle = \langle \psi_n^0 | \psi_n^1 \rangle = 0$, as long as ψ_n^1 has no ψ_n^0 component.

If $l = n$, the left side is zero, and we recover Equation 6.9; if $l \neq n$, we get

$$(E_l^0 - E_n^0)c_l^{(n)} = -\langle\psi_l^0|H'|\psi_n^0\rangle,$$

or

$$c_m^{(n)} = \frac{\langle\psi_m^0|H'|\psi_n^0\rangle}{E_n^0 - E_m^0}, \tag{6.12}$$

so

$$\boxed{\psi_n^1 = \sum_{m\neq n} \frac{\langle\psi_m^0|H'|\psi_n^0\rangle}{(E_n^0 - E_m^0)}\psi_m^0.} \tag{6.13}$$

Notice that the denominator is safe (since there is no coefficient with $m = n$) *as long as the unperturbed energy spectrum is nondegenerate*. But if two different unperturbed states share the same energy, we're in serious trouble (we divided by zero to get Equation 6.12); in that case we need **degenerate perturbation theory**, which I'll come to in Section 6.2.

That completes first-order perturbation theory: The first-order correction to the energy, E_n^1, is given by Equation 6.9, and the first-order correction to the wave function, ψ_n^1, is given by Equation 6.13. I should warn you that whereas perturbation theory often yields surprisingly accurate energies (that is, $E_n^0 + E_n^1$ is quite close to the exact value E_n), the wave functions are notoriously poor.

***Problem 6.1** Suppose we put a delta-function bump in the center of the infinite square well:

$$H' = \alpha\delta(x - a/2),$$

where α is a constant.

(a) Find the first-order correction to the allowed energies. Explain why the energies are not perturbed for even n.

(b) Find the first three nonzero terms in the expansion (Equation 6.13) of the correction to the ground state, ψ_1^1.

***Problem 6.2** For the harmonic oscillator $[V(x) = (1/2)kx^2]$, the allowed energies are

$$E_n = (n + 1/2)\hbar\omega, \quad (n = 0, 1, 2, \ldots),$$

where $\omega = \sqrt{k/m}$ is the classical frequency. Now suppose the spring constant increases slightly: $k \to (1 + \epsilon)k$. (Perhaps we cool the spring, so it becomes less flexible.)

(a) Find the *exact* new energies (trivial, in this case). Expand your formula as a power series in ϵ, up to second order.

(b) Now calculate the first-order perturbation in the energy, using Equation 6.9. What is H' here? Compare your result with part (a). *Hint:* It is not necessary—in fact, it is not *permitted*—to calculate a single integral in doing this problem.

Problem 6.3 Two identical bosons are placed in an infinite square well (Equation 2.19). They interact weakly with one another, via the potential

$$V(x_1, x_2) = -aV_0\delta(x_1 - x_2)$$

(where V_0 is a constant with the dimensions of energy, and a is the width of the well).

(a) First, ignoring the interaction between the particles, find the ground state and the first excited state—both the wave functions and the associated energies.

(b) Use first-order perturbation theory to estimate the effect of the particle-particle interaction on the energies of the ground state and the first excited state.

6.1.3 Second-Order Energies

Proceeding as before, we take the inner product of the *second* order equation (Equation 6.8) with ψ_n^0:

$$\langle\psi_n^0|H^0\psi_n^2\rangle + \langle\psi_n^0|H'\psi_n^1\rangle = E_n^0\langle\psi_n^0|\psi_n^2\rangle + E_n^1\langle\psi_n^0|\psi_n^1\rangle + E_n^2\langle\psi_n^0|\psi_n^0\rangle.$$

Again, we exploit the hermiticity of H^0:

$$\langle\psi_n^0|H^0\psi_n^2\rangle = \langle H^0\psi_n^0|\psi_n^2\rangle = E_n^0\langle\psi_n^0|\psi_n^2\rangle,$$

so the first term on the left cancels the first term on the right. Meanwhile, $\langle\psi_n^0|\psi_n^0\rangle = 1$, and we are left with a formula for E_n^2:

$$E_n^2 = \langle\psi_n^0|H'|\psi_n^1\rangle - E_n^1\langle\psi_n^0|\psi_n^1\rangle. \qquad [6.14]$$

But

$$\langle\psi_n^0|\psi_n^1\rangle = \sum_{m\neq n} c_m^{(n)}\langle\psi_n^0|\psi_m^0\rangle = 0,$$

(because the sum excludes $m = n$, and all the others are orthogonal), so

$$E_n^2 = \langle\psi_n^0|H'|\psi_n^1\rangle = \sum_{m\neq n} c_m^{(n)}\langle\psi_n^0|H'|\psi_m^0\rangle = \sum_{m\neq n}\frac{\langle\psi_m^0|H'|\psi_n^0\rangle\langle\psi_n^0|H'|\psi_m^0\rangle}{E_n^0 - E_m^0},$$

or, finally,

$$\boxed{E_n^2 = \sum_{m\neq n}\frac{|\langle\psi_m^0|H'|\psi_n^0\rangle|^2}{E_n^0 - E_m^0}.} \qquad [6.15]$$

This is the fundamental result of second-order perturbation theory.

We could go on to calculate the second-order correction to the wave function (ψ_n^2), the third-order correction to the energy, and so on, but in practice Equation 6.15 is ordinarily as far as it is useful to pursue this method.[5]

∗Problem 6.4

(a) Find the second-order correction to the energies (E_n^2) for the potential in Problem 6.1. *Comment:* You can sum the series explicitly, obtaining $-2m(\alpha/\pi\hbar n)^2$ for odd n.

(b) Calculate the second-order correction to the ground state energy (E_0^2) for the potential in Problem 6.2. Check that your result is consistent with the exact solution.

∗∗Problem 6.5 Consider a charged particle in the one-dimensional harmonic oscillator potential. Suppose we turn on a weak electric field (E), so that the potential energy is shifted by an amount $H' = -qEx$.

---

[5]In the short-hand notation $V_{mn} \equiv \langle\psi_m^0|H'|\psi_n^0\rangle$, $\Delta_{mn} \equiv E_m^0 - E_n^0$, the first three corrections to the nth energy are

$$E_n^1 = V_{nn},\quad E_n^2 = \sum_{m\neq n}\frac{|V_{nm}|^2}{\Delta_{nm}},\quad E_n^3 = \sum_{l,m\neq n}\frac{V_{nl}V_{lm}V_{mn}}{\Delta_{nl}\Delta_{nm}} - V_{nn}\sum_{m\neq n}\frac{|V_{nm}|^2}{\Delta_{nm}^2}.$$

The third order correction is given in Landau and Lifschitz, *Quantum Mechanics: Non-Relativistic Theory*, 3rd ed., Pergamon, Oxford (1977), page 136; the fourth and fifth orders (together with a powerful general technique for obtaining the higher orders) are developed by Nicholas Wheeler, *Higher-Order Spectral Perturbation* (unpublished Reed College report, 2000). Illuminating alternative formulations of time-independent perturbation theory include the Delgarno-Lewis method and the closely related "logarithmic" perturbation theory (see, for example, T. Imbo and U. Sukhatme, *Am. J. Phys.* **52**, 140 (1984), for LPT, and H. Mavromatis, *Am. J. Phys.* **59**, 738 (1991), for Delgarno-Lewis).

(a) Show that there is no first-order change in the energy levels, and calculate the second-order correction. *Hint:* See Problem 3.33.

(b) The Schrödinger equation can be solved directly in this case, by a change of variables: $x' \equiv x - (qE/m\omega^2)$. Find the exact energies, and show that they are consistent with the perturbation theory approximation.

6.2 DEGENERATE PERTURBATION THEORY

If the unperturbed states are degenerate—that is, if two (or more) distinct states (ψ_a^0 and ψ_b^0) share the same energy—then ordinary perturbation theory fails: $c_a^{(b)}$ (Equation 6.12) and E_a^2 (Equation 6.15) blow up (unless, perhaps, the numerator vanishes, $\langle\psi_a^0|H'|\psi_b^0\rangle = 0$—a loophole that will be important to us later on). In the degenerate case, therefore, there is no reason to trust even the *first*-order correction to the energy (Equation 6.9), and we must look for some other way to handle the problem.

6.2.1 Two-Fold Degeneracy

Suppose that

$$H^0\psi_a^0 = E^0\psi_a^0, \quad H^0\psi_b^0 = E^0\psi_b^0, \quad \langle\psi_a^0|\psi_b^0\rangle = 0, \qquad [6.16]$$

with ψ_a^0 and ψ_b^0 both normalized. Note that any linear combination of these states,

$$\psi^0 = \alpha\psi_a^0 + \beta\psi_b^0, \qquad [6.17]$$

is still an eigenstate of H^0, with the same eigenvalue E^0:

$$H^0\psi^0 = E^0\psi^0. \qquad [6.18]$$

Typically, the perturbation (H') will "break" (or "lift") the degeneracy: As we increase λ (from 0 to 1), the common unperturbed energy E^0 splits into two (Figure 6.4). Going the other direction, when we turn *off* the perturbation, the "upper" state reduces down to *one* linear combination of ψ_a^0 and ψ_b^0, and the "lower" state reduces to some *orthogonal* linear combination, but we don't know a priori *what* these **"good"** linear combinations will be. For this reason we can't even calculate the *first*-order energy (Equation 6.9)—we don't know what unperturbed states to use.

For the moment, therefore, let's just write the "good" unperturbed states in generic form (Equation 6.17), keeping α and β adjustable. We want to solve the Schrödinger equation,

$$H\psi = E\psi, \qquad [6.19]$$

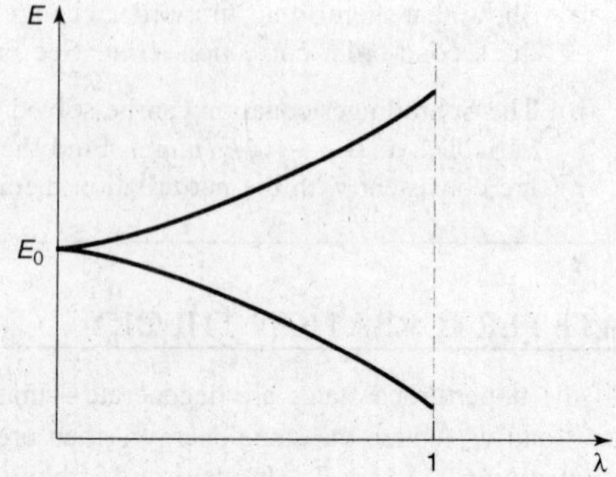

FIGURE 6.4: "Lifting" of a degeneracy by a perturbation.

with $H = H^0 + \lambda H'$ and

$$E = E^0 + \lambda E^1 + \lambda^2 E^2 + \cdots, \quad \psi = \psi^0 + \lambda\psi^1 + \lambda^2\psi^2 + \cdots. \qquad [6.20]$$

Plugging these into Equation 6.19, and collecting like powers of λ (as before) we find

$$H^0\psi^0 + \lambda(H'\psi^0 + H^0\psi^1) + \cdots = E^0\psi^0 + \lambda(E^1\psi^0 + E^0\psi^1) + \cdots.$$

But $H^0\psi^0 = E^0\psi^0$ (Equation 6.18), so the first terms cancel; at order λ^1 we have

$$H^0\psi^1 + H'\psi^0 = E^0\psi^1 + E^1\psi^0. \qquad [6.21]$$

Taking the inner product with ψ_a^0:

$$\langle\psi_a^0|H^0\psi^1\rangle + \langle\psi_a^0|H'\psi^0\rangle = E^0\langle\psi_a^0|\psi^1\rangle + E^1\langle\psi_a^0|\psi^0\rangle.$$

Because H^0 is hermitian, the first term on the left cancels the first term on the right. Putting in Equation 6.17 and exploiting the orthonormality condition (Equation 6.16), we obtain

$$\alpha\langle\psi_a^0|H'|\psi_a^0\rangle + \beta\langle\psi_a^0|H'|\psi_b^0\rangle = \alpha E^1,$$

or, more compactly,

$$\alpha W_{aa} + \beta W_{ab} = \alpha E^1, \qquad [6.22]$$

where

$$W_{ij} \equiv \langle\psi_i^0|H'|\psi_j^0\rangle, \quad (i, j = a, b). \qquad [6.23]$$

Similarly, the inner product with ψ_b^0 yields

$$\alpha W_{ba} + \beta W_{bb} = \beta E^1. \qquad [6.24]$$

Notice that the W's are (in principle) *known*—they are just the "matrix elements" of H', with respect to the unperturbed wave functions ψ_a^0 and ψ_b^0. Multiplying Equation 6.24 by W_{ab}, and using Equation 6.22 to eliminate βW_{ab}, we find:

$$\alpha[W_{ab}W_{ba} - (E^1 - W_{aa})(E^1 - W_{bb})] = 0. \qquad [6.25]$$

If α is *not* zero, Equation 6.25 yields an equation for E^1:

$$(E^1)^2 - E^1(W_{aa} + W_{bb}) + (W_{aa}W_{bb} - W_{ab}W_{ba}) = 0. \qquad [6.26]$$

Invoking the quadratic formula, and noting (from Equation 6.23) that $W_{ba} = W_{ab}^*$, we conclude that

$$\boxed{E_\pm^1 = \frac{1}{2}\left[W_{aa} + W_{bb} \pm \sqrt{(W_{aa} - W_{bb})^2 + 4|W_{ab}|^2}\right].} \qquad [6.27]$$

This is the fundamental result of degenerate perturbation theory; the two roots correspond to the two perturbed energies.

But what if α *is* zero? In that case $\beta = 1$, Equation 6.22 says $W_{ab} = 0$, and Equation 6.24 gives $E^1 = W_{bb}$. This is actually included in the general result (Equation 6.27), with the minus sign (the plus sign corresponds to $\alpha = 1$, $\beta = 0$). What's more, the *answers*,

$$E_+^1 = W_{aa} = \langle\psi_a^0|H'|\psi_a^0\rangle, \quad E_-^1 = W_{bb} = \langle\psi_b^0|H'|\psi_b^0\rangle,$$

are precisely what we would have obtained using *non*degenerate perturbation theory (Equation 6.9). We have simply been *lucky*: The states ψ_a^0 and ψ_b^0 were *already* the "good" linear combinations. Obviously, it would be greatly to our advantage if we could somehow *guess* the "good" states right from the start—then we could go ahead and use *non*degenerate perturbation theory. As it turns out, we can very often do this by exploiting the following theorem:

> **Theorem:** Let A be a hermitian operator that commutes with H^0 and H'. If ψ_a^0 and ψ_b^0 (the degenerate eigenfunctions of H^0) are also eigenfunctions of A, with distinct eigenvalues,
>
> $$A\psi_a^0 = \mu\psi_a^0, \quad A\psi_b^0 = \nu\psi_b^0, \quad \text{and } \mu \neq \nu,$$

then $W_{ab} = 0$ (and hence ψ_a^0 and ψ_b^0 are the "good" states to use in perturbation theory).

Proof: By assumption, $[A, H'] = 0$, so

$$\begin{aligned}\langle\psi_a^0|[A, H']\psi_b^0\rangle &= 0 \\ &= \langle\psi_a^0|AH'\psi_b^0\rangle - \langle\psi_a^0|H'A\psi_b^0\rangle \\ &= \langle A\psi_a^0|H'\psi_b^0\rangle - \langle\psi_a^0|H'\nu\psi_b^0\rangle \\ &= (\mu - \nu)\langle\psi_a^0|H'\psi_b^0\rangle = (\mu - \nu)W_{ab}.\end{aligned}$$

But $\mu \neq \nu$, so $W_{ab} = 0$. QED

Moral: If you're faced with degenerate states, look around for some hermitian operator A that commutes with H^0 and H'; pick as your unperturbed states ones that are simultaneously eigenfunctions of H^0 and A. Then use *ordinary* first-order perturbation theory. If you can't find such an operator, you'll have to resort to Equation 6.27, but in practice this is seldom necessary.

Problem 6.6 Let the two "good" unperturbed states be

$$\psi_\pm^0 = \alpha_\pm\psi_a^0 + \beta_\pm\psi_b^0,$$

where $\alpha_\pm$ and $\beta_\pm$ are determined (up to normalization) by Equation 6.22 (or Equation 6.24). Show explicitly that

(a) $\psi_\pm^0$ are orthogonal ($\langle\psi_+^0|\psi_-^0\rangle = 0$);

(b) $\langle\psi_+^0|H'|\psi_-^0\rangle = 0$;

(c) $\langle\psi_\pm^0|H'|\psi_\pm^0\rangle = E_\pm^1$, with $E_\pm^1$ given by Equation 6.27.

Problem 6.7 Consider a particle of mass m that is free to move in a one-dimensional region of length L that closes on itself (for instance, a bead that slides frictionlessly on a circular wire of circumference L, as in Problem 2.46).

(a) Show that the stationary states can be written in the form

$$\psi_n(x) = \frac{1}{\sqrt{L}}e^{2\pi inx/L}, \quad (-L/2 < x < L/2),$$

where $n = 0, \pm 1, \pm 2, \ldots$, and the allowed energies are

$$E_n = \frac{2}{m}\left(\frac{n\pi\hbar}{L}\right)^2.$$

Notice that—with the exception of the ground state ($n = 0$)—these are all doubly degenerate.

(b) Now suppose we introduce the perturbation

$$H' = -V_0 e^{-x^2/a^2},$$

where $a \ll L$. (This puts a little "dimple" in the potential at $x = 0$, as though we bent the wire slightly to make a "trap.") Find the first-order correction to E_n, using Equation 6.27. *Hint:* To evaluate the integrals, exploit the fact that $a \ll L$ to extend the limits from $\pm L/2$ to $\pm\infty$; after all, H' is essentially zero outside $-a < x < a$.

(c) What are the "good" linear combinations of ψ_n and ψ_{-n}, for this problem? Show that with these states you get the first-order correction using Equation 6.9.

(d) Find a hermitian operator A that fits the requirements of the theorem, and show that the simultaneous eigenstates of H^0 and A are precisely the ones you used in (c).

6.2.2 Higher-Order Degeneracy

In the previous section I assumed the degeneracy was two-fold, but it is easy to see how the method generalizes. Rewrite Equations 6.22 and 6.24 in matrix form:

$$\begin{pmatrix} W_{aa} & W_{ab} \\ W_{ba} & W_{bb} \end{pmatrix}\begin{pmatrix} \alpha \\ \beta \end{pmatrix} = E^1\begin{pmatrix} \alpha \\ \beta \end{pmatrix}. \qquad [6.28]$$

Evidently the E^1's are nothing but the *eigenvalues* of the W-matrix; Equation 6.26 is the characteristic equation for this matrix, and the "good" linear combinations of the unperturbed states are the eigenvectors of $\mathbf{W}$.

In the case of n-fold degeneracy, we look for the eigenvalues of the $n \times n$ matrix

$$W_{ij} = \langle\psi_i^0|H'|\psi_j^0\rangle. \qquad [6.29]$$

In the language of linear algebra, finding the "good" unperturbed wave functions amounts to constructing a basis in the degenerate subspace that *diagonalizes* the matrix **W**. Once again, if you can think of an operator A that *commutes* with H', and use the simultaneous eigenfunctions of A and H^0, then the W matrix will *automatically* be diagonal, and you won't have to fuss with solving the characteristic equation.[6] (If you're nervous about my casual generalization from 2-fold degeneracy to n-fold degeneracy, work Problem 6.10.)

Example 6.2 Consider the three-dimensional infinite cubical well (Problem 4.2):

$$V(x, y, z) = \begin{cases} 0, & \text{if } 0 < x < a, 0 < y < a, \text{ and } 0 < z < a; \\ \infty & \text{otherwise.} \end{cases} \tag{6.30}$$

The stationary states are

$$\psi^0_{n_x n_y n_z}(x, y, z) = \left(\frac{2}{a}\right)^{3/2} \sin\left(\frac{n_x \pi}{a} x\right) \sin\left(\frac{n_y \pi}{a} y\right) \sin\left(\frac{n_z \pi}{a} z\right), \tag{6.31}$$

where n_x, n_y, and n_z are positive integers. The corresponding allowed energies are

$$E^0_{n_x n_y n_z} = \frac{\pi^2 \hbar^2}{2ma^2}(n_x^2 + n_y^2 + n_z^2). \tag{6.32}$$

Notice that the ground state (ψ_{111}) is nondegenerate; its energy is

$$E^0_0 \equiv 3\frac{\pi^2 \hbar^2}{2ma^2}. \tag{6.33}$$

But the first excited state is (triply) degenerate:

$$\psi_a \equiv \psi_{112}, \quad \psi_b \equiv \psi_{121}, \text{ and } \psi_c \equiv \psi_{211}, \tag{6.34}$$

all share the same energy

$$E^0_1 \equiv 3\frac{\pi^2 \hbar^2}{ma^2}. \tag{6.35}$$

Now let's introduce the perturbation

$$H' = \begin{cases} V_0, & \text{if } 0 < x < a/2 \text{ and } 0 < y < a/2; \\ 0, & \text{otherwise.} \end{cases} \tag{6.36}$$

[6] Degenerate perturbation theory amounts to diagonalization of the degenerate part of the Hamiltonian. The diagonalization of matrices (and simultaneous diagonalizability of commuting matrices) is discussed in the Appendix (Section A.5).

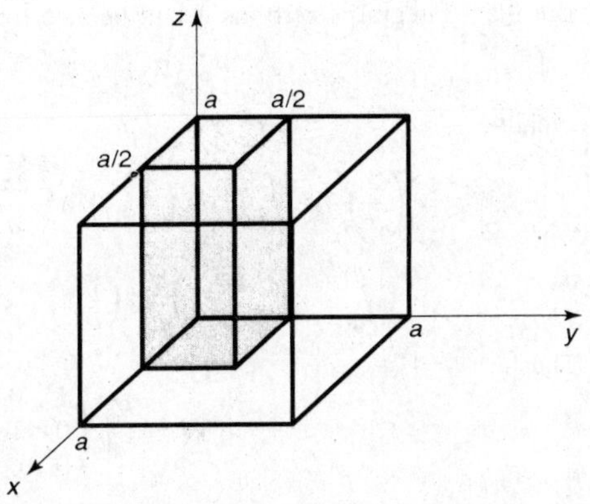

FIGURE 6.5: The perturbation increases the potential by an amount V_0 in the shaded sector.

This raises the potential by an amount V_0 in one quarter of the box (see Figure 6.5). The first-order correction to the ground state energy is given by Equation 6.9:

$$\begin{aligned} E_0^1 &= \langle \psi_{111} | H' | \psi_{111} \rangle \\ &= \left(\frac{2}{a}\right)^3 V_0 \int_0^{a/2} \sin^2\left(\frac{\pi}{a}x\right) dx \int_0^{a/2} \sin^2\left(\frac{\pi}{a}y\right) dy \int_0^a \sin^2\left(\frac{\pi}{a}z\right) dz \\ &= \frac{1}{4} V_0, \end{aligned} \quad [6.37]$$

which is pretty much what we would expect.

For the first excited state we need the full machinery of degenerate perturbation theory. The first step is to construct the matrix $\mathbf{W}$. The diagonal elements are the same as for the ground state (except that the argument of one of the sines is doubled); you can check for yourself that

$$W_{aa} = W_{bb} = W_{cc} = \frac{1}{4} V_0.$$

The off-diagonal elements are more interesting:

$$\begin{aligned} W_{ab} = \left(\frac{2}{a}\right)^3 V_0 \int_0^{a/2} \sin^2\left(\frac{\pi}{a}x\right) dx \\ \times \int_0^{a/2} \sin\left(\frac{\pi}{a}y\right) \sin\left(\frac{2\pi}{a}y\right) dy \int_0^a \sin\left(\frac{2\pi}{a}z\right) \sin\left(\frac{\pi}{a}z\right) dz. \end{aligned}$$

But the z integral is zero (as it will be also for W_{ac}), so

$$W_{ab} = W_{ac} = 0.$$

Finally,

$$W_{bc} = \left(\frac{2}{a}\right)^3 V_0 \int_0^{a/2} \sin\left(\frac{\pi}{a}x\right)\sin\left(\frac{2\pi}{a}x\right)\,dx$$
$$\times \int_0^{a/2} \sin\left(\frac{2\pi}{a}y\right)\sin\left(\frac{\pi}{a}y\right)\,dy \int_0^{a} \sin^2\left(\frac{\pi}{a}z\right)\,dz = \frac{16}{9\pi^2}V_0.$$

Thus

$$\mathbf{W} = \frac{V_0}{4}\begin{pmatrix} 1 & 0 & 0 \\ 0 & 1 & \kappa \\ 0 & \kappa & 1 \end{pmatrix} \qquad [6.38]$$

where $\kappa \equiv (8/3\pi)^2 \approx 0.7205$.

The characteristic equation for $\mathbf{W}$ (or rather, for $4\mathbf{W}/V_0$, which is easier to work with) is

$$(1-w)^3 - \kappa^2(1-w) = 0,$$

and the eigenvalues are

$$w_1 = 1; \quad w_2 = 1+\kappa \approx 1.705; \quad w_3 = 1-\kappa \approx 0.2795.$$

To first order in λ, then,

$$E_1(\lambda) = \begin{cases} E_1^0 + \lambda V_0/4, \\ E_1^0 + \lambda(1+\kappa)V_0/4, \\ E_1^0 + \lambda(1-\kappa)V_0/4, \end{cases} \qquad [6.39]$$

where E_1^0 is the (common) unperturbed energy (Equation 6.35). The perturbation lifts the degeneracy, splitting E_1^0 into three distinct energy levels (see Figure 6.6). Notice that if we had naively applied *non*degenerate perturbation theory to this problem, we would have concluded that the first-order correction (Equation 6.9) is the same for all three states, and equal to $V_0/4$—which is actually correct only for the middle state.

Meanwhile, the "good" unperturbed states are linear combinations of the form

$$\psi^0 = \alpha\psi_a + \beta\psi_b + \gamma\psi_c, \qquad [6.40]$$

where the coefficients (α, β, and γ) form the eigenvectors of the matrix $\mathbf{W}$:

$$\begin{pmatrix} 1 & 0 & 0 \\ 0 & 1 & \kappa \\ 0 & \kappa & 1 \end{pmatrix}\begin{pmatrix} \alpha \\ \beta \\ \gamma \end{pmatrix} = w\begin{pmatrix} \alpha \\ \beta \\ \gamma \end{pmatrix}$$

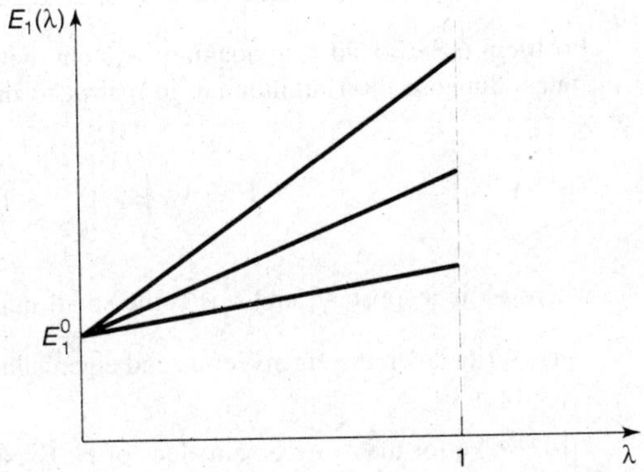

FIGURE 6.6: Lifting of the degeneracy in Example 6.2 (Equation 6.39).

For $w = 1$ we get $\alpha = 1$, $\beta = \gamma = 0$; for $w = 1 \pm \kappa$ we get $\alpha = 0$, $\beta = \pm\gamma = 1/\sqrt{2}$. (I normalized them as I went along.) Thus the "good" states are[7]

$$\psi^0 = \begin{cases} \psi_a, \\ (\psi_b + \psi_c)/\sqrt{2}, \\ (\psi_b - \psi_c)/\sqrt{2}. \end{cases} \tag{6.41}$$

Problem 6.8 Suppose we perturb the infinite cubical well (Equation 6.30) by putting a delta function "bump" at the point $(a/4, a/2, 3a/4)$:

$$H' = a^3 V_0 \delta(x - a/4)\delta(y - a/2)\delta(z - 3a/4).$$

Find the first-order corrections to the energy of the ground state and the (triply degenerate) first excited states.

---

[7]We might have guessed this result right from the start by noting that the operator P_{xy}, which interchanges x and y, commutes with H'. Its eigenvalues are $+1$ (for functions that are *even* under the interchange), and -1 (for functions that are odd). In this case ψ_a is *already* even, $(\psi_b + \psi_c)$ is even, and $(\psi_b - \psi_c)$ is odd. This is not quite conclusive, since any linear combination of the even states would still be even. But if we also use the operator Q, which takes z to $a - z$, and note that ψ_a is an eigenfunction with eigenvalue -1, whereas the other two are eigenfunctions with eigenvalue $+1$, the ambiguity is resolved. Here the operators P_{xy} and Q together play the role of A in the theorem of Section 6.2.1.

*__Problem 6.9__ Consider a quantum system with just *three* linearly independent states. Suppose the Hamiltonian, in matrix form, is

$$\mathbf{H} = V_0 \begin{pmatrix} (1-\epsilon) & 0 & 0 \\ 0 & 1 & \epsilon \\ 0 & \epsilon & 2 \end{pmatrix},$$

where V_0 is a constant, and ϵ is some small number ($\epsilon \ll 1$).

(a) Write down the eigenvectors and eigenvalues of the *unperturbed* Hamiltonian ($\epsilon = 0$).

(b) Solve for the *exact* eigenvalues of $\mathbf{H}$. Expand each of them as a power series in ϵ, up to second order.

(c) Use first- and second-order *non*degenerate perturbation theory to find the approximate eigenvalue for the state that grows out of the nondegenerate eigenvector of H^0. Compare the exact result, from (a).

(d) Use *degenerate* perturbation theory to find the first-order correction to the two initially degenerate eigenvalues. Compare the exact results.

Problem 6.10 In the text I asserted that the first-order corrections to an n-fold degenerate energy are the eigenvalues of the W matrix, and I justified this claim as the "natural" generalization of the case $n = 2$. *Prove* it, by reproducing the steps in Section 6.2.1, starting with

$$\psi^0 = \sum_{j=1}^{n} \alpha_j \psi_j^0$$

(generalizing Equation 6.17), and ending by showing that the analog to Equation 6.22 can be interpreted as the eigenvalue equation for the matrix $\mathbf{W}$.

6.3 THE FINE STRUCTURE OF HYDROGEN

In our study of the hydrogen atom (Section 4.2) we took the Hamiltonian to be

$$H = -\frac{\hbar^2}{2m}\nabla^2 - \frac{e^2}{4\pi\epsilon_0}\frac{1}{r} \qquad [6.42]$$

(electron kinetic energy plus coulombic potential energy). But this is not quite the whole story. We have already learned how to correct for the motion of the

TABLE 6.1: Hierarchy of corrections to the Bohr energies of hydrogen.

Bohr energies:	of order	$\alpha^2 mc^2$
Fine structure:	of order	$\alpha^4 mc^2$
Lamb shift:	of order	$\alpha^5 mc^2$
Hyperfine splitting:	of order	$(m/m_p)\alpha^4 mc^2$

nucleus: Just replace m by the reduced mass (Problem 5.1). More significant is the so-called **fine structure**, which is actually due to two distinct mechanisms: a **relativistic correction**, and **spin-orbit coupling**. Compared to the Bohr energies (Equation 4.70), fine structure is a tiny perturbation—smaller by a factor of α^2, where

$$\alpha \equiv \frac{e^2}{4\pi\epsilon_0\hbar c} \cong \frac{1}{137.036} \qquad [6.43]$$

is the famous **fine structure constant**. Smaller still (by another factor of α) is the **Lamb shift**, associated with the quantization of the electric field, and smaller by yet another order of magnitude is the **hyperfine structure**, which is due to the magnetic interaction between the dipole moments of the electron and the proton. This hierarchy is summarized in Table 6.1. In the present section we will analyze the fine structure of hydrogen, as an application of time-independent perturbation theory.

Problem 6.11

(a) Express the Bohr energies in terms of the fine structure constant and the rest energy (mc^2) of the electron.

(b) Calculate the fine structure constant from first principles (i.e., without recourse to the empirical values of ϵ_0, e, $\hbar$, and c). *Comment:* The fine structure constant is undoubtedly the most fundamental pure (dimensionless) number in all of physics. It relates the basic constants of electromagnetism (the charge of the electron), relativity (the speed of light), and quantum mechanics (Planck's constant). If you can solve part (b), you have the most certain Nobel Prize in history waiting for you. But I wouldn't recommend spending a lot of time on it right now; many smart people have tried, and all (so far) have failed.

6.3.1 The Relativistic Correction

The first term in the Hamiltonian is supposed to represent kinetic energy:

$$T = \frac{1}{2}mv^2 = \frac{p^2}{2m}, \qquad [6.44]$$

and the canonical substitution $\mathbf{p} \rightarrow (\hbar/i)\nabla$ yields the operator

$$T = -\frac{\hbar^2}{2m}\nabla^2. \qquad [6.45]$$

But Equation 6.44 is the *classical* expression for kinetic energy; the *relativistic* formula is

$$T = \frac{mc^2}{\sqrt{1-(v/c)^2}} - mc^2. \qquad [6.46]$$

The first term is the *total* relativistic energy (not counting *potential* energy, which we aren't concerned with at the moment), and the second term is the *rest* energy—the *difference* is the energy attributable to motion.

We need to express T in terms of the (relativistic) momentum,

$$p = \frac{mv}{\sqrt{1-(v/c)^2}}, \qquad [6.47]$$

instead of velocity. Notice that

$$p^2c^2 + m^2c^4 = \frac{m^2v^2c^2 + m^2c^4[1-(v/c)^2]}{1-(v/c)^2} = \frac{m^2c^4}{1-(v/c)^2} = (T+mc^2)^2,$$

so

$$T = \sqrt{p^2c^2 + m^2c^4} - mc^2. \qquad [6.48]$$

This relativistic equation for kinetic energy reduces (of course) to the classical result (Equation 6.44), in the nonrelativistic limit $p \ll mc$; expanding in powers of the small number (p/mc), we have

$$T = mc^2\left[\sqrt{1+\left(\frac{p}{mc}\right)^2} - 1\right] = mc^2\left[1 + \frac{1}{2}\left(\frac{p}{mc}\right)^2 - \frac{1}{8}\left(\frac{p}{mc}\right)^4 \cdots - 1\right]$$

$$= \frac{p^2}{2m} - \frac{p^4}{8m^3c^2} + \cdots. \qquad [6.49]$$

The lowest-order[8] relativistic correction to the Hamiltonian is evidently

$$H_r' = -\frac{p^4}{8m^3c^2}. \qquad [6.50]$$

---

[8]The kinetic energy of the electron in hydrogen is on the order of 10 eV, which is miniscule compared to its rest energy (511,000 eV), so the hydrogen atom is basically nonrelativistic, and we can afford to keep only the lowest-order correction. In Equation 6.49, p is the *relativistic* momentum (Equation 6.47), *not* the classical momentum mv. It is the former that we now associate with the quantum operator $-i\hbar\nabla$, in Equation 6.50.

In first-order perturbation theory, the correction to E_n is given by the expectation value of H' in the unperturbed state (Equation 6.9):

$$E_r^1 = \langle H_r' \rangle = -\frac{1}{8m^3c^2}\langle \psi | p^4 \psi \rangle = -\frac{1}{8m^3c^2}\langle p^2\psi | p^2\psi \rangle. \tag{6.51}$$

Now, the Schrödinger equation (for the unperturbed states) says

$$p^2\psi = 2m(E - V)\psi, \tag{6.52}$$

and hence[9]

$$E_r^1 = -\frac{1}{2mc^2}\langle (E - V)^2 \rangle = -\frac{1}{2mc^2}[E^2 - 2E\langle V \rangle + \langle V^2 \rangle]. \tag{6.53}$$

So far this is entirely general; but we're interested in hydrogen, for which $V(r) = -(1/4\pi\epsilon_0)e^2/r$:

$$E_r^1 = -\frac{1}{2mc^2}\left[E_n^2 + 2E_n\left(\frac{e^2}{4\pi\epsilon_0}\right)\left\langle\frac{1}{r}\right\rangle + \left(\frac{e^2}{4\pi\epsilon_0}\right)^2\left\langle\frac{1}{r^2}\right\rangle\right], \tag{6.54}$$

where E_n is the Bohr energy of the state in question.

To complete the job, we need the expectation values of $1/r$ and $1/r^2$, in the (unperturbed) state ψ_{nlm} (Equation 4.2.1). The first is easy (see Problem 6.12):

$$\left\langle\frac{1}{r}\right\rangle = \frac{1}{n^2 a}, \tag{6.55}$$

where a is the Bohr radius (Equation 4.72). The second is not so simple to derive (see Problem 6.33), but the answer is[10]

$$\left\langle\frac{1}{r^2}\right\rangle = \frac{1}{(l + 1/2)n^3a^2}. \tag{6.56}$$

It follows that

$$E_r^1 = -\frac{1}{2mc^2}\left[E_n^2 + 2E_n\left(\frac{e^2}{4\pi\epsilon_0}\right)\frac{1}{n^2a} + \left(\frac{e^2}{4\pi\epsilon_0}\right)^2\frac{1}{(l+1/2)n^3a^2}\right],$$

---

[9]There is some sleight-of-hand in this maneuver, which exploits the hermiticity of p^2 and of $(E - V)$. In truth, the operator p^4 is *not* hermitian for states with $l = 0$ (see Problem 6.15), and the applicability of perturbation theory to Equation 6.50 is therefore called into question (for the case $l = 0$). Fortunately, the *exact* solution is available; it can be obtained by using the (relativistic) Dirac equation in place of the (nonrelativistic) Schrödinger equation, and it confirms the results we obtain here by less rigorous means (see Problem 6.19).

[10]The general formula for the expectation value of *any* power of r is given in Hans A. Bethe and Edwin E. Salpeter, *Quantum Mechanics of One- and Two-Electron Atoms*, Plenum, New York (1977), p. 17.

or, eliminating a (using Equation 4.72) and expressing everything in terms of E_n (using Equation 4.70):

$$E_r^1 = -\frac{(E_n)^2}{2mc^2}\left[\frac{4n}{l+1/2} - 3\right]. \qquad [6.57]$$

Evidently the relativistic correction is smaller than E_n, by a factor of about $E_n/mc^2 = 2 \times 10^{-5}$.

You might have noticed that I used *non*degenerate perturbation theory in this calculation (Equation 6.51), in spite of the fact that the hydrogen atom is highly degenerate. But the perturbation is spherically symmetrical, so it commutes with L^2 and L_z. Moreover, the eigenfunctions of these operators (taken together) have distinct eigenvalues for the n^2 states with a given E_n. Luckily, then, the wave functions ψ_{nlm} *are* the "good" states for this problem (or, as we say, n, l, and m are the **good quantum numbers**), so as it happens the use of nondegenerate perturbation theory was legitimate (see the "Moral" to Section 6.2.1).

$*$**Problem 6.12** Use the virial theorem (Problem 4.40) to prove Equation 6.55.

Problem 6.13 In Problem 4.43 you calculated the expectation value of r^s in the state ψ_{321}. Check your answer for the special cases $s = 0$ (trivial), $s = -1$ (Equation 6.55), $s = -2$ (Equation 6.56), and $s = -3$ (Equation 6.64). Comment on the case $s = -7$.

$**$**Problem 6.14** Find the (lowest-order) relativistic correction to the energy levels of the one-dimensional harmonic oscillator. *Hint:* Use the technique in Example 2.5.

$***$**Problem 6.15** Show that p^2 is hermitian, but p^4 is *not*, for hydrogen states with $l = 0$. *Hint:* For such states ψ is independent of θ and ϕ, so

$$p^2 = -\frac{\hbar^2}{r^2}\frac{d}{dr}\left(r^2\frac{d}{dr}\right)$$

(Equation 4.13). Using integration by parts, show that

$$\langle f|p^2 g\rangle = -4\pi\hbar^2\left(r^2 f\frac{dg}{dr} - r^2 g\frac{df}{dr}\right)\Bigg|_0^\infty + \langle p^2 f|g\rangle.$$

Check that the boundary term vanishes for ψ_{n00}, which goes like

$$\psi_{n00} \sim \frac{1}{\sqrt{\pi}(na)^{3/2}}\exp(-r/na)$$

near the origin. Now do the same for p^4, and show that the boundary terms do *not* vanish. In fact:

$$\langle\psi_{n00}|p^4\psi_{m00}\rangle = \frac{8\hbar^4}{a^4}\frac{(n-m)}{(nm)^{5/2}} + \langle p^4\psi_{n00}|\psi_{m00}\rangle.$$

6.3.2 Spin-Orbit Coupling

Imagine the electron in orbit around the nucleus; from the *electron's* point of view, the proton is circling around *it* (Figure 6.7). This orbiting positive charge sets up a magnetic field **B**, in the electron frame, which exerts a torque on the spinning electron, tending to align its magnetic moment ($\boldsymbol{\mu}$) along the direction of the field. The Hamiltonian (Equation 4.157) is

$$H = -\boldsymbol{\mu} \cdot \mathbf{B}. \qquad [6.58]$$

To begin with, we need to figure out the magnetic field of the proton (**B**) and the dipole moment of the electron ($\boldsymbol{\mu}$).

The magnetic field of the proton. If we picture the proton (from the electron's perspective) as a continuous current loop (Figure 6.7), its magnetic field can be calculated from the Biot-Savart law:

$$B = \frac{\mu_0 I}{2r},$$

with an effective current $I = e/T$, where e is the charge of the proton and T is the period of the orbit. On the other hand, the orbital angular momentum of the *electron* (in the rest frame of the *nucleus*) is $L = rmv = 2\pi mr^2/T$. Moreover, **B** and **L** point in the same direction (up, in Figure 6.7), so

$$\mathbf{B} = \frac{1}{4\pi\epsilon_0}\frac{e}{mc^2r^3}\mathbf{L}. \qquad [6.59]$$

(I used $c = 1/\sqrt{\epsilon_0\mu_0}$ to eliminate μ_0 in favor of ϵ_0.)

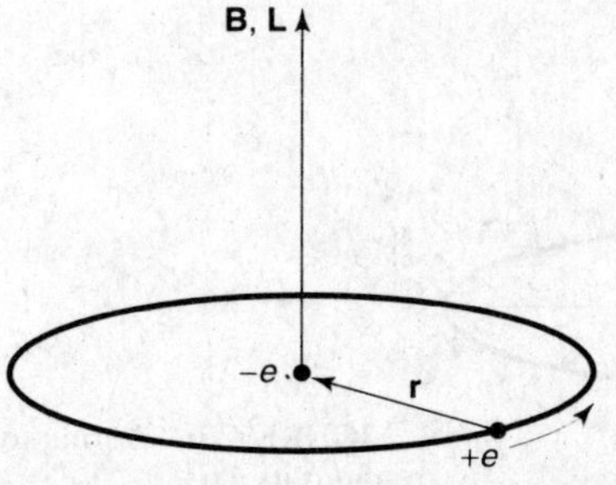

FIGURE 6.7: Hydrogen atom, from the electron's perspective.

The magnetic dipole moment of the electron. The magnetic dipole moment of a spinning charge is related to its (spin) angular momentum; the proportionality factor is the gyromagnetic ratio (which we already encountered in Section 4.4.2). Let's derive it, this time, using classical electrodynamics. Consider first a charge q smeared out around a ring of radius r, which rotates about the axis with period T (Figure 6.8). The magnetic dipole moment of the ring is defined as the current (q/T) times the area (πr^2):

$$\mu = \frac{q\pi r^2}{T}.$$

If the mass of the ring is m, its angular momentum is the moment of inertia (mr^2) times the angular velocity $(2\pi/T)$:

$$S = \frac{2\pi m r^2}{T}.$$

The gyromagnetic ratio for this configuration is evidently $\mu/S = q/2m$. Notice that it is independent of r (and T). If I had some more complicated object, such as a sphere (all I require is that it be a figure of revolution, rotating about its axis), I could calculate $\boldsymbol{\mu}$ and $\mathbf{S}$ by chopping it into little rings, and adding up their contributions. As long as the mass and the charge are distributed in the same manner (so that the charge-to-mass ratio is uniform), the gyromagnetic ratio will be the same for each ring, and hence also for the object as a whole. Moreover, the directions of $\boldsymbol{\mu}$ and $\mathbf{S}$ are the same (or opposite, if the charge is negative), so

$$\boldsymbol{\mu} = \left(\frac{q}{2m}\right)\mathbf{S}.$$

That was a purely *classical* calculation, however; as it turns out the electron's magnetic moment is *twice* the classical value:

$$\boldsymbol{\mu}_e = -\frac{e}{m}\mathbf{S}. \qquad [6.60]$$

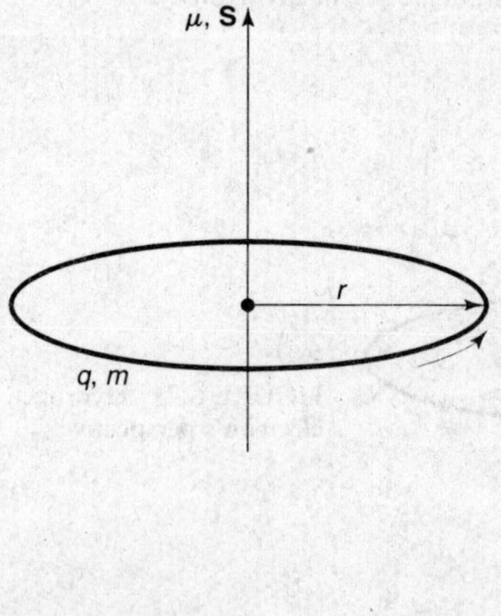

FIGURE 6.8: A ring of charge, rotating about its axis.

The "extra" factor of 2 was explained by Dirac, in his relativistic theory of the electron.[11]

Putting all this together, we have

$$H = \left(\frac{e^2}{4\pi\epsilon_0}\right)\frac{1}{m^2c^2r^3}\mathbf{S}\cdot\mathbf{L}.$$

But there is a serious fraud in this calculation: I did the analysis in the rest frame of the electron, but that's *not an inertial system*—it *accelerates*, as the electron orbits around the nucleus. You can get away with this if you make an appropriate kinematic correction, known as the **Thomas precession.**[12] In this context it throws in a factor of 1/2:[13]

$$H'_{\mathrm{so}} = \left(\frac{e^2}{8\pi\epsilon_0}\right)\frac{1}{m^2c^2r^3}\mathbf{S}\cdot\mathbf{L}. \qquad [6.61]$$

This is the **spin-orbit interaction**; apart from two corrections (the modified gyromagnetic ratio for the electron and the Thomas precession factor—which, coincidentally, exactly cancel one another) it is just what you would expect on the basis of a naive classical model. Physically, it is due to the torque exerted on the magnetic dipole moment of the spinning electron, by the magnetic field of the proton, in the electron's instantaneous rest frame.

Now the quantum mechanics. In the presence of spin-orbit coupling, the Hamiltonian no longer commutes with **L** and **S**, so the spin and orbital angular momenta are not separately conserved (see Problem 6.16). However, H'_{so} *does* commute with L^2, S^2 and the *total* angular momentum

$$\mathbf{J} \equiv \mathbf{L} + \mathbf{S}, \qquad [6.62]$$

---

[11] We have already noted that it can be dangerous to picture the electron as a spinning sphere (see Problem 4.25), and it is not too surprising that the naive classical model gets the gyromagnetic ratio wrong. The *deviation* from the classical expectation is known as the **g-factor**: $\mu = g(q/2m)\mathbf{S}$. Thus the g-factor of the electron, in Dirac's theory, is exactly 2. But **quantum electrodynamics** reveals tiny corrections to this: g_e is actually $2 + (\alpha/\pi) + \ldots = 2.002\ldots$. The calculation and measurement (which agree to exquisite precision) of the so-called **anomalous magnetic moment** of the electron were among the greatest achievements of twentieth-century physics.

[12] One way of thinking of it is that the electron is continually stepping from one inertial system to another; Thomas precession amounts to the cumulative effect of all these Lorentz transformations. We could avoid the whole problem, of course, by staying in the *lab* frame, in which the nucleus is at rest. In that case the field of the proton is purely *electric*, and you might wonder why it exerts any torque on the electron. Well, the fact is that a moving *magnetic* dipole acquires an *electric* dipole moment, and in the lab frame the spin-orbit coupling is due to the interaction of the *electric* field of the nucleus with the *electric* dipole moment of the electron. Because this analysis requires more sophisticated electrodynamics, it seems best to adopt the electron's perspective, where the physical mechanism is more transparent.

[13] More precisely, Thomas precession subtracts 1 from the g factor. See R. R. Haar and L. J. Curtis, *Am. J. Phys.*, **55**, 1044 (1987).

and hence these quantities *are* conserved (Equation 3.71). To put it another way, the eigenstates of L_z and S_z are not "good" states to use in perturbation theory, but the eigenstates of L^2, S^2, J^2, and J_z *are*. Now

$$J^2 = (\mathbf{L}+\mathbf{S})\cdot(\mathbf{L}+\mathbf{S}) = L^2 + S^2 + 2\mathbf{L}\cdot\mathbf{S},$$

so

$$\mathbf{L}\cdot\mathbf{S} = \frac{1}{2}(J^2 - L^2 - S^2), \tag{6.63}$$

and therefore the eigenvalues of $\mathbf{L}\cdot\mathbf{S}$ are

$$\frac{\hbar^2}{2}[j(j+1) - l(l+1) - s(s+1)].$$

In this case, of course, $s = 1/2$. Meanwhile, the expectation value of $1/r^3$ (see Problem 6.35(c)) is

$$\left\langle \frac{1}{r^3} \right\rangle = \frac{1}{l(l+1/2)(l+1)n^3a^3}, \tag{6.64}$$

and we conclude that

$$E^1_{\text{so}} = \langle H'_{\text{so}} \rangle = \frac{e^2}{8\pi\epsilon_0}\frac{1}{m^2c^2}\frac{(\hbar^2/2)[j(j+1) - l(l+1) - 3/4]}{l(l+1/2)(l+1)n^3a^3},$$

or, expressing it all in terms of E_n:[14]

$$E^1_{\text{so}} = \frac{(E_n)^2}{mc^2}\left\{\frac{n[j(j+1) - l(l+1) - 3/4]}{l(l+1/2)(l+1)}\right\}. \tag{6.65}$$

It is remarkable, considering the totally different physical mechanisms involved, that the relativistic correction and the spin-orbit coupling are of the same order (E_n^2/mc^2). Adding them together, we get the complete fine-structure formula (see Problem 6.17):

$$E^1_{\text{fs}} = \frac{(E_n)^2}{2mc^2}\left(3 - \frac{4n}{j+1/2}\right). \tag{6.66}$$

[14]Once again, the case $l = 0$ is problematic, since we are ostensibly dividing by zero. On the other hand, the numerator is *also* zero, since in this case $j = s$, so Equation 6.65 is indeterminate. On physical grounds there shouldn't be any spin-orbit coupling when $l = 0$. One way to resolve the ambiguity is to introduce the so-called **Darwin term** (see, for instance, G. K. Woodgate, *Elementary Atomic Structure*, 2nd ed., Oxford (1983), p. 63). Serendipitously, even though both the relativistic correction (Equation 6.57) and the spin-orbit coupling (Equation 6.65) are questionable in the case $l = 0$, their *sum* (Equation 6.66) is correct for *all* l (see Problem 6.19).

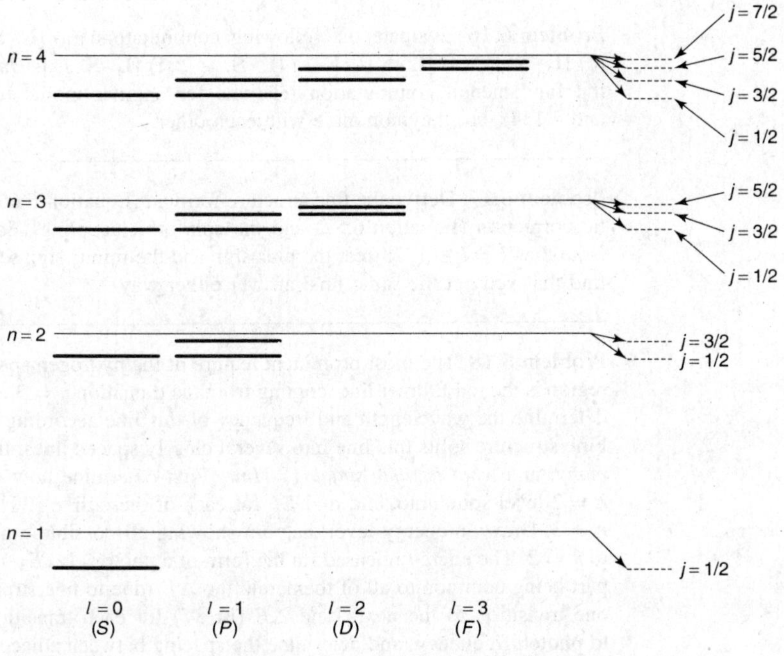

FIGURE 6.9: Energy levels of hydrogen, including fine structure (not to scale).

Combining this with the Bohr formula, we obtain the grand result for the energy levels of hydrogen, with fine structure included:

$$\boxed{E_{nj} = -\frac{13.6\,\text{eV}}{n^2}\left[1 + \frac{\alpha^2}{n^2}\left(\frac{n}{j+1/2} - \frac{3}{4}\right)\right].} \qquad [6.67]$$

Fine structure breaks the degeneracy in l (that is, for a given n, the different allowed values of l do not all carry the same energy), but it still preserves degeneracy in j (see Figure 6.9). The z-component eigenvalues for orbital and spin angular momentum (m_l and m_s) are no longer "good" quantum numbers—the stationary states are linear combinations of states with different values of these quantities; the "good" quantum numbers are n, l, s, j, and m_j.[15]

[15]To write $|j\, m_j\rangle$ (for given l and s) as a linear combination of $|l\, m_l\rangle|s\, m_s\rangle$ we would use the appropriate Clebsch-Gordan coefficients (Equation 4.185).

Problem 6.16 Evaluate the following commutators: (a) $[\mathbf{L}\cdot\mathbf{S}, \mathbf{L}]$, (b) $[\mathbf{L}\cdot\mathbf{S}, \mathbf{S}]$, (c) $[\mathbf{L}\cdot\mathbf{S}, \mathbf{J}]$, (d) $[\mathbf{L}\cdot\mathbf{S}, L^2]$, (e) $[\mathbf{L}\cdot\mathbf{S}, S^2]$, (f) $[\mathbf{L}\cdot\mathbf{S}, J^2]$. *Hint:* **L** and **S** satisfy the fundamental commutation relations for angular momentum (Equations 4.99 and 4.134), but they commute with each other.

∗**Problem 6.17** Derive the fine structure formula (Equation 6.66) from the relativistic correction (Equation 6.57) and the spin-orbit coupling (Equation 6.65). *Hint:* Note that $j = l \pm 1/2$; treat the plus sign and the minus sign separately, and you'll find that you get the same final answer either way.

∗∗**Problem 6.18** The most prominent feature of the hydrogen spectrum in the visible region is the red Balmer line, coming from the transition $n = 3$ to $n = 2$. First of all, determine the wavelength and frequency of this line according to the Bohr theory. Fine structure splits this line into several closely spaced lines; the question is: *How many*, and *what is their spacing*? *Hint:* First determine how many sublevels the $n = 2$ level splits into, and find E^1_{fs} for each of these, in eV. Then do the same for $n = 3$. Draw an energy level diagram showing all possible transitions from $n = 3$ to $n = 2$. The energy released (in the form of a photon) is $(E_3 - E_2) + \Delta E$, the first part being common to all of them, and the ΔE (due to fine structure) varying from one transition to the next. Find ΔE (in eV) for each transition. Finally, convert to photon frequency, and determine the spacing between adjacent spectral lines (in Hz)—*not* the frequency interval between each line and the *unperturbed* line (which is, of course, unobservable), but the frequency interval between each line and the *next* one. Your final answer should take the form: "The red Balmer line splits into (???) lines. In order of increasing frequency, they come from the transitions (1) $j = $ (???) to $j = $ (???), (2) $j = $ (???) to $j = $ (???), ... The frequency spacing between line (1) and line (2) is (???) Hz, the spacing between line (2) and line (3) is (???) Hz, ..."

Problem 6.19 The *exact* fine-structure formula for hydrogen (obtained from the Dirac equation without recourse to perturbation theory) is[16]

$$E_{nj} = mc^2\left\{\left[1 + \left(\frac{\alpha}{n - (j + 1/2) + \sqrt{(j + 1/2)^2 - \alpha^2}}\right)^2\right]^{-1/2} - 1\right\}.$$

Expand to order α^4 (noting that $\alpha \ll 1$), and show that you recover Equation 6.67.

[16]Bethe and Salpeter (footnote 10), page 238.

6.4 THE ZEEMAN EFFECT

When an atom is placed in a uniform external magnetic field $\mathbf{B}_{\text{ext}}$, the energy levels are shifted. This phenomenon is known as the **Zeeman effect**. For a single electron, the perturbation is

$$H'_Z = -(\boldsymbol{\mu}_l + \boldsymbol{\mu}_s) \cdot \mathbf{B}_{\text{ext}}, \tag{6.68}$$

where

$$\boldsymbol{\mu}_s = -\frac{e}{m}\mathbf{S} \tag{6.69}$$

is the magnetic dipole moment associated with electron spin, and

$$\boldsymbol{\mu}_l = -\frac{e}{2m}\mathbf{L} \tag{6.70}$$

is the dipole moment associated with orbital motion.[17] Thus

$$H'_Z = \frac{e}{2m}(\mathbf{L} + 2\mathbf{S}) \cdot \mathbf{B}_{\text{ext}}. \tag{6.71}$$

The nature of the Zeeman splitting depends critically on the strength of the external field in comparison with the *internal* field (Equation 6.59) that gives rise to spin-orbit coupling. If $B_{\text{ext}} \ll B_{\text{int}}$, then fine structure dominates, and H'_Z can be treated as a small perturbation, whereas if $B_{\text{ext}} \gg B_{\text{int}}$, then the Zeeman effect dominates, and fine structure becomes the perturbation. In the intermediate zone, where the two fields are comparable, we need the full machinery of degenerate perturbation theory, and it is necessary to diagonalize the relevant portion of the Hamiltonian "by hand." In the following sections we shall explore each of these regimes briefly, for the case of hydrogen.

Problem 6.20 Use Equation 6.59 to estimate the internal field in hydrogen, and characterize quantitatively a "strong" and "weak" Zeeman field.

6.4.1 Weak-Field Zeeman Effect

If $B_{\text{ext}} \ll B_{\text{int}}$, fine structure dominates (Equation 6.67); the "good" quantum numbers are n, l, j, and m_j (but not m_l and m_s, because—in the presence of spin-orbit coupling—**L** and **S** are not separately conserved).[18] In first-order perturbation

[17] The gyromagnetic ratio for *orbital* motion is just the classical value $(q/2m)$—it is only for *spin* that there is an "extra" factor of 2.

[18] In this problem we have a perturbation (Zeeman splitting) piled on top of a perturbation (fine structure). The "good" quantum numbers are those appropriate to the dominant perturbation—in this case the fine structure. The secondary perturbation (Zeeman splitting) lifts the remaining degeneracy in J_z, which plays here the role of the operator A in the theorem of Section 6.2.1. Technically, J_z does not commute with H'_Z, but it does in the time average sense of Equation 6.73.

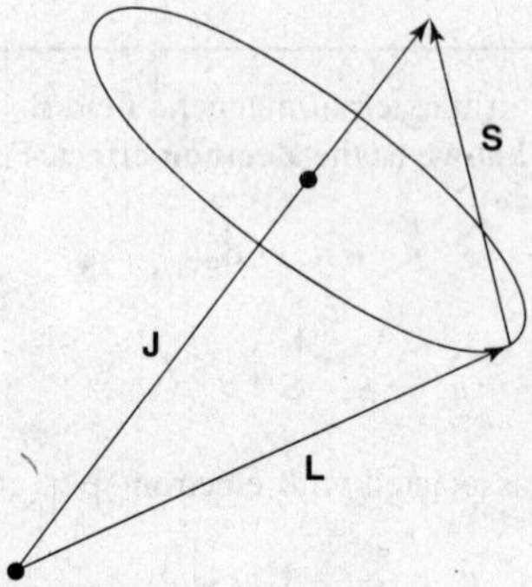

FIGURE 6.10: In the presence of spin-orbit coupling, L and S are not separately conserved; they precess about the fixed total angular momentum, J.

theory, the Zeeman correction to the energy is

$$E_Z^1 = \langle nljm_j | H_Z' | nljm_j \rangle = \frac{e}{2m} \mathbf{B}_{\text{ext}} \cdot \langle \mathbf{L} + 2\mathbf{S} \rangle. \tag{6.72}$$

Now $\mathbf{L} + 2\mathbf{S} = \mathbf{J} + \mathbf{S}$. Unfortunately, we do not immediately know the expectation value of $\mathbf{S}$. But we can figure it out, as follows: The total angular momentum $\mathbf{J} = \mathbf{L} + \mathbf{S}$ is constant (Figure 6.10); $\mathbf{L}$ and $\mathbf{S}$ precess rapidly about this fixed vector. In particular, the (time) *average* value of $\mathbf{S}$ is just its projection along $\mathbf{J}$:

$$\mathbf{S}_{\text{ave}} = \frac{(\mathbf{S} \cdot \mathbf{J})}{J^2} \mathbf{J}. \tag{6.73}$$

But $\mathbf{L} = \mathbf{J} - \mathbf{S}$, so $L^2 = J^2 + S^2 - 2\mathbf{J} \cdot \mathbf{S}$, and hence

$$\mathbf{S} \cdot \mathbf{J} = \frac{1}{2}(J^2 + S^2 - L^2) = \frac{\hbar^2}{2}[j(j+1) + s(s+1) - l(l+1)], \tag{6.74}$$

from which it follows that

$$\langle \mathbf{L} + 2\mathbf{S} \rangle = \left\langle \left(1 + \frac{\mathbf{S} \cdot \mathbf{J}}{J^2}\right) \mathbf{J} \right\rangle = \left[1 + \frac{j(j+1) - l(l+1) + 3/4}{2j(j+1)}\right] \langle \mathbf{J} \rangle. \tag{6.75}$$

The term in square brackets is known as the **Landé g-factor**, g_J.

We may as well choose the z-axis to lie along $\mathbf{B}_{\text{ext}}$; then

$$E_Z^1 = \mu_B g_J B_{\text{ext}} m_j, \tag{6.76}$$

where

$$\mu_B \equiv \frac{e\hbar}{2m} = 5.788 \times 10^{-5} \text{ eV/T} \tag{6.77}$$

is the so-called **Bohr magneton**. The *total* energy is the sum of the fine-structure part (Equation 6.67) and the Zeeman contribution (Equation 6.76). For example, the ground state ($n = 1$, $l = 0$, $j = 1/2$, and therefore $g_J = 2$) splits into two levels:

$$-13.6 \text{ eV}(1 + \alpha^2/4) \pm \mu_B B_{\text{ext}}, \qquad [6.78]$$

with the plus sign for $m_j = 1/2$, and minus for $m_j = -1/2$. These energies are plotted (as functions of B_{ext}) in Figure 6.11.

∗**Problem 6.21** Consider the (eight) $n = 2$ states, $|2ljm_j\rangle$. Find the energy of each state, under weak-field Zeeman splitting, and construct a diagram like Figure 6.11 to show how the energies evolve as B_{ext} increases. Label each line clearly, and indicate its slope.

6.4.2 Strong-Field Zeeman Effect

If $B_{\text{ext}} \gg B_{\text{int}}$, the Zeeman effect dominates;[19] with B_{ext} in the z direction, the "good" quantum numbers are now n, l, m_l, and m_s (but not j and m_j because—in the presence of the external torque—the total angular momentum is not conserved, whereas L_z and S_z *are*). The Zeeman Hamiltonian is

$$H'_Z = \frac{e}{2m} B_{\text{ext}}(L_z + 2S_z),$$

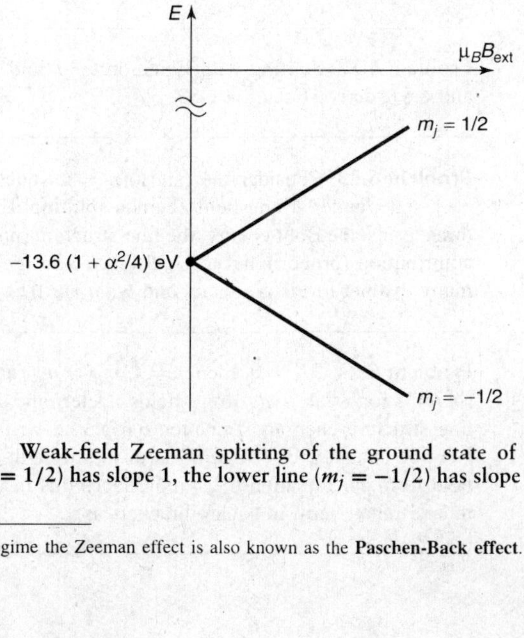

FIGURE 6.11: Weak-field Zeeman splitting of the ground state of hydrogen; the upper line ($m_j = 1/2$) has slope 1, the lower line ($m_j = -1/2$) has slope −1.

---

[19]In this regime the Zeeman effect is also known as the **Paschen-Back effect**.

and the "unperturbed" energies are

$$E_{nm_l m_s} = -\frac{13.6 \text{ eV}}{n^2} + \mu_B B_{\text{ext}}(m_l + 2m_s). \qquad [6.79]$$

That's the *answer*, if we ignore fine structure completely. But we can do better.

In first-order perturbation theory the fine-structure correction to these levels is

$$E^1_{\text{fs}} = \langle n l\, m_l\, m_s | (H'_r + H'_{\text{so}}) | n l\, m_l\, m_s \rangle. \qquad [6.80]$$

The relativistic contribution is the same as before (Equation 6.57); for the spin-orbit term (Equation 6.61) we need

$$\langle \mathbf{S} \cdot \mathbf{L} \rangle = \langle S_x \rangle \langle L_x \rangle + \langle S_y \rangle \langle L_y \rangle + \langle S_z \rangle \langle L_y \rangle = \hbar^2 m_l m_s \qquad [6.81]$$

(note that $\langle S_x \rangle = \langle S_y \rangle = \langle L_x \rangle = \langle L_y \rangle = 0$ for eigenstates of S_z and L_z). Putting all this together (Problem 6.22), we conclude that

$$E^1_{\text{fs}} = \frac{13.6 \text{ eV}}{n^3} \alpha^2 \left\{ \frac{3}{4n} - \left[\frac{l(l+1) - m_l m_s}{l(l+1/2)(l+1)} \right] \right\}. \qquad [6.82]$$

(The term in square brackets is indeterminate for $l = 0$; its correct value in this case is 1—see Problem 6.24.) The *total* energy is the sum of the Zeeman part (Equation 6.79) and the fine structure contribution (Equation 6.82).

Problem 6.22 Starting with Equation 6.80, and using Equations 6.57, 6.61, 6.64, and 6.81, derive Equation 6.82.

****Problem 6.23** Consider the (eight) $n = 2$ states, $|2\, l\, m_l\, m_s\rangle$. Find the energy of each state, under strong-field Zeeman splitting. Express each answer as the sum of three terms: the Bohr energy, the fine-structure (proportional to α^2), and the Zeeman contribution (proportional to $\mu_B B_{\text{ext}}$). If you ignore fine structure altogether, how many distinct levels are there, and what are their degeneracies?

Problem 6.24 If $l = 0$, then $j = s$, $m_j = m_s$, and the "good" states are the same ($|n\, m_s\rangle$) for weak *and* strong fields. Determine E^1_Z (from Equation 6.72) and the fine structure energies (Equation 6.67), and write down the general result for the $l = 0$ Zeeman effect—*regardless* of the strength of the field. Show that the strong-field formula (Equation 6.82) reproduces this result, provided that we interpret the indeterminate term in square brackets as 1.

6.4.3 Intermediate-Field Zeeman Effect

In the intermediate regime neither H'_Z nor H'_{fs} dominates, and we must treat the two on an equal footing, as perturbations to the Bohr Hamiltonian (Equation 6.42):

$$H' = H'_Z + H'_{fs}. \tag{6.83}$$

I'll confine my attention here to the case $n = 2$, and use as the basis for degenerate perturbation theory the states characterized by l, j, and m_j.[20] Using the Clebsch-Gordan coefficients (Problem 4.51 or Table 4.8) to express $|j\,m_j\rangle$ as a linear combination of $|l\,m_l\rangle|s\,m_s\rangle$, we have:

$$l=0\begin{cases}\psi_1 \equiv |\frac{1}{2}\,\frac{1}{2}\rangle = |0\,0\rangle|\frac{1}{2}\,\frac{1}{2}\rangle,\\[4pt] \psi_2 \equiv |\frac{1}{2}\,\frac{-1}{2}\rangle = |0\,0\rangle|\frac{1}{2}\,\frac{-1}{2}\rangle,\end{cases}$$

$$l=1\begin{cases}\psi_3 \equiv |\frac{3}{2}\,\frac{3}{2}\rangle = |1\,1\rangle|\frac{1}{2}\,\frac{1}{2}\rangle,\\[4pt] \psi_4 \equiv |\frac{3}{2}\,\frac{-3}{2}\rangle = |1\,-1\rangle|\frac{1}{2}\,\frac{-1}{2}\rangle,\\[4pt] \psi_5 \equiv |\frac{3}{2}\,\frac{1}{2}\rangle = \sqrt{2/3}|1\,0\rangle|\frac{1}{2}\,\frac{1}{2}\rangle + \sqrt{1/3}|1\,1\rangle|\frac{1}{2}\,\frac{-1}{2}\rangle,\\[4pt] \psi_6 \equiv |\frac{1}{2}\,\frac{1}{2}\rangle = -\sqrt{1/3}|1\,0\rangle|\frac{1}{2}\,\frac{1}{2}\rangle + \sqrt{2/3}|1\,1\rangle|\frac{1}{2}\,\frac{-1}{2}\rangle,\\[4pt] \psi_7 \equiv |\frac{3}{2}\,\frac{-1}{2}\rangle = \sqrt{1/3}|1\,-1\rangle|\frac{1}{2}\,\frac{1}{2}\rangle + \sqrt{2/3}|1\,0\rangle|\frac{1}{2}\,\frac{-1}{2}\rangle,\\[4pt] \psi_8 \equiv |\frac{1}{2}\,\frac{-1}{2}\rangle = -\sqrt{2/3}|1\,-1\rangle|\frac{1}{2}\,\frac{1}{2}\rangle + \sqrt{1/3}|1\,0\rangle|\frac{1}{2}\,\frac{-1}{2}\rangle.\end{cases}$$

In this basis the nonzero matrix elements of H'_{fs} are all on the diagonal, and given by Equation 6.66; H'_Z has four off-diagonal elements, and the complete matrix $-\mathbf{W}$ is (see Problem 6.25):

$$\begin{pmatrix} 5\gamma-\beta & 0 & 0 & 0 & 0 & 0 & 0 & 0\\ 0 & 5\gamma+\beta & 0 & 0 & 0 & 0 & 0 & 0\\ 0 & 0 & \gamma-2\beta & 0 & 0 & 0 & 0 & 0\\ 0 & 0 & 0 & \gamma+2\beta & 0 & 0 & 0 & 0\\ 0 & 0 & 0 & 0 & \gamma-\frac{2}{3}\beta & \frac{\sqrt{2}}{3}\beta & 0 & 0\\ 0 & 0 & 0 & 0 & \frac{\sqrt{2}}{3}\beta & 5\gamma-\frac{1}{3}\beta & 0 & 0\\ 0 & 0 & 0 & 0 & 0 & 0 & \gamma+\frac{2}{3}\beta & \frac{\sqrt{2}}{3}\beta\\ 0 & 0 & 0 & 0 & 0 & 0 & \frac{\sqrt{2}}{3}\beta & 5\gamma+\frac{1}{3}\beta \end{pmatrix}$$

[20] You can use l, m_l, m_s states if you prefer—this makes the matrix elements of H'_Z easier, but those of H'_{fs} more difficult; the W matrix will be more complicated, but its eigenvalues (which are independent of basis) are the same either way.

where

$$\gamma \equiv (\alpha/8)^2 13.6 \text{ eV} \quad \text{and} \quad \beta \equiv \mu_B B_{\text{ext}}.$$

The first four eigenvalues are already displayed along the diagonal; it remains only to find the eigenvalues of the two 2 × 2 blocks. The characteristic equation for the first of these is

$$\lambda^2 - \lambda(6\gamma - \beta) + \left(5\gamma^2 - \frac{11}{3}\gamma\beta\right) = 0,$$

and the quadratic formula gives the eigenvalues:

$$\lambda_{\pm} = -3\gamma + (\beta/2) \pm \sqrt{4\gamma^2 + (2/3)\gamma\beta + (\beta^2/4)}. \qquad [6.84]$$

The eigenvalues of the second block are the same, but with the sign of β reversed. The eight energies are listed in Table 6.2, and plotted against B_{ext} in Figure 6.12. In the zero-field limit ($\beta = 0$) they reduce to the fine-structure values; for weak fields ($\beta \ll \gamma$) they reproduce what you got in Problem 6.21; for strong fields ($\beta \gg \gamma$) we recover the results of Problem 6.23 (note the convergence to five distinct energy levels, at very high fields, as predicted in Problem 6.23).

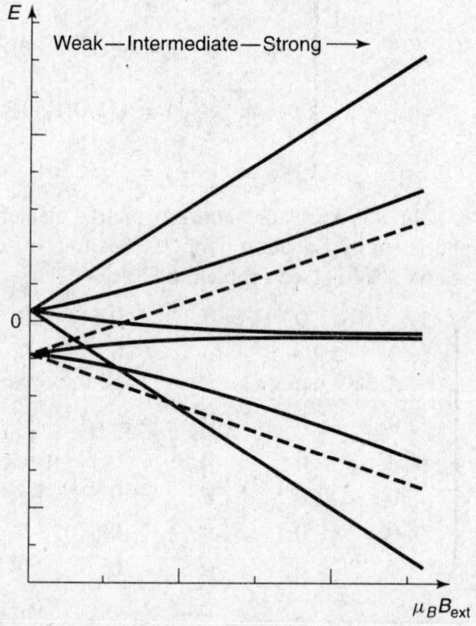

FIGURE 6.12: Zeeman splitting of the $n = 2$ states of hydrogen, in the weak, intermediate, and strong field regimes.

TABLE 6.2: Energy levels for the $n = 2$ states of hydrogen, with fine structure and Zeeman splitting.

ϵ_1	$=$	$E_2 - 5\gamma + \beta$
ϵ_2	$=$	$E_2 - 5\gamma - \beta$
ϵ_3	$=$	$E_2 - \gamma + 2\beta$
ϵ_4	$=$	$E_2 - \gamma - 2\beta$
ϵ_5	$=$	$E_2 - 3\gamma + \beta/2 + \sqrt{4\gamma^2 + (2/3)\gamma\beta + \beta^2/4}$
ϵ_6	$=$	$E_2 - 3\gamma + \beta/2 - \sqrt{4\gamma^2 + (2/3)\gamma\beta + \beta^2/4}$
ϵ_7	$=$	$E_2 - 3\gamma - \beta/2 + \sqrt{4\gamma^2 - (2/3)\gamma\beta + \beta^2/4}$
ϵ_8	$=$	$E_2 - 3\gamma - \beta/2 - \sqrt{4\gamma^2 - (2/3)\gamma\beta + \beta^2/4}$

Problem 6.25 Work out the matrix elements of H'_Z and H'_{fs}, and construct the W-matrix given in the text, for $n = 2$.

$***$**Problem 6.26** Analyze the Zeeman effect for the $n = 3$ states of hydrogen, in the weak, strong, and intermediate field regimes. Construct a table of energies (analogous to Table 6.2), plot them as functions of the external field (as in Figure 6.12), and check that the intermediate-field results reduce properly in the two limiting cases.

6.5 HYPERFINE SPLITTING

The proton itself constitutes a magnetic dipole, though its dipole moment is much smaller than the electron's because of the mass in the denominator (Equation 6.60):

$$\boldsymbol{\mu}_p = \frac{g_p e}{2m_p}\mathbf{S}_p, \quad \boldsymbol{\mu}_e = -\frac{e}{m_e}\mathbf{S}_e. \qquad [6.85]$$

(The proton is a composite structure, made up of three quarks, and its gyromagnetic ratio is not as simple as the electron's—hence the explicit g-factor (g_p), whose measured value is 5.59, as opposed to 2.00 for the electron.) According to classical electrodynamics, a dipole $\boldsymbol{\mu}$ sets up a magnetic field[21]

$$\mathbf{B} = \frac{\mu_0}{4\pi r^3}[3(\boldsymbol{\mu}\cdot\hat{r})\hat{r} - \boldsymbol{\mu}] + \frac{2\mu_0}{3}\boldsymbol{\mu}\delta^3(\mathbf{r}). \qquad [6.86]$$

[21] If you are unfamiliar with the delta function term in Equation 6.86, you can derive it by treating the dipole as a spinning charged spherical shell, in the limit as the radius goes to zero and the charge goes to infinity (with $\boldsymbol{\mu}$ held constant). See D. J. Griffiths, *Am. J. Phys.*, **50**, 698 (1982).

So the Hamiltonian of the electron, in the magnetic field due to the proton's magnetic dipole moment, is (Equation 6.58)

$$H'_{\text{hf}} = \frac{\mu_0 g_p e^2}{8\pi m_p m_e} \frac{[3(\mathbf{S}_p \cdot \hat{r})(\mathbf{S}_e \cdot \hat{r}) - \mathbf{S}_p \cdot \mathbf{S}_e]}{r^3} + \frac{\mu_0 g_p e^2}{3m_p m_e} \mathbf{S}_p \cdot \mathbf{S}_e \delta^3(\mathbf{r}). \qquad [6.87]$$

According to perturbation theory, the first-order correction to the energy (Equation 6.9) is the expectation value of the perturbing Hamiltonian:

$$E^1_{\text{hf}} = \frac{\mu_0 g_p e^2}{8\pi m_p m_e} \left\langle \frac{3(\mathbf{S}_p \cdot \hat{r})(\mathbf{S}_e \cdot \hat{r}) - \mathbf{S}_p \cdot \mathbf{S}_e}{r^3} \right\rangle + \frac{\mu_0 g_p e^2}{3m_p m_e} \langle \mathbf{S}_p \cdot \mathbf{S}_e \rangle |\psi(0)|^2. \qquad [6.88]$$

In the ground state (or any other state for which $l = 0$) the wave function is spherically symmetrical, and the first expectation value vanishes (see Problem 6.27). Meanwhile, from Equation 4.80 we find that $|\psi_{100}(0)|^2 = 1/(\pi a^3)$, so

$$E^1_{\text{hf}} = \frac{\mu_0 g_p e^2}{3\pi m_p m_e a^3} \langle \mathbf{S}_p \cdot \mathbf{S}_e \rangle, \qquad [6.89]$$

in the ground state. This is called **spin-spin coupling**, because it involves the dot product of two spins (contrast spin-orbit coupling, which involves $\mathbf{S} \cdot \mathbf{L}$).

In the presence of spin-spin coupling, the individual spin angular momenta are no longer conserved; the "good" states are eigenvectors of the *total* spin,

$$\mathbf{S} \equiv \mathbf{S}_e + \mathbf{S}_p. \qquad [6.90]$$

As before, we square this out to get

$$\mathbf{S}_p \cdot \mathbf{S}_e = \frac{1}{2}(S^2 - S_e^2 - S_p^2). \qquad [6.91]$$

But the electron and proton both have spin 1/2, so $S_e^2 = S_p^2 = (3/4)\hbar^2$. In the triplet state (spins "parallel") the total spin is 1, and hence $S^2 = 2\hbar^2$; in the singlet state the total spin is 0, and $S^2 = 0$. Thus

$$E^1_{\text{hf}} = \frac{4g_p \hbar^4}{3m_p m_e^2 c^2 a^4} \begin{cases} +1/4, & \text{(triplet);} \\ -3/4, & \text{(singlet).} \end{cases} \qquad [6.92]$$

Spin-spin coupling breaks the spin degeneracy of the ground state, lifting the triplet configuration and depressing the singlet (see Figure 6.13). The energy gap is evidently

$$\Delta E = \frac{4g_p \hbar^4}{3m_p m_e^2 c^2 a^4} = 5.88 \times 10^{-6} \text{ eV}. \qquad [6.93]$$

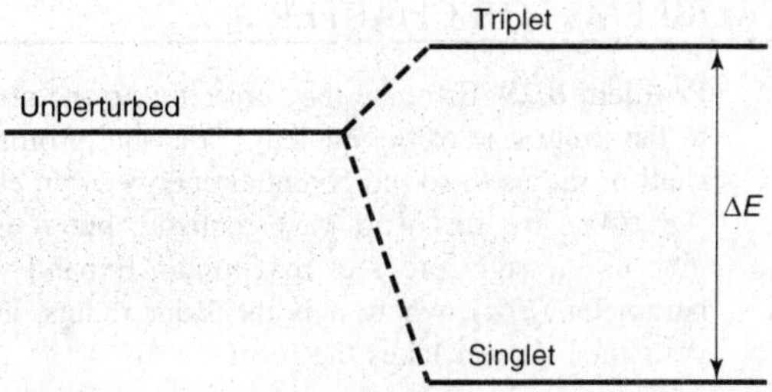

FIGURE 6.13: Hyperfine splitting in the ground state of hydrogen.

The frequency of the photon emitted in a transition from the triplet to the singlet state is

$$\nu = \frac{\Delta E}{h} = 1420 \text{ MHz}, \tag{6.94}$$

and the corresponding wavelength is $c/\nu = 21$ cm, which falls in the microwave region. This famous **21-centimeter line** is among the most pervasive and ubiquitous forms of radiation in the universe.

Problem 6.27 Let **a** and **b** be two constant vectors. Show that

$$\int (\mathbf{a} \cdot \hat{r})(\mathbf{b} \cdot \hat{r}) \sin\theta \, d\theta \, d\phi = \frac{4\pi}{3}(\mathbf{a} \cdot \mathbf{b}) \tag{6.95}$$

(the integration is over the usual range: $0 < \theta < \pi$, $0 < \phi < 2\pi$). Use this result to demonstrate that

$$\left\langle \frac{3(\mathbf{S}_p \cdot \hat{r})(\mathbf{S}_e \cdot \hat{r}) - \mathbf{S}_p \cdot \mathbf{S}_e}{r^3} \right\rangle = 0,$$

for states with $l = 0$. *Hint:* $\hat{r} = \sin\theta\cos\phi\hat{\imath} + \sin\theta\sin\phi\hat{\jmath} + \cos\theta\hat{k}$.

Problem 6.28 By appropriate modification of the hydrogen formula, determine the hyperfine splitting in the ground state of (a) **muonic hydrogen** (in which a muon—same charge and g-factor as the electron, but 207 times the mass—substitutes for the electron), (b) **positronium** (in which a positron—same mass and g-factor as the electron, but opposite charge—substitutes for the proton), and (c) **muonium** (in which an anti-muon—same mass and g-factor as a muon, but opposite charge—substitutes for the proton). *Hint:* Don't forget to use the reduced mass (Problem 5.1) in calculating the "Bohr radius" of these exotic "atoms." Incidentally, the answer you get for positronium (4.85×10^{-4} eV) is quite far from the experimental value (8.41×10^{-4} eV); the large discrepancy is due to pair annihilation ($e^+ + e^- \rightarrow \gamma + \gamma$), which contributes an extra $(3/4)\Delta E$, and does not occur (of course) in ordinary hydrogen, muonic hydrogen, or muonium.

FURTHER PROBLEMS FOR CHAPTER 6

Problem 6.29 Estimate the correction to the ground state energy of hydrogen due to the finite size of the nucleus. Treat the proton as a uniformly charged spherical shell of radius b, so the potential energy of an electron inside the shell is *constant:* $-e^2/(4\pi\epsilon_0 b)$; this isn't very realistic, but it is the simplest model, and it will give us the right order of magnitude. Expand your result in powers of the small parameter (b/a), where a is the Bohr radius, and keep only the leading term, so your final answer takes the form

$$\frac{\Delta E}{E} = A(b/a)^n.$$

Your business is to determine the constant A and the power n. Finally, put in $b \approx 10^{-15}$ m (roughly the radius of the proton) and work out the actual number. How does it compare with fine structure and hyperfine structure?

Problem 6.30 Consider the isotropic three-dimensional harmonic oscillator (Problem 4.38). Discuss the effect (in first order) of the perturbation

$$H' = \lambda x^2 yz$$

(for some constant λ) on

(a) the ground state;

(b) the (triply degenerate) first excited state. *Hint:* Use the answers to Problems 2.12 and 3.33.

$***$**Problem 6.31 Van der Waals interaction.** Consider two atoms a distance R apart. Because they are electrically neutral you might suppose there would be no force between them, but if they are polarizable there is in fact a weak attraction. To model this system, picture each atom as an electron (mass m, charge $-e$) attached by a spring (spring constant k) to the nucleus (charge $+e$), as in Figure 6.14. We'll assume the nuclei are heavy, and essentially motionless. The Hamiltonian for the unperturbed system is

$$H^0 = \frac{1}{2m}p_1^2 + \frac{1}{2}kx_1^2 + \frac{1}{2m}p_2^2 + \frac{1}{2}kx_2^2. \qquad [6.96]$$

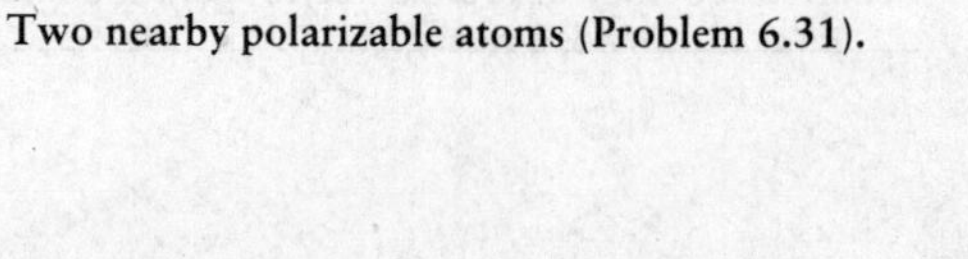

FIGURE 6.14: Two nearby polarizable atoms (Problem 6.31).

The Coulomb interaction between the atoms is

$$H' = \frac{1}{4\pi\epsilon_0}\left(\frac{e^2}{R} - \frac{e^2}{R+x_1} - \frac{e^2}{R-x_2} + \frac{e^2}{R+x_1-x_2}\right). \quad [6.97]$$

(a) Explain Equation 6.97. Assuming that $|x_1|$ and $|x_2|$ are both much less than R, show that

$$H' \cong -\frac{e^2 x_1 x_2}{2\pi\epsilon_0 R^3}. \quad [6.98]$$

(b) Show that the total Hamiltonian (Equation 6.96 plus Equation 6.98) separates into two harmonic oscillator Hamiltonians:

$$H = \left[\frac{1}{2m}p_+^2 + \frac{1}{2}\left(k - \frac{e^2}{4\pi\epsilon_0 R^3}\right)x_+^2\right] + \left[\frac{1}{2m}p_-^2 + \frac{1}{2}\left(k + \frac{e^2}{4\pi\epsilon_0 R^3}\right)x_-^2\right], \quad [6.99]$$

under the change of variables

$$x_\pm \equiv \frac{1}{\sqrt{2}}(x_1 \pm x_2), \quad \text{which entails } p_\pm = \frac{1}{\sqrt{2}}(p_1 \pm p_2). \quad [6.100]$$

(c) The ground state energy for this Hamiltonian is evidently

$$E = \frac{1}{2}\hbar(\omega_+ + \omega_-), \quad \text{where } \omega_\pm = \sqrt{\frac{k \mp (e^2/4\pi\epsilon_0 R^3)}{m}}. \quad [6.101]$$

Without the Coulomb interaction it would have been $E_0 = \hbar\omega_0$, where $\omega_0 = \sqrt{k/m}$. Assuming that $k \gg (e^2/4\pi\epsilon_0 R^3)$, show that

$$\Delta V \equiv E - E_0 \cong -\frac{\hbar}{8m^2\omega_0^3}\left(\frac{e^2}{4\pi\epsilon_0}\right)^2 \frac{1}{R^6}. \quad [6.102]$$

Conclusion: There is an attractive potential between the atoms, proportional to the inverse sixth power of their separation. This is the **van der Waals interaction** between two neutral atoms.

(d) Now do the same calculation using second-order perturbation theory. *Hint:* The unperturbed states are of the form $\psi_{n_1}(x_1)\psi_{n_2}(x_2)$, where $\psi_n(x)$ is a one-particle oscillator wave function with mass m and spring constant k; ΔV is the second-order correction to the ground state energy, for the perturbation in Equation 6.98 (notice that the *first*-order correction is zero).

$**$**Problem 6.32** Suppose the Hamiltonian H, for a particular quantum system, is a function of some parameter λ; let $E_n(\lambda)$ and $\psi_n(\lambda)$ be the eigenvalues and eigenfunctions of $H(\lambda)$. The **Feynman-Hellmann theorem**[22] states that

$$\frac{\partial E_n}{\partial \lambda} = \left\langle \psi_n \left| \frac{\partial H}{\partial \lambda} \right| \psi_n \right\rangle \qquad [6.103]$$

(assuming either that E_n is nondegenerate, or—if degenerate—that the ψ_n's are the "good" linear combinations of the degenerate eigenfunctions).

(a) Prove the Feynman-Hellmann theorem. *Hint:* Use Equation 6.9.

(b) Apply it to the one-dimensional harmonic oscillator, (i) using $\lambda = \omega$ (this yields a formula for the expectation value of V), (ii) using $\lambda = \hbar$ (this yields $\langle T \rangle$), and (iii) using $\lambda = m$ (this yields a relation between $\langle T \rangle$ and $\langle V \rangle$). Compare your answers to Problem 2.12, and the virial theorem predictions (Problem 3.31).

****Problem 6.33** The Feynman-Hellmann theorem (Problem 6.32) can be used to determine the expectation values of $1/r$ and $1/r^2$ for hydrogen.[23] The effective Hamiltonian for the radial wave functions is (Equation 4.53)

$$H = -\frac{\hbar^2}{2m}\frac{d^2}{dr^2} + \frac{\hbar^2}{2m}\frac{l(l+1)}{r^2} - \frac{e^2}{4\pi\epsilon_0}\frac{1}{r},$$

and the eigenvalues (expressed in terms of l)[24] are (Equation 4.70)

$$E_n = -\frac{me^4}{32\pi^2\epsilon_0^2\hbar^2(j_{\max} + l + 1)^2}.$$

(a) Use $\lambda = e$ in the Feynman-Hellmann theorem to obtain $\langle 1/r \rangle$. Check your result against Equation 6.55.

(b) Use $\lambda = l$ to obtain $\langle 1/r^2 \rangle$. Check your answer with Equation 6.56.

*****Problem 6.34** Prove **Kramers' relation**:[25]

$$\frac{s+1}{n^2}\langle r^s \rangle - (2s+1)a\langle r^{s-1} \rangle + \frac{s}{4}[(2l+1)^2 - s^2]a^2\langle r^{s-2} \rangle = 0, \qquad [6.104]$$

---

[22] Feynman obtained Equation 6.103 while working on his undergraduate thesis at MIT (R. P. Feynman, *Phys. Rev.* **56**, 340, 1939); Hellmann's work was published four years earlier in an obscure Russian journal.

[23] C. Sánchez del Rio, *Am. J. Phys.* **50**, 556 (1982); H. S. Valk, *Am. J. Phys.* **54**, 921 (1986).

[24] In part (b) we treat l as a continuous variable; n becomes a function of l, according to Equation 4.67, because $j_{\max}$, which must be an integer, is fixed. To avoid confusion, I have eliminated n, to reveal the dependence on l explicitly.

[25] This is also known as the (second) **Pasternack relation**. See H. Beker, *Am. J. Phys.* **65**, 1118 (1997). For a proof based on the Feynman-Hellmann theorem (Problem 6.32), see S. Balasubramanian, *Am. J. Phys.* **68**, 959 (2000).

which relates the expectation values of r to three different powers (s, $s-1$, and $s-2$), for an electron in the state ψ_{nlm} of hydrogen. *Hint:* Rewrite the radial equation (Equation 4.53) in the form

$$u'' = \left[\frac{l(l+1)}{r^2} - \frac{2}{ar} + \frac{1}{n^2a^2}\right] u,$$

and use it to express $\int (ur^s u'')dr$ in terms of $\langle r^s \rangle$, $\langle r^{s-1} \rangle$, and $\langle r^{s-2} \rangle$. Then use integration by parts to reduce the second derivative. Show that $\int (ur^s u')\,dr = -(s/2)\langle r^{s-1}\rangle$, and $\int (u' r^s u')\,dr = -[2/(s+1)]\int (u'' r^{s+1} u')\,dr$. Take it from there.

Problem 6.35

(a) Plug $s = 0$, $s = 1$, $s = 2$, and $s = 3$ into Kramers' relation (Equation 6.104) to obtain formulas for $\langle r^{-1}\rangle$, $\langle r\rangle$, $\langle r^2\rangle$, and $\langle r^3\rangle$. Note that you could continue indefinitely, to find *any* positive power.

(b) In the *other* direction, however, you hit a snag. Put in $s = -1$, and show that all you get is a relation between $\langle r^{-2}\rangle$ and $\langle r^{-3}\rangle$.

(c) But if you can get $\langle r^{-2}\rangle$ by some *other* means, you can apply the Kramers' relation to obtain the rest of the negative powers. Use Equation 6.56 (which is derived in Problem 6.33) to determine $\langle r^{-3}\rangle$, and check your answer against Equation 6.64.

*****Problem 6.36** When an atom is placed in a uniform external electric field $\mathbf{E}_{\text{ext}}$, the energy levels are shifted—a phenomenon known as the **Stark effect** (it is the electrical analog to the Zeeman effect). In this problem we analyze the Stark effect for the $n = 1$ and $n = 2$ states of hydrogen. Let the field point in the z direction, so the potential energy of the electron is

$$H'_S = eE_{\text{ext}}z = eE_{\text{ext}}r\cos\theta.$$

Treat this as a perturbation on the Bohr Hamiltonian (Equation 6.42). (Spin is irrelevant to this problem, so ignore it, and neglect the fine structure.)

(a) Show that the ground state energy is not affected by this perturbation, in first order.

(b) The first excited state is 4-fold degenerate: ψ_{200}, ψ_{211}, ψ_{210}, ψ_{21-1}. Using degenerate perturbation theory, determine the first-order corrections to the energy. Into how many levels does E_2 split?

(c) What are the "good" wave functions for part (b)? Find the expectation value of the electric dipole moment ($\mathbf{p}_e = -e\mathbf{r}$) in each of these "good" states.

Notice that the results are independent of the applied field—evidently hydrogen in its first excited state can carry a *permanent* electric dipole moment.

Hint: There are a lot of integrals in this problem, but almost all of them are zero. So study each one carefully, before you do any calculations: If the ϕ integral vanishes, there's not much point in doing the r and θ integrals! *Partial answer:* $W_{13} = W_{31} = -3eaE_{\text{ext}}$; all other elements are zero.

$* * *$**Problem 6.37** Consider the Stark effect (Problem 6.36) for the $n = 3$ states of hydrogen. There are initially nine degenerate states, ψ_{3lm} (neglecting spin, as before), and we turn on an electric field in the z direction.

(a) Construct the 9×9 matrix representing the perturbing Hamiltonian. *Partial answer:* $\langle 300|z|310\rangle = -3\sqrt{6}a$, $\langle 310|z|320\rangle = -3\sqrt{3}a$, $\langle 31\pm1|z|32\pm1\rangle = -(9/2)a$.

(b) Find the eigenvalues, and their degeneracies.

Problem 6.38 Calculate the wavelength, in centimeters, of the photon emitted under a hyperfine transition in the ground state ($n = 1$) of **deuterium**. Deuterium is "heavy" hydrogen, with an extra neutron in the nucleus; the proton and neutron bind together to form a **deuteron**, with spin 1 and magnetic moment

$$\boldsymbol{\mu}_d = \frac{g_d e}{2m_d}\mathbf{S}_d;$$

the deuteron g-factor is 1.71.

$* * *$**Problem 6.39** In a crystal, the electric field of neighboring ions perturbs the energy levels of an atom. As a crude model, imagine that a hydrogen atom is surrounded by three pairs of point charges, as shown in Figure 6.15. (Spin is irrelevant to this problem, so ignore it.)

(a) Assuming that $r \ll d_1$, $r \ll d_2$, and $r \ll d_3$, show that

$$H' = V_0 + 3(\beta_1 x^2 + \beta_2 y^2 + \beta_3 z^2) - (\beta_1 + \beta_2 + \beta_3)r^2,$$

where

$$\beta_i \equiv -\frac{e}{4\pi\epsilon_0}\frac{q_i}{d_i^3}, \quad \text{and } V_0 = 2(\beta_1 d_1^2 + \beta_2 d_2^2 + \beta_3 d_3^2).$$

(b) Find the lowest-order correction to the ground state energy.

(c) Calculate the first-order corrections to the energy of the first excited states ($n = 2$). Into how many levels does this four-fold degenerate system split, (i) in the case of **cubic symmetry**, $\beta_1 = \beta_2 = \beta_3$; (ii) in the case of **tetragonal symmetry**, $\beta_1 = \beta_2 \neq \beta_3$; (iii) in the general case of **orthorhombic symmetry** (all three different)?

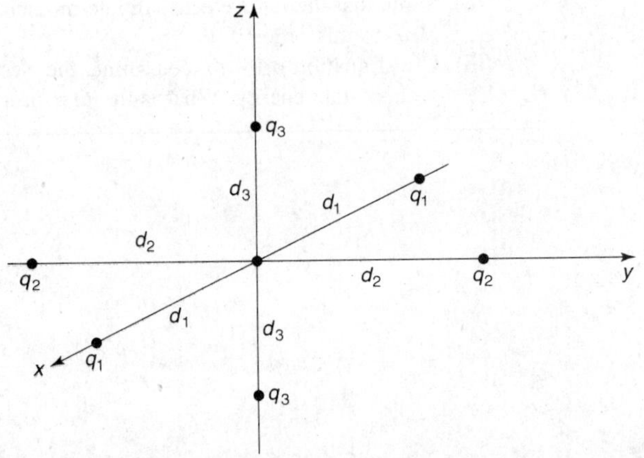

FIGURE 6.15: Hydrogen atom surrounded by six point charges (crude model for a crystal lattice); Problem 6.39.

∗∗∗**Problem 6.40** Sometimes it is possible to solve Equation 6.10 directly, without having to expand ψ_n^1 in terms of the unperturbed wave functions (Equation 6.11). Here are two particularly nice examples.

(a) **Stark effect in the ground state of hydrogen.**

(i) Find the first-order correction to the ground state of hydrogen in the presence of a uniform external electric field E_{ext} (the Stark effect—see Problem 6.36). *Hint:* Try a solution of the form

$$(A + Br + Cr^2)e^{-r/a} \cos\theta;$$

your problem is to find the constants A, B, and C that solve Equation 6.10.

(ii) Use Equation 6.14 to determine the second-order correction to the ground state energy (the first-order correction is zero, as you found in Problem 6.36(a)). *Answer:* $-m(3a^2eE_{\text{ext}}/2\hbar)^2$.

(b) If the proton had an *electric* dipole moment p, the potential energy of the electron in hydrogen would be perturbed in the amount

$$H' = -\frac{ep\cos\theta}{4\pi\epsilon_0 r^2}.$$

(i) Solve Equation 6.10 for the first-order correction to the ground state wave function.

(ii) Show that the *total* electric dipole moment of the atom is (surprisingly) *zero*, to this order.

(iii) Use Equation 6.14 to determine the second-order correction to the ground state energy. What is the *first*-order correction?

CHAPTER 7

THE VARIATIONAL PRINCIPLE

7.1 THEORY

Suppose you want to calculate the ground state energy, E_{gs}, for a system described by the Hamiltonian H, but you are unable to solve the (time-independent) Schrödinger equation. The **variational principle** will get you an *upper bound* for E_{gs}, which is sometimes all you need, and often, if you're clever about it, very close to the exact value. Here's how it works: Pick *any normalized function* ψ *whatsoever*; I claim that

$$\boxed{E_{gs} \leq \langle\psi|H|\psi\rangle \equiv \langle H\rangle.} \qquad [7.1]$$

That is, the expectation value of H, in the (presumably incorrect) state ψ is certain to *overestimate* the ground state energy. Of course, if ψ just happens to be one of the *excited* states, then *obviously* $\langle H\rangle$ exceeds E_{gs}; the point is that the same holds for any ψ whatsoever.

Proof: Since the (unknown) eigenfunctions of H form a complete set, we can express ψ as a linear combination of them:[1]

$$\psi = \sum_n c_n\psi_n, \quad \text{with} \quad H\psi_n = E_n\psi_n.$$

[1] If the Hamiltonian admits scattering states, as well as bound states, then we'll need an integral as well as a sum, but the argument is unchanged.

Since ψ is normalized,

$$1 = \langle\psi|\psi\rangle = \left\langle \sum_m c_m \psi_m \middle| \sum_n c_n \psi_n \right\rangle = \sum_m \sum_n c_m^* c_n \langle\psi_m|\psi_n\rangle = \sum_n |c_n|^2,$$

(assuming the eigenfunctions themselves have been orthonormalized: $\langle\psi_m|\psi_n\rangle = \delta_{mn}$). Meanwhile,

$$\langle H\rangle = \left\langle \sum_m c_m \psi_m \middle| H \sum_n c_n \psi_n \right\rangle = \sum_m \sum_n c_m^* E_n c_n \langle\psi_m|\psi_n\rangle = \sum_n E_n |c_n|^2.$$

But the ground state energy is, by definition, the *smallest* eigenvalue, so $E_{\text{gs}} \le E_n$, and hence

$$\langle H\rangle \ge E_{\text{gs}} \sum_n |c_n|^2 = E_{\text{gs}},$$

which is what we were trying to prove.

Example 7.1 Suppose we want to find the ground state energy for the one-dimensional harmonic oscillator:

$$H = -\frac{\hbar^2}{2m}\frac{d^2}{dx^2} + \frac{1}{2}m\omega^2 x^2.$$

Of course, we already know the *exact* answer, in this case (Equation 2.61): $E_{\text{gs}} = (1/2)\hbar\omega$; but this makes it a good test of the method. We might pick as our "trial" wave function the Gaussian,

$$\psi(x) = Ae^{-bx^2}, \tag{7.2}$$

where b is a constant, and A is determined by normalization:

$$1 = |A|^2 \int_{-\infty}^{\infty} e^{-2bx^2}\,dx = |A|^2\sqrt{\frac{\pi}{2b}} \Rightarrow A = \left(\frac{2b}{\pi}\right)^{1/4}. \tag{7.3}$$

Now

$$\langle H\rangle = \langle T\rangle + \langle V\rangle, \tag{7.4}$$

where, in this case,

$$\langle T\rangle = -\frac{\hbar^2}{2m}|A|^2 \int_{-\infty}^{\infty} e^{-bx^2}\frac{d^2}{dx^2}\left(e^{-bx^2}\right)dx = \frac{\hbar^2 b}{2m}, \tag{7.5}$$

and

$$\langle V \rangle = \frac{1}{2}m\omega^2|A|^2 \int_{-\infty}^{\infty} e^{-2bx^2} x^2\,dx = \frac{m\omega^2}{8b},$$

so

$$\langle H \rangle = \frac{\hbar^2 b}{2m} + \frac{m\omega^2}{8b}. \qquad [7.6]$$

According to Equation 7.1, this exceeds E_{gs} *for any* b; to get the *tightest* bound, let's *minimize* $\langle H \rangle$:

$$\frac{d}{db}\langle H \rangle = \frac{\hbar^2}{2m} - \frac{m\omega^2}{8b^2} = 0 \Rightarrow b = \frac{m\omega}{2\hbar}.$$

Putting this back into $\langle H \rangle$, we find

$$\langle H \rangle_{\min} = \frac{1}{2}\hbar\omega. \qquad [7.7]$$

In this case we hit the ground state energy right on the nose—because (obviously) I "just happened" to pick a trial function with precisely the form of the *actual* ground state (Equation 2.59). But the gaussian is very easy to work with, so it's a popular trial function, even when it bears little resemblance to the true ground state.

Example 7.2 Suppose we're looking for the ground state energy of the delta-function potential:

$$H = -\frac{\hbar^2}{2m}\frac{d^2}{dx^2} - \alpha\delta(x).$$

Again, we already know the exact answer (Equation 2.129): $E_{gs} = -m\alpha^2/2\hbar^2$. As before, we'll use a gaussian trial function (Equation 7.2). We've already determined the normalization, and calculated $\langle T \rangle$; all we need is

$$\langle V \rangle = -\alpha|A|^2 \int_{-\infty}^{\infty} e^{-2bx^2}\delta(x)\,dx = -\alpha\sqrt{\frac{2b}{\pi}}.$$

Evidently

$$\langle H \rangle = \frac{\hbar^2 b}{2m} - \alpha\sqrt{\frac{2b}{\pi}}, \qquad [7.8]$$

and we know that this exceeds E_{gs} for all b. Minimizing it,

$$\frac{d}{db}\langle H \rangle = \frac{\hbar^2}{2m} - \frac{\alpha}{\sqrt{2\pi b}} = 0 \Rightarrow b = \frac{2m^2\alpha^2}{\pi\hbar^4}.$$

So

$$\langle H \rangle_{\min} = -\frac{m\alpha^2}{\pi\hbar^2}, \qquad [7.9]$$

which is indeed somewhat higher than E_{gs}, since $\pi > 2$.

I said you can use *any* (normalized) trial function ψ whatsoever, and this is true in a sense. However, for *discontinuous* functions it takes some fancy footwork to assign a sensible meaning to the second derivative (which you need, in order to calculate $\langle T \rangle$). Continuous functions with kinks in them are fair game, however, as long as you are careful; the next example shows how to handle them.[2]

Example 7.3 Find an upper bound on the ground state energy of the one-dimensional infinite square well (Equation 2.19), using the "triangular" trial wave function (Figure 7.1):[3]

$$\psi(x) = \begin{cases} Ax, & \text{if } 0 \le x \le a/2, \\ A(a-x), & \text{if } a/2 \le x \le a, \\ 0, & \text{otherwise,} \end{cases} \qquad [7.10]$$

where A is determined by normalization:

$$1 = |A|^2 \left[\int_0^{a/2} x^2\,dx + \int_{a/2}^{a} (a-x)^2\,dx \right] = |A|^2 \frac{a^3}{12} \Rightarrow A = \frac{2}{a}\sqrt{\frac{3}{a}}. \qquad [7.11]$$

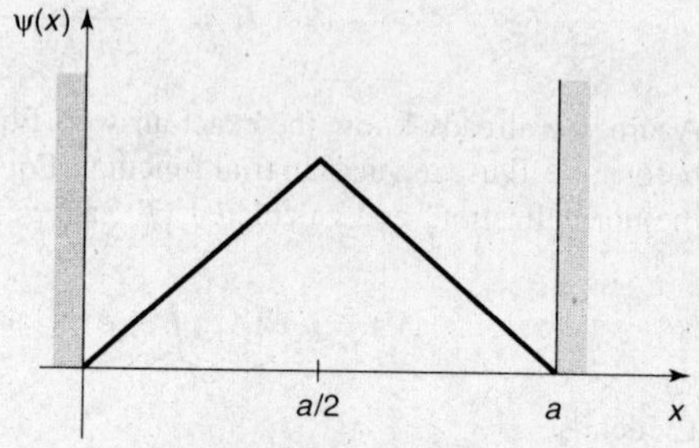

FIGURE 7.1: Triangular trial wave function for the infinite square well (Equation 7.10).

---

[2]For a collection of interesting examples see W. N. Mei, *Int. J. Educ. Sci. Tech.* **30**, 513 (1999).

[3]There is no point in trying a function (such as the gaussian) that "leaks" outside the well, because you'll get $\langle V \rangle = \infty$, and Equation 7.1 tells you nothing.

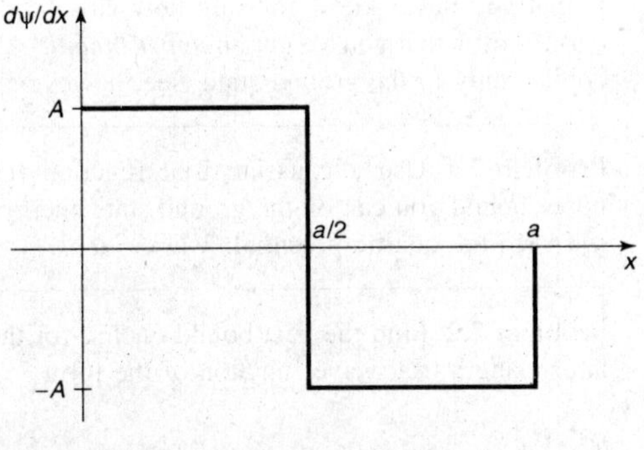

FIGURE 7.2: Derivative of the wave function in Figure 7.1.

In this case

$$\frac{d\psi}{dx} = \begin{cases} A, & \text{if } 0 < x < a/2, \\ -A, & \text{if } a/2 < x < a, \\ 0, & \text{otherwise,} \end{cases}$$

as indicated in Figure 7.2. Now, the derivative of a step function is a delta function (see Problem 2.24(b)):

$$\frac{d^2\psi}{dx^2} = A\delta(x) - 2A\delta(x - a/2) + A\delta(x - a), \qquad [7.12]$$

and hence

$$\langle H \rangle = -\frac{\hbar^2 A}{2m} \int [\delta(x) - 2\delta(x - a/2) + \delta(x - a)]\psi(x)\, dx$$

$$= -\frac{\hbar^2 A}{2m}[\psi(0) - 2\psi(a/2) + \psi(a)] = \frac{\hbar^2 A^2 a}{2m} = \frac{12\hbar^2}{2ma^2}. \qquad [7.13]$$

The exact ground state energy is $E_{gs} = \pi^2\hbar^2/2ma^2$ (Equation 2.27), so the theorem works ($12 > \pi^2$).

The variational principle is extraordinarily powerful, and embarrassingly easy to use. What a physical chemist does to find the ground state energy of some complicated molecule is write down a trial wave function with a large number of adjustable parameters, calculate $\langle H \rangle$, and tweak the parameters to get the lowest possible value. Even if ψ has little resemblance to the true wave function, you often get miraculously accurate values for E_{gs}. Naturally, if you have some way of guessing a *realistic* ψ, so much the better. The only *trouble* with the method is that you never know for sure how close you are to the target—all you can be *certain* of is that you've got an *upper bound*.[4] Moreover, as it stands, the technique applies only to the ground state (see, however, Problem 7.4).[5]

***Problem 7.1** Use a gaussian trial function (Equation 7.2) to obtain the lowest upper bound you can on the ground state energy of (a) the linear potential: $V(x) = \alpha|x|$; (b) the quartic potential: $V(x) = \alpha x^4$.

****Problem 7.2** Find the best bound on E_{gs} for the one-dimensional harmonic oscillator using a trial wave function of the form

$$\psi(x) = \frac{A}{x^2 + b^2},$$

where A is determined by normalization and b is an adjustable parameter.

Problem 7.3 Find the best bound on E_{gs} for the delta-function potential $V(x) = -\alpha\delta(x)$, using a triangular trial function (Equation 7.10, only centered at the origin). This time a is an adjustable parameter.

Problem 7.4

(a) Prove the following corollary to the variational principle: If $\langle\psi|\psi_{gs}\rangle = 0$, then $\langle H\rangle \geq E_{fe}$, where E_{fe} is the energy of the first excited state.

Thus, if we can find a trial function that is orthogonal to the exact ground state, we can get an upper bound on the *first excited state*. In general, it's difficult to be sure that ψ is orthogonal to ψ_{gs}, since (presumably) we don't *know* the latter. However, if the potential $V(x)$ is an *even* function of x, then the ground state is likewise even, and hence any *odd* trial function will automatically meet the condition for the corollary.

(b) Find the best bound on the first excited state of the one-dimensional harmonic oscillator using the trial function

$$\psi(x) = Axe^{-bx^2}.$$

---

[4] In practice this isn't much of a limitation, and there are sometimes ways of estimating the accuracy. The ground state helium has been calculated to many significant digits in this way—see for example G. W. Drake *et al.*, *Phys. Rev. A* **65**, 054501 (2002) or Vladimir I. Korobov, *Phys. Rev. A* **66**, 024501 (2002).

[5] For a systematic extension of the variational principle to the calculation of excited state energies see, for example, Linus Pauling and E. Bright Wilson, *Introduction to Quantum Mechanics, With Applications to Chemistry*, McGraw-Hill, New York (1935, paperback edition 1985), Section 26.

Problem 7.5

(a) Use the variational principle to prove that first-order non-degenerate perturbation theory always *overestimates* (or at any rate never *underestimates*) the ground state energy.

(b) In view of (a), you would expect that the *second*-order correction to the ground state is always negative. Confirm that this is indeed the case, by examining Equation 6.15.

7.2 THE GROUND STATE OF HELIUM

The helium atom (Figure 7.3) consists of two electrons in orbit around a nucleus containing two protons (also some neutrons, which are irrelevant to our purpose). The Hamiltonian for this system (ignoring fine structure and smaller corrections) is:

$$H = -\frac{\hbar^2}{2m}(\nabla_1^2 + \nabla_2^2) - \frac{e^2}{4\pi\epsilon_0}\left(\frac{2}{r_1} + \frac{2}{r_2} - \frac{1}{|\mathbf{r}_1 - \mathbf{r}_2|}\right). \quad [7.14]$$

Our problem is to calculate the ground state energy, E_{gs}. Physically, this represents the amount of energy it would take to strip off both electrons. (Given E_{gs} it is easy to figure out the "ionization energy" required to remove a *single* electron—see Problem 7.6.) The ground state energy of helium has been measured to great precision in the laboratory:

$$E_{gs} = -78.975 \text{ eV} \quad \text{(experimental)}. \quad [7.15]$$

This is the number we would like to reproduce theoretically.

It is curious that such a simple and important problem has no known exact solution.[6] The trouble comes from the electron-electron repulsion,

$$V_{ee} = \frac{e^2}{4\pi\epsilon_0}\frac{1}{|\mathbf{r}_1 - \mathbf{r}_2|}. \quad [7.16]$$

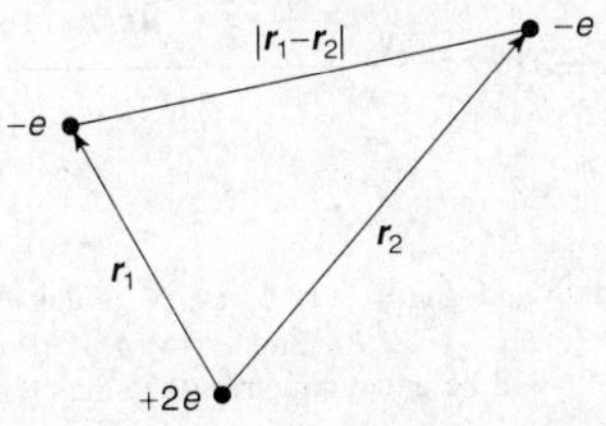

FIGURE 7.3: The helium atom.

[6]There do exist exactly soluble three-body problems with many of the qualitative features of helium, but using non-coulombic potentials (see Problem 7.17).

If we ignore this term altogether, H splits into two independent hydrogen Hamiltonians (only with a nuclear charge of $2e$, instead of e); the exact solution is just the product of hydrogenic wave functions:

$$\psi_0(\mathbf{r}_1, \mathbf{r}_2) \equiv \psi_{100}(\mathbf{r}_1)\psi_{100}(\mathbf{r}_2) = \frac{8}{\pi a^3} e^{-2(r_1+r_2)/a}, \qquad [7.17]$$

and the energy is $8E_1 = -109$ eV (Equation 5.31).[7] This is a long way from -79 eV, but it's a start.

To get a better approximation for E_{gs} we'll apply the variational principle, using ψ_0 as the trial wave function. This is a particularly convenient choice because it's an eigenfunction of *most* of the Hamiltonian:

$$H\psi_0 = (8E_1 + V_{ee})\psi_0. \qquad [7.18]$$

Thus

$$\langle H \rangle = 8E_1 + \langle V_{ee} \rangle, \qquad [7.19]$$

where[8]

$$\langle V_{ee} \rangle = \left(\frac{e^2}{4\pi\epsilon_0}\right)\left(\frac{8}{\pi a^3}\right)^2 \int \frac{e^{-4(r_1+r_2)/a}}{|\mathbf{r}_1 - \mathbf{r}_2|} d^3\mathbf{r}_1\, d^3\mathbf{r}_2. \qquad [7.20]$$

I'll do the $\mathbf{r}_2$ integral first; for this purpose $\mathbf{r}_1$ is fixed, and we may as well orient the $\mathbf{r}_2$ coordinate system so that the polar axis lies along $\mathbf{r}_1$ (see Figure 7.4). By the law of cosines,

$$|\mathbf{r}_1 - \mathbf{r}_2| = \sqrt{r_1^2 + r_2^2 - 2r_1 r_2 \cos\theta_2}, \qquad [7.21]$$

and hence

$$I_2 \equiv \int \frac{e^{-4r_2/a}}{|\mathbf{r}_1 - \mathbf{r}_2|} d^3 r_2 = \int \frac{e^{-4r_2/a}}{\sqrt{r_1^2 + r_2^2 - 2r_1 r_2 \cos\theta_2}} r_2^2 \sin\theta_2\, dr_2 d\theta_2 d\phi_2. \qquad [7.22]$$

The ϕ_2 integral is trivial (2π); the θ_2 integral is

$$\int_0^{\pi} \frac{\sin\theta_2}{\sqrt{r_1^2 + r_2^2 - 2r_1 r_2 \cos\theta_2}} d\theta_2 = \left. \frac{\sqrt{r_1^2 + r_2^2 - 2r_1 r_2 \cos\theta_2}}{r_1 r_2} \right|_0^{\pi}$$

---

[7] Here a is the ordinary Bohr radius and $E_n = -13.6/n^2$ eV is the nth Bohr energy; recall that for a nucleus with atomic number Z, $E_n \to Z^2 E_n$ and $a \to a/Z$ (Problem 4.16). The spin configuration associated with Equation 7.17 will be antisymmetric (the singlet).

[8] You can, if you like, interpret Equation 7.20 as first-order perturbation theory, with $H' = V_{ee}$. However, I regard this as a misuse of the method, since the perturbation is comparable in size to the unperturbed potential. I prefer, therefore, to think of it as a variational calculation, in which we are looking for an upper bound on E_{gs}.

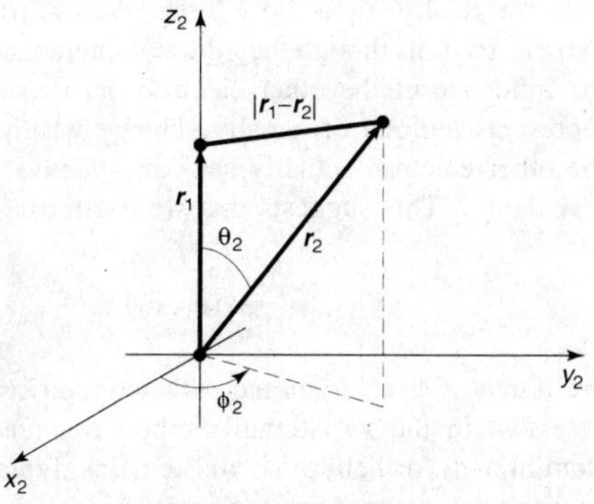

FIGURE 7.4: Choice of coordinates for the r_2-integral (Equation 7.20).

$$= \frac{1}{r_1 r_2}\left(\sqrt{r_1^2 + r_2^2 + 2r_1 r_2} - \sqrt{r_1^2 + r_2^2 - 2r_1 r_2}\right)$$

$$= \frac{1}{r_1 r_2}\left[(r_1 + r_2) - |r_1 - r_2|\right] = \begin{cases} 2/r_1, & \text{if } r_2 < r_1, \\ 2/r_2, & \text{if } r_2 > r_1. \end{cases} \tag{7.23}$$

Thus

$$I_2 = 4\pi\left(\frac{1}{r_1}\int_0^{r_1} e^{-4r_2/a} r_2^2\, dr_2 + \int_{r_1}^{\infty} e^{-4r_2/a} r_2\, dr_2\right)$$

$$= \frac{\pi a^3}{8r_1}\left[1 - \left(1 + \frac{2r_1}{a}\right)e^{-4r_1/a}\right]. \tag{7.24}$$

It follows that $\langle V_{ee}\rangle$ is equal to

$$\left(\frac{e^2}{4\pi\epsilon_0}\right)\left(\frac{8}{\pi a^3}\right)\int\left[1 - \left(1 + \frac{2r_1}{a}\right)e^{-4r_1/a}\right]e^{-4r_1/a} r_1 \sin\theta_1\, dr_1 d\theta_1 d\phi_1.$$

The angular integrals are easy (4π), and the r_1 integral becomes

$$\int_0^{\infty}\left[re^{-4r/a} - \left(r + \frac{2r^2}{a}\right)e^{-8r/a}\right]dr = \frac{5a^2}{128}.$$

Finally, then,

$$\langle V_{ee}\rangle = \frac{5}{4a}\left(\frac{e^2}{4\pi\epsilon_0}\right) = -\frac{5}{2}E_1 = 34 \text{ eV}, \tag{7.25}$$

and therefore

$$\langle H\rangle = -109 \text{ eV} + 34 \text{ eV} = -75 \text{ eV}. \tag{7.26}$$

Not bad (remember, the experimental value is -79 eV). But we can do better.

We need to think up a more realistic trial function than ψ_0 (which treats the two electrons as though they did not interact at all). Rather than completely *ignoring* the influence of the other electron, let us say that, on the average, each electron represents a cloud of negative charge which partially *shields* the nucleus, so that the other electron actually sees an *effective* nuclear charge (Z) that is somewhat *less* than 2. This suggests that we use a trial function of the form

$$\psi_1(\mathbf{r}_1, \mathbf{r}_2) \equiv \frac{Z^3}{\pi a^3} e^{-Z(r_1+r_2)/a}. \qquad [7.27]$$

We'll treat Z as a variational parameter, picking the value that minimizes H. (Please note that in the variational method we *never touch the Hamiltonian itself*—the Hamiltonian for helium is, and remains, Equation 7.14. But it's fine to *think* about approximating the Hamiltonian *as a way of motivating the choice of the trial wave function.*

This wave function is an eigenstate of the "unperturbed" Hamiltonian (neglecting electron repulsion), only with Z, instead of 2, in the Coulomb terms. With this in mind, we rewrite H (Equation 7.14) as follows:

$$H = -\frac{\hbar^2}{2m}(\nabla_1^2 + \nabla_2^2) - \frac{e^2}{4\pi\epsilon_0}\left(\frac{Z}{r_1} + \frac{Z}{r_2}\right)$$
$$+ \frac{e^2}{4\pi\epsilon_0}\left(\frac{(Z-2)}{r_1} + \frac{(Z-2)}{r_2} + \frac{1}{|\mathbf{r}_1 - \mathbf{r}_2|}\right). \qquad [7.28]$$

The expectation value of H is evidently

$$\langle H \rangle = 2Z^2 E_1 + 2(Z-2)\left(\frac{e^2}{4\pi\epsilon_0}\right)\left\langle \frac{1}{r} \right\rangle + \langle V_{ee} \rangle. \qquad [7.29]$$

Here $\langle 1/r \rangle$ is the expectation value of $1/r$ in the (one-particle) hydrogenic ground state ψ_{100} (but with nuclear charge Z); according to Equation 6.55,

$$\left\langle \frac{1}{r} \right\rangle = \frac{Z}{a}. \qquad [7.30]$$

The expectation value of V_{ee} is the same as before (Equation 7.25), except that instead of $Z = 2$ we now want *arbitrary* Z—so we multiply a by $2/Z$:

$$\langle V_{ee} \rangle = \frac{5Z}{8a}\left(\frac{e^2}{4\pi\epsilon_0}\right) = -\frac{5Z}{4} E_1. \qquad [7.31]$$

Putting all this together, we find

$$\langle H \rangle = \left[2Z^2 - 4Z(Z-2) - (5/4)Z\right] E_1 = [-2Z^2 + (27/4)Z]E_1. \qquad [7.32]$$

According to the variational principle, this quantity exceeds E_{gs} for *any* value of Z. The *lowest* upper bound occurs when $\langle H \rangle$ is minimized:

$$\frac{d}{dZ}\langle H \rangle = [-4Z + (27/4)]E_1 = 0,$$

from which it follows that

$$Z = \frac{27}{16} = 1.69. \qquad [7.33]$$

This seems reasonable; it tells us that the other electron partially screens the nucleus, reducing its effective charge from 2 down to about 1.69. Putting in this value for Z, we find

$$\langle H \rangle = \frac{1}{2}\left(\frac{3}{2}\right)^6 E_1 = -77.5 \text{ eV}. \qquad [7.34]$$

The ground state of helium has been calculated with great precision in this way, using increasingly complicated trial wave functions, with more and more adjustable parameters.[9] But we're within 2% of the correct answer, and, frankly, at this point my own interest in the problem begins to wane.[10]

Problem 7.6 Using $E_{gs} = -79.0$ eV for the ground state energy of helium, calculate the ionization energy (the energy required to remove just *one* electron). *Hint:* First calculate the ground state energy of the helium ion, He^+, with a single electron orbiting the nucleus; then subtract the two energies.

∗**Problem 7.7** Apply the techniques of this Section to the H^- and Li^+ ions (each has two electrons, like helium, but nuclear charges $Z = 1$ and $Z = 3$, respectively). Find the effective (partially shielded) nuclear charge, and determine the best upper bound on E_{gs}, for each case. *Comment:* In the case of H^- you should find that $\langle H \rangle > -13.6$ eV, which would appear to indicate that there is no bound state at all, since it would be energetically favorable for one electron to fly off, leaving behind a neutral hydrogen atom. This is not entirely surprising, since the electrons are less strongly attracted to the nucleus than they are in helium, and the electron repulsion tends to break the atom apart. However, it turns out to be incorrect. With a more sophisticated trial wave function (see Problem 7.18) it can be shown that $E_{gs} < -13.6$ eV, and hence that a bound state *does* exist. It's only *barely* bound,

---

[9]The classic studies are E. A. Hylleraas, *Z. Phys.* **65**, 209 (1930); C. L. Pekeris, *Phys. Rev.* **115**, 1216 (1959). For more recent work see footnote 4.

[10]The first excited state of helium can be calculated in much the same way, using a trial wave function orthogonal to the ground state. See P. J. E. Peebles, *Quantum Mechanics*, Princeton U.P., Princeton, NJ (1992), Section 40.

however, and there are no excited bound states,[11] so H^- has no discrete spectrum (all transitions are to and from the continuum). As a result, it is difficult to study in the laboratory, although it exists in great abundance on the surface of the sun.[12]

7.3 THE HYDROGEN MOLECULE ION

Another classic application of the variational principle is to the hydrogen molecule ion, H_2^+, consisting of a single electron in the Coulomb field of two protons (Figure 7.5). I shall assume for the moment that the protons are fixed in position, a specified distance R apart, although one of the most interesting byproducts of the calculation is going to be the actual *value* of R. The Hamiltonian is

$$H = -\frac{\hbar^2}{2m}\nabla^2 - \frac{e^2}{4\pi\epsilon_0}\left(\frac{1}{r_1} + \frac{1}{r_2}\right), \quad [7.35]$$

where r_1 and r_2 are the distances to the electron from the respective protons. As always, our strategy will be to guess a reasonable trial wave function, and invoke the variational principle to get a bound on the ground state energy. (Actually, our main interest is in finding out whether this system bonds at *all*—that is, whether its energy is less than that of a neutral hydrogen atom plus a free proton. If our trial wave function indicates that there *is* a bound state, a *better* trial function can only make the bonding even stronger.)

To construct the trial wave function, imagine that the ion is formed by taking a hydrogen atom in its ground state (Equation 4.80),

$$\psi_0(\mathbf{r}) = \frac{1}{\sqrt{\pi a^3}} e^{-r/a}, \quad [7.36]$$

bringing the second proton in from "infinity," and nailing it down a distance R away. If R is substantially greater than the Bohr radius, the electron's wave function probably isn't changed very much. But we would like to treat the two protons on

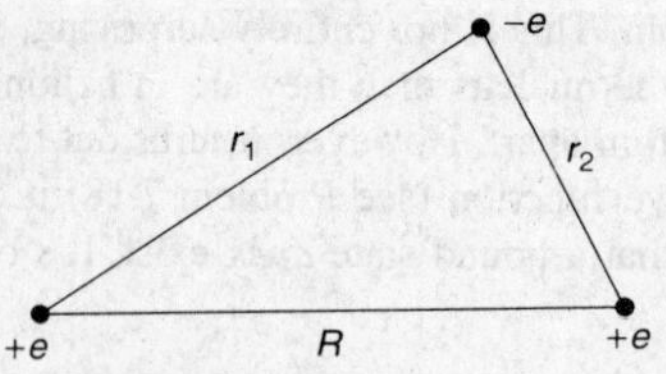

FIGURE 7.5: The hydrogen molecule ion, H_2^+.

[11] Robert N. Hill, *J. Math. Phys.* **18**, 2316 (1977).

[12] For further discussion see Hans A. Bethe and Edwin E. Salpeter, *Quantum Mechanics of One- and Two-Electron Atoms*, Plenum, New York (1977), Section 34.

an equal footing, so that the electron has the same probability of being associated with either one. This suggests that we consider a trial function of the form

$$\psi = A\left[\psi_0(r_1) + \psi_0(r_2)\right]. \qquad [7.37]$$

(Quantum chemists call this the **LCAO** technique, because we are expressing the *molecular* wave function as a **l**inear **c**ombination of **a**tomic **o**rbitals.)

Our first task is to *normalize* the trial function:

$$1 = \int |\psi|^2 d^3\mathbf{r} = |A|^2 \left[\int |\psi_0(r_1)|^2 d^3\mathbf{r} + \int |\psi_0(r_2)|^2 d^3\mathbf{r} + 2\int \psi_0(r_1)\psi_0(r_2)\, d^3\mathbf{r}\right]. \qquad [7.38]$$

The first two integrals are 1 (since ψ_0 itself is normalized); the third is more tricky. Let

$$I \equiv \langle\psi_0(r_1)|\psi_0(r_2)\rangle = \frac{1}{\pi a^3}\int e^{-(r_1+r_2)/a}\, d^3\mathbf{r}. \qquad [7.39]$$

Picking coordinates so that proton 1 is at the origin and proton 2 is on the z axis at the point R (Figure 7.6), we have

$$r_1 = r \quad \text{and} \quad r_2 = \sqrt{r^2 + R^2 - 2rR\cos\theta}, \qquad [7.40]$$

and therefore

$$I = \frac{1}{\pi a^3}\int e^{-r/a} e^{-\sqrt{r^2+R^2-2rR\cos\theta}/a}\, r^2 \sin\theta\, dr\, d\theta\, d\phi. \qquad [7.41]$$

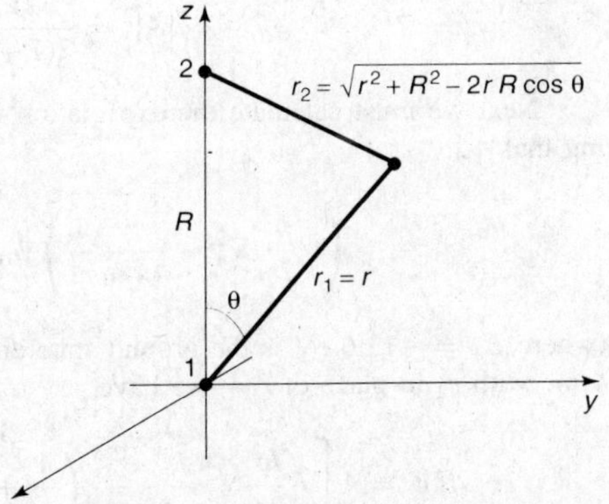

FIGURE 7.6: Coordinates for the calculation of *I* (Equation 7.39).

The ϕ integral is trivial (2π). To do the θ integral, let

$$y \equiv \sqrt{r^2 + R^2 - 2rR\cos\theta}, \quad \text{so that } d(y^2) = 2y\,dy = 2rR\sin\theta\,d\theta.$$

Then

$$\int_0^{\pi} e^{-\sqrt{r^2+R^2-2rR\cos\theta}/a}\sin\theta\,d\theta = \frac{1}{rR}\int_{|r-R|}^{r+R} e^{-y/a}y\,dy$$
$$= -\frac{a}{rR}\left[e^{-(r+R)/a}(r+R+a) - e^{-|r-R|/a}(|r-R|+a)\right].$$

The r integral is now straightforward:

$$I = \frac{2}{a^2R}\left[-e^{-R/a}\int_0^{\infty}(r+R+a)e^{-2r/a}r\,dr + e^{-R/a}\int_0^R (R-r+a)r\,dr\right.$$
$$\left. + e^{R/a}\int_R^{\infty}(r-R+a)e^{-2r/a}r\,dr\right].$$

Evaluating the integrals, we find (after some algebraic simplification),

$$I = e^{-R/a}\left[1 + \left(\frac{R}{a}\right) + \frac{1}{3}\left(\frac{R}{a}\right)^2\right]. \tag{7.42}$$

I is called an **overlap** integral; it measures the amount by which $\psi_0(r_1)$ overlaps $\psi_0(r_2)$ (notice that it goes to 1 as $R \to 0$, and to 0 as $R \to \infty$). In terms of I, the normalization factor (Equation 7.38) is

$$|A|^2 = \frac{1}{2(1+I)}. \tag{7.43}$$

Next we must calculate the expectation value of H in the trial state ψ. Noting that

$$\left(-\frac{\hbar^2}{2m}\nabla^2 - \frac{e^2}{4\pi\epsilon_0}\frac{1}{r_1}\right)\psi_0(r_1) = E_1\psi_0(r_1)$$

(where $E_1 = -13.6$ eV is the ground state energy of atomic hydrogen)—and the same with r_2 in place of r_1—we have

$$H\psi = A\left[-\frac{\hbar^2}{2m}\nabla^2 - \frac{e^2}{4\pi\epsilon_0}\left(\frac{1}{r_1} + \frac{1}{r_2}\right)\right][\psi_0(r_1) + \psi_0(r_2)]$$
$$= E_1\psi - A\left(\frac{e^2}{4\pi\epsilon_0}\right)\left[\frac{1}{r_2}\psi_0(r_1) + \frac{1}{r_1}\psi_0(r_2)\right].$$

It follows that

$$\langle H\rangle = E_1 - 2|A|^2\left(\frac{e^2}{4\pi\epsilon_0}\right)\left[\langle\psi_0(r_1)\left|\frac{1}{r_2}\right|\psi_0(r_1)\rangle + \langle\psi_0(r_1)\left|\frac{1}{r_1}\right|\psi_0(r_2)\rangle\right]. \quad [7.44]$$

I'll let you calculate the two remaining quantities, the so-called **direct integral**,

$$D \equiv a\langle\psi_0(r_1)\left|\frac{1}{r_2}\right|\psi_0(r_1)\rangle, \quad [7.45]$$

and the **exchange integral**,

$$X \equiv a\langle\psi_0(r_1)\left|\frac{1}{r_1}\right|\psi_0(r_2)\rangle. \quad [7.46]$$

The results (see Problem 7.8) are

$$D = \frac{a}{R} - \left(1 + \frac{a}{R}\right)e^{-2R/a}, \quad [7.47]$$

and

$$X = \left(1 + \frac{R}{a}\right)e^{-R/a}. \quad [7.48]$$

Putting all this together, and recalling (Equations 4.70 and 4.72) that $E_1 = -(e^2/4\pi\epsilon_0)(1/2a)$, we conclude:

$$\langle H\rangle = \left[1 + 2\frac{(D+X)}{(1+I)}\right]E_1. \quad [7.49]$$

According to the variational principle, the ground state energy is *less* than $\langle H\rangle$. Of course, this is only the *electron's* energy—there is also potential energy associated with the proton-proton repulsion:

$$V_{pp} = \frac{e^2}{4\pi\epsilon_0}\frac{1}{R} = -\frac{2a}{R}E_1. \quad [7.50]$$

Thus the *total* energy of the system, in units of $-E_1$, and expressed as a function of $x \equiv R/a$, is less than

$$F(x) = -1 + \frac{2}{x}\left\{\frac{(1-(2/3)x^2)e^{-x} + (1+x)e^{-2x}}{1 + (1 + x + (1/3)x^2)e^{-x}}\right\}. \quad [7.51]$$

This function is plotted in Figure 7.7. Evidently bonding *does* occur, for there exists a region in which the graph goes below -1, indicating that the energy is less than that of a neutral atom plus a free proton (-13.6 eV). It's a covalent bond, with the electron shared equally by the two protons. The equilibrium separation of the protons is about 2.4 Bohr radii, or 1.3 Å (the experimental value is 1.06 Å). The

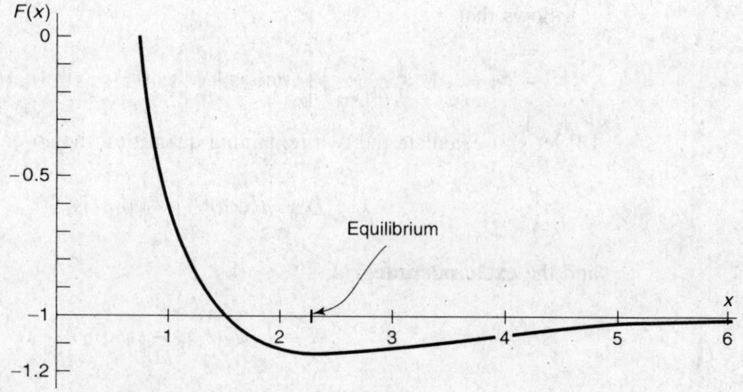

FIGURE 7.7: Plot of the function $F(x)$, Equation 7.51, showing existence of a bound state (x is the distance between the protons, in units of the Bohr radius).

calculated binding energy is 1.8 eV, whereas the experimental value is 2.8 eV (the variational principle, as always, *over*estimates the ground state energy—and hence *under*estimates the strength of the bond—but never mind: The essential point was to see whether binding occurs at all; a better variational function can only make the potential well even deeper.

***Problem 7.8** Evaluate D and X (Equations 7.45 and 7.46). Check your answers against Equations 7.47 and 7.48.

****Problem 7.9** Suppose we used a *minus* sign in our trial wave function (Equation 7.37):

$$\psi = A[\psi_0(r_1) - \psi_0(r_2)]. \quad [7.52]$$

Without doing any new integrals, find $F(x)$ (the analog to Equation 7.51) for this case, and construct the graph. Show that there is no evidence of bonding.[13] (Since the variational principle only gives an *upper bound*, this doesn't *prove* that bonding cannot occur for such a state, but it certainly doesn't look promising). *Comment:* Actually, any function of the form

$$\psi = A[\psi_0(r_1) + e^{i\phi}\psi_0(r_2)]. \quad [7.53]$$

[13] Bonding occurs when the electron "prefers" to be between the protons, attracting them inward. But the odd linear combination (Equation 7.52) has a *node* at the center, so it's not surprising that this configuration drives the protons apart.

has the desired property that the electron is equally likely to be associated with either proton. However, since the Hamiltonian (Equation 7.35) is invariant under the interchange P: $r_1 \leftrightarrow r_2$, its eigenfunctions can be chosen to be simultaneously eigenfunctions of P. The plus sign (Equation 7.37) goes with the eigenvalue $+1$, and the minus sign (Equation 7.52) with the eigenvalue -1; nothing is to be gained by considering the ostensibly more general case (Equation 7.53), though you're welcome to try it, if you're interested.

$***$**Problem 7.10** The second derivative of $F(x)$, at the equilibrium point, can be used to estimate the natural frequency of vibration (ω) of the two protons in the hydrogen molecule ion (see Section 2.3). If the ground state energy ($\hbar\omega/2$) of this oscillator exceeds the binding energy of the system, it will fly apart. Show that in fact the oscillator energy is small enough that this will *not* happen, and estimate how many bound vibrational levels there are. *Note:* You're not going to be able to obtain the position of the minimum—still less the second derivative at that point—analytically. Do it numerically, on a computer.

FURTHER PROBLEMS FOR CHAPTER 7

Problem 7.11

(a) Use a trial wave function of the form

$$\psi(x) = \begin{cases} A\cos(\pi x/a), & \text{if } (-a/2 < x < a/2), \\ 0 & \text{otherwise} \end{cases}$$

to obtain a bound on the ground state energy of the one-dimensional harmonic oscillator. What is the "best" value of a? Compare $\langle H\rangle_{\text{min}}$ with the exact energy. *Note:* This trial function has a "kink" in it (a discontinuous derivative) at $\pm a/2$; do you need to take account of this, as I did in Example 7.3?

(b) Use $\psi(x) = B\sin(\pi x/a)$ on the interval $(-a, a)$ to obtain a bound on the first excited state. Compare the exact answer.

$**$**Problem 7.12**

(a) Generalize Problem 7.2, using the trial wave function[14]

$$\psi(x) = \frac{A}{(x^2+b^2)^n},$$

[14] W. N. Mei, *Int. J. Educ. Sci. Tech.* **27**, 285 (1996).

for arbitrary n. *Partial answer:* The best value of b is given by

$$b^2 = \frac{\hbar}{m\omega}\left[\frac{n(4n-1)(4n-3)}{2(2n+1)}\right]^{1/2}.$$

(b) Find the least upper bound on the first excited state of the harmonic oscillator using a trial function of the form

$$\psi(x) = \frac{Bx}{(x^2+b^2)^n}.$$

Partial answer: The best value of b is given by

$$b^2 = \frac{\hbar}{m\omega}\left[\frac{n(4n-5)(4n-3)}{2(2n+1)}\right]^{1/2}.$$

(c) Notice that the bounds approach the exact energies as $n \to \infty$. Why is that? *Hint:* Plot the trial wave functions for $n = 2$, $n = 3$, and $n = 4$, and compare them with the true wave functions (Equations 2.59 and 2.62). To do it analytically, start with the identity

$$e^z = \lim_{n\to\infty}\left(1+\frac{z}{n}\right)^n.$$

Problem 7.13 Find the lowest bound on the ground state of hydrogen you can get using a gaussian trial wave function

$$\psi(\mathbf{r}) = Ae^{-br^2},$$

where A is determined by normalization and b is an adjustable parameter. *Answer:* -11.5 eV.

****Problem 7.14** If the photon had a nonzero mass ($m_\gamma \neq 0$), the Coulomb potential would be replaced by the **Yukawa potential**,

$$V(\mathbf{r}) = -\frac{e^2}{4\pi\epsilon_0}\frac{e^{-\mu r}}{r}, \qquad [7.54]$$

where $\mu = m_\gamma c/\hbar$. With a trial wave function of your own devising, estimate the binding energy of a "hydrogen" atom with this potential. Assume $\mu a \ll 1$, and give your answer correct to order $(\mu a)^2$.

Problem 7.15 Suppose you're given a quantum system whose Hamiltonian H_0 admits just two eigenstates, ψ_a (with energy E_a), and ψ_b (with energy E_b). They are orthogonal, normalized, and nondegenerate (assume E_a is the smaller of the two energies). Now we turn on a perturbation H', with the following matrix elements:

$$\langle\psi_a|H'|\psi_a\rangle = \langle\psi_b|H'|\psi_b\rangle = 0; \quad \langle\psi_a|H'|\psi_b\rangle = \langle\psi_b|H'|\psi_a\rangle = h, \qquad [7.55]$$

where h is some specified constant.

(a) Find the exact eigenvalues of the perturbed Hamiltonian.

(b) Estimate the energies of the perturbed system using second-order perturbation theory.

(c) Estimate the ground state energy of the perturbed system using the variational principle, with a trial function of the form

$$\psi = (\cos\phi)\psi_a + (\sin\phi)\psi_b, \qquad [7.56]$$

where ϕ is an adjustable parameter. *Note:* Writing the linear combination in this way is just a neat way to guarantee that ψ is normalized.

(d) Compare your answers to (a), (b), and (c). Why is the variational principle so accurate, in this case?

Problem 7.16 As an explicit example of the method developed in Problem 7.15, consider an electron at rest in a uniform magnetic field $\mathbf{B} = B_z\hat{k}$, for which the Hamiltonian is (Equation 4.158):

$$H_0 = \frac{eB_z}{m}S_z. \qquad [7.57]$$

The eigenspinors, χ_a and χ_b, and the corresponding energies, E_a and E_b, are given in Equation 4.161. Now we turn on a perturbation, in the form of a uniform field in the x direction:

$$H' = \frac{eB_x}{m}S_x. \qquad [7.58]$$

(a) Find the matrix elements of H', and confirm that they have the structure of Equation 7.55. What is h?

(b) Using your result in Problem 7.15(b), find the new ground state energy, in second-order perturbation theory.

(c) Using your result in Problem 7.15(c), find the variational principle bound on the ground state energy.

∗∗∗**Problem 7.17** Although the Schrödinger equation for helium itself cannot be solved exactly, there exist "helium-like" systems that do admit exact solutions.

A simple example[15] is "rubber-band helium," in which the Coulomb forces are replaced by Hooke's law forces:

$$H = -\frac{\hbar^2}{2m}(\nabla_1^2 + \nabla_2^2) + \frac{1}{2}m\omega^2(r_1^2 + r_2^2) - \frac{\lambda}{4}m\omega^2|\mathbf{r}_1 - \mathbf{r}_2|^2. \quad [7.59]$$

(a) Show that the change of variables from $\mathbf{r}_1$, $\mathbf{r}_2$, to

$$\mathbf{u} \equiv \frac{1}{\sqrt{2}}(\mathbf{r}_1 + \mathbf{r}_2), \quad \mathbf{v} \equiv \frac{1}{\sqrt{2}}(\mathbf{r}_1 - \mathbf{r}_2), \quad [7.60]$$

turns the Hamiltonian into two independent three-dimensional harmonic oscillators:

$$H = \left[-\frac{\hbar^2}{2m}\nabla_u^2 + \frac{1}{2}m\omega^2 u^2\right] + \left[-\frac{\hbar^2}{2m}\nabla_v^2 + \frac{1}{2}(1-\lambda)m\omega^2 v^2\right]. \quad [7.61]$$

(b) What is the *exact* ground state energy for this system?

(c) If we didn't know the exact solution, we might be inclined to apply the method of Section 7.2 to the Hamiltonian in its original form (Equation 7.59). Do so (but don't bother with shielding). How does your result compare with the exact answer? *Answer:* $\langle H \rangle = 3\hbar\omega(1 - \lambda/4)$.

*****Problem 7.18** In Problem 7.7 we found that the trial wave function with shielding (Equation 7.27), which worked well for helium, is inadequate to confirm the existence of a bound state for the negative hydrogen ion. Chandrasekhar[16] used a trial wave function of the form

$$\psi(\mathbf{r}_1, \mathbf{r}_2) \equiv A[\psi_1(r_1)\psi_2(r_2) + \psi_2(r_1)\psi_1(r_2)], \quad [7.62]$$

where

$$\psi_1(r) \equiv \sqrt{\frac{Z_1^3}{\pi a^3}}e^{-Z_1 r/a}, \quad \text{and} \quad \psi_2(r) \equiv \sqrt{\frac{Z_2^3}{\pi a^3}}e^{-Z_2 r/a}. \quad [7.63]$$

In effect, he allowed two *different* shielding factors, suggesting that one electron is relatively close to the nucleus, and the other is farther out. (Because electrons are identical particles, the spatial wave function must be symmetrized with respect to interchange. The *spin* state—which is irrelevant to the calculation—is evidently

[15] For a more sophisticated model, see R. Crandall, R. Whitnell, and R. Bettega, *Am. J. Phys.* **52**, 438 (1984).

[16] S. Chandrasekhar, *Astrophys. J.* **100**, 176 (1944).

antisymmetric.) Show that by astute choice of the adjustable parameters Z_1 and Z_2 you can get $\langle H\rangle$ less than -13.6 eV. *Answer:*

$$\langle H\rangle = \frac{E_1}{x^6+y^6}\left(-x^8+2x^7+\frac{1}{2}x^6y^2-\frac{1}{2}x^5y^2-\frac{1}{8}x^3y^4+\frac{11}{8}xy^6-\frac{1}{2}y^8\right),$$

where $x \equiv Z_1+Z_2$ and $y \equiv 2\sqrt{Z_1Z_2}$. Chandrasekhar used $Z_1 = 1.039$ (since this is larger than 1, the motivating interpretation as an effective nuclear charge cannot be sustained, but never mind—it's still an acceptable trial wave function) and $Z_2 = 0.283$.

Problem 7.19 The fundamental problem in harnessing nuclear fusion is getting the two particles (say, two deuterons) close enough together for the attractive (but short-range) nuclear force to overcome the Coulomb repulsion. The "bulldozer" method is to heat the particles up to fantastic temperatures, and allow the random collisions to bring them together. A more exotic proposal is **muon catalysis**, in which we construct a "hydrogen molecule ion," only with deuterons in place of protons, and a *muon* in place of the electron. Predict the equilibrium separation distance between the deuterons in such a structure, and explain why muons are superior to electrons for this purpose.[17]

∗∗∗**Problem 7.20 Quantum dots.** Consider a particle constrained to move in two dimensions in the cross-shaped region shown in Figure 7.8. The "arms" of the cross continue out to infinity. The potential is zero within the cross, and infinite in the shaded areas outside. Surprisingly, this configuration admits a positive-energy bound state.[18]

(a) Show that the lowest energy that can propagate off to infinity is

$$E_{\text{threshold}} = \frac{\pi^2\hbar^2}{8ma^2};$$

any solution with energy *less* than that has to be a bound state. *Hint:* Go way out one arm (say, $x \gg a$), and solve the Schrödinger equation by separation of variables; if the wave function propagates out to infinity, the dependence on x must take the form $\exp(ik_x x)$ with $k_x > 0$.

---

[17] The classic paper on muon-catalyzed fusion is J. D. Jackson, *Phys. Rev.* **106**, 330 (1957); for a more recent popular review, see J. Rafelski and S. Jones, *Scientific American*, November 1987, page 84.

[18] This model is taken from R. L. Schult *et al.*, *Phys. Rev. B* **39**, 5476 (1989). In the presence of quantum tunneling a classically bound state becomes unbound; this is the reverse: A classically *un*bound state is quantum mechanically *bound*.

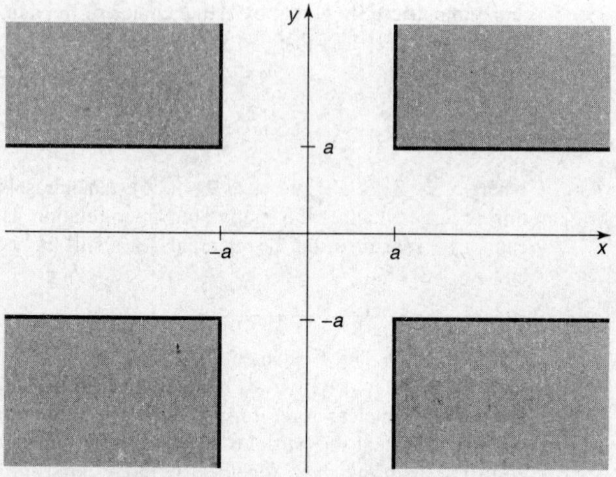

FIGURE 7.8: The cross-shaped region for Problem 7.20.

(b) Now use the variational principle to show that the ground state has energy less than $E_{\text{threshold}}$. Use the following trial wave function (suggested by Krishna Rajagopal):

$$\psi(x, y) = A \begin{cases} (1 - |xy|/a^2)e^{-\alpha}, & |x| \leq a \text{ and } |y| \leq a \\ (1 - |x|/a)e^{-\alpha|y|/a}, & |x| \leq a \text{ and } |y| > a \\ (1 - |y|/a)e^{-\alpha|x|/a}, & |x| > a \text{ and } |y| \leq a \\ 0, & \text{elsewhere.} \end{cases}$$

Normalize it to determine A, and calculate the expectation value of H. *Answer:*

$$\langle H \rangle = \frac{3\hbar^2}{ma^2}\left(\frac{\alpha^2 + 2\alpha + 3}{6 + 11\alpha}\right).$$

Now minimize with respect to α, and show that the result is less than $E_{\text{threshold}}$. *Hint:* Take full advantage of the symmetry of the problem—you only need to integrate over 1/8 of the open region, since the other 7 integrals will be the same. Note however that whereas the trial wave function is continuous, its *derivatives* are *not*—there are "roof-lines" at $x = 0$, $y = 0$, $x = \pm a$, and $y = \pm a$, where you will need to exploit the technique of Example 7.3.

CHAPTER 8

THE WKB APPROXIMATION

The **WKB** (Wentzel, Kramers, Brillouin)[1] method is a technique for obtaining approximate solutions to the time-independent Schrödinger equation in one dimension (the same basic idea can be applied to many other differential equations, and to the radial part of the Schrödinger equation in three dimensions). It is particularly useful in calculating bound state energies and tunneling rates through potential barriers.

The essential idea is as follows: Imagine a particle of energy E moving through a region where the potential $V(x)$ is *constant*. If $E > V$, the wave function is of the form

$$\psi(x) = Ae^{\pm ikx}, \quad \text{with } k \equiv \sqrt{2m(E - V)}/\hbar.$$

The plus sign indicates that the particle is traveling to the right, and the minus sign means it is going to the left (the general solution, of course, is a linear combination of the two). The wave function is oscillatory, with fixed wavelength ($\lambda = 2\pi/k$) and unchanging amplitude (A). Now suppose that $V(x)$ is *not* constant, but varies rather slowly in comparison to λ, so that over a region containing many full wavelengths the potential is *essentially* constant. Then it is reasonable to suppose that ψ remains *practically* sinusoidal, except that the wavelength and the amplitude change slowly with x. This is the inspiration behind the WKB approximation. In effect, it identifies two different levels of x-dependence: rapid oscillations, *modulated* by gradual variation in amplitude and wavelength.

[1] In Holland it's KWB, in France it's BWK, and in England it's JWKB (for Jeffreys).

By the same token, if $E < V$ (and V is constant), then ψ is exponential:

$$\psi(x) = Ae^{\pm\kappa x}, \quad \text{with } \kappa \equiv \sqrt{2m(V-E)}/\hbar.$$

And if $V(x)$ is *not* constant, but varies slowly in comparison with $1/\kappa$, the solution remains *practically* exponential, except that A and κ are now slowly-varying functions of x.

Now, there is one place where this whole program is bound to fail, and that is in the immediate vicinity of a classical **turning point**, where $E \approx V$. For here λ (or $1/\kappa$) goes to infinity, and $V(x)$ can hardly be said to vary "slowly" in comparison. As we shall see, a proper handling of the turning points is the most difficult aspect of the WKB approximation, though the final results are simple to state and easy to implement.

8.1 THE "CLASSICAL" REGION

The Schrödinger equation,

$$-\frac{\hbar^2}{2m}\frac{d^2\psi}{dx^2} + V(x)\psi = E\psi,$$

can be rewritten in the following way:

$$\frac{d^2\psi}{dx^2} = -\frac{p^2}{\hbar^2}\psi, \quad [8.1]$$

where

$$p(x) \equiv \sqrt{2m[E - V(x)]} \quad [8.2]$$

is the classical formula for the momentum of a particle with total energy E and potential energy $V(x)$. For the moment, I'll assume that $E > V(x)$, so that $p(x)$ is *real*; we call this the "classical" region, for obvious reasons—classically the particle is *confined* to this range of x (see Figure 8.1). In general, ψ is some complex function; we can express it in terms of its *amplitude*, $A(x)$, and its *phase*, $\phi(x)$—both of which are *real*:

$$\psi(x) = A(x)e^{i\phi(x)}. \quad [8.3]$$

Using a prime to denote the derivative with respect to x, we find:

$$\frac{d\psi}{dx} = (A' + iA\phi')e^{i\phi},$$

and

$$\frac{d^2\psi}{dx^2} = [A'' + 2iA'\phi' + iA\phi'' - A(\phi')^2]e^{i\phi}. \quad [8.4]$$

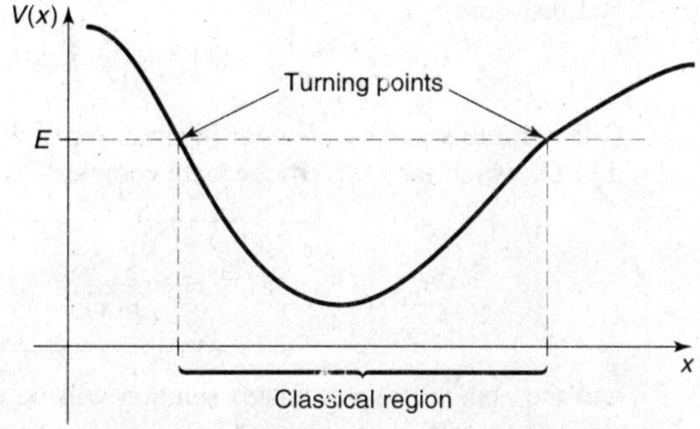

FIGURE 8.1: Classically, the particle is confined to the region where $E \geq V(x)$.

Putting this into Equation 8.1:

$$A'' + 2iA'\phi' + iA\phi'' - A(\phi')^2 = -\frac{p^2}{\hbar^2}A. \quad [8.5]$$

This is equivalent to two *real* equations, one for the real part and one for the imaginary part:

$$A'' - A(\phi')^2 = -\frac{p^2}{\hbar^2}A, \quad \text{or} \quad A'' = A\left[(\phi')^2 - \frac{p^2}{\hbar^2}\right], \quad [8.6]$$

and

$$2A'\phi' + A\phi'' = 0, \quad \text{or} \quad \left(A^2\phi'\right)' = 0. \quad [8.7]$$

Equations 8.6 and 8.7 are entirely equivalent to the original Schrödinger equation. The second one is easily solved:

$$A^2\phi' = C^2, \quad \text{or} \quad A = \frac{C}{\sqrt{\phi'}}, \quad [8.8]$$

where C is a (real) constant. The first one (Equation 8.6) cannot be solved in general—so here comes the approximation: *We assume that the amplitude A varies slowly*, so that the A'' term is negligible. (More precisely, we assume that A''/A is much less than both $(\phi')^2$ and $p^2/\hbar^2$.) In that case we can drop the left side of Equation 8.6, and we are left with

$$(\phi')^2 = \frac{p^2}{\hbar^2}, \quad \text{or} \quad \frac{d\phi}{dx} = \pm\frac{p}{\hbar},$$

and therefore

$$\phi(x) = \pm \frac{1}{\hbar} \int p(x)\,dx. \tag{8.9}$$

(I'll write this as an *indefinite* integral, for now—any constants can be absorbed into C, which may thereby become complex.) It follows that

$$\boxed{\psi(x) \cong \frac{C}{\sqrt{p(x)}} e^{\pm \frac{i}{\hbar} \int p(x)\,dx},} \tag{8.10}$$

and the general (approximate) solution will be a linear combination of two such terms, one with each sign.

Notice that

$$|\psi(x)|^2 \cong \frac{|C|^2}{p(x)}, \tag{8.11}$$

which says that the probability of finding the particle at point x is inversely proportional to its (classical) momentum (and hence its velocity) at that point. This is exactly what you would expect—the particle doesn't spend long in the places where it is moving rapidly, so the probability of getting caught there is small. In fact, the WKB approximation is sometimes *derived* by starting with this "semi-classical" observation, instead of by dropping the A'' term in the differential equation. The latter approach is cleaner mathematically, but the former offers a more plausible physical rationale.

Example 8.1 Potential well with two vertical walls. Suppose we have an infinite square well with a bumpy bottom (Figure 8.2):

$$V(x) = \begin{cases} \text{some specified function,} & \text{if } 0 < x < a, \\ \infty, & \text{otherwise.} \end{cases} \tag{8.12}$$

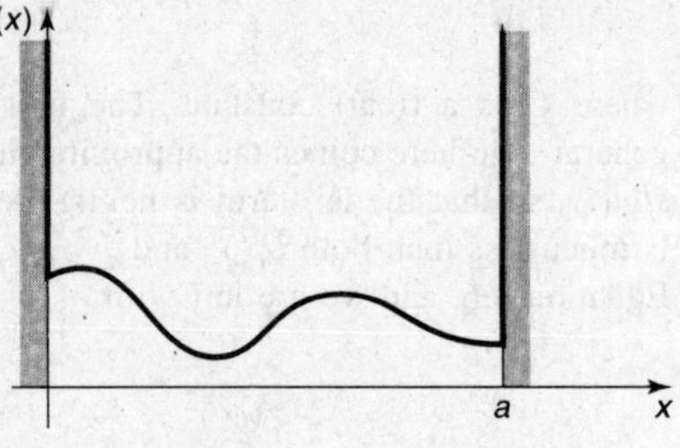

FIGURE 8.2: Infinite square well with a bumpy bottom.

Inside the well (assuming $E > V(x)$ throughout) we have

$$\psi(x) \cong \frac{1}{\sqrt{p(x)}}\left[C_+ e^{i\phi(x)} + C_- e^{-i\phi(x)}\right],$$

or, more conveniently,

$$\psi(x) \cong \frac{1}{\sqrt{p(x)}}\left[C_1 \sin\phi(x) + C_2 \cos\phi(x)\right], \tag{8.13}$$

where (exploiting the freedom noted earlier to impose a convenient lower limit on the integral)

$$\phi(x) = \frac{1}{\hbar}\int_0^x p(x')\,dx'. \tag{8.14}$$

Now $\psi(x)$ must go to zero at $x = 0$, and therefore (since $\phi(0) = 0$) $C_2 = 0$. Also, $\psi(x)$ goes to zero at $x = a$, so

$$\phi(a) = n\pi \quad (n = 1, 2, 3, \ldots). \tag{8.15}$$

Conclusion:

$$\boxed{\int_0^a p(x)\,dx = n\pi\hbar.} \tag{8.16}$$

This quantization condition determines the (approximate) allowed energies.

For instance, if the well has a *flat* bottom ($V(x) = 0$), then $p(x) = \sqrt{2mE}$ (a constant), and Equation 8.16 says $pa = n\pi\hbar$, or

$$E_n = \frac{n^2\pi^2\hbar^2}{2ma^2},$$

which is the old formula for the energy levels of the infinite square well (Equation 2.27). In this case the WKB approximation yields the *exact* answer (the amplitude of the true wave function is *constant*, so dropping A'' cost us nothing).

∗Problem 8.1 Use the WKB approximation to find the allowed energies (E_n) of an infinite square well with a "shelf," of height V_0 extending half-way across (Figure 6.3):

$$V(x) = \begin{cases} V_0, & \text{if } 0 < x < a/2, \\ 0, & \text{if } a/2 < x < a, \\ \infty, & \text{otherwise.} \end{cases}$$

Express your answer in terms of V_0 and $E_n^0 \equiv (n\pi\hbar)^2/2ma^2$ (the nth allowed energy for the infinite square well with *no* shelf). Assume that $E_1^0 > V_0$, but do *not* assume that $E_n \gg V_0$. Compare your result with what we got in Example 6.1 using first-order perturbation theory. Note that they are in agreement if either V_0 is very small (the perturbation theory regime) or n is very large (the WKB—semi-classical—regime).

∗∗Problem 8.2 An illuminating alternative derivation of the WKB formula (Equation 8.10) is based on an expansion in powers of $\hbar$. Motivated by the free-particle wave function, $\psi = A\exp(\pm ipx/\hbar)$, we write

$$\psi(x) = e^{if(x)/\hbar},$$

where $f(x)$ is some *complex* function. (Note that there is no loss of generality here—*any* nonzero function can be written in this way.)

(a) Put this into Schrödinger's equation (in the form of Equation 8.1), and show that

$$i\hbar f'' - (f')^2 + p^2 = 0.$$

(b) Write $f(x)$ as a power series in $\hbar$:

$$f(x) = f_0(x) + \hbar f_1(x) + \hbar^2 f_2(x) + \cdots,$$

and, collecting like powers of $\hbar$, show that

$$(f_0')^2 = p^2, \quad if_0'' = 2f_0'f_1', \quad if_1'' = 2f_0'f_2' + (f_1')^2, \quad \text{etc.}$$

(c) Solve for $f_0(x)$ and $f_1(x)$, and show that—to first order in $\hbar$—you recover Equation 8.10.

Note: The logarithm of a negative number is defined by $\ln(-z) = \ln(z) + in\pi$, where n is an odd integer. If this formula is new to you, try exponentiating both sides, and you'll see where it comes from.

8.2 TUNNELING

So far, I have assumed that $E > V$, so $p(x)$ is real. But we can easily write down the corresponding result in the *non*classical region ($E < V$)—it's the same as before (Equation 8.10), only now $p(x)$ is *imaginary*:[2]

$$\psi(x) \cong \frac{C}{\sqrt{|p(x)|}} e^{\pm \frac{1}{\hbar} \int |p(x)|\, dx}. \tag{8.17}$$

Consider, for example, the problem of scattering from a rectangular barrier with a bumpy top (Figure 8.3). To the left of the barrier ($x < 0$),

$$\psi(x) = Ae^{ikx} + Be^{-ikx}, \tag{8.18}$$

where A is the incident amplitude, B is the reflected amplitude, and $k \equiv \sqrt{2mE}/\hbar$ (see Section 2.5). To the right of the barrier ($x > a$),

$$\psi(x) = Fe^{ikx}; \tag{8.19}$$

F is the transmitted amplitude, and the transmission probability is

$$T = \frac{|F|^2}{|A|^2}. \tag{8.20}$$

In the tunneling region ($0 \leq x \leq a$), the WKB approximation gives

$$\psi(x) \cong \frac{C}{\sqrt{|p(x)|}} e^{\frac{1}{\hbar}\int_0^x |p(x')|\, dx'} + \frac{D}{\sqrt{|p(x)|}} e^{-\frac{1}{\hbar}\int_0^x |p(x')|\, dx'}. \tag{8.21}$$

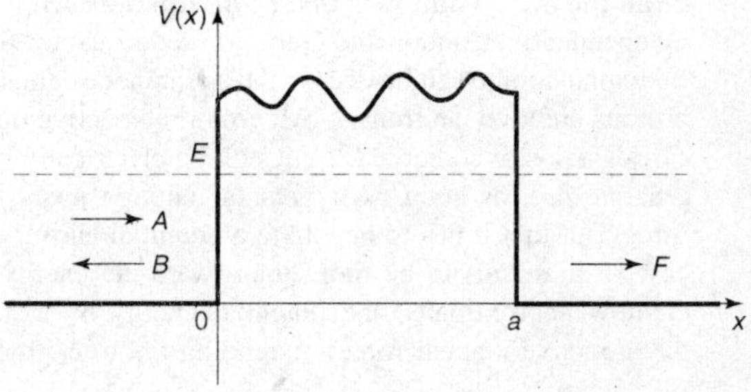

FIGURE 8.3: Scattering from a rectangular barrier with a bumpy top.

[2]In this case the wave function is *real*, and the analogs to Equations 8.6 and 8.7 do not follow *necessarily* from Equation 8.5, although they are still *sufficient*. If this bothers you, study the alternative derivation in Problem 8.2.

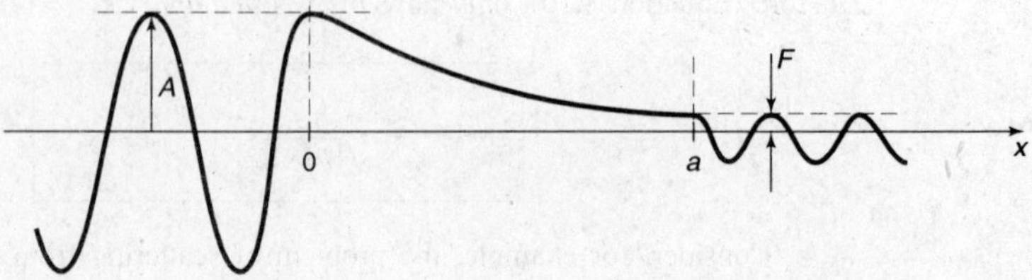

FIGURE 8.4: Qualitative structure of the wave function, for scattering from a high, broad barrier.

If the barrier is very high and/or very wide (which is to say, if the probability of tunneling is small), then the coefficient of the exponentially *increasing* term (C) must be small (in fact, it would be *zero* if the barrier were *infinitely* broad), and the wave function looks something like[3] Figure 8.4. The relative amplitudes of the incident and transmitted waves are determined essentially by the total decrease of the exponential over the nonclassical region:

$$\frac{|F|}{|A|} \sim e^{-\frac{1}{\hbar}\int_0^a |p(x')|\,dx'},$$

so that

$$T \cong e^{-2\gamma}, \quad \text{with } \gamma \equiv \frac{1}{\hbar}\int_0^a |p(x)|\,dx. \qquad [8.22]$$

Example 8.2 Gamow's theory of alpha decay.[4] In 1928, George Gamow (and, independently, Condon and Gurney) used Equation 8.22 to provide the first successful explanation of alpha decay (the spontaneous emission of an alpha-particle—two protons and two neutrons—by certain radioactive nuclei).[5] Since the alpha particle carries a positive charge ($2e$), it will be electrically repelled by the leftover nucleus (charge Ze), as soon as it gets far enough away to escape the nuclear binding force. But first it has to negotiate a potential barrier that was already known (in the case of uranium) to be more than twice the energy of the emitted alpha particle. Gamow approximated the potential energy by a finite square well (representing the attractive nuclear force), extending out to r_1 (the radius of the nucleus), joined

[3]This heuristic argument can be made more rigorous—see Problem 8.10.

[4]For a more complete discussion, and alternative formulations, see Barry R. Holstein, *Am. J. Phys.* **64**, 1061 (1996).

[5]For an interesting brief history see Eugen Merzbacher, "The Early History of Quantum Tunneling," *Physics Today*, August 2002, p. 44.

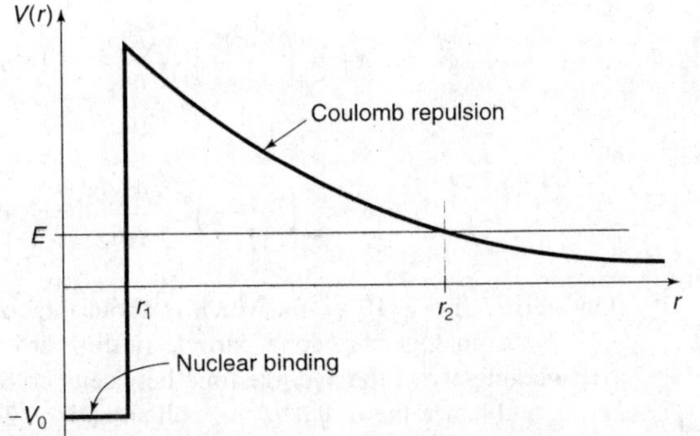

FIGURE 8.5: Gamow's model for the potential energy of an alpha particle in a radioactive nucleus.

to a repulsive coulombic tail (Figure 8.5), and identified the escape mechanism as quantum tunneling (this was, by the way, the first time that quantum mechanics had been applied to nuclear physics).

If E is the energy of the emitted alpha particle, the outer turning point (r_2) is determined by

$$\frac{1}{4\pi\epsilon_0}\frac{2Ze^2}{r_2} = E. \tag{8.23}$$

The exponent γ (Equation 8.22) is evidently[6]

$$\gamma = \frac{1}{\hbar}\int_{r_1}^{r_2}\sqrt{2m\left(\frac{1}{4\pi\epsilon_0}\frac{2Ze^2}{r} - E\right)}\,dr = \frac{\sqrt{2mE}}{\hbar}\int_{r_1}^{r_2}\sqrt{\frac{r_2}{r} - 1}\,dr.$$

The integral can be done by substitution (let $r \equiv r_2 \sin^2 u$), and the result is

$$\gamma = \frac{\sqrt{2mE}}{\hbar}\left[r_2\left(\frac{\pi}{2} - \sin^{-1}\sqrt{\frac{r_1}{r_2}}\right) - \sqrt{r_1(r_2 - r_1)}\right]. \tag{8.24}$$

Typically, $r_1 \ll r_2$, and we can simplify this result using the small angle approximation ($\sin\epsilon \cong \epsilon$):

$$\gamma \cong \frac{\sqrt{2mE}}{\hbar}\left[\frac{\pi}{2}r_2 - 2\sqrt{r_1 r_2}\right] = K_1\frac{Z}{\sqrt{E}} - K_2\sqrt{Zr_1}, \tag{8.25}$$

[6] In this case the potential does not drop to zero on the left side of the barrier (moreover, this is really a three-dimensional problem), but the essential idea, contained in Equation 8.22, is all we really need.

where

$$K_1 \equiv \left(\frac{e^2}{4\pi\epsilon_0}\right)\frac{\pi\sqrt{2m}}{\hbar} = 1.980 \text{ MeV}^{1/2}, \qquad [8.26]$$

and

$$K_2 \equiv \left(\frac{e^2}{4\pi\epsilon_0}\right)^{1/2}\frac{4\sqrt{m}}{\hbar} = 1.485 \text{ fm}^{-1/2}. \qquad [8.27]$$

[One fermi (fm) is 10^{-15} m, which is about the size of a typical nucleus.]

If we imagine the alpha particle rattling around inside the nucleus, with an average velocity v, the average time between "collisions" with the "wall" is about $2r_1/v$, and hence the *frequency* of collisions is $v/2r_1$. The probability of escape at each collision is $e^{-2\gamma}$, so the probability of emission, per unit time, is $(v/2r_1)e^{-2\gamma}$, and hence the **lifetime** of the parent nucleus is about

$$\tau = \frac{2r_1}{v}e^{2\gamma}. \qquad [8.28]$$

Unfortunately, we don't know v—but it hardly matters, for the exponential factor varies over a *fantastic* range (twenty-five orders of magnitude), as we go from one radioactive nucleus to another; relative to this the variation in v is pretty insignificant. In particular, if you plot the *logarithm* of the experimentally measured lifetime against $1/\sqrt{E}$, the result is a beautiful straight line (Figure 8.6),[7] just as you would expect from Equations 8.25 and 8.28.

$*$**Problem 8.3** Use Equation 8.22 to calculate the approximate transmission probability for a particle of energy E that encounters a finite square barrier of height $V_0 > E$ and width $2a$. Compare your answer with the exact result (Problem 2.33), to which it should reduce in the WKB regime $T \ll 1$.

$**$**Problem 8.4** Calculate the lifetimes of U^{238} and Po^{212}, using Equations 8.25 and 8.28. *Hint:* The density of nuclear matter is relatively constant (i.e., the same for all nuclei), so $(r_1)^3$ is proportional to A (the number of neutrons plus protons). Empirically,

$$r_1 \cong (1.07 \text{ fm})A^{1/3}. \qquad [8.29]$$

[7] From David Park, *Introduction to the Quantum Theory*, 3rd ed., McGraw-Hill (1992); it was adapted from I. Perlman and J. O. Rasmussen, "Alpha Radioactivity," *Encyclopedia of Physics*, Vol. **42**, Springer (1957). This material is reproduced with permission of The McGraw-Hill Companies.

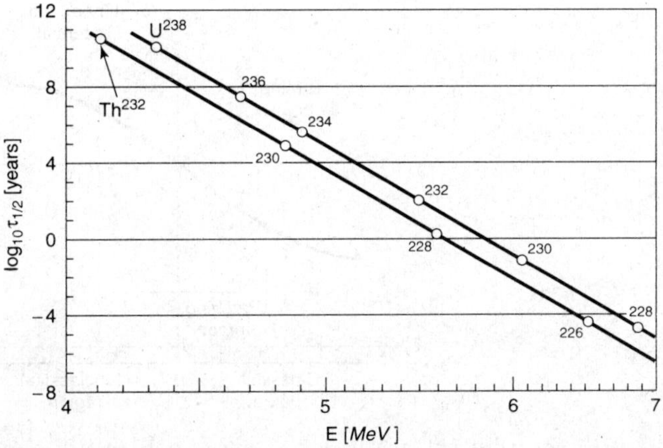

FIGURE 8.6: Graph of the logarithm of the lifetime versus $1/\sqrt{E}$ (where E is the energy of the emitted alpha particle), for uranium and thorium.

The energy of the emitted alpha particle can be deduced by using Einstein's formula ($E = mc^2$):

$$E = m_p c^2 - m_d c^2 - m_\alpha c^2, \tag{8.30}$$

where m_p is the mass of the parent nucleus, m_d is the mass of the daughter nucleus, and m_α is the mass of the alpha particle (which is to say, the He^4 nucleus). To figure out what the daughter nucleus is, note that the alpha particle carries off two protons and two neutrons, so Z decreases by 2 and A by 4. Look up the relevant nuclear masses. To estimate v, use $E = (1/2)m_\alpha v^2$; this ignores the (negative) potential energy inside the nucleus, and surely *underestimates* v, but it's about the best we can do at this stage. Incidentally, the experimental lifetimes are 6×10^9 yrs and 0.5 μs, respectively.

8.3 THE CONNECTION FORMULAS

In the discussion so far I have assumed that the "walls" of the potential well (or the barrier) are *vertical*, so that the "exterior" solution is simple, and the boundary conditions trivial. As it turns out, our main results (Equations 8.16 and 8.22) are reasonably accurate even when the edges are not so abrupt (indeed, in Gamow's theory they were applied to just such a case). Nevertheless, it is of some interest to study more closely what happens to the wave function at a turning point ($E = V$), where the "classical" region joins the "nonclassical" region, and the WKB

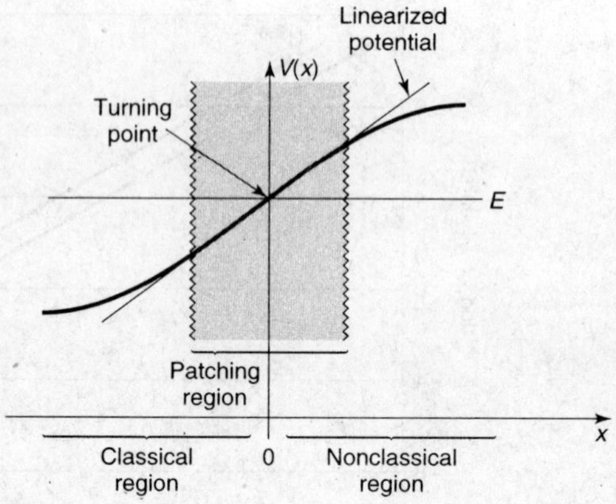

FIGURE 8.7: Enlarged view of the right-hand turning point.

approximation itself breaks down. In this section I'll treat the bound state problem (Figure 8.1); you get to do the scattering problem for yourself (Problem 8.10).[8]

For simplicity, let's shift the axes over so that the right-hand turning point occurs at $x = 0$ (Figure 8.7). In the WKB approximation, we have

$$\psi(x) \cong \begin{cases} \dfrac{1}{\sqrt{p(x)}}\left[Be^{\frac{i}{\hbar}\int_x^0 p(x')\,dx'} + Ce^{-\frac{i}{\hbar}\int_x^0 p(x')\,dx'}\right], & \text{if } x < 0. \\ \dfrac{1}{\sqrt{|p(x)|}}De^{-\frac{1}{\hbar}\int_0^x |p(x')|\,dx'}, & \text{if } x > 0. \end{cases} \qquad [8.31]$$

(Assuming $V(x)$ remains greater than E for *all* $x > 0$, we can exclude the positive exponent in this region, because it blows up as $x \to \infty$.) Our task is to join the two solutions at the boundary. But there is a serious difficulty here: In the WKB approximation, ψ goes to *infinity* at the turning point (where $p(x) \to 0$). The *true* wave function, of course, has no such wild behavior—as anticipated, the WKB method simply fails in the vicinity of a turning point. And yet, it is precisely the boundary conditions at the turning points that determine the allowed energies. What we need to do, then, is *splice* the two WKB solutions together, using a "patching" wave function that straddles the turning point.

Since we only need the patching wave function (ψ_p) in the neighborhood of the origin, we'll *approximate the potential by a straight line*:

$$V(x) \cong E + V'(0)x, \qquad [8.32]$$

[8] *Warning:* The following argument is quite technical, and you may wish to skip it on a first reading.

and solve the Schrödinger for this linearized V:

$$-\frac{\hbar^2}{2m}\frac{d^2\psi_p}{dx^2} + [E + V'(0)x]\psi_p = E\psi_p,$$

or

$$\frac{d^2\psi_p}{dx^2} = \alpha^3 x\psi_p, \quad [8.33]$$

where

$$\alpha \equiv \left[\frac{2m}{\hbar^2}V'(0)\right]^{1/3}. \quad [8.34]$$

The α's can be absorbed into the independent variable by defining

$$z \equiv \alpha x, \quad [8.35]$$

so that

$$\frac{d^2\psi_p}{dz^2} = z\psi_p, \quad [8.36]$$

This is **Airy's equation**, and the solutions are called **Airy functions**.[9] Since the Airy equation is a second-order differential equation, there are two linearly independent Airy functions, $Ai(z)$ and $Bi(z)$.

TABLE 8.1: Some properties of the Airy functions.

Differential Equation:	$\frac{d^2y}{dz^2} = zy.$
Solutions:	Linear combinations of Airy Functions, $Ai(z)$ and $Bi(z)$.
Integral Representation:	$Ai(z) = \frac{1}{\pi}\int_0^\infty \cos\left(\frac{s^3}{3} + sz\right)ds,$
	$Bi(z) = \frac{1}{\pi}\int_0^\infty \left[e^{-\frac{s^3}{3}+sz} + \sin\left(\frac{s^3}{3} + sz\right)\right]ds.$
Asymptotic Forms:	
$Ai(z) \sim \frac{1}{2\sqrt{\pi}z^{1/4}}e^{-\frac{2}{3}z^{3/2}}$, $Bi(z) \sim \frac{1}{\sqrt{\pi}z^{1/4}}e^{\frac{2}{3}z^{3/2}}$; $z \gg 0$;	$Ai(z) \sim \frac{1}{\sqrt{\pi}(-z)^{1/4}}\sin\left[\frac{2}{3}(-z)^{3/2} + \frac{\pi}{4}\right]$, $Bi(z) \sim \frac{1}{\sqrt{\pi}(-z)^{1/4}}\cos\left[\frac{2}{3}(-z)^{3/2} + \frac{\pi}{4}\right]$; $z \ll 0$.

[9] *Classically*, a linear potential means a constant force, and hence a constant acceleration—the simplest nontrivial motion possible, and the *starting* point for elementary mechanics. It is ironic that the same potential in *quantum* mechanics gives rise to unfamiliar transcendental functions, and plays only a peripheral role in the theory.

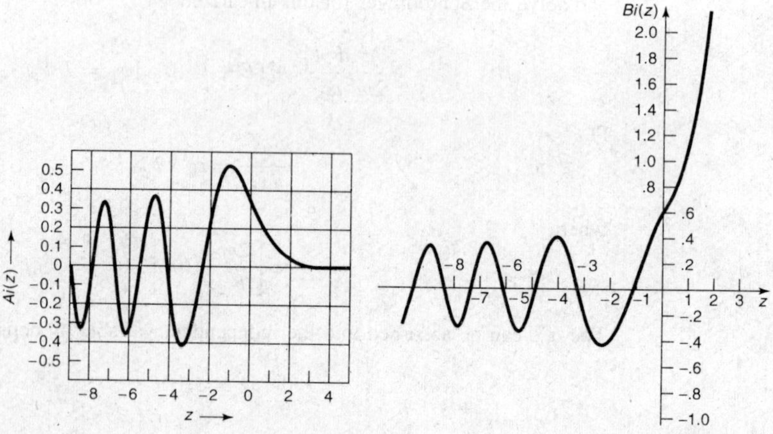

FIGURE 8.8: Graph of the Airy functions.

They are related to Bessel functions of order 1/3; some of their properties are listed in Table 8.1 and they are plotted in Figure 8.8. Evidently the patching wave function is a linear combination of $Ai(z)$ and $Bi(z)$:

$$\psi_p(x) = aAi(\alpha x) + bBi(\alpha x), \tag{8.37}$$

for appropriate constants a and b.

Now ψ_p is the (approximate) wave function in the neighborhood of the origin; our job is to match it to the WKB solutions in the overlap regions on either side (see Figure 8.9). These overlap zones are close enough to the turning point that the linearized potential is reasonably accurate (so that ψ_p is a good approximation to the true wave function), and yet far enough away from the turning point that the WKB approximation is reliable.[10] In the overlap regions Equation 8.32 holds, and therefore (in the notation of Equation 8.34)

$$p(x) \cong \sqrt{2m(E - E - V'(0)x)} = \hbar\alpha^{3/2}\sqrt{-x}. \tag{8.38}$$

In particular, in overlap region 2,

$$\int_0^x |p(x')|\,dx' \cong \hbar\alpha^{3/2}\int_0^x \sqrt{x'}\,dx' = \frac{2}{3}\hbar(\alpha x)^{3/2},$$

[10]This is a delicate double constraint, and it is possible to concoct potentials so pathological that no such overlap region exists. However, in practical applications this seldom occurs. See Problem 8.8.

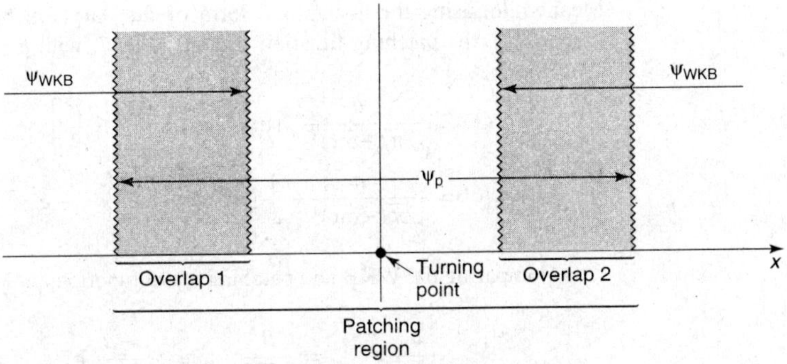

FIGURE 8.9: Patching region and the two overlap zones.

and therefore the WKB wave function (Equation 8.31) can be written as

$$\psi(x) \cong \frac{D}{\sqrt{\hbar}\alpha^{3/4}x^{1/4}} e^{-\frac{2}{3}(\alpha x)^{3/2}}. \qquad [8.39]$$

Meanwhile, using the large-z asymptotic forms[11] of the Airy functions (from Table 8.1), the patching wave function (Equation 8.37) in overlap region 2 becomes

$$\psi_p(x) \cong \frac{a}{2\sqrt{\pi}(\alpha x)^{1/4}} e^{-\frac{2}{3}(\alpha x)^{3/2}} + \frac{b}{\sqrt{\pi}(\alpha x)^{1/4}} e^{\frac{2}{3}(\alpha x)^{3/2}}. \qquad [8.40]$$

Comparing the two solutions, we see that

$$a = \sqrt{\frac{4\pi}{\alpha\hbar}} D, \quad \text{and} \quad b = 0. \qquad [8.41]$$

Now we go back and repeat the procedure for overlap region 1. Once again, $p(x)$ is given by Equation 8.38, but this time x is *negative*, so

$$\int_x^0 p(x')\,dx' \cong \frac{2}{3}\hbar(-\alpha x)^{3/2} \qquad [8.42]$$

and the WKB wave function (Equation 8.31) is

$$\psi(x) \cong \frac{1}{\sqrt{\hbar}\alpha^{3/4}(-x)^{1/4}} \left[Be^{i\frac{2}{3}(-\alpha x)^{3/2}} + Ce^{-i\frac{2}{3}(-\alpha x)^{3/2}} \right]. \qquad [8.43]$$

[11] At first glance it seems absurd to use a *large-z* approximation in this region, which after all is supposed to be reasonably close to the turning point at $z = 0$ (so that the linear approximation to the potential is valid). But notice that the argument here is αx, and if you study the matter carefully (see Problem 8.8) you will find that there *is* (typically) a region in which αx is large, but at the same time it is reasonable to approximate $V(x)$ by a straight line.

Meanwhile, using the asymptotic form of the Airy function for large *negative* z (Table 8.1), the patching function (Equation 8.37, with $b = 0$) reads

$$\begin{aligned}\psi_p(x) &\cong \frac{a}{\sqrt{\pi}(-\alpha x)^{1/4}} \sin\left[\frac{2}{3}(-\alpha x)^{3/2} + \frac{\pi}{4}\right] \\ &= \frac{a}{\sqrt{\pi}(-\alpha x)^{1/4}} \frac{1}{2i}\left[e^{i\pi/4} e^{i\frac{2}{3}(-\alpha x)^{3/2}} - e^{-i\pi/4} e^{-i\frac{2}{3}(-\alpha x)^{3/2}}\right]. \end{aligned} \quad [8.44]$$

Comparing the WKB and patching wave functions in overlap region 1, we find

$$\frac{a}{2i\sqrt{\pi}} e^{i\pi/4} = \frac{B}{\sqrt{\hbar\alpha}} \quad \text{and} \quad \frac{-a}{2i\sqrt{\pi}} e^{-i\pi/4} = \frac{C}{\sqrt{\hbar\alpha}},$$

or, putting in Equation 8.41 for a:

$$B = -ie^{i\pi/4} D, \quad \text{and} \quad C = ie^{-i\pi/4} D. \quad [8.45]$$

These are the so-called **connection formulas**, joining the WKB solutions at either side of the turning point. We're done with the patching wave function now—its only purpose was to bridge the gap. Expressing everything in terms of the one normalization constant D, and shifting the turning point back from the origin to an arbitrary point x_2, the WKB wave function (Equation 8.31) becomes

$$\psi(x) \cong \begin{cases} \dfrac{2D}{\sqrt{p(x)}} \sin\left[\dfrac{1}{\hbar}\displaystyle\int_x^{x_2} p(x')\,dx' + \dfrac{\pi}{4}\right], & \text{if } x < x_2; \\ \dfrac{D}{\sqrt{|p(x)|}} \exp\left[-\dfrac{1}{\hbar}\displaystyle\int_{x_2}^{x} |p(x')|\,dx'\right], & \text{if } x > x_2. \end{cases} \quad [8.46]$$

Example 8.3 **Potential well with one vertical wall.** Imagine a potential well that has one vertical side (at $x = 0$) and one sloping side (Figure 8.10). In this case $\psi(0) = 0$, so Equation 8.46 says

$$\frac{1}{\hbar}\int_0^{x_2} p(x)\,dx + \frac{\pi}{4} = n\pi, \quad (n = 1, 2, 3, \ldots),$$

or

$$\boxed{\int_0^{x_2} p(x)\,dx = \left(n - \frac{1}{4}\right)\pi\hbar.} \quad [8.47]$$

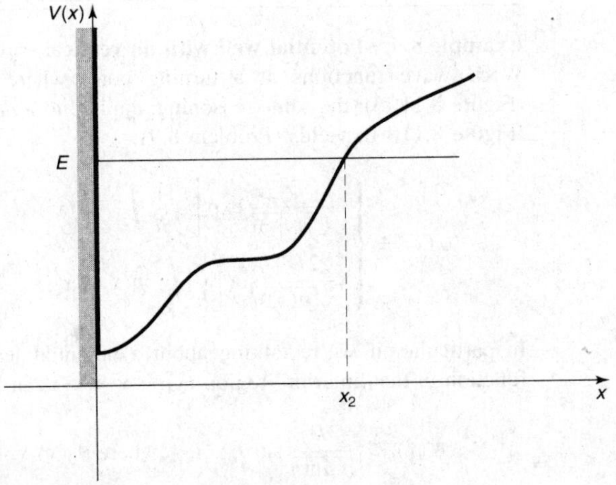

FIGURE 8.10: **Potential well with one vertical wall.**

For instance, consider the "half-harmonic oscillator,"

$$V(x) = \begin{cases} \frac{1}{2}m\omega^2 x^2, & \text{if } x > 0, \\ 0, & \text{otherwise.} \end{cases} \quad [8.48]$$

In this case

$$p(x) = \sqrt{2m[E - (1/2)m\omega^2 x^2]} = m\omega\sqrt{x_2^2 - x^2},$$

where

$$x_2 = \frac{1}{\omega}\sqrt{\frac{2E}{m}}$$

is the turning point. So

$$\int_0^{x_2} p(x)\,dx = m\omega \int_0^{x_2} \sqrt{x_2^2 - x^2}\,dx = \frac{\pi}{4} m\omega x_2^2 = \frac{\pi E}{2\omega},$$

and the quantization condition (Equation 8.47) yields

$$E_n = \left(2n - \frac{1}{2}\right)\hbar\omega = \left(\frac{3}{2}, \frac{7}{2}, \frac{11}{2}, \ldots\right)\hbar\omega. \quad [8.49]$$

In this particular case the WKB approximation actually delivers the *exact* allowed energies (which are precisely the *odd* energies of the *full* harmonic oscillator—see Problem 2.42).

Example 8.4 Potential well with no vertical walls. Equation 8.46 connects the WKB wave functions at a turning point where the potential slopes *upward* (Figure 8.11(a)); the same reasoning, applied to a *downward*-sloping turning point (Figure 8.11(b)), yields (Problem 8.9)

$$\psi(x) \cong \begin{cases} \dfrac{D'}{\sqrt{|p(x)|}} \exp\left[-\dfrac{1}{\hbar}\displaystyle\int_x^{x_1} |p(x')|\,dx'\right], & \text{if } x < x_1; \\ \dfrac{2D'}{\sqrt{p(x)}} \sin\left[\dfrac{1}{\hbar}\displaystyle\int_{x_1}^{x} p(x')\,dx' + \dfrac{\pi}{4}\right], & \text{if } x > x_1. \end{cases} \tag{8.50}$$

In particular, if we're talking about a potential *well* (Figure 8.11(c)), the wave function in the "interior" region ($x_1 < x < x_2$) can be written *either* as

$$\psi(x) \cong \frac{2D}{\sqrt{p(x)}} \sin\theta_2(x), \quad \text{where } \theta_2(x) \equiv \frac{1}{\hbar}\int_x^{x_2} p(x')\,dx' + \frac{\pi}{4},$$

(Equation 8.46), *or* as

$$\psi(x) \cong \frac{-2D'}{\sqrt{p(x)}} \sin\theta_1(x), \quad \text{where } \theta_1(x) \equiv -\frac{1}{\hbar}\int_{x_1}^{x} p(x')\,dx' - \frac{\pi}{4},$$

(Equation 8.50). Evidently the arguments of the sine functions must be equal, modulo π:[12] $\theta_2 = \theta_1 + n\pi$, from which it follows that

$$\boxed{\int_{x_1}^{x_2} p(x)\,dx = \left(n - \frac{1}{2}\right)\pi\hbar, \quad \text{with } n = 1, 2, 3, \ldots.} \tag{8.51}$$

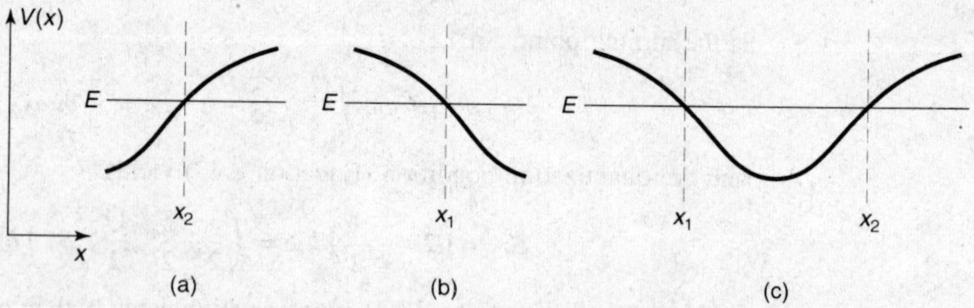

FIGURE 8.11: Upward-sloping and downward-sloping turning points.

[12] *Not* 2π—an overall minus sign can be absorbed into the normalization factors D and D'.

This quantization condition determines the allowed energies for the "typical" case of a potential well with two sloping sides. Notice that it differs from the formulas for two vertical walls (Equation 8.16) or one vertical wall (Equation 8.47) only in the number that is subtracted from n (0, 1/4, or 1/2). Since the WKB approximation works best in the semi-classical (large n) regime, the distinction is more in appearance than in substance. In any event, the result is extraordinarily powerful, for it enables us to calculate (approximate) allowed energies *without ever solving the Schrödinger equation*, by simply evaluating one integral. The wave function itself has dropped out of sight.

∗∗**Problem 8.5** Consider the quantum mechanical analog to the classical problem of a ball (mass m) bouncing elastically on the floor.[13]

(a) What is the potential energy, as a function of height x above the floor? (For negative x, the potential is *infinite*—the ball can't get there at all.)

(b) Solve the Schrödinger equation for this potential, expressing your answer in terms of the appropriate Airy function (note that $Bi(z)$ blows up for large z, and must therefore be rejected). Don't bother to normalize $\psi(x)$.

(c) Using $g = 9.80\ \text{m/s}^2$ and $m = 0.100$ kg, find the first four allowed energies, in joules, correct to three significant digits. *Hint:* See Milton Abramowitz and Irene A. Stegun, *Handbook of Mathematical Functions*, Dover, New York (1970), page 478; the notation is defined on page 450.

(d) What is the ground state energy, in eV, of an *electron* in this gravitational field? How high off the ground is this electron, on the average? *Hint:* Use the virial theorem to determine $\langle x \rangle$.

∗**Problem 8.6** Analyze the bouncing ball (Problem 8.5) using the WKB approximation.

(a) Find the allowed energies, E_n, in terms of m, g, and $\hbar$.

(b) Now put in the particular values given in Problem 8.5(c), and compare the WKB approximation to the first four energies with the "exact" results.

(c) About how large would the quantum number n have to be to give the ball an average height of, say, 1 meter above the ground?

---

[13] For more on the quantum bouncing ball see J. Gea-Banacloche, *Am. J. Phys.* **67**, 776 (1999) and N. Wheeler, "Classical/quantum dynamics in a uniform gravitational field," unpublished Reed College report (2002). This may sound like an awfully artificial problem, but the experiment has actually been done, using neutrons (V. V. Nesvizhevsky *et al.*, Nature **415**, 297 (2002)).

***Problem 8.7** Use the WKB approximation to find the allowed energies of the harmonic oscillator.

Problem 8.8 Consider a particle of mass m in the nth stationary state of the harmonic oscillator (angular frequency ω).

(a) Find the turning point, x_2.

(b) How far (d) could you go *above* the turning point before the error in the linearized potential (Equation 8.32, but with the turning point at x_2) reaches 1%? That is, if

$$\frac{V(x_2+d) - V_{\text{lin}}(x_2+d)}{V(x_2)} = 0.01,$$

what is d?

(c) The asymptotic form of $Ai(z)$ is accurate to 1% as long as $z \geq 5$. For the d in part (b), determine the smallest n such that $\alpha d \geq 5$. (For any n larger than this there exists an overlap region in which the linearized potential is good to 1% *and* the large-z form of the Airy function is good to 1%.)

****Problem 8.9** Derive the connection formulas at a downward-sloping turning point, and confirm Equation 8.50.

*****Problem 8.10** Use appropriate connection formulas to analyze the problem of scattering from a barrier with sloping walls (Figure 8.12). *Hint:* Begin by writing the WKB wave function in the form

$$\psi(x) \cong \begin{cases} \dfrac{1}{\sqrt{p(x)}}\left[Ae^{\frac{i}{\hbar}\int_x^{x_1} p(x')\,dx'} + Be^{-\frac{i}{\hbar}\int_x^{x_1} p(x')\,dx'}\right], & (x < x_1); \\ \dfrac{1}{\sqrt{|p(x)|}}\left[Ce^{\frac{1}{\hbar}\int_{x_1}^{x} |p(x')|\,dx'} + De^{-\frac{1}{\hbar}\int_{x_1}^{x} |p(x')|\,dx'}\right], & (x_1 < x < x_2); \\ \dfrac{1}{\sqrt{p(x)}}\left[Fe^{\frac{i}{\hbar}\int_{x_2}^{x} p(x')\,dx'}\right], & (x > x_2). \end{cases} \quad [8.52]$$

Do *not* assume $C = 0$. Calculate the tunneling probability, $T = |F|^2/|A|^2$, and show that your result reduces to Equation 8.22 in the case of a broad, high barrier.

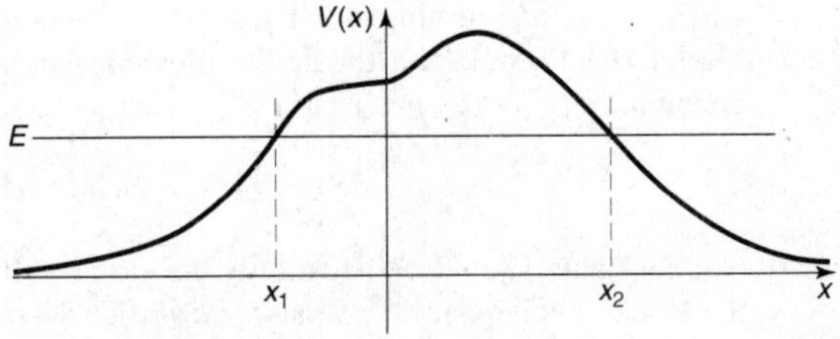

FIGURE 8.12: Barrier with sloping walls.

FURTHER PROBLEMS FOR CHAPTER 8

∗∗**Problem 8.11** Use the WKB approximation to find the allowed energies of the general power-law potential:

$$V(x) = \alpha |x|^{\nu},$$

where ν is a positive number. Check your result for the case $\nu = 2$. *Answer:*[14]

$$E_n = \alpha \left[(n - 1/2)\hbar \sqrt{\frac{\pi}{2m\alpha}} \frac{\Gamma\left(\frac{1}{\nu} + \frac{3}{2}\right)}{\Gamma\left(\frac{1}{\nu} + 1\right)} \right]^{\left(\frac{2\nu}{\nu+2}\right)} \qquad [8.53]$$

∗∗**Problem 8.12** Use the WKB approximation to find the bound state energy for the potential in Problem 2.51. Compare the exact answer. *Answer:* $-[(9/8) - (1/\sqrt{2})]\,\hbar^2 a^2/m$.

Problem 8.13 For spherically symmetrical potentials we can apply the WKB approximation to the radial part (Equation 4.37). In the case $l = 0$ it is reasonable[15] to use Equation 8.47 in the form

$$\int_0^{r_0} p(r)\, dr = (n - 1/4)\pi\hbar, \qquad [8.54]$$

[14] As always, the WKB result is most accurate in the semi-classical (large n) regime. In particular, Equation 8.53 is not very good for the ground state ($n = 1$). See W. N. Mei, *Am. J. Phys.* **66**, 541 (1998).

[15] Application of the WKB approximation to the radial equation raises some delicate and subtle problems, which I will not go into here. The classic paper on the subject is R. Langer, *Phys. Rev.* **51**, 669 (1937).

where r_0 is the turning point (in effect, we treat $r = 0$ as an infinite wall). Exploit this formula to estimate the allowed energies of a particle in the logarithmic potential

$$V(r) = V_0 \ln(r/a)$$

(for constants V_0 and a). Treat only the case $l = 0$. Show that the spacing between the levels is independent of mass. *Partial answer:*

$$E_{n+1} - E_n = V_0 \ln\left(\frac{n+3/4}{n-1/4}\right).$$

$**$**Problem 8.14** Use the WKB approximation in the form

$$\int_{r_1}^{r_2} p(r)\, dr = (n - 1/2)\pi\hbar \qquad [8.55]$$

to estimate the bound state energies for hydrogen. Don't forget the centrifugal term in the effective potential (Equation 4.38). The following integral may help:

$$\int_a^b \frac{1}{x}\sqrt{(x-a)(b-x)}\, dx = \frac{\pi}{2}(\sqrt{b} - \sqrt{a})^2. \qquad [8.56]$$

Note that you recover the Bohr levels when $n \gg l$ and $n \gg 1/2$. *Answer:*

$$E_{nl} \cong \frac{-13.6 \text{ eV}}{[n - (1/2) + \sqrt{l(l+1)}]^2}. \qquad [8.57]$$

$***$**Problem 8.15** Consider the case of a symmetrical double well, such as the one pictured in Figure 8.13. We are interested in bound states with $E < V(0)$.

(a) Write down the WKB wave functions in regions (i) $x > x_2$, (ii) $x_1 < x < x_2$, and (iii) $0 < x < x_1$. Impose the appropriate connection formulas at x_1 and x_2 (this has already been done, in Equation 8.46, for x_2; you will have to work out x_1 for yourself), to show that

$$\psi(x) \cong \begin{cases} \dfrac{D}{\sqrt{|p(x)|}} \exp\left[-\dfrac{1}{\hbar}\displaystyle\int_{x_2}^{x} |p(x')| dx'\right], & \text{(i)} \\ \dfrac{2D}{\sqrt{p(x)}} \sin\left[\dfrac{1}{\hbar}\displaystyle\int_{x}^{x_2} p(x')\, dx' + \dfrac{\pi}{4}\right], & \text{(ii)} \\ \dfrac{D}{\sqrt{|p(x)|}} \left[2\cos\theta\, e^{\frac{1}{\hbar}\int_x^{x_1} |p(x')| dx'} + \sin\theta\, e^{-\frac{1}{\hbar}\int_x^{x_1} |p(x')| dx'}\right], & \text{(iii)} \end{cases}$$

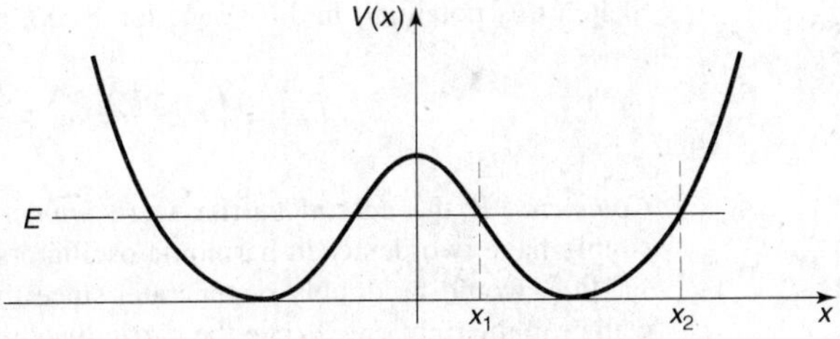

FIGURE 8.13: Symmetrical double well; Problem 8.15.

where

$$\theta \equiv \frac{1}{\hbar}\int_{x_1}^{x_2} p(x)\, dx. \qquad [8.58]$$

(b) Because $V(x)$ is symmetric, we need only consider even (+) and odd (−) wave functions. In the former case $\psi'(0) = 0$, and in the latter case $\psi(0) = 0$. Show that this leads to the following quantization condition:

$$\tan\theta = \pm 2e^{\phi}, \qquad [8.59]$$

where

$$\phi \equiv \frac{1}{\hbar}\int_{-x_1}^{x_1} |p(x')|\, dx'. \qquad [8.60]$$

Equation 8.59 determines the (approximate) allowed energies (note that E comes into x_1 and x_2, so θ and ϕ are both functions of E).

(c) We are particularly interested in a high and/or broad central barrier, in which case ϕ is large, and e^{ϕ} is *huge*. Equation 8.59 then tells us that θ must be very close to a half-integer multiple of π. With this in mind, write $\theta = (n + 1/2)\pi + \epsilon$, where $|\epsilon| \ll 1$, and show that the quantization condition becomes

$$\theta \cong \left(n + \frac{1}{2}\right)\pi \mp \frac{1}{2}e^{-\phi}. \qquad [8.61]$$

(d) Suppose each well is a parabola:[16]

$$V(x) = \begin{cases} \frac{1}{2}m\omega^2(x+a)^2, & \text{if } x < 0, \\ \frac{1}{2}m\omega^2(x-a)^2, & \text{if } x > 0. \end{cases} \qquad [8.62]$$

[16]Even if $V(x)$ is not strictly parabolic in each well, this calculation of θ, and hence the result (Equation 8.63) will be *approximately* correct, in the sense discussed at the beginning of Section 2.3, with $\omega \equiv \sqrt{V''(x_0)/m}$, where x_0 is the position of the minimum.

Sketch this potential, find θ (Equation 8.58), and show that

$$E_n^{\pm} \cong \left(n + \frac{1}{2}\right)\hbar\omega \mp \frac{\hbar\omega}{2\pi} e^{-\phi}. \qquad [8.63]$$

Comment: If the central barrier were *impenetrable* ($\phi \to \infty$), we would simply have two detached harmonic oscillators, and the energies, $E_n = (n + 1/2)\hbar\omega$, would be doubly degenerate, since the particle could be in the left well or in the right one. When the barrier becomes *finite* (putting the two wells into "communication"), the degeneracy is lifted. The even states (ψ_n^+) have slightly *lower* energy, and the odd ones (ψ_n^-) have slightly higher energy.

(e) Suppose the particle starts out in the *right* well—or, more precisely, in a state of the form

$$\Psi(x, 0) = \frac{1}{\sqrt{2}}(\psi_n^+ + \psi_n^-),$$

which, assuming the phases are picked in the "natural" way, will be concentrated in the right well. Show that it oscillates back and forth between the wells, with a period

$$\tau = \frac{2\pi^2}{\omega} e^{\phi}. \qquad [8.64]$$

(f) Calculate ϕ, for the specific potential in part (d), and show that for $V(0) \gg E$, $\phi \sim m\omega a^2/\hbar$.

Problem 8.16 Tunneling in the Stark effect. When you turn on an external electric field, the electron in an atom can, in principle, tunnel out, ionizing the atom. *Question:* Is this likely to happen in a typical Stark effect experiment? We can estimate the probability using a crude one-dimensional model, as follows. Imagine a particle in a very deep finite square well (Section 2.6).

(a) What is the energy of the ground state, measured up from the bottom of the well? Assume $V_0 \gg \hbar^2/ma^2$. *Hint:* This is just the ground state energy of the *infinite* square well (of width $2a$).

(b) Now introduce a perturbation $H' = -\alpha x$ (for an electron in an electric field $\mathbf{E} = -E_{\text{ext}}\hat{\imath}$ we would have $\alpha = eE_{\text{ext}}$). Assume it is relatively weak ($\alpha a \ll \hbar^2/ma^2$). Sketch the total potential, and note that the particle can now tunnel out, in the direction of positive x.

(c) Calculate the tunneling factor γ (Equation 8.22), and estimate the time it would take for the particle to escape (Equation 8.28). *Answer:* $\gamma = \sqrt{8mV_0^3}/3\alpha\hbar$, $\tau = (8ma^2/\pi\hbar)e^{2\gamma}$.

(d) Put in some reasonable numbers: $V_0 = 20$ eV (typical binding energy for an outer electron), $a = 10^{-10}$ m (typical atomic radius), $E_{\text{ext}} = 7 \times 10^6$ V/m (strong laboratory field), e and m the charge and mass of the electron. Calculate τ, and compare it to the age of the universe.

Problem 8.17 About how long would it take for a can of beer at room temperature to topple over spontaneously, as a result of quantum tunneling? *Hint:* Treat it as a uniform cylinder of mass m, radius R, and length h. As the can tips, let x be the height of the center above its equilibrium position ($h/2$). The potential energy is mgx, and it topples when x reaches the critical value $x_0 = \sqrt{R^2 + (h/2)^2} - h/2$. Calculate the tunneling probability (Equation 8.22), for $E = 0$. Use Equation 8.28, with the thermal energy ($(1/2)mv^2 = (1/2)k_BT$) to estimate the velocity. Put in reasonable numbers, and give your final answer in years.[17]

---

[17] R. E. Crandall, *Scientific American*, February 1997, p. 74.

CHAPTER 9

TIME-DEPENDENT PERTURBATION THEORY

Up to this point, practically everything we have done belongs to the subject that might properly be called **quantum statics**, in which the potential energy function is *independent of time*: $V(\mathbf{r}, t) = V(\mathbf{r})$. In that case the (time-dependent) Schrödinger equation,

$$H\Psi = i\hbar\frac{\partial \Psi}{\partial t},$$

can be solved by separation of variables:

$$\Psi(\mathbf{r}, t) = \psi(\mathbf{r})e^{-iEt/\hbar},$$

where $\psi(\mathbf{r})$ satisfies the time-*in*dependent Schrödinger equation,

$$H\psi = E\psi.$$

Because the time dependence of separable solutions is carried by the exponential factor ($e^{-iEt/\hbar}$), which cancels out when we construct the physically relevant quantity $|\Psi|^2$, all probabilities and expectation values are constant in time. By forming *linear combinations* of these stationary states we obtain wave functions with more interesting time dependence, but even then the possible values of the energy, and their respective probabilities, are constant.

If we want to allow for **transitions** (**quantum jumps**, as they are sometimes called) between one energy level and another, we must introduce a *time-dependent* potential (**quantum dynamics**). There are precious few exactly solvable problems in quantum dynamics. However, if the time-dependent portion of the Hamiltonian is small compared to the time-independent part, it can be treated as a perturbation. My purpose in this chapter is to develop time-dependent perturbation theory, and study its most important application: the emission or absorption of radiation by an atom.

9.1 TWO-LEVEL SYSTEMS

To begin with, let us suppose that there are just *two* states of the (unperturbed) system, ψ_a and ψ_b. They are eigenstates of the unperturbed Hamiltonian, H^0:

$$H^0\psi_a = E_a\psi_a, \quad \text{and} \quad H^0\psi_b = E_b\psi_b, \tag{9.1}$$

and they are orthonormal:

$$\langle\psi_a|\psi_b\rangle = \delta_{ab}. \tag{9.2}$$

Any state can be expressed as a linear combination of them; in particular,

$$\Psi(0) = c_a\psi_a + c_b\psi_b. \tag{9.3}$$

The states ψ_a and ψ_b might be position-space wave functions, or spinors, or something more exotic—it doesn't matter; it is the *time* dependence that concerns us here, so when I write $\Psi(t)$, I simply mean the state of the system at time t. In the absence of any perturbation, each component evolves with its characteristic exponential factor:

$$\Psi(t) = c_a\psi_a e^{-iE_at/\hbar} + c_b\psi_b e^{-iE_bt/\hbar}. \tag{9.4}$$

We say that $|c_a|^2$ is the "probability that the particle is in state ψ_a"—by which we *really* mean the probability that a measurement of the energy would yield the value E_a. Normalization of Ψ requires, of course, that

$$|c_a|^2 + |c_b|^2 = 1. \tag{9.5}$$

9.1.1 The Perturbed System

Now suppose we turn on a time-dependent perturbation, $H'(t)$. Since ψ_a and ψ_b constitute a complete set, the wave function $\Psi(t)$ can still be expressed as a linear combination of them. The only difference is that c_a and c_b are now *functions of* t:

$$\Psi(t) = c_a(t)\psi_a e^{-iE_at/\hbar} + c_b(t)\psi_b e^{-iE_bt/\hbar}. \tag{9.6}$$

(I could absorb the exponential factors into $c_a(t)$ and $c_b(t)$, and some people prefer to do it this way, but I think it is nicer to keep visible that part of the time dependence that would be present even *without* the perturbation.) The whole problem is to determine c_a and c_b, as functions of time. If, for example, the particle started out in the state ψ_a ($c_a(0) = 1$, $c_b(0) = 0$), and at some later time t_1 we find that $c_a(t_1) = 0$, $c_b(t_1) = 1$, we shall report that the system underwent a transition from ψ_a to ψ_b.

We solve for $c_a(t)$ and $c_b(t)$ by demanding that $\Psi(t)$ satisfy the time-dependent Schrödinger equation,

$$H\Psi = i\hbar\frac{\partial\Psi}{\partial t}, \quad \text{where } H = H^0 + H'(t). \tag{9.7}$$

From Equations 9.6 and 9.7, we find:

$$c_a[H^0\psi_a]e^{-iE_at/\hbar} + c_b[H^0\psi_b]e^{-iE_bt/\hbar} + c_a[H'\psi_a]e^{-iE_at/\hbar} + c_b[H'\psi_b]e^{-iE_bt/\hbar}$$
$$= i\hbar\Big[\dot{c}_a\psi_a e^{-iE_at/\hbar} + \dot{c}_b\psi_b e^{-iE_bt/\hbar}$$
$$+ c_a\psi_a\left(-\frac{iE_a}{\hbar}\right)e^{-iE_at/\hbar} + c_b\psi_b\left(-\frac{iE_b}{\hbar}\right)e^{-iE_bt/\hbar}\Big].$$

In view of Equation 9.1, the first two terms on the left cancel the last two terms on the right, and hence

$$c_a[H'\psi_a]e^{-iE_at/\hbar} + c_b[H'\psi_b]e^{-iE_bt/\hbar} = i\hbar\left[\dot{c}_a\psi_a e^{-iE_at/\hbar} + \dot{c}_b\psi_b e^{-iE_bt/\hbar}\right]. \tag{9.8}$$

To isolate $\dot{c}_a$, we use the standard trick: Take the inner product with ψ_a, and exploit the orthogonality of ψ_a and ψ_b (Equation 9.2):

$$c_a\langle\psi_a|H'|\psi_a\rangle e^{-iE_at/\hbar} + c_b\langle\psi_a|H'|\psi_b\rangle e^{-iE_bt/\hbar} = i\hbar\dot{c}_a e^{-iE_at/\hbar}.$$

For short, we define

$$H'_{ij} \equiv \langle\psi_i|H'|\psi_j\rangle; \tag{9.9}$$

note that the hermiticity of H' entails $H'_{ji} = (H'_{ij})^*$. Multiplying through by $-(i/\hbar)e^{iE_at/\hbar}$, we conclude that:

$$\dot{c}_a = -\frac{i}{\hbar}\left[c_a H'_{aa} + c_b H'_{ab} e^{-i(E_b-E_a)t/\hbar}\right]. \tag{9.10}$$

Similarly, the inner product with ψ_b picks out $\dot{c}_b$:

$$c_a\langle\psi_b|H'|\psi_a\rangle e^{-iE_at/\hbar} + c_b\langle\psi_b|H'|\psi_b\rangle e^{-iE_bt/\hbar} = i\hbar\dot{c}_b e^{-iE_bt/\hbar},$$

and hence

$$\dot{c}_b = -\frac{i}{\hbar}\left[c_b H'_{bb} + c_a H'_{ba} e^{i(E_b-E_a)t/\hbar}\right]. \tag{9.11}$$

Equations 9.10 and 9.11 determine $c_a(t)$ and $c_b(t)$; taken together, they are completely equivalent to the (time-dependent) Schrödinger equation, for a two-level system. Typically, the diagonal matrix elements of H' vanish (see Problem 9.4 for the general case):

$$H'_{aa} = H'_{bb} = 0. \qquad [9.12]$$

If so, the equations simplify:

$$\boxed{\dot{c}_a = -\frac{i}{\hbar} H'_{ab} e^{-i\omega_0 t} c_b, \quad \dot{c}_b = -\frac{i}{\hbar} H'_{ba} e^{i\omega_0 t} c_a,} \qquad [9.13]$$

where

$$\omega_0 \equiv \frac{E_b - E_a}{\hbar}. \qquad [9.14]$$

(I'll assume that $E_b \geq E_a$, so $\omega_0 \geq 0$.)

***Problem 9.1** A hydrogen atom is placed in a (time-dependent) electric field $\mathbf{E} = E(t)\hat{k}$. Calculate all four matrix elements H'_{ij} of the perturbation $H' = eEz$ between the ground state ($n = 1$) and the (quadruply degenerate) first excited states ($n = 2$). Also show that $H'_{ii} = 0$ for all five states. *Note:* There is only one integral to be done here, if you exploit oddness with respect to z; only one of the $n = 2$ states is "accessible" from the ground state by a perturbation of this form, and therefore the system functions as a two-state configuration—assuming transitions to higher excited states can be ignored.

***Problem 9.2** Solve Equation 9.13 for the case of a *time-independent* perturbation, assuming that $c_a(0) = 1$ and $c_b(0) = 0$. Check that $|c_a(t)|^2 + |c_b(t)|^2 = 1$. *Comment:* Ostensibly, this system oscillates between "pure ψ_a" and "some ψ_b." Doesn't this contradict my general assertion that no transitions occur for time-independent perturbations? No, but the reason is rather subtle: In this case ψ_a and ψ_b are not, and never were, eigenstates of the Hamiltonian—a measurement of the energy *never* yields E_a or E_b. In time-dependent perturbation theory we typically contemplate turning *on* the perturbation for a while, and then turning it *off* again, in order to examine the system. At the beginning, and at the end, ψ_a and ψ_b are eigenstates of the exact Hamiltonian, and only in this context does it make sense to say that the system underwent a transition from one to the other. For the present problem, then, assume that the perturbation was turned on at time $t = 0$, and off again at time t—this doesn't affect the *calculations*, but it allows for a more sensible interpretation of the result.

****Problem 9.3** Suppose the perturbation takes the form of a delta function (in time):

$$H' = U\delta(t);$$

assume that $U_{aa} = U_{bb} = 0$, and let $U_{ab} = U_{ba}^* \equiv \alpha$. If $c_a(-\infty) = 1$ and $c_b(-\infty) = 0$, find $c_a(t)$ and $c_b(t)$, and check that $|c_a(t)|^2 + |c_b(t)|^2 = 1$. What is the net probability ($P_{a\to b}$ for $t \to \infty$) that a transition occurs? *Hint:* You might want to treat the delta function as the limit of a sequence of rectangles. *Answer:* $P_{a\to b} = \sin^2(|\alpha|/\hbar)$.

9.1.2 Time-Dependent Perturbation Theory

So far, everything is *exact*: We have made no assumption about the *size* of the perturbation. But if H' is "small," we can solve Equation 9.13 by a process of successive approximations, as follows. Suppose the particle starts out in the lower state:

$$c_a(0) = 1, \quad c_b(0) = 0. \tag{9.15}$$

If there were *no perturbation at all*, they would stay this way forever:

Zeroth Order:

$$c_a^{(0)}(t) = 1, \quad c_b^{(0)}(t) = 0. \tag{9.16}$$

(I'll use a superscript in parentheses to indicate the order of the approximation.)

To calculate the first-order approximation, we insert the zeroth-order values on the right side of Equation 9.13:

First Order:

$$\frac{dc_a^{(1)}}{dt} = 0 \Rightarrow c_a^{(1)}(t) = 1;$$
$$\frac{dc_b^{(1)}}{dt} = -\frac{i}{\hbar} H'_{ba} e^{i\omega_0 t} \Rightarrow c_b^{(1)} = -\frac{i}{\hbar}\int_0^t H'_{ba}(t') e^{i\omega_0 t'}\, dt'. \tag{9.17}$$

Now we insert *these* expressions on the right, to obtain the *second*-order approximation:

Second Order:

$$\frac{dc_a^{(2)}}{dt} = -\frac{i}{\hbar} H'_{ab} e^{-i\omega_0 t}\left(-\frac{i}{\hbar}\right)\int_0^t H'_{ba}(t') e^{i\omega_0 t'}\, dt' \Rightarrow$$
$$c_a^{(2)}(t) = 1 - \frac{1}{\hbar^2}\int_0^t H'_{ab}(t') e^{-i\omega_0 t'}\left[\int_0^{t'} H'_{ba}(t'') e^{i\omega_0 t''}\, dt''\right] dt', \tag{9.18}$$

while c_b is unchanged ($c_b^{(2)}(t) = c_b^{(1)}(t)$). (Notice that $c_a^{(2)}(t)$ *includes* the zeroth-order term; the second-order *correction* would be the integral part alone.)

In principle, we could continue this ritual indefinitely, always inserting the nth-order approximation into the right side of Equation 9.13, and solving for the $(n+1)$th order. The zeroth order contains *no* factors of H', the first-order correction contains *one* factor of H', the second-order correction has *two* factors of H', and so on.[1] The error in the first-order approximation is evident in the fact that $|c_a^{(1)}(t)|^2 + |c_b^{(1)}(t)|^2 \neq 1$ (the *exact* coefficients must, of course, obey Equation 9.5). However, $|c_a^{(1)}(t)|^2 + |c_b^{(1)}(t)|^2$ *is* equal to 1 *to first order in* H', which is all we can expect from a first-order approximation. And the same goes for the higher orders.

****Problem 9.4** Suppose you *don't* assume $H'_{aa} = H'_{bb} = 0$.

(a) Find $c_a(t)$ and $c_b(t)$ in first-order perturbation theory, for the case $c_a(0) = 1$, $c_b(0) = 0$. Show that $|c_a^{(1)}(t)|^2 + |c_b^{(1)}(t)|^2 = 1$, to first order in H'.

(b) There is a nicer way to handle this problem. Let

$$d_a \equiv e^{\frac{i}{\hbar}\int_0^t H'_{aa}(t')dt'} c_a, \quad d_b \equiv e^{\frac{i}{\hbar}\int_0^t H'_{bb}(t')dt'} c_b. \tag{9.19}$$

Show that

$$\dot{d}_a = -\frac{i}{\hbar} e^{i\phi} H'_{ab} e^{-i\omega_0 t} d_b; \quad \dot{d}_b = -\frac{i}{\hbar} e^{-i\phi} H'_{ba} e^{i\omega_0 t} d_a, \tag{9.20}$$

where

$$\phi(t) \equiv \frac{1}{\hbar}\int_0^t [H'_{aa}(t') - H'_{bb}(t')]\, dt'. \tag{9.21}$$

So the equations for d_a and d_b are identical in structure to Equation 9.13 (with an extra factor $e^{i\phi}$ tacked onto H').

(c) Use the method in part (b) to obtain $c_a(t)$ and $c_b(t)$ in first-order perturbation theory, and compare your answer to (a). Comment on any discrepancies.

***Problem 9.5** Solve Equation 9.13 to second order in perturbation theory, for the general case $c_a(0) = a$, $c_b(0) = b$.

****Problem 9.6** Calculate $c_a(t)$ and $c_b(t)$, to second order, for a time-independent perturbation (Problem 9.2). Compare your answer with the exact result.

---

[1] Notice that c_a is modified in every *even* order, and c_b in every *odd* order; this would not be true if the perturbation included diagonal terms, or if the system started out in a linear combination of the two states.

9.1.3 Sinusoidal Perturbations

Suppose the perturbation has sinusoidal time dependence:

$$H'(\mathbf{r}, t) = V(\mathbf{r})\cos(\omega t), \tag{9.22}$$

so that

$$H'_{ab} = V_{ab}\cos(\omega t), \tag{9.23}$$

where

$$V_{ab} \equiv \langle \psi_a | V | \psi_b \rangle. \tag{9.24}$$

(As before, I'll assume the diagonal matrix elements vanish, since this is almost always the case in practice.) To first order (from now on we'll work *exclusively* in first order, and I'll dispense with the superscripts) we have (Equation 9.17):

$$\begin{aligned} c_b(t) &\cong -\frac{i}{\hbar} V_{ba} \int_0^t \cos(\omega t') e^{i\omega_0 t'}\, dt' = -\frac{iV_{ba}}{2\hbar}\int_0^t \left[e^{i(\omega_0+\omega)t'} + e^{i(\omega_0-\omega)t'}\right] dt' \\ &= -\frac{V_{ba}}{2\hbar}\left[\frac{e^{i(\omega_0+\omega)t}-1}{\omega_0+\omega} + \frac{e^{i(\omega_0-\omega)t}-1}{\omega_0-\omega}\right]. \end{aligned} \tag{9.25}$$

That's the *answer*, but it's a little cumbersome to work with. Things simplify substantially if we restrict our attention to driving frequencies (ω) that are very close to the transition frequency (ω_0), so that the second term in the square brackets dominates; specifically, we assume:

$$\omega_0 + \omega \gg |\omega_0 - \omega|. \tag{9.26}$$

This is not much of a limitation, since perturbations at *other* frequencies have a negligible probability of causing a transition anyway.[2] Dropping the first term, we have

$$\begin{aligned} c_b(t) &\cong -\frac{V_{ba}}{2\hbar}\frac{e^{i(\omega_0-\omega)t/2}}{\omega_0-\omega}\left[e^{i(\omega_0-\omega)t/2} - e^{-i(\omega_0-\omega)t/2}\right] \\ &= -i\frac{V_{ba}}{\hbar}\frac{\sin[(\omega_0-\omega)t/2]}{\omega_0-\omega}e^{i(\omega_0-\omega)t/2}. \end{aligned} \tag{9.27}$$

The **transition probability**—the probability that a particle which started out in the state ψ_a will be found, at time t, in the state ψ_b—is

$$\boxed{P_{a\to b}(t) = |c_b(t)|^2 \cong \frac{|V_{ab}|^2}{\hbar^2}\frac{\sin^2[(\omega_0-\omega)t/2]}{(\omega_0-\omega)^2}.} \tag{9.28}$$

---

[2] In the following sections we will be applying this theory to the case of *light*, for which $\omega \sim 10^{15}$ s^{-1}, so the denominator in *both* terms is huge, except (for the second one) in the neighborhood of ω_0.

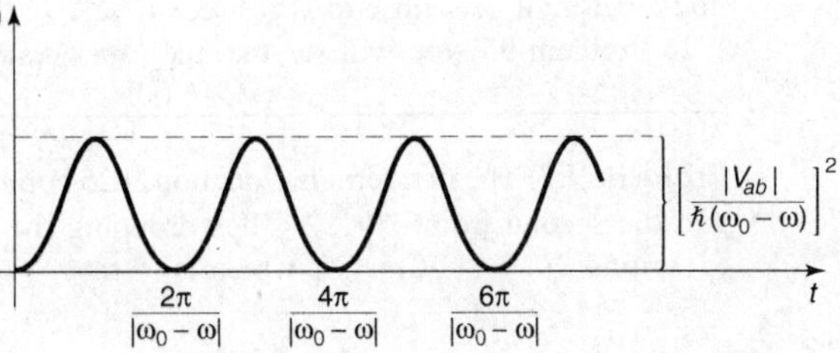

FIGURE 9.1: Transition probability as a function of time, for a sinusoidal perturbation (Equation 9.28).

The most remarkable feature of this result is that, as a function of time, the transition probability *oscillates* sinusoidally (Figure 9.1). After rising to a maximum of $|V_{ab}|^2/\hbar^2(\omega_0 - \omega)^2$—necessarily much less than 1, else the assumption that the perturbation is "small" would be invalid—it drops back down to zero! At times $t_n = 2n\pi/|\omega_0 - \omega|$, where $n = 1, 2, 3, \ldots$, the particle is *certain* to be back in the lower state. If you want to maximize your chances of provoking a transition, you should *not* keep the perturbation on for a long period; you do better to *turn it off* after a time $\pi/|\omega_0 - \omega|$, and hope to "catch" the system in the upper state. In Problem 9.7 it is shown that this "flopping" is not an artifact of perturbation theory—it occurs also in the exact solution, though the flopping *frequency* is modified somewhat.

As I noted earlier, the probability of a transition is greatest when the driving frequency is close to the "natural" frequency, ω_0. This is illustrated in Figure 9.2, where $P_{a\to b}$ is plotted as a function of ω. The peak has a height of $(|V_{ab}|t/2\hbar)^2$ and a width $4\pi/t$; evidently it gets higher and narrower as time goes on. (Ostensibly, the maximum increases without limit. However, the perturbation assumption breaks

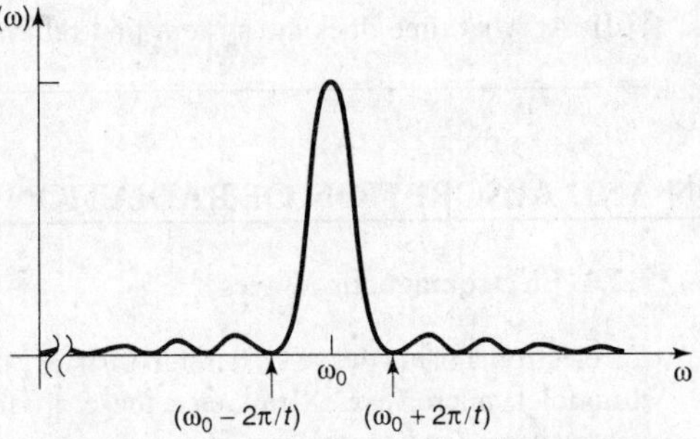

FIGURE 9.2: Transition probability as a function of driving frequency (Equation 9.28).

down before it gets close to 1, so we can believe the result only for relatively small t. In Problem 9.7 you will see that the *exact* result never exceeds 1.)

∗∗Problem 9.7 The first term in Equation 9.25 comes from the $e^{i\omega t}/2$ part of $\cos(\omega t)$, and the second from $e^{-i\omega t}/2$. Thus dropping the first term is formally equivalent to writing $H' = (V/2)e^{-i\omega t}$, which is to say,

$$H'_{ba} = \frac{V_{ba}}{2}e^{-i\omega t}, \quad H'_{ab} = \frac{V_{ab}}{2}e^{i\omega t}. \tag{9.29}$$

(The latter is required to make the Hamiltonian matrix hermitian—or, if you prefer, to pick out the dominant term in the formula analogous to Equation 9.25 for $c_a(t)$.) Rabi noticed that if you make this so-called **rotating wave approximation** at the *beginning* of the calculation, Equation 9.13 can be solved exactly, with no need for perturbation theory, and no assumption about the strength of the field.

(a) Solve Equation 9.13 in the rotating wave approximation (Equation 9.29), for the usual initial conditions: $c_a(0) = 1$, $c_b(0) = 0$. Express your results ($c_a(t)$ and $c_b(t)$) in terms of the **Rabi flopping frequency**,

$$\omega_r \equiv \frac{1}{2}\sqrt{(\omega - \omega_0)^2 + (|V_{ab}|/\hbar)^2}. \tag{9.30}$$

(b) Determine the transition probability, $P_{a\to b}(t)$, and show that it never exceeds 1. Confirm that $|c_a(t)|^2 + |c_b(t)|^2 = 1$.

(c) Check that $P_{a\to b}(t)$ reduces to the perturbation theory result (Equation 9.28) when the perturbation is "small," and state precisely what small *means* in this context, as a constraint on V.

(d) At what time does the system first return to its initial state?

9.2 EMISSION AND ABSORPTION OF RADIATION

9.2.1 Electromagnetic Waves

An electromagnetic wave (I'll refer to it as "light," though it could be infrared, ultraviolet, microwave, X-ray, etc.; these differ only in their frequencies) consists of transverse (and mutually perpendicular) oscillating electric and magnetic fields (Figure 9.3). An atom, in the presence of a passing light wave, responds primarily to the electric component. If the wavelength is long (compared to the size of the

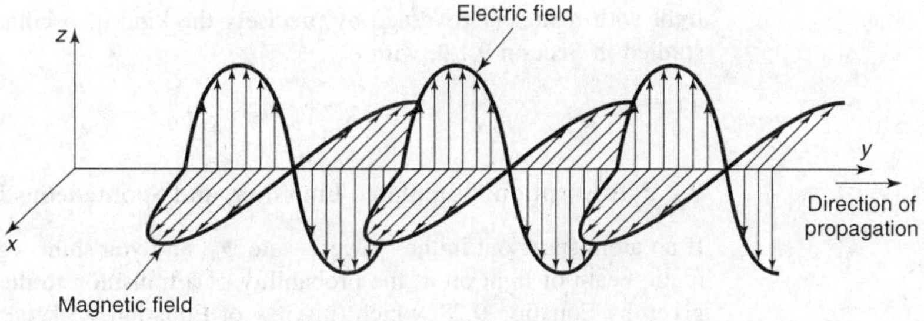

FIGURE 9.3: An electromagnetic wave.

atom), we can ignore the *spatial* variation in the field;[3] the atom, then, is exposed to a sinusoidally oscillating electric field

$$\mathbf{E} = E_0 \cos(\omega t)\,\hat{k} \qquad [9.31]$$

(for the moment I'll assume the light is monochromatic, and polarized along the z direction). The perturbing Hamiltonian is[4]

$$H' = -qE_0 z\cos(\omega t), \qquad [9.32]$$

where q is the charge of the electron.[5] Evidently[6]

$$H'_{ba} = -\wp E_0 \cos(\omega t), \quad \text{where } \wp \equiv q\langle\psi_b|z|\psi_a\rangle. \qquad [9.33]$$

Typically, ψ is an even or odd function of z; in either case $z|\psi|^2$ is odd, and integrates to zero (see Problem 9.1 for some examples). This licenses our usual assumption that the diagonal matrix elements of H' vanish. Thus the interaction of

[3]For visible light $\lambda \sim 5000$Å, while the diameter of an atom is around 1 Å, so this approximation is reasonable; but it would *not* be for X-rays. Problem 9.21 explores the effect of spatial variation of the field.

[4]The energy of a charge q in a static field $\mathbf{E}$ is $-q\int \mathbf{E}\cdot d\mathbf{r}$. You may well object to the use of an electro*static* formula for a manifestly time-dependent field. I am implicitly assuming that the period of oscillation is long compared to the time it takes the charge to move around (within the atom).

[5]As usual, we assume the nucleus is heavy and stationary; it is the wave function of the *electron* that concerns us.

[6]The letter $\wp$ is supposed to remind you of **electric dipole moment** (for which, in electrodynamics, the letter p is customarily used—in this context it is rendered as a squiggly $\wp$ to avoid confusion with momentum). Actually, $\wp$ is the off-diagonal matrix element of the z component of the dipole moment operator, $q\mathbf{r}$. Because of its association with electric dipole moments, radiation governed by Equation 9.33 is called **electric dipole radiation**; it is overwhelmingly the dominant kind, at least in the visible region. See Problem 9.21 for generalizations and terminology.

light with matter is governed by precisely the kind of oscillatory perturbation we studied in Section 9.1.3, with

$$V_{ba} = -\wp E_0. \quad [9.34]$$

9.2.2 Absorption, Stimulated Emission, and Spontaneous Emission

If an atom starts out in the "lower" state ψ_a, and you shine a polarized monochromatic beam of light on it, the probability of a transition to the "upper" state ψ_b is given by Equation 9.28, which (in view of Equation 9.34) takes the form

$$P_{a\to b}(t) = \left(\frac{|\wp|E_0}{\hbar}\right)^2 \frac{\sin^2[(\omega_0 - \omega)t/2]}{(\omega_0 - \omega)^2}. \quad [9.35]$$

In this process, the atom absorbs energy $E_b - E_a = \hbar\omega_0$ from the electromagnetic field. We say that it has "absorbed a photon" (Figure 9.4(a)). (As I mentioned earlier, the word "photon" really belongs to **quantum electrodynamics** [the quantum theory of the electromagnetic field], whereas we are treating the field itself *classically*. But this language is convenient, as long as you don't read more into it than is really there.)

I could, of course, go back and run the whole derivation for a system that starts off in the *upper* state ($c_a(0) = 0$, $c_b(0) = 1$). Do it for yourself, if you like; it comes out *exactly the same*—except that this time we're calculating $P_{b\to a} = |c_a(t)|^2$, the probability of a transition *down* to the *lower* level:

$$P_{b\to a}(t) = \left(\frac{|\wp|E_0}{\hbar}\right)^2 \frac{\sin^2[(\omega_0 - \omega)t/2]}{(\omega_0 - \omega)^2}. \quad [9.36]$$

(It *has* to come out this way—all we're doing is switching $a \leftrightarrow b$, which substitutes $-\omega_0$ for ω_0. When we get to Equation 9.25 we now keep the *first* term, with $-\omega_0 + \omega$ in the denominator, and the rest is the same as before.) But when you stop to think of it, this is an absolutely *astonishing* result: If the particle is in the *upper* state, and you shine light on it, it can make a transition to the *lower* state, and in fact the probability of such a transition is exactly the same as for a transition *upward* from the *lower* state. This process, which was first predicted by Einstein, is called **stimulated emission**.

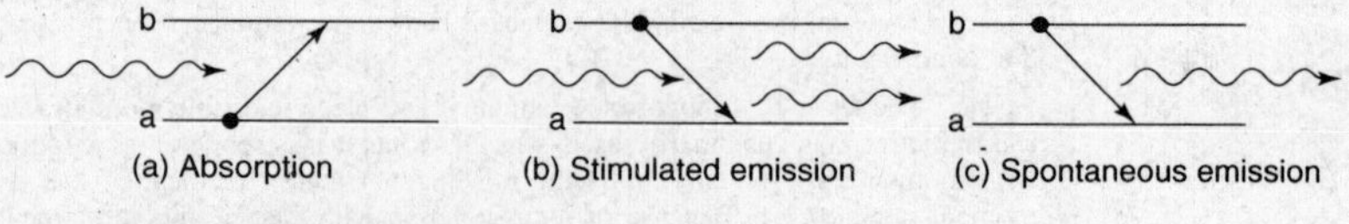

FIGURE 9.4: Three ways in which light interacts with atoms: (a) absorption, (b) stimulated emission, (c) spontaneous emission.

In the case of stimulated emission the electromagnetic field *gains* energy $\hbar\omega_0$ from the atom; we say that one photon went in and *two* photons came out—the original one that caused the transition plus another one from the transition itself (Figure 9.4(b)). This raises the possibility of *amplification*, for if I had a bottle of atoms, all in the upper state, and triggered it with a single incident photon, a chain reaction would occur, with the first photon producing 2, these 2 producing 4, and so on. We'd have an enormous number of photons coming out, all with the same frequency and at virtually the same instant. This is, of course, the principle behind the **laser** (light amplification by stimulated emission of radiation). Note that it is essential (for laser action) to get a majority of the atoms into the upper state (a so-called **population inversion**), because *absorption* (which *costs* one photon) competes with stimulated emission (which *produces* one); if you started with an even mixture of the two states, you'd get no amplification at all.

There is a third mechanism (in addition to absorption and stimulated emission) by which radiation interacts with matter; it is called **spontaneous emission**. Here an atom in the excited state makes a transition downward, with the release of a photon, but without any applied electromagnetic field to initiate the process (Figure 9.4(c)). This is the mechanism that accounts for the typical decay of an atomic excited state. At first sight it is far from clear why spontaneous emission should occur at *all*. If the atom is in a stationary state (albeit an excited one), and there is no external perturbation, it should just sit there forever. And so it *would*, if it were *really* free of all external perturbations. However, in quantum electrodynamics the fields are nonzero *even in the ground state*—just as the harmonic oscillator (for example) has nonzero energy (to wit: $\hbar\omega/2$) in its ground state. You can turn out all the lights, and cool the room down to absolute zero, but there is still some electromagnetic radiation present, and it is this "zero point" radiation that serves to catalyze spontaneous emission. When you come right down to it, there is really no such thing as *truly* spontaneous emission; it's *all* stimulated emission. The only distinction to be made is whether the field that does the stimulating is one that *you* put there, or one that *God* put there. In this sense it is exactly the reverse of the classical radiative process, in which it's all spontaneous, and there is no such thing as stimulated emission.

Quantum electrodynamics is beyond the scope of this book,[7] but there is a lovely argument, due to Einstein,[8] which interrelates the three processes (absorption, stimulated emission, and spontaneous emission). Einstein did not identify the *mechanism* responsible for spontaneous emission (perturbation by the ground-state electromagnetic field), but his results nevertheless enable us to calculate the

---

[7]For an accessible treatment see Rodney Loudon, *The Quantum Theory of Light*, 2nd ed. (Clarendon Press, Oxford, 1983).

[8]Einstein's paper was published in 1917, well before the Schrödinger equation. Quantum electrodynamics comes into the argument via the Planck blackbody formula (Equation 5.113), which dates from 1900.

spontaneous emission rate, and from that the natural lifetime of an excited atomic state.[9] Before we turn to that, however, we need to consider the response of an atom to non-monochromatic, unpolarized, incoherent electromagnetic waves coming in from all directions—such as it would encounter, for instance, if it were immersed in thermal radiation.

9.2.3 Incoherent Perturbations

The energy density in an electromagnetic wave is[10]

$$u = \frac{\epsilon_0}{2} E_0^2, \qquad [9.37]$$

where E_0 is (as before) the amplitude of the electric field. So the transition probability (Equation 9.36) is (not surprisingly) proportional to the energy density of the fields:

$$P_{b\to a}(t) = \frac{2u}{\epsilon_0 \hbar^2} |\wp|^2 \frac{\sin^2[(\omega_0 - \omega)t/2]}{(\omega_0 - \omega)^2}. \qquad [9.38]$$

But this is for a **monochromatic** wave, at a single frequency ω. In many applications the system is exposed to electromagnetic waves at a whole *range* of frequencies; in that case $u \to \rho(\omega)d\omega$, where $\rho(\omega)d\omega$ is the energy density in the frequency range $d\omega$, and the net transition probability takes the form of an integral:[11]

$$P_{b\to a}(t) = \frac{2}{\epsilon_0 \hbar^2} |\wp|^2 \int_0^\infty \rho(\omega) \left\{ \frac{\sin^2[(\omega_0 - \omega)t/2]}{(\omega_0 - \omega)^2} \right\} d\omega. \qquad [9.39]$$

---

[9] For an interesting alternative derivation using "seat-of-the-pants" quantum electrodynamics, see Problem 9.9.

[10] D. Griffiths, *Introduction to Electrodynamics,* 3rd ed. (Prentice Hall, Upper Saddle River, NJ, 1999), Section 9.2.3. In general, the energy per unit volume in electromagnetic fields is

$$u = (\epsilon_0/2)E^2 + (1/2\mu_0)B^2.$$

For electromagnetic waves, the electric and magnetic contributions are equal, so

$$u = \epsilon_0 E^2 = \epsilon_0 E_0^2 \cos^2(\omega t),$$

and the average over a full cycle is $(\epsilon_0/2)E_0^2$, since the average of $\cos^2$ (or $\sin^2$) is 1/2.

[11] Equation 9.39 assumes that the perturbations at different frequencies are *independent*, so that the total transition probability is a sum of the individual probabilities. If the different components are **coherent** (phase-correlated), then we should add *amplitudes* ($c_b(t)$), not *probabilities* ($|c_b(t)|^2$), and there will be cross-terms. For the applications we will consider the perturbations are always incoherent.

The term in curly brackets is sharply peaked about ω_0 (Figure 9.2), whereas $\rho(\omega)$ is ordinarily quite broad, so we may as well replace $\rho(\omega)$ by $\rho(\omega_0)$, and take it outside the integral:

$$P_{b\to a}(t) \cong \frac{2|\wp|^2}{\epsilon_0\hbar^2}\rho(\omega_0)\int_0^\infty \frac{\sin^2[(\omega_0-\omega)t/2]}{(\omega_0-\omega)^2}\,d\omega. \qquad [9.40]$$

Changing variables to $x \equiv (\omega_0 - \omega)t/2$, extending the limits of integration to $x = \pm\infty$ (since the integrand is essentially zero out there anyway), and looking up the definite integral

$$\int_{-\infty}^{\infty} \frac{\sin^2 x}{x^2}\,dx = \pi, \qquad [9.41]$$

we find

$$P_{b\to a}(t) \cong \frac{\pi|\wp|^2}{\epsilon_0\hbar^2}\rho(\omega_0)t. \qquad [9.42]$$

This time the transition probability is proportional to t. The bizarre "flopping" phenomenon characteristic of a monochromatic perturbation gets "washed out" when we hit the system with an incoherent spread of frequencies. In particular, the **transition rate** ($R \equiv dP/dt$) is now a *constant*:

$$R_{b\to a} = \frac{\pi}{\epsilon_0\hbar^2}|\wp|^2\rho(\omega_0). \qquad [9.43]$$

So far, we have assumed that the perturbing wave is coming in along the y direction (Figure 9.3), and polarized in the z direction. But we are interested in the case of an atom bathed in radiation coming from *all* directions, and with all possible polarizations; the energy in the fields ($\rho(\omega)$) is shared equally among these different modes. What we need, in place of $|\wp|^2$, is the *average* of $|\boldsymbol{\wp}\cdot\hat{n}|^2$, where

$$\boldsymbol{\wp} \equiv q\langle\psi_b|\mathbf{r}|\psi_a\rangle \qquad [9.44]$$

(generalizing Equation 9.33), and the average is over all polarizations and all incident directions.

The averaging can be carried out as follows: Choose spherical coordinates such that the z-axis is now along the direction of propagation (so the polarization is in the xy plane) and the (fixed) vector $\boldsymbol{\wp}$ lies in the yz plane (Figure 9.5):[12]

$$\hat{n} = \cos\phi\,\hat{\imath} + \sin\phi\,\hat{\jmath}, \quad \boldsymbol{\wp} = \wp\sin\theta\,\hat{\jmath} + \wp\cos\theta\,\hat{k}. \qquad [9.45]$$

---

[12] I'll treat $\boldsymbol{\wp}$ as though it were *real*, even though in general it will be complex. Since

$$|\boldsymbol{\wp}\cdot\hat{n}|^2 = |\mathrm{Re}(\boldsymbol{\wp})\cdot\hat{n} + i\,\mathrm{Im}(\boldsymbol{\wp})\cdot\hat{n}|^2 = |\mathrm{Re}(\boldsymbol{\wp})\cdot\hat{n}|^2 + |\mathrm{Im}(\boldsymbol{\wp})\cdot\hat{n}|^2$$

we can do the whole calculation for the real and imaginary parts separately, and simply add the results. In Equation 9.47 the absolute value sign is doing double duty, signifying both the vector magnitude *and* to the complex amplitude:

$$|\boldsymbol{\wp}|^2 = |\wp_x|^2 + |\wp_y|^2 + |\wp_z|^2.$$

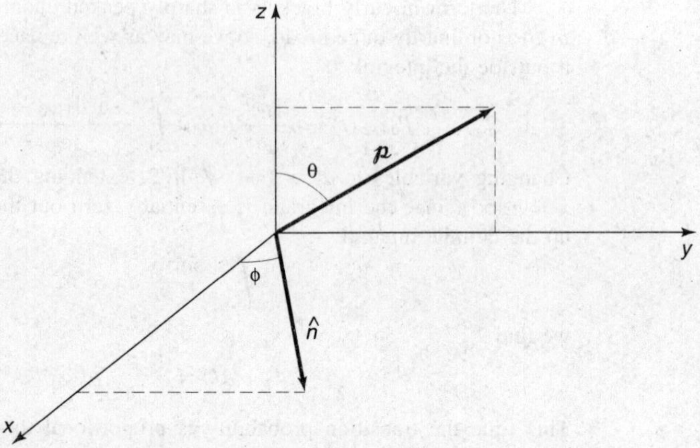

FIGURE 9.5: Axes for the averaging of $|\boldsymbol{p} \cdot \hat{n}|^2$.

Then

$$\boldsymbol{p} \cdot \hat{n} = p \sin\theta \sin\phi,$$

and

$$|\boldsymbol{p} \cdot \hat{n}|^2_{\text{ave}} = \frac{1}{4\pi}\int |\boldsymbol{p}|^2 \sin^2\theta \sin^2\phi \sin\theta \, d\theta \, d\phi$$
$$= \frac{|\boldsymbol{p}|^2}{4\pi}\int_0^{\pi} \sin^3\theta \, d\theta \int_0^{2\pi} \sin^2\phi \, d\phi = \frac{1}{3}|\boldsymbol{p}|^2. \qquad [9.46]$$

Conclusion: The transition rate for stimulated emission from state b to state a, under the influence of incoherent, unpolarized light incident from all directions, is

$$\boxed{R_{b\to a} = \frac{\pi}{3\epsilon_0\hbar^2}|\boldsymbol{p}|^2 \rho(\omega_0),} \qquad [9.47]$$

where $\boldsymbol{p}$ is the matrix element of the electric dipole moment between the two states (Equation 9.44), and $\rho(\omega_0)$ is the energy density in the fields, per unit frequency, evaluated at $\omega_0 = (E_b - E_a)/\hbar$.[13]

[13]This is a special case of Fermi's **Golden Rule** for time-dependent perturbation theory, which says that the transition rate is proportional to the square of the matrix element of the perturbing potential and to the strength of the perturbation at the transition frequency.

9.3 SPONTANEOUS EMISSION

9.3.1 Einstein's A and B Coefficients

Picture a container of atoms, N_a of them in the lower state (ψ_a), and N_b of them in the upper state (ψ_b). Let A be the spontaneous emission rate,[14] so that the number of particles leaving the upper state by this process, per unit time, is $N_b A$.[15] The transition rate for stimulated emission, as we have seen (Equation 9.47), is proportional to the energy density of the electromagnetic field: $B_{ba}\rho(\omega_0)$; the number of particles leaving the upper state by this mechanism, per unit time, is $N_b B_{ba}\rho(\omega_0)$. The absorption rate is likewise proportional to $\rho(\omega_0)$—call it $B_{ab}\rho(\omega_0)$; the number of particles per unit time *joining* the upper level is therefore $N_a B_{ab}\rho(\omega_0)$. All told, then,

$$\frac{dN_b}{dt} = -N_b A - N_b B_{ba}\rho(\omega_0) + N_a B_{ab}\rho(\omega_0). \qquad [9.48]$$

Suppose these atoms are in thermal equilibrium with the ambient field, so that the number of particles in each level is *constant*. In that case $dN_b/dt = 0$, and it follows that

$$\rho(\omega_0) = \frac{A}{(N_a/N_b)B_{ab} - B_{ba}}. \qquad [9.49]$$

On the other hand, we know from elementary statistical mechanics[16] that the number of particles with energy E, in thermal equilibrium at temperature T, is proportional to the **Boltzmann factor**, $\exp(-E/k_B T)$, so

$$\frac{N_a}{N_b} = \frac{e^{-E_a/k_B T}}{e^{-E_b/k_B T}} = e^{\hbar\omega_0/k_B T}, \qquad [9.50]$$

and hence

$$\rho(\omega_0) = \frac{A}{e^{\hbar\omega_0/k_B T}B_{ab} - B_{ba}}. \qquad [9.51]$$

But Planck's blackbody formula (Equation 5.113) tells us the energy density of thermal radiation:

$$\rho(\omega) = \frac{\hbar}{\pi^2 c^3}\frac{\omega^3}{e^{\hbar\omega/k_B T} - 1}. \qquad [9.52]$$

Comparing the two expressions, we conclude that

$$B_{ab} = B_{ba} \qquad [9.53]$$

---

[14] Normally I'd use R for a transition rate, but out of deference to *der Alte* everyone follows Einstein's notation in this context.

[15] Assume that N_a and N_b are very large, so we can treat them as continuous functions of time and ignore statistical fluctuations.

[16] See, for example, Charles Kittel and Herbert Kroemer, *Thermal Physics*, 2nd ed. (Freeman, New York, 1980), Chapter 3.

and

$$A = \frac{\omega_0^3 \hbar}{\pi^2 c^3} B_{ba}. \qquad [9.54]$$

Equation 9.53 confirms what we already knew: The transition rate for stimulated emission is the same as for absorption. But it was an astonishing result in 1917—indeed, Einstein was forced to "invent" stimulated emission in order to reproduce Planck's formula. Our present attention, however, focuses on Equation 9.54, for this tells us the spontaneous emission rate (A)—which is what we are looking for—in terms of the stimulated emission rate ($B_{ba}\rho(\omega_0)$)—which we already know. From Equation 9.47 we read off

$$B_{ba} = \frac{\pi}{3\epsilon_0 \hbar^2} |\wp|^2, \qquad [9.55]$$

and it follows that the spontaneous emission rate is

$$\boxed{A = \frac{\omega_0^3 |\wp|^2}{3\pi \epsilon_0 \hbar c^3}.} \qquad [9.56]$$

Problem 9.8 As a mechanism for downward transitions, spontaneous emission competes with thermally stimulated emission (stimulated emission for which blackbody radiation is the source). Show that at room temperature ($T = 300$ K) thermal stimulation dominates for frequencies well below 5×10^{12} Hz, whereas spontaneous emission dominates for frequencies well above 5×10^{12} Hz. Which mechanism dominates for visible light?

Problem 9.9 You could derive the spontaneous emission rate (Equation 9.56) without the detour through Einstein's A and B coefficients if you knew the ground-state energy density of the electromagnetic field, $\rho_0(\omega)$, for then it would simply be a case of stimulated emission (Equation 9.47). To do this honestly would require quantum electrodynamics, but if you are prepared to believe that the ground state consists of *one photon in each mode*, then the derivation is very simple:

(a) Replace Equation 5.111 by $N_\omega = d_k$, and deduce $\rho_0(\omega)$. (Presumably this formula breaks down at high frequency, else the total "vacuum energy" would be *infinite* ... but that's a story for a different day.)

(b) Use your result, together with Equation 9.47, to obtain the spontaneous emission rate. Compare Equation 9.56.

9.3.2 The Lifetime of an Excited State

Equation 9.56 is our fundamental result; it gives the transition rate for spontaneous emission. Suppose, now, that you have somehow pumped a large number of atoms into the excited state. As a result of spontaneous emission, this number will decrease as time goes on; specifically, in a time interval dt you will lose a fraction $A\,dt$ of them:

$$dN_b = -AN_b\,dt, \qquad [9.57]$$

(assuming there is no mechanism to replenish the supply).[17] Solving for $N_b(t)$, we find:

$$N_b(t) = N_b(0)e^{-At}; \qquad [9.58]$$

evidently the number remaining in the excited state decreases exponentially, with a time constant

$$\tau = \frac{1}{A}. \qquad [9.59]$$

We call this the **lifetime** of the state—technically, it is the time it takes for $N_b(t)$ to reach $1/e \approx 0.368$ of its initial value.

I have assumed all along that there are only *two* states for the system, but this was just for notational simplicity—the spontaneous emission formula (Equation 9.56) gives the transition rate for $\psi_b \to \psi_a$ regardless of what other states may be accessible (see Problem 9.15). Typically, an excited atom has many different **decay modes** (that is: ψ_b can decay to a large number of different lower-energy states, $\psi_{a_1}, \psi_{a_2}, \psi_{a_3}, \ldots$). In that case the transition rates *add*, and the net lifetime is

$$\tau = \frac{1}{A_1 + A_2 + A_3 + \cdots}. \qquad [9.60]$$

Example 9.1 Suppose a charge q is attached to a spring and constrained to oscillate along the x-axis. Say it starts out in the state $|n\rangle$ (Equation 2.61), and decays by spontaneous emission to state $|n'\rangle$. From Equation 9.44 we have

$$\boldsymbol{\wp} = q\langle n|x|n'\rangle\hat{\imath}.$$

You calculated the matrix elements of x back in Problem 3.33:

$$\langle n|x|n'\rangle = \sqrt{\frac{\hbar}{2m\omega}}\left(\sqrt{n'}\,\delta_{n,n'-1} + \sqrt{n}\,\delta_{n',n-1}\right),$$

---

[17] This situation is not to be confused with the case of thermal equilibrium, which we considered in the previous section. We assume here that the atoms have been lifted *out* of equilibrium, and are in the process of cascading back down to their equilibrium levels.

where ω is the natural frequency of the oscillator. (I no longer need this letter for the frequency of the stimulating radiation.) But we're talking about *emission*, so n' must be *lower* than n; for our purposes, then,

$$\boldsymbol{\wp} = q\sqrt{\frac{n\hbar}{2m\omega}}\,\delta_{n',n-1}\,\hat{\imath}. \qquad [9.61]$$

Evidently transitions occur only to states one step lower on the "ladder," and the frequency of the photon emitted is

$$\omega_0 = \frac{E_n - E_{n'}}{\hbar} = \frac{(n+1/2)\hbar\omega - (n'+1/2)\hbar\omega}{\hbar} = (n-n')\omega = \omega. \qquad [9.62]$$

Not surprisingly, the system radiates at the classical oscillator frequency. The transition rate (Equation 9.56) is

$$A = \frac{nq^2\omega^2}{6\pi\epsilon_0 mc^3}, \qquad [9.63]$$

and the lifetime of the nth stationary state is

$$\tau_n = \frac{6\pi\epsilon_0 mc^3}{nq^2\omega^2}. \qquad [9.64]$$

Meanwhile, each radiated photon carries an energy $\hbar\omega$, so the *power* radiated is $A\hbar\omega$:

$$P = \frac{q^2\omega^2}{6\pi\epsilon_0 mc^3}(n\hbar\omega),$$

or, since the energy of an oscillator in the nth state is $E = (n+1/2)\hbar\omega$,

$$P = \frac{q^2\omega^2}{6\pi\epsilon_0 mc^3}\left(E - \frac{1}{2}\hbar\omega\right). \qquad [9.65]$$

This is the average power radiated by a quantum oscillator with (initial) energy E.

For comparison, let's determine the average power radiated by a *classical* oscillator with the same energy. According to classical electrodynamics, the power radiated by an accelerating charge q is given by the **Larmor formula**:[18]

$$P = \frac{q^2a^2}{6\pi\epsilon_0 c^3}. \qquad [9.66]$$

For a harmonic oscillator with amplitude x_0, $x(t) = x_0\cos(\omega t)$, and the acceleration is $a = -x_0\omega^2\cos(\omega t)$. Averaging over a full cycle, then,

$$P = \frac{q^2x_0^2\omega^4}{12\pi\epsilon_0 c^3}.$$

[18]See, for example, Griffiths (footnote 10), Section 11.2.1.

But the *energy* of the oscillator is $E = (1/2)m\omega^2 x_0^2$, so $x_0^2 = 2E/m\omega^2$, and hence

$$P = \frac{q^2\omega^2}{6\pi\epsilon_0 mc^3}E. \qquad [9.67]$$

This is the average power radiated by a *classical* oscillator with energy E. In the classical limit ($\hbar \to 0$) the classical and quantum formulas agree;[19] however, the quantum formula (Equation 9.65) protects the ground state: If $E = (1/2)\hbar\omega$ the oscillator does not radiate.

Problem 9.10 The **half-life** ($t_{1/2}$) of an excited state is the time it would take for half the atoms in a large sample to make a transition. Find the relation between $t_{1/2}$ and τ (the "lifetime" of the state).

*****Problem 9.11** Calculate the lifetime (in *seconds*) for each of the four $n = 2$ states of hydrogen. *Hint:* You'll need to evaluate matrix elements of the form $\langle\psi_{100}|x|\psi_{200}\rangle$, $\langle\psi_{100}|y|\psi_{211}\rangle$, and so on. Remember that $x = r\sin\theta\cos\phi$, $y = r\sin\theta\sin\phi$, and $z = r\cos\theta$. Most of these integrals are zero, so scan them before you start calculating. *Answer:* 1.60×10^{-9} seconds for all except ψ_{200}, which is infinite.

9.3.3 Selection Rules

The calculation of spontaneous emission rates has been reduced to a matter of evaluating matrix elements of the form

$$\langle\psi_b|\mathbf{r}|\psi_a\rangle.$$

As you will have discovered if you worked Problem 9.11 (if you *didn't*, go back right now and *do* so!), these quantities are very often *zero*, and it would be helpful to know in advance when this is going to happen, so we don't waste a lot of time evaluating unnecessary integrals. Suppose we are interested in systems like hydrogen, for which the Hamiltonian is spherically symmetrical. In that case we may specify the states with the usual quantum numbers n, l, and m, and the matrix elements are

$$\langle n'l'm'|\mathbf{r}|nlm\rangle.$$

[19] In fact, if we express P in terms of the energy *above the ground state*, the two formulas are identical.

Clever exploitation of the angular momentum commutation relations and the hermiticity of the angular momentum operators yields a set of powerful constraints on this quantity.

Selection rules involving m and m': Consider first the commutators of L_z with x, y, and z, which we worked out in Chapter 4 (see Equation 4.122):

$$[L_z, x] = i\hbar y, \quad [L_z, y] = -i\hbar x, \quad [L_z, z] = 0. \qquad [9.68]$$

From the third of these it follows that

$$0 = \langle n'l'm'|[L_z, z]|nlm\rangle = \langle n'l'm'|(L_z z - zL_z)|nlm\rangle$$
$$= \langle n'l'm'|[(m'\hbar)z - z(m\hbar)]|nlm\rangle = (m' - m)\hbar\langle n'l'm'|z|nlm\rangle.$$

Conclusion:

$$\text{Either } m' = m, \quad \text{or else } \langle n'l'm'|z|nlm\rangle = 0. \qquad [9.69]$$

So unless $m' = m$, the matrix elements of z are always zero.

Meanwhile, from the commutator of L_z with x we get

$$\langle n'l'm'|[L_z, x]|nlm\rangle = \langle n'l'm'|(L_z x - xL_z)|nlm\rangle$$
$$= (m' - m)\hbar\langle n'l'm'|x|nlm\rangle = i\hbar\langle n'l'm'|y|nlm\rangle.$$

Conclusion:

$$(m' - m)\langle n'l'm'|x|nlm\rangle = i\langle n'l'm'|y|nlm\rangle. \qquad [9.70]$$

So you never have to compute matrix elements of y—you can always get them from the corresponding matrix elements of x.

Finally, the commutator of L_z with y yields

$$\langle n'l'm'|[L_z, y]|nlm\rangle = \langle n'l'm'|(L_z y - yL_z)|nlm\rangle$$
$$= (m' - m)\hbar\langle n'l'm'|y|nlm\rangle = -i\hbar\langle n'l'm'|x|nlm\rangle.$$

Conclusion:

$$(m' - m)\langle n'l'm'|y|nlm\rangle = -i\langle n'l'm'|x|nlm\rangle. \qquad [9.71]$$

In particular, combining Equations 9.70 and 9.71,

$$(m' - m)^2\langle n'l'm'|x|nlm\rangle = i(m' - m)\langle n'l'm'|y|nlm\rangle = \langle n'l'm'|x|nlm\rangle,$$

and hence:

$$\text{Either } (m' - m)^2 = 1, \quad \text{or else } \langle n'l'm'|x|nlm\rangle = \langle n'l'm'|y|nlm\rangle = 0. \qquad [9.72]$$

From Equations 9.69 and 9.72 we obtain the **selection rule** for m:

$$\boxed{\text{No transitions occur unless } \Delta m = \pm 1 \text{ or } 0.} \tag{9.73}$$

This is an easy result to understand, if you remember that the photon carries spin 1, and hence *its* value of m is 1, 0, or -1;[20] conservation of (the z component of) angular momentum requires that the atom give up whatever the photon takes away.

Selection rules involving l and l': In Problem 9.12 you are asked to derive the following commutation relation:

$$\left[L^2, [L^2, \mathbf{r}]\right] = 2\hbar^2(\mathbf{r}L^2 + L^2\mathbf{r}). \tag{9.74}$$

As before, we sandwich this commutator between $\langle n'l'm'|$ and $|nlm\rangle$ to derive the selection rule:

$$\begin{aligned}
\langle n'l'm'|[L^2, [L^2, \mathbf{r}]]|nlm\rangle &= 2\hbar^2\langle n'l'm'|(\mathbf{r}L^2 + L^2\mathbf{r})|nlm\rangle \\
= 2\hbar^4[l(l+1) + l'(l'+1)]\langle n'l'm'|\mathbf{r}|nlm\rangle &= \langle n'l'm'|(L^2[L^2, \mathbf{r}] - [L^2, \mathbf{r}]L^2)|nlm\rangle \\
&= \hbar^2[l'(l'+1) - l(l+1)]\langle n'l'm'|[L^2, \mathbf{r}]|nlm\rangle \\
&= \hbar^2[l'(l'+1) - l(l+1)]\langle n'l'm'|(L^2\mathbf{r} - \mathbf{r}L^2)|nlm\rangle \\
&= \hbar^4[l'(l'+1) - l(l+1)]^2\langle n'l'm'|\mathbf{r}|nlm\rangle.
\end{aligned} \tag{9.75}$$

Conclusion:

$$\begin{aligned}
\text{Either } 2[l(l+1) + l'(l'+1)] &= [l'(l'+1) - l(l+1)]^2 \\
\text{or else } \quad \langle n'l'm'|\mathbf{r}|nlm\rangle &= 0.
\end{aligned} \tag{9.76}$$

But

$$[l'(l'+1) - l(l+1)] = (l' + l + 1)(l' - l)$$

and

$$2[l(l+1) + l'(l'+1)] = (l' + l + 1)^2 + (l' - l)^2 - 1,$$

so the first condition in Equation 9.76 can be written in the form

$$[(l' + l + 1)^2 - 1][(l' - l)^2 - 1] = 0. \tag{9.77}$$

---

[20] When the polar axis is along the direction of propagation, the middle value does not occur, and if you are only interested in the *number* of linearly independent photon states, the answer is 2, not 3. However, in this case the photon need not be going in the z direction, and all three values are possible.

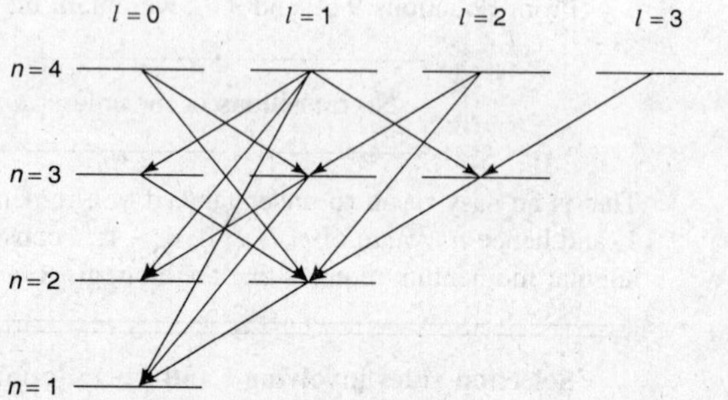

FIGURE 9.6: Allowed decays for the first four Bohr levels in hydrogen.

The first factor *cannot* be zero (unless $l' = l = 0$—this loophole is closed in Problem 9.13), so the condition simplifies to $l' = l \pm 1$. Thus we obtain the selection rule for l:

$$\text{No transitions occur unless } \Delta l = \pm 1. \qquad [9.78]$$

Again, this result (though far from trivial to *derive*) is easy to *interpret*: The photon carries spin 1, so the rules for addition of angular momentum would allow $l' = l + 1$, $l' = l$, or $l' = l - 1$ (for electric dipole radiation the middle possibility—though permitted by conservation of angular momentum—does not occur).

Evidently not all transitions to lower-energy states can proceed by spontaneous emission; some are forbidden by the selection rules. The scheme of allowed transitions for the first four Bohr levels in hydrogen is shown in Figure 9.6. Note that the $2S$ state (ψ_{200}) is "stuck": It cannot decay, because there is no lower-energy state with $l = 1$. It is called a **metastable** state, and its lifetime is indeed much longer than that of, for example, the $2P$ states (ψ_{211}, ψ_{210}, and ψ_{21-1}). Metastable states do eventually decay, by collisions, or by what are (misleadingly) called **forbidden** transitions (Problem 9.21), or by multiphoton emission.

***Problem 9.12** Prove the commutation relation in Equation 9.74. *Hint:* First show that

$$[L^2, z] = 2i\hbar(xL_y - yL_x - i\hbar z).$$

Use this, and the fact that $\mathbf{r} \cdot \mathbf{L} = \mathbf{r} \cdot (\mathbf{r} \times \mathbf{p}) = 0$, to demonstrate that

$$[L^2, [L^2, z]] = 2\hbar^2(zL^2 + L^2 z).$$

The generalization from z to $\mathbf{r}$ is trivial.

Problem 9.13 Close the "loophole" in Equation 9.78 by showing that if $l' = l = 0$ then $\langle n'l'm'|\mathbf{r}|nlm\rangle = 0$.

$**$**Problem 9.14** An electron in the $n = 3$, $l = 0$, $m = 0$ state of hydrogen decays by a sequence of (electric dipole) transitions to the ground state.

(a) What decay routes are open to it? Specify them in the following way:

$$|300\rangle \rightarrow |nlm\rangle \rightarrow |n'l'm'\rangle \rightarrow \cdots \rightarrow |100\rangle.$$

(b) If you had a bottle full of atoms in this state, what fraction of them would decay via each route?

(c) What is the lifetime of this state? *Hint:* Once it's made the first transition, it's no longer in the state $|300\rangle$, so only the first step in each sequence is relevant in computing the lifetime. When there is more than one decay route open, the transition rates add.

FURTHER PROBLEMS FOR CHAPTER 9

$**$**Problem 9.15** Develop time-dependent perturbation theory for a multilevel system, starting with the generalization of Equations 9.1 and 9.2:

$$H_0\psi_n = E_n\psi_n, \quad \langle\psi_n|\psi_m\rangle = \delta_{nm}. \qquad [9.79]$$

At time $t = 0$ we turn on a perturbation $H'(t)$, so that the total Hamiltonian is

$$H = H_0 + H'(t). \qquad [9.80]$$

(a) Generalize Equation 9.6 to read

$$\Psi(t) = \sum c_n(t)\psi_n e^{-iE_nt/\hbar}, \qquad [9.81]$$

and show that

$$\dot{c}_m = -\frac{i}{\hbar}\sum_n c_n H'_{mn} e^{i(E_m-E_n)t/\hbar}, \qquad [9.82]$$

where

$$H'_{mn} \equiv \langle\psi_m|H'|\psi_n\rangle. \qquad [9.83]$$

(b) If the system starts out in the state ψ_N, show that (in first-order perturbation theory)

$$c_N(t) \cong 1 - \frac{i}{\hbar}\int_0^t H'_{NN}(t')\,dt'. \qquad [9.84]$$

and

$$c_m(t) \cong -\frac{i}{\hbar}\int_0^t H'_{mN}(t')e^{i(E_m-E_N)t'/\hbar}\,dt', \qquad (m \neq N). \tag{9.85}$$

(c) For example, suppose H' is *constant* (except that it was turned on at $t = 0$, and switched off again at some later time t). Find the probability of transition from state N to state M $(M \neq N)$, as a function of t. *Answer:*

$$4|H'_{MN}|^2 \frac{\sin^2[(E_N - E_M)t/2\hbar]}{(E_N - E_M)^2}. \tag{9.86}$$

(d) Now suppose H' is a sinusoidal function of time: $H' = V\cos(\omega t)$. Making the usual assumptions, show that transitions occur only to states with energy $E_M = E_N \pm \hbar\omega$, and the transition probability is

$$P_{N\to M} = |V_{MN}|^2 \frac{\sin^2[(E_N - E_M \pm \hbar\omega)t/2\hbar]}{(E_N - E_M \pm \hbar\omega)^2}. \tag{9.87}$$

(e) Suppose a multilevel system is immersed in incoherent electromagnetic radiation. Using Section 9.2.3 as a guide, show that the transition rate for stimulated emission is given by the same formula (Equation 9.47) as for a two-level system.

Problem 9.16 For the examples in Problem 9.15(c) and (d), calculate $c_m(t)$, to first order. Check the normalization condition:

$$\sum_m |c_m(t)|^2 = 1, \tag{9.88}$$

and comment on any discrepancy. Suppose you wanted to calculate the probability of *remaining* in the original state ψ_N; would you do better to use $|c_N(t)|^2$, or $1 - \sum_{m\neq N}|c_m(t)|^2$?

Problem 9.17 A particle starts out (at time $t = 0$) in the Nth state of the infinite square well. Now the "floor" of the well rises temporarily (maybe water leaks in, and then drains out again), so that the potential inside is uniform but time dependent: $V_0(t)$, with $V_0(0) = V_0(T) = 0$.

(a) Solve for the *exact* $c_m(t)$, using Equation 9.82, and show that the wave function changes *phase*, but no transitions occur. Find the phase change, $\phi(T)$, in terms of the function $V_0(t)$.

(b) Analyze the same problem in first-order perturbation theory, and compare your answers.

Comment: The same result holds *whenever* the perturbation simply adds a constant (constant in x, that is, not in t) to the potential; it has nothing to do with the infinite square well, as such. Compare Problem 1.8.

∗**Problem 9.18** A particle of mass m is initially in the ground state of the (one-dimensional) infinite square well. At time $t = 0$ a "brick" is dropped into the well, so that the potential becomes

$$V(x) = \begin{cases} V_0, & \text{if } 0 \leq x \leq a/2, \\ 0, & \text{if } a/2 < x \leq a, \\ \infty, & \text{otherwise,} \end{cases}$$

where $V_0 \ll E_1$. After a time T, the brick is removed, and the energy of the particle is measured. Find the probability (in first-order perturbation theory) that the result is now E_2.

Problem 9.19 We have encountered stimulated emission, (stimulated) absorption, and spontaneous emission. How come there is no such thing as spontaneous *absorption*?

∗∗∗**Problem 9.20 Magnetic resonance.** A spin-1/2 particle with gyromagnetic ratio γ, at rest in a static magnetic field $B_0\hat{k}$, precesses at the Larmor frequency $\omega_0 = \gamma B_0$ (Example 4.3). Now we turn on a small transverse radiofrequency (rf) field, $B_{\rm rf}[\cos(\omega t)\,\hat{\imath} - \sin(\omega t)\,\hat{\jmath}]$, so that the total field is

$$\mathbf{B} = B_{\rm rf}\cos(\omega t)\,\hat{\imath} - B_{\rm rf}\sin(\omega t)\,\hat{\jmath} + B_0\,\hat{k}. \qquad [9.89]$$

(a) Construct the 2×2 Hamiltonian matrix (Equation 4.158) for this system.

(b) If $\chi(t) = \begin{pmatrix} a(t) \\ b(t) \end{pmatrix}$ is the spin state at time t, show that

$$\dot{a} = \frac{i}{2}\left(\Omega e^{i\omega t} b + \omega_0 a\right); \quad \dot{b} = \frac{i}{2}\left(\Omega e^{-i\omega t} a - \omega_0 b\right), \qquad [9.90]$$

where $\Omega \equiv \gamma B_{\rm rf}$ is related to the strength of the rf field.

(c) Find the general solution for $a(t)$ and $b(t)$, in terms of their initial values a_0 and b_0. *Answer:*

$$a(t) = \left\{a_0\cos(\omega' t/2) + \frac{i}{\omega'}\left[a_0(\omega_0 - \omega) + b_0\Omega\right]\sin(\omega' t/2)\right\} e^{i\omega t/2}$$

$$b(t) = \left\{b_0\cos(\omega' t/2) + \frac{i}{\omega'}\left[b_0(\omega - \omega_0) + a_0\Omega\right]\sin(\omega' t/2)\right\} e^{-i\omega t/2}$$

where

$$\omega' \equiv \sqrt{(\omega - \omega_0)^2 + \Omega^2}. \qquad [9.91]$$

(d) If the particle starts out with spin up (i.e., $a_0 = 1$, $b_0 = 0$), find the probability of a transition to spin down, as a function of time. *Answer:* $P(t) = \{\Omega^2/[(\omega - \omega_0)^2 + \Omega^2]\} \sin^2(\omega' t/2)$.

(e) Sketch the **resonance curve**,

$$P(\omega) = \frac{\Omega^2}{(\omega - \omega_0)^2 + \Omega^2}, \qquad [9.92]$$

as a function of the driving frequency ω (for fixed ω_0 and Ω). Note that the maximum occurs at $\omega = \omega_0$. Find the "full width at half maximum," $\Delta\omega$.

(f) Since $\omega_0 = \gamma B_0$, we can use the experimentally observed resonance to determine the magnetic dipole moment of the particle. In a **nuclear magnetic resonance** (nmr) experiment the g-factor of the proton is to be measured, using a static field of 10,000 gauss and an rf field of amplitude 0.01 gauss. What will the resonant frequency be? (See Section 6.5 for the magnetic moment of the proton.) Find the width of the resonance curve. (Give your answers in Hz.)

*****Problem 9.21** In Equation 9.31 I assumed that the atom is so small (in comparison to the wavelength of light) that spatial variations in the field can be ignored. The *true* electric field would be

$$\mathbf{E}(\mathbf{r}, t) = \mathbf{E}_0 \cos(\mathbf{k} \cdot \mathbf{r} - \omega t). \qquad [9.93]$$

If the atom is centered at the origin, then $\mathbf{k} \cdot \mathbf{r} \ll 1$ over the relevant volume ($|\mathbf{k}| = 2\pi/\lambda$, so $\mathbf{k} \cdot \mathbf{r} \sim r/\lambda \ll 1$), and that's why we could afford to drop this term. Suppose we keep the first-order correction:

$$\mathbf{E}(\mathbf{r}, t) = \mathbf{E}_0[\cos(\omega t) + (\mathbf{k} \cdot \mathbf{r}) \sin(\omega t)]. \qquad [9.94]$$

The first term gives rise to the **allowed** (**electric dipole**) transitions we considered in the text; the second leads to so-called **forbidden** (**magnetic dipole** and **electric quadrupole**) transitions (higher powers of $\mathbf{k} \cdot \mathbf{r}$ lead to even *more* "forbidden" transitions, associated with higher multipole moments).[21]

(a) Obtain the spontaneous emission rate for forbidden transitions (don't bother to average over polarization and propagation directions, though this should really be done to complete the calculation). *Answer:*

$$R_{b\to a} = \frac{q^2\omega^5}{\pi\epsilon_0\hbar c^5} |\langle a|(\hat{n} \cdot \mathbf{r})(\hat{k} \cdot \mathbf{r})|b\rangle|^2. \qquad [9.95]$$

[21] For a systematic treatment (including the role of the magnetic field) see David Park, *Introduction to the Quantum Theory*, 3rd ed. (McGraw-Hill, New York, 1992), Chapter 11.

(b) Show that for a one-dimensional oscillator the forbidden transitions go from level n to level $n - 2$, and the transition rate (suitably averaged over $\hat{n}$ and $\hat{k}$) is

$$R = \frac{\hbar q^2 \omega^3 n(n-1)}{15\pi \epsilon_0 m^2 c^5}. \qquad [9.96]$$

(*Note:* Here ω is the frequency of the *photon*, not the oscillator.) Find the *ratio* of the "forbidden" rate to the "allowed" rate, and comment on the terminology.

(c) Show that the $2S \to 1S$ transition in hydrogen is not possible even by a "forbidden" transition. (As it turns out, this is true for all the higher multipoles as well; the dominant decay is in fact by two-photon emission, and the lifetime is about a tenth of a second.[22])

*****Problem 9.22** Show that the spontaneous emission rate (Equation 9.56) for a transition from n, l to n', l' in hydrogen is

$$\frac{e^2 \omega^3 I^2}{3\pi \epsilon_0 \hbar c^3} \times \begin{cases} \dfrac{l+1}{2l+1}, & \text{if } l' = l+1, \\ \dfrac{l}{2l-1} & \text{if } l' = l-1, \end{cases} \qquad [9.97]$$

where

$$I \equiv \int_0^\infty r^3 R_{nl}(r) R_{n'l'}(r)\, dr. \qquad [9.98]$$

(The atom starts out with a specific value of m, and it goes to *any* of the states m' consistent with the selection rules: $m' = m+1, m$, or $m-1$. Notice that the answer is independent of m.) *Hint:* First calculate all the nonzero matrix elements of x, y, and z between $|nlm\rangle$ and $|n'l'm'\rangle$ for the case $l' = l+1$. From these, determine the quantity

$$|\langle n', l+1, m+1|\mathbf{r}|nlm\rangle|^2 + |\langle n', l+1, m|\mathbf{r}|nlm\rangle|^2 + |\langle n', l+1, m-1|\mathbf{r}|nlm\rangle|^2.$$

Then do the same for $l' = l - 1$.

---

[22]See Masataka Mizushima, *Quantum Mechanics of Atomic Spectra and Atomic Structure*, Benjamin, New York (1970), Section 5.6.

CHAPTER 10

THE ADIABATIC APPROXIMATION

10.1 THE ADIABATIC THEOREM

10.1.1 Adiabatic Processes

Imagine a perfect pendulum, with no friction or air resistance, oscillating back and forth in a vertical plane. If you grab the support and shake it in a jerky manner the bob will swing around chaotically. But if you *very gently and steadily* move the support (Figure 10.1), the pendulum will continue to swing in a nice smooth way, in the same plane (or one parallel to it), with the same amplitude. This *gradual change of the external conditions* defines an **adiabatic** process. Notice that there are two characteristic times involved: T_i, the "internal" time, representing the motion of the system itself (in this case the period of the pendulum's oscillations), and T_e, the "external" time, over which the parameters of the system change appreciably (if the pendulum were mounted on a vibrating platform, for example, T_e would be the period of the *platform's* motion). An adiabatic process is one for which $T_e \gg T_i$.[1]

The basic strategy for analyzing an adiabatic process is first to solve the problem with the external parameters held *constant*, and only at the *end* of the calculation allow them to vary (slowly) with time. For example, the classical period of a pendulum of (fixed) length L is $2\pi\sqrt{L/g}$; if the length is now gradually *changing*, the period will presumably be $2\pi\sqrt{L(t)/g}$. A more subtle example occurred in our discussion of the hydrogen molecule ion (Section 7.3). We began

[1]For an interesting discussion of classical adiabatic processes, see Frank S. Crawford, *Am. J. Phys.* **58**, 337 (1990).

FIGURE 10.1: Adiabatic motion: If the case is transported very gradually, the pendulum inside keeps swinging with the same amplitude, in a plane parallel to the original one.

by assuming that the nuclei were *at rest*, a fixed distance R apart, and we solved for the motion of the electron. Once we had found the ground state energy of the system as a function of R, we located the equilibrium separation and from the curvature of the graph we obtained the frequency of vibration of the nuclei (Problem 7.10). In molecular physics this technique (beginning with nuclei at rest, calculating electronic wave functions, and using these to obtain information about the positions and—relatively sluggish—motion of the nuclei) is known as the **Born-Oppenheimer approximation**.

In quantum mechanics, the essential content of the **adiabatic approximation** can be cast in the form of a theorem. Suppose the Hamiltonian changes *gradually* from some initial form H^i to some final form H^f. The **adiabatic theorem** states that if the particle was initially in the nth eigenstate of H^i, it will be carried (under the Schrödinger equation) into the nth eigenstate of H^f. (I assume that the spectrum is discrete and nondegenerate throughout the transition from H^i to H^f, so there is no ambiguity about the ordering of the states; these conditions can be relaxed, given a suitable procedure for "tracking" the eigenfunctions, but I'm not going to pursue that here.)

For example, suppose we prepare a particle in the ground state of the infinite square well (Figure 10.2(a)):

$$\psi^i(x) = \sqrt{\frac{2}{a}} \sin\left(\frac{\pi}{a}x\right). \tag{10.1}$$

If we now gradually move the right wall out to $2a$, the adiabatic theorem says that the particle will end up in the ground state of the expanded well (Figure 10.2(b)):

$$\psi^f(x) = \sqrt{\frac{1}{a}} \sin\left(\frac{\pi}{2a}x\right), \tag{10.2}$$

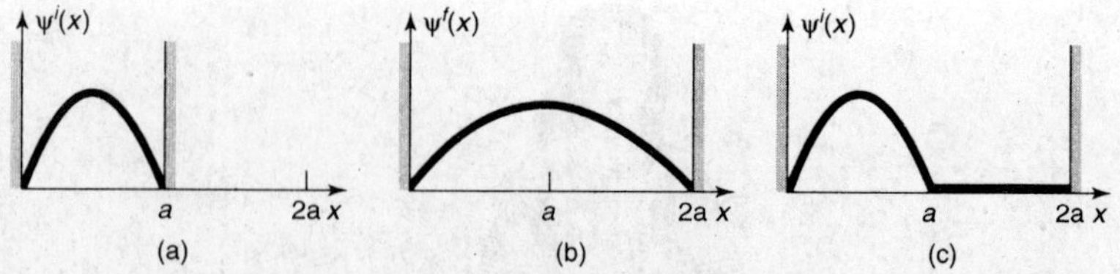

FIGURE 10.2: (a) Particle starts out in the ground state of the infinite square well. (b) If the wall moves *slowly*, the particle remains in the ground state. (c) If the wall moves *rapidly*, the particle is left (momentarily) in its initial state.

(apart, perhaps, from a phase factor). Notice that we're not talking about a *small* change in the Hamiltonian (as in perturbation theory)—this one is *huge*. All we require is that it happen *slowly*. Energy is not conserved here: Whoever is moving the wall is extracting energy from the system, just like the piston on a slowly expanding cylinder of gas. By contrast, if the well expands *suddenly*, the resulting state is still $\psi^i(x)$ (Figure 10.2(c)), which is a complicated linear combination of eigenstates of the new Hamiltonian (Problem 2.38). In this case energy *is* conserved (at least, its *expectation value* is); as in the *free* expansion of a gas (into a vacuum) when the barrier is suddenly removed, no work is done.

∗ ∗ ∗**Problem 10.1** The case of an infinite square well whose right wall expands at a *constant* velocity (v) can be solved *exactly*.[2] A complete set of solutions is

$$\Phi_n(x, t) \equiv \sqrt{\frac{2}{w}} \sin\left(\frac{n\pi}{w}x\right) e^{i(mvx^2 - 2E_n^i at)/2\hbar w}, \qquad [10.3]$$

where $w(t) \equiv a + vt$ is the (instantaneous) width of the well and $E_n^i \equiv n^2\pi^2\hbar^2/2ma^2$ is the nth allowed energy of the *original* well (width a). The *general* solution is a linear combination of the Φ's:

$$\Psi(x, t) = \sum_{n=1}^{\infty} c_n \Phi_n(x, t); \qquad [10.4]$$

the coefficients c_n are *independent of* t.

(a) Check that Equation 10.3 satisfies the time-dependent Schrödinger equation, with the appropriate boundary conditions.

---

[2] S. W. Doescher and M. H. Rice, *Am. J. Phys.* **37**, 1246 (1969).

(b) Suppose a particle starts out ($t = 0$) in the ground state of the initial well:

$$\Psi(x,0) = \sqrt{\frac{2}{a}}\sin\left(\frac{\pi}{a}x\right).$$

Show that the expansion coefficients can be written in the form

$$c_n = \frac{2}{\pi}\int_0^{\pi} e^{-i\alpha z^2}\sin(nz)\sin(z)\,dz, \qquad [10.5]$$

where $\alpha \equiv mva/2\pi^2\hbar$ is a dimensionless measure of the speed with which the well expands. (Unfortunately, this integral cannot be evaluated in terms of elementary functions.)

(c) Suppose we allow the well to expand to twice its original width, so the "external" time is given by $w(T_e) = 2a$. The "internal" time is the *period* of the time-dependent exponential factor in the (initial) ground state. Determine T_e and T_i, and show that the adiabatic regime corresponds to $\alpha \ll 1$, so that $\exp(-i\alpha z^2) \cong 1$ over the domain of integration. Use this to determine the expansion coefficients, c_n. Construct $\Psi(x,t)$, and confirm that it is consistent with the adiabatic theorem.

(d) Show that the phase factor in $\Psi(x,t)$ can be written in the form

$$\theta(t) = -\frac{1}{\hbar}\int_0^t E_1(t')\,dt', \qquad [10.6]$$

where $E_n(t) \equiv n^2\pi^2\hbar^2/2mw^2$ is the *instantaneous* eigenvalue, at time t. Comment on this result.

10.1.2 Proof of the Adiabatic Theorem

The adiabatic theorem is simple to state, and it *sounds* plausible, but it is not easy to prove.[3] If the Hamiltonian is *independent* of time, then a particle which starts out in the nth eigenstate,[4] ψ_n,

$$H\psi_n = E_n\psi_n, \qquad [10.7]$$

[3] The theorem is usually attributed to Ehrenfest, who studied adiabatic processes in early versions of the quantum theory. The first proof in modern quantum mechanics was given by Born and Fock, *Zeit. f. Physik* **51**, 165 (1928). Other proofs will be found in Messiah, *Quantum Mechanics*, Wiley, New York (1962), Vol. II, Chapter XVII, Section 12, J-T Hwang and Philip Pechukas, *J. Chem. Phys.* **67**, 4640, 1977, and Gasiorowicz, *Quantum Physics*, Wiley, New York (1974), Chapter 22, Problem 6. The argument given here follows B. H. Bransden and C. J. Joachain, *Introduction to Quantum Mechanics*, 2nd ed., Addison-Wesley, Boston, MA (2000), Section 9.4.

[4] I'll suppress the dependence on position (or spin, etc.); in this argument only the time dependence is at issue.

remains in the nth eigenstate, simply picking up a phase factor:

$$\Psi_n(t) = \psi_n e^{-iE_n t/\hbar}. \qquad [10.8]$$

If the Hamiltonian *changes* with time, then the eigenfunctions and eigenvalues are themselves time-dependent:

$$H(t)\psi_n(t) = E_n(t)\psi_n(t), \qquad [10.9]$$

but they still constitute (at any particular instant) an orthonormal set

$$\langle\psi_n(t)|\psi_m(t)\rangle = \delta_{nm}, \qquad [10.10]$$

and they are complete, so the general solution to the time-dependent Schrödinger equation

$$i\hbar\frac{\partial}{\partial t}\Psi(t) = H(t)\Psi(t) \qquad [10.11]$$

can be expressed as a linear combination of them:

$$\Psi(t) = \sum_n c_n(t)\psi_n(t)e^{i\theta_n(t)}, \qquad [10.12]$$

where

$$\theta_n(t) \equiv -\frac{1}{\hbar}\int_0^t E_n(t')\,dt' \qquad [10.13]$$

generalizes the "standard" phase factor to the case where E_n varies with time. (As usual, I could have included it in the coefficient $c_n(t)$, but it is convenient to factor out this portion of the time dependence, since it would be present even for a time-*in*dependent Hamiltonian.)

Substituting Equation 10.12 into Equation 10.11 we obtain

$$i\hbar\sum_n\left[\dot{c}_n\psi_n + c_n\dot{\psi}_n + ic_n\psi_n\dot{\theta}_n\right]e^{i\theta_n} = \sum_n c_n(H\psi_n)e^{i\theta_n} \qquad [10.14]$$

(I use a dot to denote the time derivative). In view of Equations 10.9 and 10.13 the last two terms cancel, leaving

$$\sum_n \dot{c}_n\psi_n e^{i\theta_n} = -\sum_n c_n\dot{\psi}_n e^{i\theta_n}. \qquad [10.15]$$

Taking the inner product with ψ_m, and invoking the orthonormality of the instantaneous eigenfunctions (Equation 10.10),

$$\sum_n \dot{c}_n\delta_{mn}e^{i\theta_n} = -\sum_n c_n\langle\psi_m|\dot{\psi}_n\rangle e^{i\theta_n},$$

or

$$\dot{c}_m(t) = -\sum_n c_n \langle \psi_m | \dot{\psi}_n \rangle e^{i(\theta_n - \theta_m)}. \qquad [10.16]$$

Now, differentiating Equation 10.9 with respect to time yields

$$\dot{H}\psi_n + H\dot{\psi}_n = \dot{E}_n\psi_n + E_n\dot{\psi}_n,$$

and hence (again taking the inner product with ψ_m)

$$\langle \psi_m | \dot{H} | \psi_n \rangle + \langle \psi_m | H | \dot{\psi}_n \rangle = \dot{E}_n \delta_{mn} + E_n \langle \psi_m | \dot{\psi}_n \rangle. \qquad [10.17]$$

Exploiting the hermiticity of H to write $\langle \psi_m | H | \dot{\psi}_n \rangle = E_m \langle \psi_m | \dot{\psi}_n \rangle$, it follows that for $n \neq m$

$$\langle \psi_m | \dot{H} | \psi_n \rangle = (E_n - E_m) \langle \psi_m | \dot{\psi}_n \rangle. \qquad [10.18]$$

Putting this into Equation 10.16 (and assuming, remember, that the energies are nondegenerate) we conclude that

$$\dot{c}_m(t) = -c_m \langle \psi_m | \dot{\psi}_m \rangle - \sum_{n \neq m} c_n \frac{\langle \psi_m | \dot{H} | \psi_n \rangle}{E_n - E_m} e^{(-i/\hbar) \int_0^t [E_n(t') - E_m(t')] dt'}. \qquad [10.19]$$

This result is *exact*. Now comes the adiabatic approximation: Assume that $\dot{H}$ is extremely small, and drop the second term,[5] leaving

$$\dot{c}_m(t) = -c_m \langle \psi_m | \dot{\psi}_m \rangle, \qquad [10.20]$$

with the solution

$$c_m(t) = c_m(0) e^{i\gamma_m(t)}, \qquad [10.21]$$

where[6]

$$\gamma_m(t) \equiv i \int_0^t \left\langle \psi_m(t') \middle| \frac{\partial}{\partial t'} \psi_m(t') \right\rangle dt'. \qquad [10.22]$$

In particular, if the particle starts out in the nth eigenstate (which is to say, if $c_n(0) = 1$, and $c_m(0) = 0$ for $m \neq n$), then (Equation 10.12)

$$\boxed{\Psi_n(t) = e^{i\theta_n(t)} e^{i\gamma_n(t)} \psi_n(t),} \qquad [10.23]$$

so it remains in the nth eigenstate (of the evolving Hamiltonian), picking up only a couple of phase factors. QED

[5]Rigorous justification of this step is not trivial. See A. C. Aguiar Pinto *et al.*, *Am. J. Phys.* **68**, 955 (2000).

[6]Notice that γ is *real*, since the normalization of ψ_m entails $(d/dt)\langle \psi_m | \psi_m \rangle = \langle \psi_m | \dot{\psi}_m \rangle + \langle \dot{\psi}_m | \psi_m \rangle = 2\text{Re}\left(\langle \psi_m | \dot{\psi}_m \rangle\right) = 0$.

Example 10.1 Imagine an electron (charge $-e$, mass m) at rest at the origin, in the presence of a magnetic field whose *magnitude* (B_0) is constant, but whose *direction* sweeps out a cone, of opening angle α, at constant angular velocity ω (Figure 10.3):

$$\mathbf{B}(t) = B_0[\sin\alpha\cos(\omega t)\hat{\imath} + \sin\alpha\sin(\omega t)\hat{\jmath} + \cos\alpha\hat{k}]. \tag{10.24}$$

The Hamiltonian (Equation 4.158) is

$$H(t) = \frac{e}{m}\mathbf{B}\cdot\mathbf{S} = \frac{e\hbar B_0}{2m}[\sin\alpha\cos(\omega t)\sigma_x + \sin\alpha\sin(\omega t)\sigma_y + \cos\alpha\sigma_z]$$
$$= \frac{\hbar\omega_1}{2}\begin{pmatrix} \cos\alpha & e^{-i\omega t}\sin\alpha \\ e^{i\omega t}\sin\alpha & -\cos\alpha \end{pmatrix}, \tag{10.25}$$

where

$$\omega_1 \equiv \frac{eB_0}{m}. \tag{10.26}$$

The normalized eigenspinors of $H(t)$ are

$$\chi_+(t) = \begin{pmatrix} \cos(\alpha/2) \\ e^{i\omega t}\sin(\alpha/2) \end{pmatrix}, \tag{10.27}$$

and

$$\chi_-(t) = \begin{pmatrix} e^{-i\omega t}\sin(\alpha/2) \\ -\cos(\alpha/2) \end{pmatrix}; \tag{10.28}$$

they represent spin up and spin down, respectively, *along the instantaneous direction of* $\mathbf{B}(t)$ (see Problem 4.30). The corresponding eigenvalues are

$$E_\pm = \pm\frac{\hbar\omega_1}{2}. \tag{10.29}$$

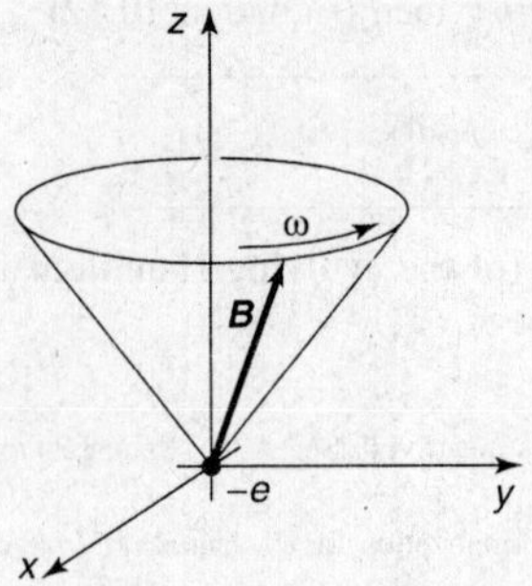

FIGURE 10.3: Magnetic field sweeps around in a cone, at angular velocity ω (Equation 10.24).

Suppose the electron starts out with spin up, along $\mathbf{B}(0)$:[7]

$$\chi(0) = \begin{pmatrix} \cos(\alpha/2) \\ \sin(\alpha/2) \end{pmatrix}. \qquad [10.30]$$

The exact solution to the time-dependent Schrödinger equation is (Problem 10.2):

$$\chi(t) = \begin{pmatrix} \left[\cos(\lambda t/2) - i\frac{(\omega_1 - \omega)}{\lambda}\sin(\lambda t/2)\right]\cos(\alpha/2)e^{-i\omega t/2} \\ \left[\cos(\lambda t/2) - i\frac{(\omega_1 + \omega)}{\lambda}\sin(\lambda t/2)\right]\sin(\alpha/2)e^{+i\omega t/2} \end{pmatrix}, \qquad [10.31]$$

where

$$\lambda \equiv \sqrt{\omega^2 + \omega_1^2 - 2\omega\omega_1\cos\alpha}. \qquad [10.32]$$

Or, expressing it as a linear combination of χ_+ and χ_-:

$$\chi(t) = \left[\cos\left(\frac{\lambda t}{2}\right) - i\frac{(\omega_1 - \omega\cos\alpha)}{\lambda}\sin\left(\frac{\lambda t}{2}\right)\right]e^{-i\omega t/2}\chi_+(t)$$
$$+ i\left[\frac{\omega}{\lambda}\sin\alpha\sin\left(\frac{\lambda t}{2}\right)\right]e^{+i\omega t/2}\chi_-(t). \qquad [10.33]$$

Evidently the (exact) probability of a transition to spin down (along the current direction of $\mathbf{B}$) is

$$|\langle\chi(t)|\chi_-(t)\rangle|^2 = \left[\frac{\omega}{\lambda}\sin\alpha\sin\left(\frac{\lambda t}{2}\right)\right]^2. \qquad [10.34]$$

The adiabatic theorem says that this transition probability should vanish in the limit $T_e \gg T_i$, where T_e is the characteristic time for changes in the Hamiltonian (in this case, $1/\omega$) and T_i is the characteristic time for changes in the wave function (in this case, $\hbar/(E_+ - E_-) = 1/\omega_1$). Thus the adiabatic approximation means $\omega \ll \omega_1$: The field rotates slowly, in comparison with the phase of the (unperturbed) wave functions. In the adiabatic regime $\lambda \cong \omega_1$, and therefore

$$|\langle\chi(t)|\chi_-(t)\rangle|^2 \cong \left[\frac{\omega}{\omega_1}\sin\alpha\sin\left(\frac{\lambda t}{2}\right)\right]^2 \to 0, \qquad [10.35]$$

as advertised. The magnetic field leads the electron around by its nose, with the spin always pointing in the direction of $\mathbf{B}$. By contrast, if $\omega \gg \omega_1$ then $\lambda \cong \omega$, and the system bounces back and forth between spin up and spin down (Figure 10.4).

---

[7] This is essentially the same as Problem 9.20, except that now the electron starts out with spin up along $\mathbf{B}$, whereas in Equation 9.20(d) it started out with spin up along z.

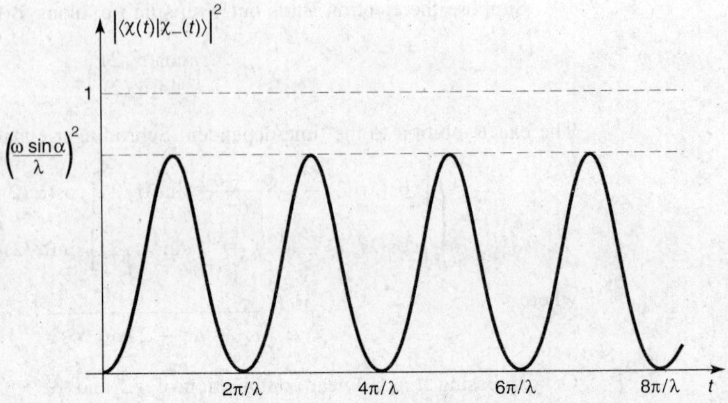

FIGURE 10.4: Plot of the transition probability, Equation 10.34, in the *non*adiabatic regime ($\omega \gg \omega_1$).

$**$**Problem 10.2** Check that Equation 10.31 satisfies the time-dependent Schrödinger equation for the Hamiltonian in Equation 10.25. Also confirm Equation 10.33, and show that the sum of the squares of the coefficients is 1, as required for normalization.

10.2 BERRY'S PHASE

10.2.1 Nonholonomic Processes

Let's go back to the classical model I used (in Section 10.1.1) to develop the notion of an adiabatic process: a perfectly frictionless pendulum, whose support is carried around from place to place. I claimed that as long as the motion of the support is *very slow*, compared to the period of the pendulum (so that the pendulum executes many oscillations before the support has moved appreciably), it will continue to swing in the same plane (or one parallel to it), with the same amplitude (and, of course, the same frequency).

But what if I took this ideal pendulum up to the North Pole, and set it swinging—say, in the direction of Portland (Figure 10.5). For the moment, pretend the earth is not rotating. Very gently (that is, *adiabatically*), I carry it down the longitude line passing through Portland, to the equator. At this point it is swinging north-south. Now I carry it (still swinging north-south) part way around the equator. And finally, I take it back up to the North Pole, along the new longitude line.

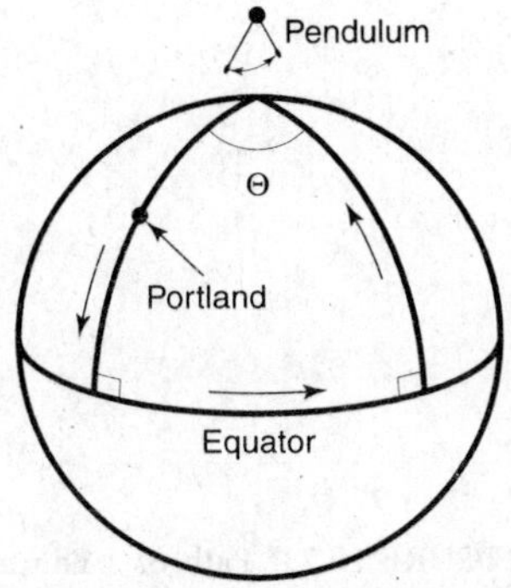

FIGURE 10.5: Itinerary for adiabatic transport of a pendulum on the surface of the earth.

It is clear that the pendulum will no longer be swinging in the same plane as it was when I set out—indeed, the new plane makes an angle Θ with the old one, where Θ is the angle between the southbound and the northbound longitude lines.

As it happens, Θ is equal to the *solid angle* (Ω) subtended (at the center of the earth) by the path around which I carried the pendulum. For this path surrounds a fraction $\Theta/2\pi$ of the northern hemisphere, so its area is $A = (1/2)(\Theta/2\pi)4\pi R^2 = \Theta R^2$ (where R is the radius of the earth), and hence

$$\Theta = A/R^2 \equiv \Omega. \tag{10.36}$$

This is a particularly nice way to express the answer, because it turns out to be independent of the *shape* of the path (Figure 10.6).[8]

Incidentally, the **Foucault pendulum** is an example of precisely this sort of adiabatic transport around a closed loop on a sphere—only this time instead of *me*

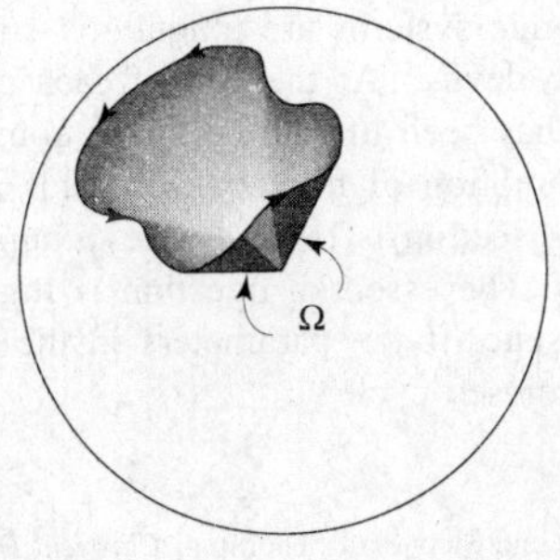

FIGURE 10.6: Arbitrary path on the surface of a sphere, subtending a solid angle Ω.

[8]You can prove this for yourself, if you are interested. Think of the circuit as being made up of tiny segments of great circles (geodesics on the sphere); the pendulum makes a fixed angle with each geodesic segment, so the net angular deviation is related to the sum of the vertex angles of the spherical polygon.

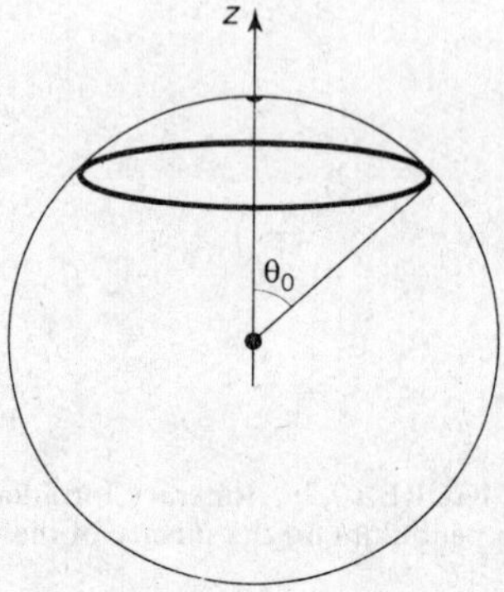

FIGURE 10.7: Path of a Foucault pendulum, in the course of one day.

carrying the pendulum around, I let the *rotation of the earth* do the job. The solid angle subtended by a latitude line θ_0 (Figure 10.7) is

$$\Omega = \int \sin\theta \, d\theta d\phi = 2\pi(-\cos\theta)\Big|_0^{\theta_0} = 2\pi(1 - \cos\theta_0). \quad [10.37]$$

Relative to the earth (which has meanwhile turned through an angle of 2π), the daily precession of the Foucault pendulum is $2\pi\cos\theta_0$—a result that is ordinarily obtained by appeal to Coriolis forces in the rotating reference frame,[9] but is seen in this context to admit a purely *geometrical* interpretation.

A system such as this, which does not return to its original state when transported around a closed loop, is said to be **nonholonomic**. (The "transport" in question need not involve physical *motion*: What we have in mind is that the parameters of the system are changed in some fashion that eventually returns them to their initial values.) Nonholonomic systems are ubiquitous—in a sense, every cyclical engine is a nonholonomic device: At the end of each cycle the car has moved forward a bit, or a weight has been lifted slightly, or something. The idea has even been applied to the locomotion of microbes in fluids at low Reynolds number.[10] My project for the next section is to study the *quantum mechanics of nonholonomic adiabatic processes*. The essential question is this: How does the final state differ from the initial state, if the parameters in the Hamiltonian are carried adiabatically around some closed cycle?

---

[9]See, for example, Jerry B. Marion and Stephen T. Thornton, *Classical Dynamics of Particles and Systems*, 4th ed., Saunders, Fort Worth, TX (1995), Example 10.5. Geographers measure latitude (λ) up from the equator, rather than down from the pole, so $\cos\theta_0 = \sin\lambda$.

[10]The pendulum example is an application of **Hannay's angle**, which is the classical analog to Berry's phase. For a collection of papers on both subjects, see Alfred Shapere and Frank Wilczek, eds., *Geometric Phases in Physics*, World Scientific, Singapore (1989).

10.2.2 Geometric Phase

In Section 10.1.2 I showed that a particle which starts out in the nth eigenstate of $H(0)$ remains, under adiabatic conditions, in the nth eigenstate of $H(t)$, picking up only a time-dependent phase factor. Specifically, its wave function is (Equation 10.23)

$$\Psi_n(t) = e^{i[\theta_n(t)+\gamma_n(t)]}\psi_n(t), \qquad [10.38]$$

where

$$\theta_n(t) \equiv -\frac{1}{\hbar}\int_0^t E_n(t')\,dt' \qquad [10.39]$$

is the **dynamic phase** (generalizing the usual factor $\exp(-iE_nt/\hbar)$ to the case where E_n is a function of time), and

$$\gamma_n(t) \equiv i\int_0^t \left\langle \psi_n(t') \left| \frac{\partial}{\partial t'}\psi_n(t') \right. \right\rangle dt' \qquad [10.40]$$

is the so-called **geometric phase.**

Now $\psi_n(t)$ depends on t because there is some parameter $R(t)$ in the Hamiltonian that is changing with time. (In Problem 10.1, $R(t)$ would be the width of the expanding square well.) Thus

$$\frac{\partial\psi_n}{\partial t} = \frac{\partial\psi_n}{\partial R}\frac{dR}{dt}, \qquad [10.41]$$

so

$$\gamma_n(t) = i\int_0^t \left\langle \psi_n \left| \frac{\partial\psi_n}{\partial R} \right. \right\rangle \frac{dR}{dt'}\,dt' = i\int_{R_i}^{R_f} \left\langle \psi_n \left| \frac{\partial\psi_n}{\partial R} \right. \right\rangle dR, \qquad [10.42]$$

where R_i and R_f are the initial and final values of $R(t)$. In particular, if the Hamiltonian returns to its original form after time T, so that $R_f = R_i$, then $\gamma_n(T) = 0$—nothing very interesting *there*!

However, I assumed (in Equation 10.41) that there is only *one* parameter in the Hamiltonian that is changing. Suppose there are N of them: $R_1(t)$, $R_2(t)$, ..., $R_N(t)$; in that case

$$\frac{\partial\psi_n}{\partial t} = \frac{\partial\psi_n}{\partial R_1}\frac{dR_1}{dt} + \frac{\partial\psi_n}{\partial R_2}\frac{dR_2}{dt} + \cdots + \frac{\partial\psi_n}{\partial R_N}\frac{dR_N}{dt} = (\nabla_R\psi_n)\cdot\frac{d\mathbf{R}}{dt}, \qquad [10.43]$$

where $\mathbf{R} \equiv (R_1, R_n, \ldots, R_N)$, and ∇_R is the gradient with respect to these parameters. This time we have

$$\gamma_n(t) = i\int_{\mathbf{R}_i}^{\mathbf{R}_f} \langle\psi_n|\nabla_R\psi_n\rangle\cdot d\mathbf{R}, \qquad [10.44]$$

and if the Hamiltonian returns to its original form after a time T, the net geometric phase change is

$$\boxed{\gamma_n(T) = i \oint \langle \psi_n | \nabla_R \psi_n \rangle \cdot d\mathbf{R}.} \tag{10.45}$$

This is a *line* integral around a closed loop in parameter-space, and it is *not*, in general, zero. Equation 10.45 was first obtained by Michael Berry, in 1984,[11] and $\gamma_n(T)$ is called **Berry's phase**. Notice that $\gamma_n(T)$ depends *only on the path taken*, not on how *fast* that path is traversed (provided, of course, that it is slow enough to validate the adiabatic hypothesis). By contrast, the accumulated *dynamic* phase,

$$\theta_n(T) = -\frac{1}{\hbar}\int_0^T E_n(t')\,dt',$$

depends critically on the elapsed time.

We are accustomed to thinking that the phase of the wave function is arbitrary—physical quantities involve $|\Psi|^2$, and the phase factor cancels out. For this reason, most people assumed until recently that the geometric phase was of no conceivable physical significance—after all, the phase of $\psi_n(t)$ itself is arbitrary. It was Berry's insight that if you carry the Hamiltonian around a closed *loop*, bringing it back to its original form, the relative phase at the beginning and the end of the process is *not* arbitrary, and can actually be measured.

For example, suppose we take a beam of particles (all in the state Ψ), and split it in two, so that one beam passes through an adiabatically changing potential, while the other does not. When the two beams are recombined, the total wave function has the form

$$\Psi = \frac{1}{2}\Psi_0 + \frac{1}{2}\Psi_0 e^{i\Gamma}, \tag{10.46}$$

where Ψ_0 is the "direct" beam wave function, and Γ is the *extra* phase (in part dynamic, and in part geometric) acquired by the beam subjected to the varying H. In this case

$$|\Psi|^2 = \frac{1}{4}|\Psi_0|^2\left(1+e^{i\Gamma}\right)\left(1+e^{-i\Gamma}\right)$$

$$= \frac{1}{2}|\Psi_0|^2(1+\cos\Gamma) = |\Psi_0|^2\cos^2(\Gamma/2). \tag{10.47}$$

So by looking for points of constructive and destructive interference (where Γ is an even or odd multiple of π, respectively), one can easily measure Γ. (Berry, and

[11] M. V. Berry, *Proc. R. Soc. Lond.* A **392**, 45 (1984), reprinted in Wilczek and Shapere (footnote 10). It is astonishing, in retrospect, that this result escaped notice for sixty years.

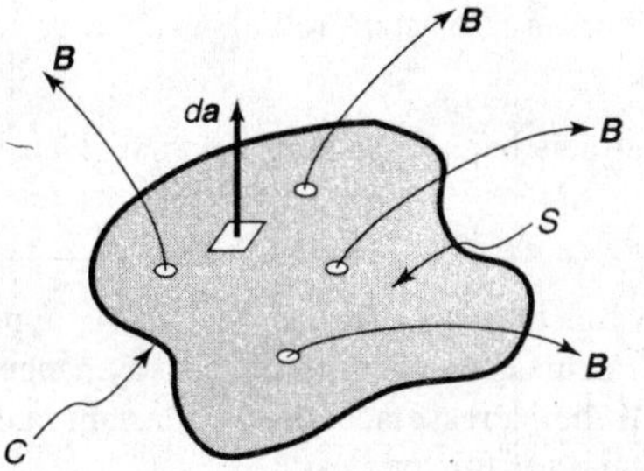

FIGURE 10.8: Magnetic flux through a surface S bounded by the closed curve C.

other early writers, worried that the geometric phase might be swamped by a larger dynamic phase, but it has proved possible to arrange things so as to separate out the two contributions.)

When the parameter space is three dimensional, $\mathbf{R} = (R_1, R_2, R_3)$, Berry's formula (Equation 10.45) is reminiscent of the expression for **magnetic flux** in terms of the vector potential $\mathbf{A}$. The flux, Φ, through a surface S bounded by a curve C (Figure 10.8), is

$$\Phi \equiv \int_S \mathbf{B} \cdot d\mathbf{a}. \qquad [10.48]$$

If we write the magnetic field in terms of the vector potential ($\mathbf{B} = \nabla \times \mathbf{A}$), and apply Stokes' theorem:

$$\Phi = \int_S (\nabla \times \mathbf{A}) \cdot d\mathbf{a} = \oint_C \mathbf{A} \cdot d\mathbf{r}. \qquad [10.49]$$

Thus Berry's phase can be thought of as the "flux" of a "magnetic field"

$$\text{“}\mathbf{B}\text{”} = i \nabla_R \times \langle \psi_n | \nabla_R \psi_n \rangle, \qquad [10.50]$$

through the (closed loop) trajectory in parameter-space. To put it the other way around, in the three-dimensional case Berry's phase can be written as a surface integral,

$$\gamma_n(T) = i \int [\nabla_R \times \langle \psi_n | \nabla_R \psi_n \rangle] \cdot d\mathbf{a}. \qquad [10.51]$$

The magnetic analogy can be carried much further, but for our purposes Equation 10.51 is merely a convenient alternative expression for $\gamma_n(T)$.

∗Problem 10.3

(a) Use Equation 10.42 to calculate the geometric phase change when the infinite square well expands adiabatically from width w_1 to width w_2. Comment on this result.

(b) If the expansion occurs at a constant rate ($dw/dt = v$), what is the dynamic phase change for this process?

(c) If the well now contracts back to its original size, what is Berry's phase for the cycle?

Problem 10.4 The delta function well (Equation 2.114) supports a single bound state (Equation 2.129). Calculate the geometric phase change when α gradually increases from α_1 to α_2. If the increase occurs at a constant rate ($d\alpha/dt = c$), what is the dynamic phase change for this process?

Problem 10.5 Show that if $\psi_n(t)$ is *real*, the geometric phase vanishes. (Problems 10.3 and 10.4 are examples of this.) You might try to beat the rap by tacking an unnecessary (but perfectly legal) phase factor onto the eigenfunctions: $\psi'_n(t) \equiv e^{i\phi_n}\psi_n(t)$, where $\phi_n(\mathbf{R})$ is an arbitrary (real) function. Try it. You'll get a nonzero geometric phase, all right, but note what happens when you put it back into Equation 10.23. And for a *closed* loop it gives *zero*. *Moral:* For nonzero Berry's phase, you need (i) more than one time-dependent parameter in the Hamiltonian, and (ii) a Hamiltonian that yields nontrivially complex eigenfunctions.

Example 10.2 The classic example of Berry's phase is an electron at the origin, subjected to a magnetic field of constant magnitude but changing direction. Consider first the special case (analyzed in Example 10.1) in which $\mathbf{B}(t)$ precesses around at a constant angular velocity ω, making a fixed angle α with the z axis. The *exact* solution (for an electron that starts out with "spin up" along $\mathbf{B}$) is given by Equation 10.33. In the adiabatic regime, $\omega \ll \omega_1$,

$$\lambda = \omega_1\sqrt{1 - 2\frac{\omega}{\omega_1}\cos\alpha + \left(\frac{\omega}{\omega_1}\right)^2} \cong \omega_1\left(1 - \frac{\omega}{\omega_1}\cos\alpha\right) = \omega_1 - \omega\cos\alpha, \quad [10.52]$$

and Equation 10.33 becomes

$$\chi(t) \cong e^{-i\omega_1 t/2}e^{i(\omega\cos\alpha)t/2}e^{-i\omega t/2}\chi_+(t) + i\left[\frac{\omega}{\omega_1}\sin\alpha\sin\left(\frac{\omega_1 t}{2}\right)\right]e^{+i\omega t/2}\chi_-(t). \quad [10.53]$$

As $\omega/\omega_1 \to 0$ the second term drops out completely, and the result matches the expected adiabatic form (Equation 10.23). The dynamic phase is

$$\theta_+(t) = -\frac{1}{\hbar}\int_0^t E_+(t')\,dt' = -\frac{\omega_1 t}{2}, \quad ['0.54]$$

(where $E_+ = \hbar\omega_1/2$, from Equation 10.29), so the geometric phase is

$$\gamma_+(t) = (\cos\alpha - 1)\frac{\omega t}{2}. \tag{10.55}$$

For a complete cycle $T = 2\pi/\omega$, and therefore Berry's phase is

$$\gamma_+(T) = \pi(\cos\alpha - 1). \tag{10.56}$$

Now consider the more general case, in which the tip of the magnetic field vector sweeps out an *arbitrary* closed curve on the surface of a sphere of radius $r = B_0$ (Figure 10.9). The eigenstate representing spin up along $\mathbf{B}(t)$ has the form (see Problem 4.30):

$$\chi_+ = \begin{pmatrix} \cos(\theta/2) \\ e^{i\phi}\sin(\theta/2) \end{pmatrix}, \tag{10.57}$$

where θ and ϕ (the spherical coordinates of $\mathbf{B}$) are now *both* functions of time. Looking up the gradient in spherical coordinates, we find

$$\begin{aligned} \nabla\chi_+ &= \frac{\partial\chi_+}{\partial r}\hat{r} + \frac{1}{r}\frac{\partial\chi_+}{\partial\theta}\hat{\theta} + \frac{1}{r\sin\theta}\frac{\partial\chi_+}{\partial\phi}\hat{\phi} \\ &= \frac{1}{r}\begin{pmatrix} -(1/2)\sin(\theta/2) \\ (1/2)e^{i\phi}\cos(\theta/2) \end{pmatrix}\hat{\theta} + \frac{1}{r\sin\theta}\begin{pmatrix} 0 \\ ie^{i\phi}\sin(\theta/2) \end{pmatrix}\hat{\phi}. \end{aligned} \tag{10.58}$$

Hence

$$\begin{aligned} \langle\chi_+|\nabla\chi_+\rangle &= \frac{1}{2r}\left[-\sin(\theta/2)\cos(\theta/2)\,\hat{\theta} + \sin(\theta/2)\cos(\theta/2)\,\hat{\theta} + 2i\frac{\sin^2(\theta/2)}{\sin\theta}\,\hat{\phi}\right] \\ &= i\frac{\sin^2(\theta/2)}{r\sin\theta}\,\hat{\phi}. \end{aligned} \tag{10.59}$$

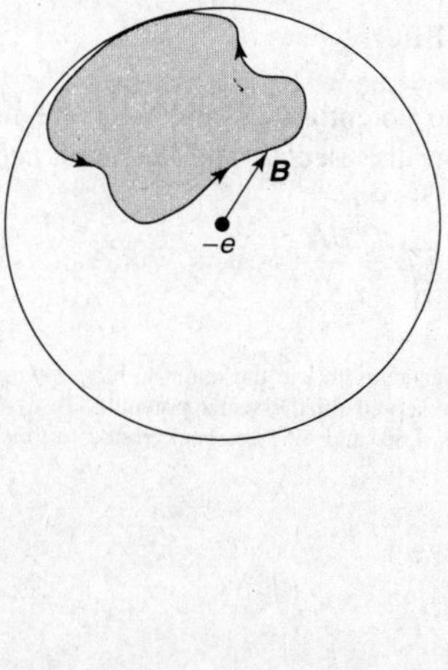

FIGURE 10.9: Magnetic field of constant magnitude but changing direction sweeps out a closed loop.

For Equation 10.51 we need the *curl* of this quantity:

$$\nabla \times \langle \chi_+ | \nabla \chi_+ \rangle = \frac{1}{r \sin\theta} \frac{\partial}{\partial \theta} \left[\sin\theta \left(\frac{i \sin^2(\theta/2)}{r \sin\theta} \right) \right] \hat{r} = \frac{i}{2r^2} \hat{r}. \qquad [10.60]$$

According to Equation 10.51, then,

$$\gamma_+(T) = -\frac{1}{2} \int \frac{1}{r^2} \hat{r} \cdot d\mathbf{a}. \qquad [10.61]$$

The integral is over the area on the sphere swept out by **B** in the course of the cycle, so $d\mathbf{a} = r^2 \, d\Omega \, \hat{r}$, and we conclude that

$$\gamma_+(T) = -\frac{1}{2} \int d\Omega = -\frac{1}{2} \Omega, \qquad [10.62]$$

where Ω is the solid angle subtended at the origin. This is a delightfully simple result, and tantalizingly reminiscent of the classical problem with which we began the discussion (transport of a frictionless pendulum around a closed path on the surface of the earth). It says that if you take a magnet, and lead the electron's spin around adiabatically in an arbitrary closed path, the net (geometric) phase change will be minus one-half the solid angle swept out by the magnetic field vector. In view of Equation 10.37, this general result is consistent with the special case (Equation 10.56), as of course it *had* to be.

***Problem 10.6** Work out the analog to Equation 10.62 for a particle of spin 1. *Answer:* $-\Omega$. (Incidentally, for spin s the result is $-s\Omega$.)

10.2.3 The Aharonov-Bohm Effect

In classical electrodynamics the potentials (φ and $\mathbf{A}$)[12] are not directly measurable—the *physical* quantities are the electric and magnetic *fields*:

$$\mathbf{E} = -\nabla\varphi - \frac{\partial \mathbf{A}}{\partial t}, \quad \mathbf{B} = \nabla \times \mathbf{A}. \qquad [10.63]$$

[12]It is customary in quantum mechanics to use the letter V for *potential energy*, but in electrodynamics the same letter is ordinarily reserved for the scalar potential. To avoid confusion I use φ for the scalar potential. See Problems 4.59, 4.60, and 4.61 for background to this section.

The fundamental laws (Maxwell's equations and the Lorentz force rule) make no reference to potentials, which are (from a logical point of view) no more than convenient but dispensable theoretical constructs. Indeed, you can with impunity *change* the potentials:

$$\varphi \to \varphi' = \varphi - \frac{\partial \Lambda}{\partial t}, \quad \mathbf{A} \to \mathbf{A}' = \mathbf{A} + \nabla\Lambda, \qquad [10.64]$$

where Λ is any function of position and time; this is called a **gauge transformation**, and it has no effect on the fields (as you can easily check using Equation 10.63).

In quantum mechanics the potentials play a more significant role, for the Hamiltonian is expressed in terms of φ and $\mathbf{A}$, not $\mathbf{E}$ and $\mathbf{B}$:

$$H = \frac{1}{2m}\left(\frac{\hbar}{i}\nabla - q\mathbf{A}\right)^2 + q\varphi. \qquad [10.65]$$

Nevertheless, the theory is still invariant under gauge transformations (see Problem 4.61), and for a long time it was taken for granted that there could be no electromagnetic influences in regions where $\mathbf{E}$ and $\mathbf{B}$ are zero—any more than there can be in the classical theory. But in 1959 Aharonov and Bohm[13] showed that the vector potential *can* affect the quantum behavior of a charged particle, even when it is moving through a region in which the field itself is zero. I'll work out a simple example first, then discuss the Aharonov-Bohm effect, and finally indicate how it all relates to Berry's phase.

Imagine a particle constrained to move in a circle of radius b (a bead on a wire ring, if you like). Along the axis runs a solenoid of radius $a < b$, carrying a steady electric current I (see Figure 10.10). If the solenoid is extremely long, the magnetic field inside it is uniform, and the field outside is zero. But the vector potential outside the solenoid is *not* zero; in fact (adopting the convenient gauge condition $\nabla \cdot \mathbf{A} = 0$),[14]

$$\mathbf{A} = \frac{\Phi}{2\pi r}\hat{\phi}, \quad (r > a), \qquad [10.66]$$

where $\Phi = \pi a^2 B$ is the **magnetic flux** through the solenoid. Meanwhile, the solenoid itself is uncharged, so the scalar potential φ is zero. In this case the Hamiltonian (Equation 10.65) becomes

$$H = \frac{1}{2m}\left[-\hbar^2\nabla^2 + q^2A^2 + 2i\hbar q\mathbf{A}\cdot\nabla\right]. \qquad [10.67]$$

---

[13] Y. Aharonov and D. Bohm, *Phys. Rev.* **115**, 485 (1959). For a significant precursor, see W. Ehrenberg and R. E. Siday, *Proc. Phys. Soc. London* **B62**, 8 (1949).

[14] See, for instance, D. J. Griffiths, *Introduction to Electrodynamics*, 3rd ed., Prentice Hall, Upper Saddle River, NJ (1999), Equation 5.71.

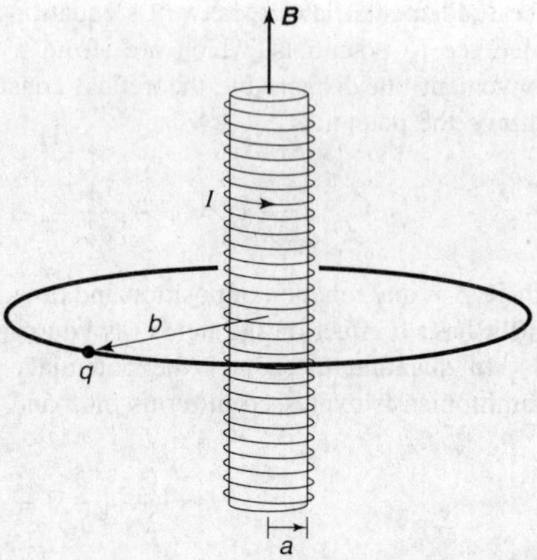

FIGURE 10.10: Charged bead on a circular ring through which a long solenoid passes.

But the wave function depends only on the azimuthal angle ϕ ($\theta = \pi/2$ and $r = b$), so $\nabla \to (\hat{\phi}/b)(d/d\phi)$, and the Schrödinger equation reads

$$\frac{1}{2m}\left[-\frac{\hbar^2}{b^2}\frac{d^2}{d\phi^2} + \left(\frac{q\Phi}{2\pi b}\right)^2 + i\frac{\hbar q\Phi}{\pi b^2}\frac{d}{d\phi}\right]\psi(\phi) = E\psi(\phi). \qquad [10.68]$$

This is a linear differential equation with constant coefficients:

$$\frac{d^2\psi}{d\phi^2} - 2i\beta\frac{d\psi}{d\phi} + \epsilon\psi = 0, \qquad [10.69]$$

where

$$\beta \equiv \frac{q\Phi}{2\pi\hbar} \quad \text{and} \quad \epsilon \equiv \frac{2mb^2E}{\hbar^2} - \beta^2. \qquad [10.70]$$

Solutions are of the form

$$\psi = Ae^{i\lambda\phi}, \qquad [10.71]$$

with

$$\lambda = \beta \pm \sqrt{\beta^2 + \epsilon} = \beta \pm \frac{b}{\hbar}\sqrt{2mE}. \qquad [10.72]$$

Continuity of $\psi(\phi)$, at $\phi = 2\pi$, requires that λ be an *integer*:

$$\beta \pm \frac{b}{\hbar}\sqrt{2mE} = n. \qquad [10.73]$$

and it follows that

$$E_n = \frac{\hbar^2}{2mb^2}\left(n - \frac{q\Phi}{2\pi\hbar}\right)^2, \quad (n = 0, \pm 1, \pm 2, \dots). \qquad [10.74]$$

The solenoid lifts the two-fold degeneracy of the bead-on-a-ring (Problem 2.46): Positive n, representing a particle traveling in the *same* direction as the current in the solenoid, has a somewhat *lower* energy (assuming q is positive) than negative n, describing a particle traveling in the *opposite* direction. More important, the allowed energies clearly depend on the field inside the solenoid, *even though the field at the location of the particle is zero!*[15]

More generally, suppose a particle is moving through a region where $\mathbf{B}$ is zero (so $\nabla \times \mathbf{A} = 0$), but $\mathbf{A}$ itself is *not*. (I'll assume that $\mathbf{A}$ is static, although the method can be generalized to time-dependent potentials.) The (time-dependent) Schrödinger equation,

$$\left[\frac{1}{2m}\left(\frac{\hbar}{i}\nabla - q\mathbf{A}\right)^2 + V\right]\Psi = i\hbar\frac{\partial\Psi}{\partial t}, \qquad [10.75]$$

with potential energy V—which may or may not include an electrical contribution $q\varphi$—can be simplified by writing

$$\Psi = e^{ig}\Psi', \qquad [10.76]$$

where

$$g(\mathbf{r}) \equiv \frac{q}{\hbar}\int_{\mathcal{O}}^{\mathbf{r}} \mathbf{A}(\mathbf{r}') \cdot d\mathbf{r}', \qquad [10.77]$$

and $\mathcal{O}$ is some (arbitrarily chosen) reference point. Note that this definition makes sense *only* when $\nabla \times \mathbf{A} = 0$ throughout the region in question—otherwise the line integral would depend on the *path* taken from $\mathcal{O}$ to $\mathbf{r}$, and hence would not define a function of $\mathbf{r}$. In terms of Ψ', the gradient of Ψ is

$$\nabla\Psi = e^{ig}(i\nabla g)\Psi' + e^{ig}(\nabla\Psi');$$

but $\nabla g = (q/\hbar)\mathbf{A}$, so

$$\left(\frac{\hbar}{i}\nabla - q\mathbf{A}\right)\Psi = \frac{\hbar}{i}e^{ig}\nabla\Psi', \qquad [10.78]$$

---

[15] It is a peculiar property of **superconducting** rings that the enclosed flux is *quantized*: $\Phi = (2\pi\hbar/q)n'$, where n' is an integer. In that case the effect is undetectable, since $E_n = (\hbar^2/2mb^2)(n+n')^2$, and $(n+n')$ is just another integer. (Incidentally, the charge q here turns out to be *twice* the charge of an electron; the superconducting electrons are locked together in pairs.) However, **flux quantization** is enforced by the *superconductor* (which induces circulating currents to make up the difference), not by the solenoid or the electromagnetic field, and it does not occur in the (nonsuperconducting) example considered here.

and it follows that

$$\left(\frac{\hbar}{i}\nabla - q\mathbf{A}\right)^2 \Psi = -\hbar^2 e^{ig}\nabla^2\Psi'. \qquad [10.79]$$

Putting this into Equation 10.75, and cancelling the common factor e^{ig}, we are left with

$$-\frac{\hbar^2}{2m}\nabla^2\Psi' + V\Psi' = i\hbar\frac{\partial\Psi'}{\partial t}. \qquad [10.80]$$

Evidently Ψ' satisfies the Schrödinger equation *without* **A**. If we can solve Equation 10.80, correcting for the presence of a (curl-free) vector potential will be trivial: Just tack on the phase factor e^{ig}.

Aharonov and Bohm proposed an experiment in which a beam of electrons is split in two, and passed either side of a long solenoid, before being recombined (Figure 10.11). The beams are kept well away from the solenoid itself, so they encounter only regions where $\mathbf{B} = 0$. But **A**, which is given by Equation 10.66, is *not* zero, and (assuming V is the same on both sides), the two beams arrive with *different phases*:

$$g = \frac{q}{\hbar}\int \mathbf{A}\cdot d\mathbf{r} = \frac{q\Phi}{2\pi\hbar}\int\left(\frac{1}{r}\hat{\phi}\right)\cdot(r\hat{\phi}\,d\phi) = \pm\frac{q\Phi}{2\hbar}. \qquad [10.81]$$

The plus sign applies to the electrons traveling in the same direction as **A**—which is to say, in the same direction as the current in the solenoid. The beams arrive out

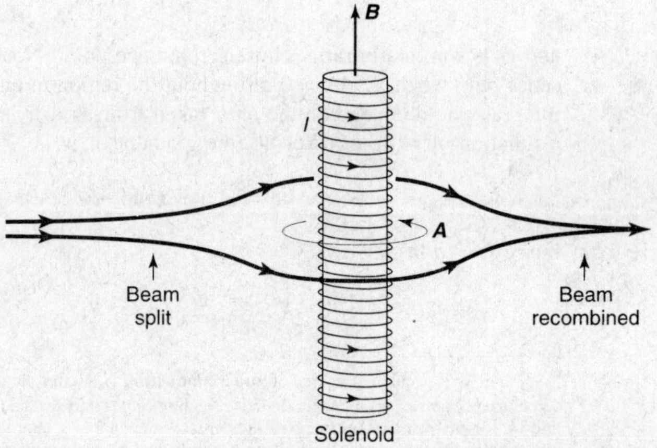

FIGURE 10.11: The Aharonov-Bohm effect: The electron beam splits, with half passing either side of a long solenoid.

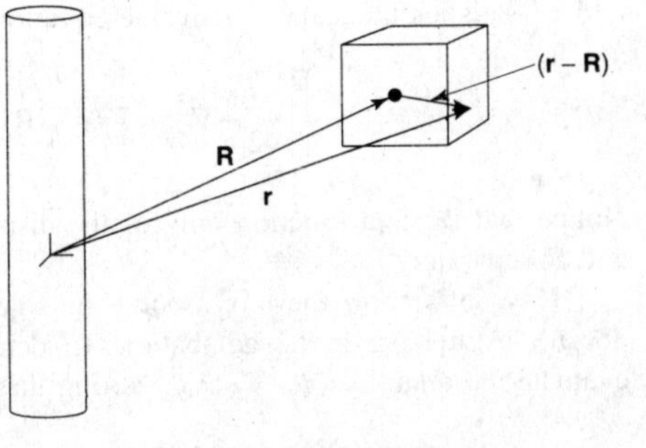

FIGURE 10.12: Particle confined to a box, by a potential $V(\mathbf{r} - \mathbf{R})$.

of phase by an amount proportional to the magnetic flux their paths encircle:

$$\text{phase difference} = \frac{q\Phi}{\hbar}. \qquad [10.82]$$

This phase shift leads to measurable interference (Equation 10.47), which has been confirmed experimentally by Chambers and others.[16]

As Berry pointed out in his first paper on the subject, the Aharonov-Bohm effect can be regarded as an example of geometric phase. Suppose the charged particle is confined to a box (which is centered at point **R** outside the solenoid) by a potential $V(\mathbf{r} - \mathbf{R})$—see Figure 10.12. (In a moment we're going to transport the box around the solenoid, so **R** will become a function of time, but for now it is just some fixed vector.) The eigenfunctions of the Hamiltonian are determined by

$$\left\{\frac{1}{2m}\left[\frac{\hbar}{i}\nabla - q\mathbf{A}(\mathbf{r})\right]^2 + V(\mathbf{r} - \mathbf{R})\right\}\psi_n = E_n\psi_n. \qquad [10.83]$$

We have already learned how to solve equations of this form: Let

$$\psi_n = e^{ig}\psi_n', \qquad [10.84]$$

where[17]

$$g \equiv \frac{q}{\hbar}\int_{\mathbf{R}}^{\mathbf{r}} \mathbf{A}(\mathbf{r}') \cdot d\mathbf{r}', \qquad [10.85]$$

---

[16] R. G. Chambers, *Phys. Rev. Lett.* **5**, 3 (1960).

[17] It is convenient to set the reference point $\mathcal{O}$ at the center of the box, for this guarantees that we recover the original phase convention when we complete the journey around the solenoid. If you use a point in fixed *space*, for example, you'll have to readjust the phase "by hand," at the far end, because the path will have wrapped around the solenoid, circling regions where the curl of **A** does not vanish. This leads to exactly the same answer, but it's a crude way to do it. In general, when choosing the phase convention for the eigenfunctions in Equation 10.9, you want to make sure that $\psi_n(x, T) = \psi_n(x, 0)$, so that no spurious phase changes are introduced.

and ψ' satisfies the same eigenvalue equation, only with $\mathbf{A} \to 0$:

$$\left[-\frac{\hbar^2}{2m}\nabla^2 + V(\mathbf{r}-\mathbf{R})\right]\psi'_n = E_n\psi'_n. \qquad [10.86]$$

Notice that ψ'_n is a function only of the displacement $(\mathbf{r}-\mathbf{R})$, not (like ψ_n) of $\mathbf{r}$ and $\mathbf{R}$ separately.

Now let's carry the box around the solenoid (in this application the process doesn't even have to be adiabatic). To determine Berry's phase we must first evaluate the quantity $\langle\psi_n|\nabla_R\psi_n\rangle$. Noting that

$$\nabla_R\psi_n = \nabla_R\left[e^{ig}\psi'_n(\mathbf{r}-\mathbf{R})\right] = -i\frac{q}{\hbar}\mathbf{A}(\mathbf{R})e^{ig}\psi'_n(\mathbf{r}-\mathbf{R}) + e^{ig}\nabla_R\psi'_n(\mathbf{r}-\mathbf{R}),$$

we find

$$\begin{aligned}\langle\psi_n|\nabla_R\psi_n\rangle &= \int e^{-ig}[\psi'_n(\mathbf{r}-\mathbf{R})]^* e^{ig}\left[-i\frac{q}{\hbar}\mathbf{A}(\mathbf{R})\psi'_n(\mathbf{r}-\mathbf{R}) + \nabla_R\psi'_n(\mathbf{r}-\mathbf{R})\right]d^3\mathbf{r} \\ &= -i\frac{q}{\hbar}\mathbf{A}(\mathbf{R}) - \int[\psi'_n(\mathbf{r}-\mathbf{R})]^*\nabla\psi'_n(\mathbf{r}-\mathbf{R})\,d^3\mathbf{r}. \end{aligned} \qquad [10.87]$$

The ∇ with no subscript denotes the gradient with respect to $\mathbf{r}$, and I used the fact that $\nabla_R = -\nabla$, when acting on a function of $(\mathbf{r}-\mathbf{R})$. But the last integral is $i/\hbar$ times the expectation value of momentum, in an eigenstate of the Hamiltonian $-(\hbar^2/2m)\nabla^2 + V$, which we know from Section 2.1 is *zero*. So

$$\langle\psi_n|\nabla_R\psi_n\rangle = -i\frac{q}{\hbar}\mathbf{A}(\mathbf{R}). \qquad [10.88]$$

Putting this into Berry's formula (Equation 10.45), we conclude that

$$\gamma_n(T) = \frac{q}{\hbar}\oint\mathbf{A}(\mathbf{R})\cdot d\mathbf{R} = \frac{q}{\hbar}\int(\nabla\times\mathbf{A})\cdot d\mathbf{a} = \frac{q\Phi}{\hbar}, \qquad [10.89]$$

which neatly confirms the Aharonov-Bohm result (Equation 10.82), and reveals that the Aharonov-Bohm effect is a particular instance of geometric phase.[18]

What are we to make of the Aharonov-Bohm effect? Evidently our classical preconceptions are simply *mistaken*: There *can* be electromagnetic effects in regions where the fields are zero. Note however that this does not make $\mathbf{A}$ itself

---

[18]Incidentally, in this case the analogy between Berry's phase and magnetic flux (Equation 10.50) is *almost* an identity: "$\mathbf{B}$" $= (q/\hbar)\mathbf{B}$.

measurable—only the enclosed *flux* comes into the final answer, and the theory remains gauge invariant.

Problem 10.7

(a) Derive Equation 10.67 from Equation 10.65.

(b) Derive Equation 10.79, starting with Equation 10.78.

FURTHER PROBLEMS FOR CHAPTER 10

∗∗∗**Problem 10.8** A particle starts out in the ground state of the infinite square well (on the interval $0 \le x \le a$). Now a wall is slowly erected, slightly off-center:[19]

$$V(x) = f(t)\delta\left(x - \frac{a}{2} - \epsilon\right),$$

where $f(t)$ rises gradually from 0 to ∞. According to the adiabatic theorem, the particle will remain in the ground state of the evolving Hamiltonian.

(a) Find (and sketch) the ground state at $t \to \infty$. *Hint:* This should be the ground state of the infinite square well with an impenetrable barrier at $a/2 + \epsilon$. Note that the particle is confined to the (slightly) larger left "half" of the well.

(b) Find the (transcendental) equation for the ground state of the Hamiltonian at time t. *Answer:*

$$z \sin z = T[\cos z - \cos(z\delta)],$$

where $z \equiv ka$, $T \equiv maf(t)/\hbar^2$, $\delta \equiv 2\epsilon/a$, and $k \equiv \sqrt{2mE}/\hbar$.

(c) Setting $\delta = 0$, solve graphically for z, and show that it goes from π to 2π as T goes from 0 to ∞. Explain this result.

(d) Now set $\delta = 0.01$ and solve numerically for z, using $T = 0$, 1, 5, 20, 100, and 1000.

(e) Find the probability P_r that the particle is in the right "half" of the well, as a function of z and δ. *Answer:* $P_r = 1/[1 + (I_+/I_-)]$, where $I_\pm \equiv \left[1 \pm \delta - (1/z)\sin\left(z(1 \pm \delta)\right)\right]\sin^2[z(1 \mp \delta)/2]$. Evaluate this expression numerically for the T's in part (d). Comment on your results.

---

[19] Julio Gea-Banacloche, *Am. J. Phys.* **70**, 307 (2002) uses a rectangular barrier; the delta-function version was suggested by M. Lakner and J. Peternelj, *Am. J. Phys.* **71**, 519 (2003).

(f) Plot the ground state wave function for those same values of T and δ. Note how it gets squeezed into the left half of the well, as the barrier grows.[20]

$***$**Problem 10.9** Suppose the one-dimensional harmonic oscillator (mass m, frequency ω) is subjected to a driving force of the form $F(t) = m\omega^2 f(t)$, where $f(t)$ is some specified function (I have factored out $m\omega^2$ for notational convenience; $f(t)$ has the dimensions of *length*). The Hamiltonian is

$$H(t) = -\frac{\hbar^2}{2m}\frac{\partial^2}{\partial x^2} + \frac{1}{2}m\omega^2 x^2 - m\omega^2 x f(t). \qquad [10.90]$$

Assume that the force was first turned on at time $t = 0$: $f(t) = 0$ for $t \le 0$. This system can be solved exactly, both in classical mechanics and in quantum mechanics.[21]

(a) Determine the *classical* position of the oscillator, assuming it started from rest at the origin ($x_c(0) = \dot{x}_c(0) = 0$). *Answer*:

$$x_c(t) = \omega \int_0^t f(t') \sin[\omega(t - t')]\, dt'. \qquad [10.91]$$

(b) Show that the solution to the (time-dependent) Schrödinger equation for this oscillator, assuming it started out in the nth state of the *undriven* oscillator ($\Psi(x, 0) = \psi_n(x)$ where $\psi_n(x)$ is given by Equation 2.61), can be written as

$$\Psi(x, t) = \psi_n(x - x_c) e^{\frac{i}{\hbar}\left[-(n+\frac{1}{2})\hbar\omega t + m\dot{x}_c(x - \frac{x_c}{2}) + \frac{m\omega^2}{2}\int_0^t f(t')x_c(t')dt'\right]}. \qquad [10.92]$$

(c) Show that the eigenfunctions and eigenvalues of $H(t)$ are

$$\psi_n(x, t) = \psi_n(x - f); \quad E_n(t) = \left(n + \frac{1}{2}\right)\hbar\omega - \frac{1}{2}m\omega^2 f^2. \qquad [10.93]$$

(d) Show that in the adiabatic approximation the classical position (Equation 10.91) reduces to $x_c(t) \cong f(t)$. State the precise criterion for adiabaticity, in this context, as a constraint on the time derivative of f. *Hint:* Write $\sin[\omega(t - t')]$ as $(1/\omega)(d/dt')\cos[\omega(t - t')]$ and use integration by parts.

---

[20] Gea-Banacloche (footnote 19) discusses the evolution of the wave function *without* assuming the adiabatic theorem, and confirms these results in the adiabatic limit.

[21] See Y. Nogami, *Am. J. Phys.* **59**, 64 (1991), and references therein.

(e) Confirm the adiabatic theorem for this example, by using the results in (c) and (d) to show that

$$\Psi(x,t) \cong \psi_n(x,t)e^{i\theta_n(t)}e^{i\gamma_n(t)}. \qquad [10.94]$$

Check that the dynamic phase has the correct form (Equation 10.39). Is the geometric phase what you would expect?

Problem 10.10 The adiabatic approximation can be regarded as the first term in an **adiabatic series** for the coefficients $c_m(t)$ in Equation 10.12. Suppose the system starts out in the nth state; in the adiabatic approximation, it *remains* in the nth state, picking up only a time-dependent geometric phase factor (Equation 10.21):

$$c_m(t) = \delta_{mn}e^{i\gamma_n(t)}.$$

(a) Substitute this into the right side of Equation 10.16 to obtain the "first correction" to adiabaticity:

$$c_m(t) = c_m(0) - \int_0^t \left\langle \psi_m(t') \middle| \frac{\partial}{\partial t'}\psi_n(t') \right\rangle e^{i\gamma_n(t')}e^{i(\theta_n(t')-\theta_m(t'))}\,dt'. \qquad [10.95]$$

This enables us to calculate transition probabilities in the *nearly* adiabatic regime. To develop the "second correction," we would insert Equation 10.95 on the right side of Equation 10.16, and so on.

(b) As an example, apply Equation 10.95 to the driven oscillator (Problem 10.9). Show that (in the near-adiabatic approximation) transitions are possible only to the two immediately adjacent levels, for which

$$c_{n+1}(t) = i\sqrt{\frac{m\omega}{2\hbar}}\sqrt{n+1}\int_0^t \dot{f}(t')e^{i\omega t'}\,dt',$$

$$c_{n-1}(t) = i\sqrt{\frac{m\omega}{2\hbar}}\sqrt{n}\int_0^t \dot{f}(t')e^{-i\omega t'}\,dt'.$$

(The transition *probabilities* are the absolute squares of these, of course.)

CHAPTER 11

SCATTERING

11.1 INTRODUCTION

11.1.1 Classical Scattering Theory

Imagine a particle incident on some scattering center (say, a proton fired at a heavy nucleus). It comes in with energy E and **impact parameter** b, and it emerges at some **scattering angle** θ—see Figure 11.1. (I'll assume for simplicity that the target is azimuthally symmetrical, so the trajectory remains in one plane, and that the target is very heavy, so the recoil is negligible.) The essential problem of classical scattering theory is this: *Given the impact parameter, calculate the scattering angle.* Ordinarily, of course, the smaller the impact parameter, the greater the scattering angle.

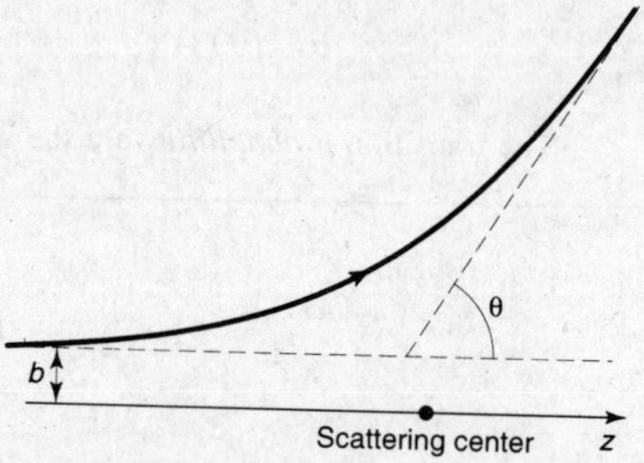

FIGURE 11.1: The classical scattering problem, showing the impact parameter b and the scattering angle θ.

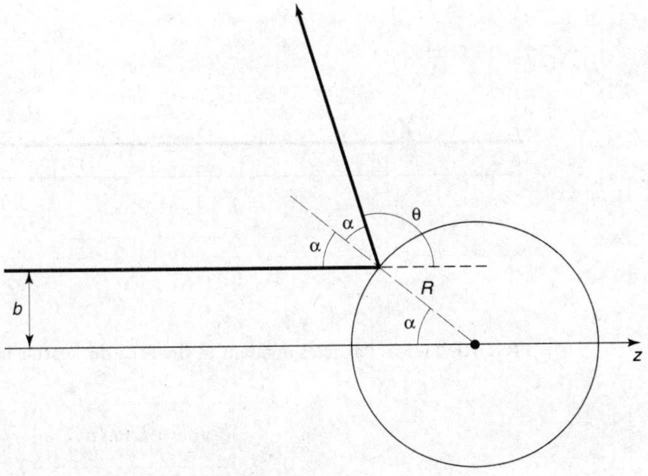

FIGURE 11.2: Elastic hard-sphere scattering.

Example 11.1 Hard-sphere scattering. Suppose the target is a billiard ball, of radius R, and the incident particle is a BB, which bounces off elastically (Figure 11.2). In terms of the angle α, the impact parameter is $b = R\sin\alpha$, and the scattering angle is $\theta = \pi - 2\alpha$, so

$$b = R\sin\left(\frac{\pi}{2} - \frac{\theta}{2}\right) = R\cos\left(\frac{\theta}{2}\right). \qquad [11.1]$$

Evidently

$$\theta = \begin{cases} 2\cos^{-1}(b/R), & \text{if } b \le R, \\ 0, & \text{if } b \ge R. \end{cases} \qquad [11.2]$$

More generally, particles incident within an infinitesimal patch of cross-sectional area $d\sigma$ will scatter into a corresponding infinitesimal solid angle $d\Omega$ (Figure 11.3). The larger $d\sigma$ is, the bigger $d\Omega$ will be; the proportionality factor, $D(\theta) \equiv d\sigma/d\Omega$, is called the **differential (scattering) cross-section**:[1]

[1] This is terrible language: D isn't a *differential*, and it isn't a cross-section. To my ear, the words "differential cross-section" would attach more naturally to $d\sigma$. But I'm afraid we're stuck with this terminology. I should also warn you that the notation $D(\theta)$ is nonstandard: Most people just call it $d\sigma/d\Omega$—which makes Equation 11.3 look like a tautology. I think it will be less confusing if we give the differential cross-section its own symbol.

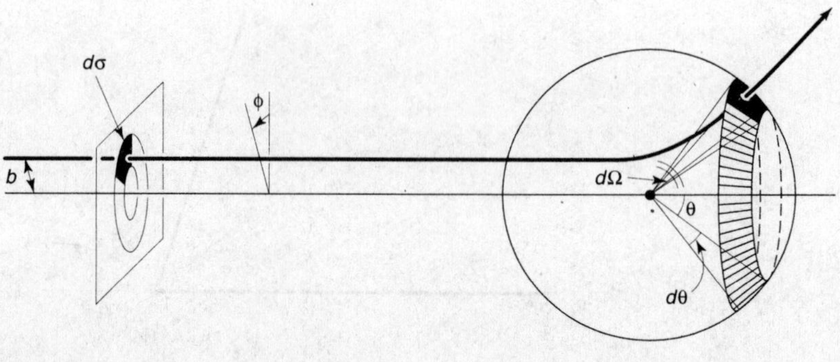

FIGURE 11.3: Particles incident in the area $d\sigma$ scatter into the solid angle $d\Omega$.

$$\boxed{d\sigma = D(\theta)\,d\Omega.} \qquad [11.3]$$

In terms of the impact parameter and the azimuthal angle ϕ, $d\sigma = b\,db\,d\phi$ and $d\Omega = \sin\theta\,d\theta\,d\phi$, so

$$D(\theta) = \frac{b}{\sin\theta}\left|\frac{db}{d\theta}\right|. \qquad [11.4]$$

(Since θ is typically a *decreasing* function of b, the derivative is actually negative—hence the absolute value sign.)

Example 11.2 Hard-sphere scattering (continued). In the case of hard-sphere scattering (Example 11.1)

$$\frac{db}{d\theta} = -\frac{1}{2}R\sin\left(\frac{\theta}{2}\right), \qquad [11.5]$$

so

$$D(\theta) = \frac{R\cos(\theta/2)}{\sin\theta}\left(\frac{R\sin(\theta/2)}{2}\right) = \frac{R^2}{4}. \qquad [11.6]$$

This example is unusual, in that the differential cross-section is independent of θ.

The **total cross-section** is the *integral* of $D(\theta)$, over all solid angles:

$$\boxed{\sigma \equiv \int D(\theta)\,d\Omega;} \qquad [11.7]$$

roughly speaking, it is the total area of incident beam that is scattered by the target. For example, in the case of hard-sphere scattering,

$$\sigma = (R^2/4)\int d\Omega = \pi R^2, \qquad [11.8]$$

which is just what we would expect: It's the cross-sectional area of the sphere; BB's incident within this area will hit the target, and those farther out will miss it completely. But the virtue of the formalism developed here is that it applies just as well to "soft" targets (such as the Coulomb field of a nucleus) that are *not* simply "hit-or-miss."

Finally, suppose we have a *beam* of incident particles, with uniform intensity (or **luminosity**, as particle physicists call it)

$$\mathcal{L} \equiv \text{number of incident particles per unit area, per unit time.} \qquad [11.9]$$

The number of particles entering area $d\sigma$ (and hence scattering into solid angle $d\Omega$), per unit time, is $dN = \mathcal{L}\, d\sigma = \mathcal{L}\, D(\theta)\, d\Omega$, so

$$D(\theta) = \frac{1}{\mathcal{L}}\frac{dN}{d\Omega}. \qquad [11.10]$$

This is often taken as the *definition* of the differential cross-section, because it makes reference only to quantities easily measured in the laboratory: If the detector accepts particles scattering into a solid angle $d\Omega$, we simply count the *number* recorded, per unit time, divide by $d\Omega$, and normalize to the luminosity of the incident beam.

$***$**Problem 11.1 Rutherford scattering.** An incident particle of charge q_1 and kinetic energy E scatters off a heavy stationary particle of charge q_2.

(a) Derive the formula relating the impact parameter to the scattering angle.[2] *Answer:* $b = (q_1 q_2/8\pi\epsilon_0 E)\cot(\theta/2)$.

(b) Determine the differential scattering cross-section. *Answer:*

$$D(\theta) = \left[\frac{q_1 q_2}{16\pi\epsilon_0 E \sin^2(\theta/2)}\right]^2. \qquad [11.11]$$

(c) Show that the total cross-section for Rutherford scattering is *infinite*. We say that the $1/r$ potential has "infinite range"; you can't escape from a Coulomb force.

---

[2]This isn't easy; and you might want to refer to a book on classical mechanics, such as Jerry B. Marion and Stephen T. Thornton, *Classical Dynamics of Particles and Systems*, 4th ed., Saunders, Fort Worth, TX (1995), Section 9.10.

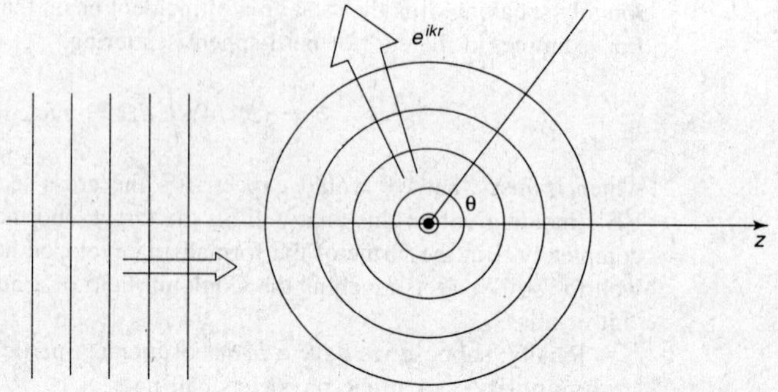

FIGURE 11.4: Scattering of waves; incoming plane wave generates outgoing spherical wave.

11.1.2 Quantum Scattering Theory

In the quantum theory of scattering, we imagine an incident *plane* wave, $\psi(z) = Ae^{ikz}$, traveling in the z direction, which encounters a scattering potential, producing an outgoing *spherical* wave (Figure 11.4).[3] That is, we look for solutions to the Schrödinger equation of the general form

$$\boxed{\psi(r, \theta) \approx A\left\{e^{ikz} + f(\theta)\frac{e^{ikr}}{r}\right\}, \quad \text{for large } r.} \qquad [11.12]$$

(The spherical wave carries a factor of $1/r$, because this portion of $|\psi|^2$ must go like $1/r^2$ to conserve probability.) The **wave number** k is related to the energy of the incident particles in the usual way:

$$k \equiv \frac{\sqrt{2mE}}{\hbar}. \qquad [11.13]$$

As before, I shall assume the target is azimuthally symmetrical; in the more general case the amplitude f of the outgoing spherical wave could depend on ϕ as well as θ.

---

[3]For the moment, there's not much *quantum* mechanics in this; what we're really talking about is the scattering of *waves*, as opposed to classical *particles*, and you could even think of Figure 11.4 as a picture of water waves encountering a rock, or (better, since we're interested in three-dimensional scattering) sound waves bouncing off a basketball. In that case we'd write the wave function in the *real* form

$$A\left[\cos(kz) + f(\theta)\cos(kr + \delta)/r\right],$$

and $f(\theta)$ would represent the amplitude of the scattered sound wave in the direction θ.

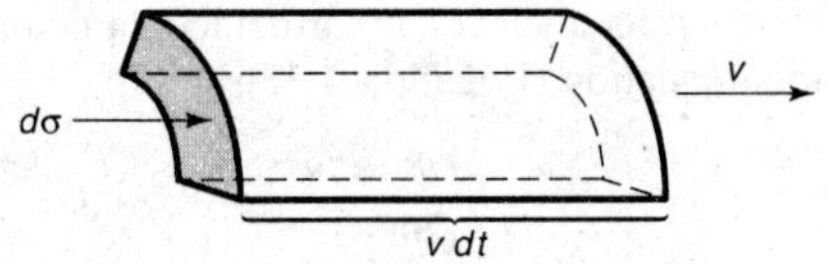

FIGURE 11.5: The volume dV of incident beam that passes through area $d\sigma$ in time dt.

The whole problem is to determine the **scattering amplitude** $f(\theta)$; it tells you the *probability of scattering in a given direction* θ, and hence is related to the differential cross-section. Indeed, the probability that the incident particle, traveling at speed v, passes through the infinitesimal area $d\sigma$, in time dt, is (see Figure 11.5)

$$dP = |\psi_{\text{incident}}|^2\, dV = |A|^2 (v\, dt)\, d\sigma.$$

But this is equal to the probability that the particle scatters into the corresponding solid angle $d\Omega$:

$$dP = |\psi_{\text{scattered}}|^2\, dV = \frac{|A|^2 |f|^2}{r^2} (v\, dt) r^2\, d\Omega,$$

from which it follows that $d\sigma = |f|^2\, d\Omega$, and hence

$$\boxed{D(\theta) = \frac{d\sigma}{d\Omega} = |f(\theta)|^2.} \qquad [11.14]$$

Evidently the differential cross-section (which is the quantity of interest to the experimentalist) is equal to the absolute square of the scattering amplitude (which is obtained by solving the Schrödinger equation). In the following sections we will study two techniques for calculating the scattering amplitude: **partial wave analysis** and the **Born approximation**.

Problem 11.2 Construct the analogs to Equation 11.12 for one-dimensional and two-dimensional scattering.

11.2 PARTIAL WAVE ANALYSIS

11.2.1 Formalism

As we found in Chapter 4, the Schrödinger equation for a spherically symmetrical potential $V(r)$ admits the separable solutions

$$\psi(r, \theta, \phi) = R(r) Y_l^m(\theta, \phi), \qquad [11.15]$$

where Y_l^m is a spherical harmonic (Equation 4.32), and $u(r) = rR(r)$ satisfies the radial equation (Equation 4.37):

$$-\frac{\hbar^2}{2m}\frac{d^2u}{dr^2} + \left[V(r) + \frac{\hbar^2}{2m}\frac{l(l+1)}{r^2}\right]u = Eu. \qquad [11.16]$$

At *very* large r the potential goes to zero, and the centrifugal contribution is negligible, so

$$\frac{d^2u}{dr^2} \approx -k^2 u.$$

The general solution is

$$u(r) = Ce^{ikr} + De^{-ikr};$$

the first term represents an *outgoing* spherical wave, and the second an *incoming* one—for the scattered wave we evidently want $D = 0$. At very large r, then,

$$R(r) \sim \frac{e^{ikr}}{r},$$

as we already deduced (on physical grounds) in the previous section (Equation 11.12).

That's for *very* large r (more precisely, for $kr \gg 1$; in optics it would be called the **radiation zone**). As in one-dimensional scattering theory, we assume that the potential is "localized," in the sense that exterior to some finite scattering region it is essentially zero (Figure 11.6). In the intermediate region (where V can be ignored but the centrifugal term cannot),[4] the radial equation becomes

$$\frac{d^2u}{dr^2} - \frac{l(l+1)}{r^2}u = -k^2u, \qquad [11.17]$$

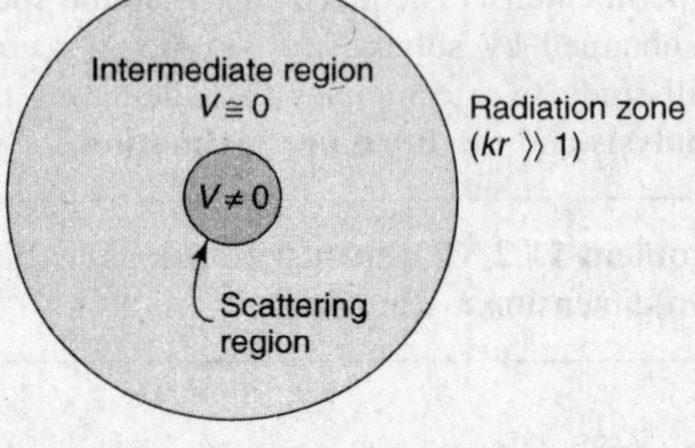

FIGURE 11.6: Scattering from a localized potential: the scattering region (darker shading), the intermediate region (lighter shading), and the radiation zone (where $kr \gg 1$).

[4]What follows does not apply to the Coulomb potential, since $1/r$ goes to zero more slowly than $1/r^2$, as $r \to \infty$, and the centrifugal term does *not* dominate in this region. In this sense the Coulomb potential is not localized, and partial wave analysis is inapplicable.

TABLE 11.1: Spherical Hankel functions, $h_l^{(1)}(x)$ and $h_l^{(2)}(x)$.

$$h_0^{(1)} = -i\frac{e^{ix}}{x} \qquad h_0^{(2)} = i\frac{e^{-ix}}{x}$$

$$h_1^{(1)} = \left(-\frac{i}{x^2} - \frac{1}{x}\right)e^{ix} \qquad h_1^{(2)} = \left(\frac{i}{x^2} - \frac{1}{x}\right)e^{-ix}$$

$$h_2^{(1)} = \left(-\frac{3i}{x^3} - \frac{3}{x^2} + \frac{i}{x}\right)e^{ix} \qquad h_2^{(2)} = \left(\frac{3i}{x^3} - \frac{3}{x^2} + \frac{i}{x}\right)e^{-ix}$$

$$\left.\begin{aligned} h_l^{(1)} &\to \frac{1}{x}(-i)^{l+1}e^{ix} \\ h_l^{(2)} &\to \frac{1}{x}(i)^{l+1}e^{-ix} \end{aligned}\right\} \text{for } x \gg 1$$

and the general solution (Equation 4.45) is a linear combination of spherical Bessel functions:

$$u(r) = Arj_l(kr) + Brn_l(kr). \qquad [11.18]$$

However, neither j_l (which is somewhat like a sine function) nor n_l (which is a sort of generalized cosine function) represents an outgoing (or an incoming) wave. What we need are the linear combinations analogous to e^{ikr} and e^{-ikr}; these are known as **spherical Hankel functions**:

$$h_l^{(1)}(x) \equiv j_l(x) + in_l(x); \quad h_l^{(2)}(x) \equiv j_l(x) - in_l(x). \qquad [11.19]$$

The first few spherical Hankel functions are listed in Table 11.1. At large r, $h_l^{(1)}(kr)$ (the "Hankel function of the first kind") goes like e^{ikr}/r, whereas $h_l^{(2)}(kr)$ (the "Hankel function of the second kind") goes like e^{-ikr}/r; for outgoing waves, then, we need *spherical Hankel functions of the first kind:*

$$R(r) \sim h_l^{(1)}(kr). \qquad [11.20]$$

Thus the exact wave function, outside the scattering region (where $V(r) = 0$), is

$$\psi(r, \theta, \phi) = A\left\{e^{ikz} + \sum_{l,m} C_{l,m}\, h_l^{(1)}(kr)\, Y_l^m(\theta, \phi)\right\}. \qquad [11.21]$$

The first term is the incident plane wave, and the sum (with expansion coefficients $C_{l,m}$) represents the scattered wave. But since we are assuming the potential is spherically symmetric, the wave function cannot depend on ϕ.[5] So only terms with

[5]There's nothing wrong with θ dependence, of course, because the incoming plane wave defines a z direction, breaking the spherical symmetry. But the *azimuthal* symmetry remains; the incident plane wave has no ϕ dependence, and there is nothing in the scattering process that could introduce any ϕ dependence in the outgoing wave.

$m = 0$ survive (remember, $Y_l^m \sim e^{im\phi}$). Now (from Equations 4.27 and 4.32)

$$Y_l^0(\theta, \phi) = \sqrt{\frac{2l+1}{4\pi}}\, P_l(\cos\theta), \qquad [11.22]$$

where P_l is the lth Legendre polynomial. It is customary to redefine the expansion coefficients, letting $C_{l,0} \equiv i^{l+1} k \sqrt{4\pi(2l+1)}\, a_l$:

$$\boxed{\psi(r, \theta) = A \left\{ e^{ikz} + k \sum_{l=0}^{\infty} i^{l+1} (2l+1)\, a_l\, h_l^{(1)}(kr)\, P_l(\cos\theta) \right\}.} \qquad [11.23]$$

You'll see in a moment why this peculiar notation is convenient; a_l is called the lth **partial wave amplitude**.

Now, for *very large* r the Hankel function goes like $(-i)^{l+1} e^{ikr}/kr$ (Table 11.1), so

$$\psi(r, \theta) \approx A \left\{ e^{ikz} + f(\theta) \frac{e^{ikr}}{r} \right\}, \qquad [11.24]$$

where

$$f(\theta) = \sum_{l=0}^{\infty} (2l+1)\, a_l\, P_l(\cos\theta). \qquad [11.25]$$

This confirms more rigorously the general structure postulated in Equation 11.12, and tells us how to compute the scattering amplitude, $f(\theta)$, in terms of the partial wave amplitudes (a_l). The differential cross-section is

$$D(\theta) = |f(\theta)|^2 = \sum_l \sum_{l'} (2l+1)(2l'+1)\, a_l^* a_{l'}\, P_l(\cos\theta)\, P_{l'}(\cos\theta), \qquad [11.26]$$

and the total cross-section is

$$\sigma = 4\pi \sum_{l=0}^{\infty} (2l+1)\, |a_l|^2. \qquad [11.27]$$

(I used the orthogonality of the Legendre polynomials, Equation 4.34, to do the angular integration.)

11.2.2 Strategy

All that remains is to determine the partial wave amplitudes, a_l, for the potential in question. This is accomplished by solving the Schrödinger equation in the *interior* region (where $V(r)$ is distinctly *non*-zero), and matching this to the exterior solution (Equation 11.23), using the appropriate boundary conditions. The only problem is that as it stands my notation is hybrid: I used *spherical* coordinates for the scattered wave, but *cartesian* coordinates for the incident wave. We need to rewrite the wave function in a more consistent notation.

Of course, e^{ikz} satisfies the Schrödinger equation with $V = 0$. On the other hand, I just argued that the *general* solution to the Schrödinger equation with $V = 0$ can be written in the form

$$\sum_{l,m} \left[A_{l,m}\, j_l(kr) + B_{l,m}\, n_l(kr)\right] Y_l^m(\theta, \phi).$$

In particular, then, it must be possible to express e^{ikz} in this way. But e^{ikz} is finite at the origin, so no Neumann functions are allowed ($n_l(kr)$ blows up at $r = 0$), and since $z = r\cos\theta$ has no ϕ dependence, only $m = 0$ terms occur. The explicit expansion of a plane wave in terms of spherical waves is known as **Rayleigh's formula**:[6]

$$e^{ikz} = \sum_{l=0}^{\infty} i^l (2l+1)\, j_l(kr) P_l(\cos\theta). \qquad [11.28]$$

Using this, the wave function in the exterior region can be expressed entirely in terms of r and θ:

$$\psi(r,\theta) = A \sum_{l=0}^{\infty} i^l (2l+1) \left[j_l(kr) + ik\, a_l\, h_l^{(1)}(kr)\right] P_l(\cos\theta). \qquad [11.29]$$

Example 11.3 Quantum hard-sphere scattering. Suppose

$$V(r) = \begin{cases} \infty, & \text{for } r \le a, \\ 0, & \text{for } r > a. \end{cases} \qquad [11.30]$$

The boundary condition, then, is

$$\psi(a,\theta) = 0, \qquad [11.31]$$

so

$$\sum_{l=0}^{\infty} i^l (2l+1) \left[j_l(ka) + ik\, a_l\, h_l^{(1)}(ka)\right] P_l(\cos\theta) = 0 \qquad [11.32]$$

for all θ, from which it follows (Problem 11.3) that

$$a_l = -i \frac{j_l(ka)}{k\, h_l^{(1)}(ka)}. \qquad [11.33]$$

---

[6] For a guide to the proof, see George Arfken and Hans-Jurgen Weber, *Mathematical Methods for Physicists*, 5th ed., Academic Press, Orlando (2000), Exercises 12.4.7 and 12.4.8.

In particular, the total cross-section is

$$\sigma = \frac{4\pi}{k^2}\sum_{l=0}^{\infty}(2l+1)\left|\frac{j_l(ka)}{h_l^{(1)}(ka)}\right|^2. \qquad [11.34]$$

That's the *exact* answer, but it's not terribly illuminating, so let's consider the limiting case of *low-energy scattering*: $ka \ll 1$. (Since $k = 2\pi/\lambda$, this amounts to saying that the wavelength is much greater than the radius of the sphere.) Referring to Table 4.4, we note that $n_l(z)$ is much larger than $j_l(z)$, for small z, so

$$\frac{j_l(z)}{h_l^{(1)}(z)} = \frac{j_l(z)}{j_l(z) + i n_l(z)} \approx -i\frac{j_l(z)}{n_l(z)}$$
$$\approx -i\frac{2^l l! z^l/(2l+1)!}{-(2l)! z^{-l-1}/2^l l!} = \frac{i}{2l+1}\left[\frac{2^l l!}{(2l)!}\right]^2 z^{2l+1}, \qquad [11.35]$$

and hence

$$\sigma \approx \frac{4\pi}{k^2}\sum_{l=0}^{\infty}\frac{1}{2l+1}\left[\frac{2^l l!}{(2l)!}\right]^4 (ka)^{4l+2}.$$

But we're assuming $ka \ll 1$, so the higher powers are negligible—in the low energy approximation the scattering is dominated by the $l = 0$ term. (This means that the differential cross-section is independent of θ, just as it was in the classical case.) Evidently

$$\sigma \approx 4\pi a^2, \qquad [11.36]$$

for low energy hard-sphere scattering. Surprisingly, the scattering cross-section is *four times* the geometrical cross-section—in fact, σ is the *total surface area of the sphere*. This "larger effective size" is characteristic of long-wavelength scattering (it would be true in optics, as well); in a sense, these waves "feel" their way around the whole sphere, whereas classical *particles* only see the head-on cross-section.

Problem 11.3 Prove Equation 11.33, starting with Equation 11.32. *Hint:* Exploit the orthogonality of the Legendre polynomials to show that the coefficients with different values of l must separately vanish.

∗∗**Problem 11.4** Consider the case of low-energy scattering from a spherical delta-function shell:

$$V(r) = \alpha\delta(r-a),$$

where α and a are constants. Calculate the scattering amplitude, $f(\theta)$, the differential cross-section, $D(\theta)$, and the total cross-section, σ. Assume $ka \ll 1$, so that only the $l = 0$ term contributes significantly. (To simplify matters, throw out all $l \neq 0$ terms right from the start.) The main problem, of course, is to determine a_0. Express your answer in terms of the dimensionless quantity $\beta \equiv 2ma\alpha/\hbar^2$. *Answer:* $\sigma = 4\pi a^2\beta^2/(1+\beta)^2$.

11.3 PHASE SHIFTS

Consider first the problem of *one*-dimensional scattering from a localized potential $V(x)$ on the half-line $x < 0$ (Figure 11.7). I'll put a "brick wall" at $x = 0$, so a wave incident from the left,

$$\psi_i(x) = Ae^{ikx} \quad (x < -a) \tag{11.37}$$

is entirely reflected

$$\psi_r(x) = Be^{-ikx} \quad (x < -a). \tag{11.38}$$

Whatever happens in the interaction region ($-a < x < 0$), the amplitude of the reflected wave has *got* to be the same as that of the incident wave, by conservation of probability. But it need not have the same *phase*. If there were no potential at all (just the wall at $x = 0$), then $B = -A$, since the total wave function (incident plus reflected) must vanish at the origin:

$$\psi_0(x) = A\left(e^{ikx} - e^{-ikx}\right) \quad (V(x) = 0). \tag{11.39}$$

If the potential is *not* zero, the wave function (for $x < -a$) takes the form

$$\psi(x) = A\left(e^{ikx} - e^{i(2\delta - kx)}\right) \quad (V(x) \neq 0). \tag{11.40}$$

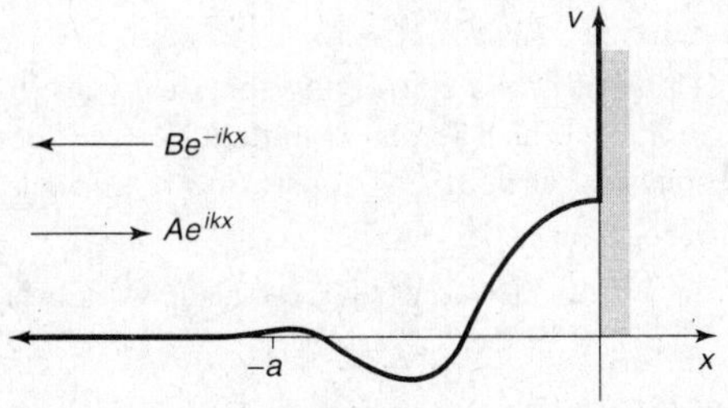

FIGURE 11.7: One-dimensional scattering from a localized potential bounded on the right by an infinite wall.

The whole theory of scattering reduces to the problem of calculating the **phase shift**[7] δ (as a function of k, and hence of the energy $E = \hbar^2 k^2/2m$), for a specified potential. We do this, of course, by solving the Schrödinger equation in the scattering region ($-a < x < 0$), and imposing appropriate boundary conditions (see Problem 11.5). The virtue of working with the phase shift (as opposed to the complex amplitude B) is that it illuminates the physics (because of conservation of probability, *all* the potential can do is shift the phase of the reflected wave) and simplifies the mathematics (trading a complex quantity—two real numbers—for a single real quantity).

Now let's return to the three-dimensional case. The incident plane wave (Ae^{ikz}) carries no angular momentum in the z direction (Rayleigh's formula contains no terms with $m \neq 0$), but it includes all values of the *total* angular momentum ($l = 0, 1, 2, \ldots$). Because angular momentum is conserved (by a spherically symmetric potential), each **partial wave** (labelled by a particular l) scatters independently, with (again) no change in amplitude[8]—only in phase. If there is no potential at all, then $\psi_0 = Ae^{ikz}$, and the lth partial wave is (Equation 11.28)

$$\psi_0^{(l)} = Ai^l(2l+1)\, j_l(kr)\, P_l(\cos\theta) \quad (V(r) = 0). \qquad [11.41]$$

But (from Equation 11.19 and Table 11.1)

$$j_l(x) = \frac{1}{2}\left[h^{(1)}(x) + h_l^{(2)}(x)\right] \approx \frac{1}{2x}\left[(-i)^{l+1}e^{ix} + i^{l+1}e^{-ix}\right] \quad (x \gg 1). \qquad [11.42]$$

So for large r

$$\psi_0^{(l)} \approx A\frac{(2l+1)}{2ikr}\left[e^{ikr} - (-1)^l e^{-ikr}\right] P_l(\cos\theta) \quad (V(r) = 0). \qquad [11.43]$$

The second term in square brackets represents an incoming spherical wave; it is unchanged when we introduce the scattering potential. The first term is the outgoing wave; it picks up a phase shift δ_l:

$$\psi^{(l)} \approx A\frac{(2l+1)}{2ikr}\left[e^{i(kr+2\delta_l)} - (-1)^l e^{-ikr}\right] P_l(\cos\theta) \quad (V(r) \neq 0). \qquad [11.44]$$

Think of it as a converging spherical wave (due exclusively to the $h_l^{(2)}$ component in e^{ikz}), which is phase shifted $2\delta_l$ (see footnote 7) and emerges as an outgoing spherical wave (the $h_l^{(1)}$ part of e^{ikz} as well as the scattered wave itself).

---

[7]The 2 in front of δ is conventional. We think of the incident wave as being phase shifted once on the way in, and again on the way out; by δ we mean the "one way" phase shift, and the *total* is therefore 2δ.

[8]One reason this subject can be so confusing is that practically everything is called an "amplitude:" $f(\theta)$ is the "scattering amplitude," a_l is the "partial wave amplitude," but the first is a function of θ, and both are complex numbers. I'm *now* talking about "amplitude" in the original sense: the (*real*, of course) height of a sinusoidal wave.

In Section 11.2.1 the whole theory was expressed in terms of the partial wave amplitudes a_l; now we have formulated it in terms of the phase shifts δ_l. There must be a connection between the two. Indeed, comparing the asymptotic (large r) form of Equation 11.23

$$\psi^{(l)} \approx A\left\{\frac{(2l+1)}{2ikr}\left[e^{ikr} - (-1)^l e^{-ikr}\right] + \frac{(2l+1)}{r} a_l\, e^{ikr}\right\} P_l(\cos\theta) \qquad [11.45]$$

with the generic expression in terms of δ_l (Equation 11.44), we find[9]

$$a_l = \frac{1}{2ik}\left(e^{2i\delta_l} - 1\right) = \frac{1}{k} e^{i\delta_l} \sin(\delta_l). \qquad [11.46]$$

It follows in particular (Equation 11.25) that

$$f(\theta) = \frac{1}{k}\sum_{l=0}^{\infty}(2l+1)e^{i\delta_l}\sin(\delta_l)\, P_l(\cos\theta) \qquad [11.47]$$

and (Equation 11.27)

$$\sigma = \frac{4\pi}{k^2}\sum_{l=0}^{\infty}(2l+1)\sin^2(\delta_l). \qquad [11.48]$$

Again, the advantage of working with phase shifts (as opposed to partial wave amplitudes) is that they are easier to interpret physically, and simpler mathematically—the phase shift formalism exploits conservation of angular momentum to reduce a complex quantity a_l (two real numbers) to a single real one δ_l.

Problem 11.5 A particle of mass m and energy E is incident from the left on the potential

$$V(x) = \begin{cases} 0, & (x < -a), \\ -V_0, & (-a \le x \le 0), \\ \infty, & (x > 0). \end{cases}$$

(a) If the incoming wave is Ae^{ikx} (where $k = \sqrt{2mE}/\hbar$), find the reflected wave. *Answer:*

$$Ae^{-2ika}\left[\frac{k - ik'\cot(k'a)}{k + ik'\cot(k'a)}\right]e^{-ikx}, \quad \text{where } k' = \sqrt{2m(E+V_0)}/\hbar.$$

[9] Although I used the asymptotic form of the wave function to draw the connection between a_l and δ_l, there is nothing approximate about the result (Equation 11.46). Both of them are *constants* (independent of r), and δ_l *means* the phase shift in the asymptotic region (where the Hankel functions have settled down to $e^{\pm ikr}/kr$).

(b) Confirm that the reflected wave has the same amplitude as the incident wave.

(c) Find the phase shift δ (Equation 11.40) for a very deep well ($E \ll V_0$). *Answer:* $\delta = -ka$.

Problem 11.6 What are the partial wave phase shifts (δ_l) for hard-sphere scattering (Example 11.3)?

Problem 11.7 Find the S-wave ($l = 0$) partial wave phase shift $\delta_0(k)$ for scattering from a delta-function shell (Problem 11.4). Assume that the radial wave function $u(r)$ goes to 0 as $r \to \infty$. *Answer:*

$$-\cot^{-1}\left[\cot(ka) + \frac{ka}{\beta \sin^2(ka)}\right], \quad \text{where } \beta \equiv \frac{2m\alpha a}{\hbar^2}$$

11.4 THE BORN APPROXIMATION

11.4.1 Integral Form of the Schrödinger Equation

The time-independent Schrödinger equation,

$$-\frac{\hbar^2}{2m}\nabla^2\psi + V\psi = E\psi, \qquad [11.49]$$

can be written more succinctly as

$$(\nabla^2 + k^2)\psi = Q, \qquad [11.50]$$

where

$$k \equiv \frac{\sqrt{2mE}}{\hbar} \quad \text{and} \quad Q \equiv \frac{2m}{\hbar^2}V\psi \qquad [11.51]$$

This has the superficial form of the **Helmholtz equation**; note, however, that the "inhomogeneous" term (Q) *itself* depends on ψ.

Suppose we could find a function $G(\mathbf{r})$ that solves the Helmholtz equation with a *delta function* "source:"

$$(\nabla^2 + k^2)G(\mathbf{r}) = \delta^3(\mathbf{r}). \qquad [11.52]$$

Then we could express ψ as an integral:

$$\psi(\mathbf{r}) = \int G(\mathbf{r} - \mathbf{r}_0)Q(\mathbf{r}_0)\, d^3\mathbf{r}_0. \qquad [11.53]$$

For it is easy to show that this satisfies Schrödinger's equation, in the form of Equation 11.50:

$$(\nabla^2 + k^2)\psi(\mathbf{r}) = \int \left[(\nabla^2 + k^2)G(\mathbf{r} - \mathbf{r}_0)\right] Q(\mathbf{r}_0)\, d^3\mathbf{r}_0$$

$$= \int \delta^3(\mathbf{r} - \mathbf{r}_0) Q(\mathbf{r}_0)\, d^3\mathbf{r}_0 = Q(\mathbf{r}).$$

$G(\mathbf{r})$ is called the **Green's function** for the Helmholtz equation. (In general, the Green's function for a linear differential equation represents the "response" to a delta-function source.)

Our first task[10] is to solve Equation 11.52 for $G(\mathbf{r})$. This is most easily accomplished by taking the Fourier transform, which turns the *differential* equation into an *algebraic* equation. Let

$$G(\mathbf{r}) = \frac{1}{(2\pi)^{3/2}} \int e^{i\mathbf{s}\cdot\mathbf{r}} g(\mathbf{s})\, d^3\mathbf{s}. \qquad [11.54]$$

Then

$$(\nabla^2 + k^2)G(\mathbf{r}) = \frac{1}{(2\pi)^{3/2}} \int \left[(\nabla^2 + k^2)e^{i\mathbf{s}\cdot\mathbf{r}}\right] g(\mathbf{s})\, d^3\mathbf{s}.$$

But

$$\nabla^2 e^{i\mathbf{s}\cdot\mathbf{r}} = -s^2 e^{i\mathbf{s}\cdot\mathbf{r}}, \qquad [11.55]$$

and (see Equation 2.144)

$$\delta^3(\mathbf{r}) = \frac{1}{(2\pi)^3} \int e^{i\mathbf{s}\cdot\mathbf{r}}\, d^3\mathbf{s}, \qquad [11.56]$$

so Equation 11.52 says

$$\frac{1}{(2\pi)^{3/2}} \int (-s^2 + k^2) e^{i\mathbf{s}\cdot\mathbf{r}} g(\mathbf{s})\, d^3\mathbf{s} = \frac{1}{(2\pi)^3} \int e^{i\mathbf{s}\cdot\mathbf{r}}\, d^3\mathbf{s}.$$

It follows[11] that

$$g(\mathbf{s}) = \frac{1}{(2\pi)^{3/2}(k^2 - s^2)}. \qquad [11.57]$$

Putting this back into Equation 11.54, we find:

$$G(\mathbf{r}) = \frac{1}{(2\pi)^3} \int e^{i\mathbf{s}\cdot\mathbf{r}} \frac{1}{(k^2 - s^2)}\, d^3\mathbf{s}. \qquad [11.58]$$

---

[10] *Warning:* You are approaching two pages of heavy analysis, including contour integration; if you wish, skip straight to the answer, Equation 11.65.

[11] This is clearly *sufficient*, but it is also *necessary*, as you can show by combining the two terms into a single integral, and using Plancherel's theorem, Equation 2.102.

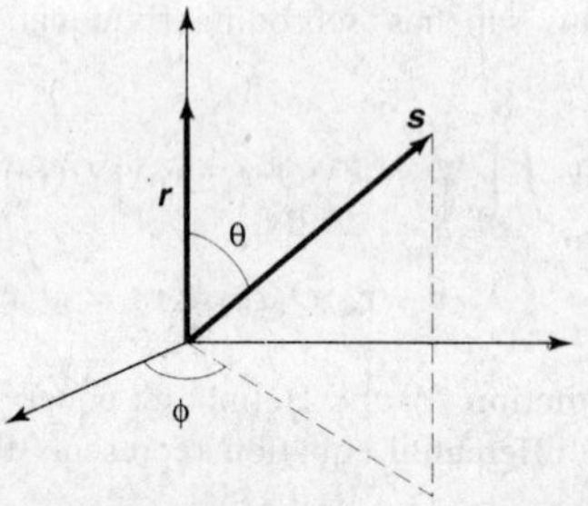

FIGURE 11.8: Convenient coordinates for the integral in Equation 11.58.

Now, **r** is *fixed*, as far as the **s** integration is concerned, so we may as well choose spherical coordinates (s, θ, ϕ) with the polar axis along **r** (Figure 11.8). Then $\mathbf{s}\cdot\mathbf{r} = sr\cos\theta$, the ϕ integral is trivial (2π), and the θ integral is

$$\int_0^{\pi} e^{isr\cos\theta}\sin\theta\, d\theta = -\left.\frac{e^{isr\cos\theta}}{isr}\right|_0^{\pi} = \frac{2\sin(sr)}{sr}. \qquad [11.59]$$

Thus

$$G(\mathbf{r}) = \frac{1}{(2\pi)^2}\frac{2}{r}\int_0^{\infty}\frac{s\sin(sr)}{k^2-s^2}ds = \frac{1}{4\pi^2 r}\int_{-\infty}^{\infty}\frac{s\sin(sr)}{k^2-s^2}ds. \qquad [11.60]$$

The remaining integral is not so simple. It pays to revert to exponential notation, and factor the denominator:

$$G(\mathbf{r}) = \frac{i}{8\pi^2 r}\left\{\int_{-\infty}^{\infty}\frac{se^{isr}}{(s-k)(s+k)}ds - \int_{-\infty}^{\infty}\frac{se^{-isr}}{(s-k)(s+k)}ds\right\}$$
$$= \frac{i}{8\pi^2 r}(I_1 - I_2) \qquad [11.61]$$

These two integrals can be evaluated using **Cauchy's integral formula**:

$$\oint \frac{f(z)}{(z-z_0)}dz = 2\pi i f(z_0), \qquad [11.62]$$

if z_0 lies within the contour (otherwise the integral is zero). In the present case the integration is along the real axis, and it passes *right over* the pole singularities at $\pm k$. We have to decide how to skirt the poles—I'll go *over* the one at $-k$ and *under* the one at $+k$ (Figure 11.9). (You're welcome to choose some *other* convention if you like—even winding seven times around each pole—you'll get a different Green's function, but, as I'll show you in a minute, they're all equally acceptable.)

For each integral in Equation 11.61 we must "close the contour" in such a way that the semicircle at infinity contributes nothing. In the case of I_1, the factor

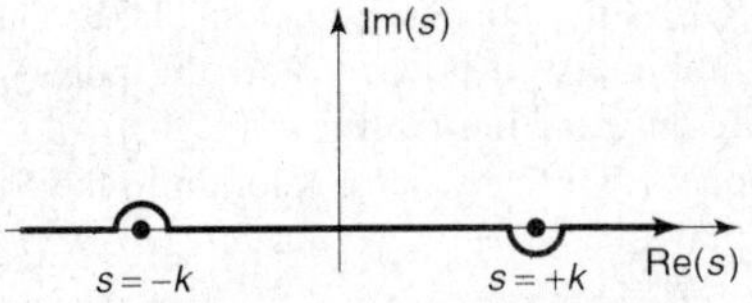

FIGURE 11.9: Skirting the poles in the contour integral (Equation 11.61).

e^{isr} goes to zero when s has a large *positive* imaginary part; for this one we close *above* (Figure 11.10(a)). The contour encloses only the singularity at $s = +k$, so

$$I_1 = \oint \left[\frac{se^{isr}}{s+k}\right]\frac{1}{s-k}\,ds = 2\pi i\left[\frac{se^{isr}}{s+k}\right]\bigg|_{s=k} = i\pi e^{ikr} \qquad [11.63]$$

In the case of I_2, the factor e^{-isr} goes to zero when s has a large *negative* imaginary part, so we close *below* (Figure 11.10(b)); this time the contour encloses the singularity at $s = -k$ (and it goes around in the *clockwise* direction, so we pick up a minus sign):

$$I_2 = -\oint \left[\frac{se^{-isr}}{s-k}\right]\frac{1}{s+k}\,ds = -2\pi i\left[\frac{se^{-isr}}{s-k}\right]\bigg|_{s=-k} = -i\pi e^{ikr} \qquad [11.64]$$

Conclusion:

$$G(\mathbf{r}) = \frac{i}{8\pi^2 r}\left[\left(i\pi e^{ikr}\right) - \left(-i\pi e^{ikr}\right)\right] = -\frac{e^{ikr}}{4\pi r}. \qquad [11.65]$$

This, finally, is the Green's function for the Helmholtz equation—the solution to Equation 11.52. (If you got lost in all that analysis, you might want to *check* the result by direct differentiation—see Problem 11.8.) Or rather, it is *a* Green's function for the Helmholtz equation, for we can add to $G(\mathbf{r})$ any function $G_0(\mathbf{r})$ that satisfies the *homogeneous* Helmholtz equation:

$$(\nabla^2 + k^2)G_0(\mathbf{r}) = 0; \qquad [11.66]$$

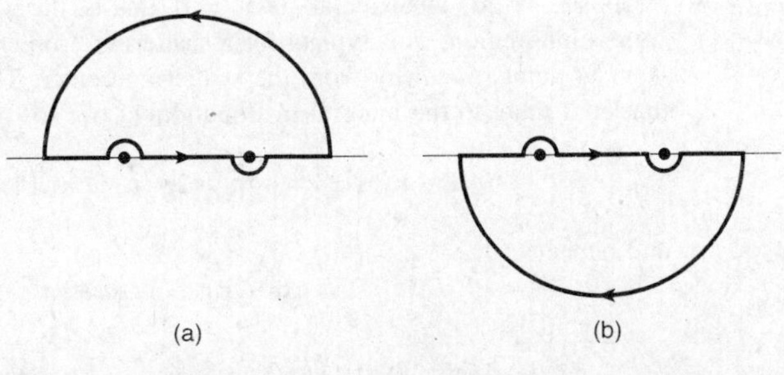

FIGURE 11.10: Closing the contour in Equations 11.63 and 11.64.

clearly, the result $(G + G_0)$ still satisfies Equation 11.52. This ambiguity corresponds precisely to the ambiguity in how to skirt the poles—a different choice amounts to picking a different function $G_0(\mathbf{r})$.

Returning to Equation 11.53, the general solution to the Schrödinger equation takes the form

$$\boxed{\psi(\mathbf{r}) = \psi_0(\mathbf{r}) - \frac{m}{2\pi\hbar^2}\int \frac{e^{ik|\mathbf{r}-\mathbf{r}_0|}}{|\mathbf{r}-\mathbf{r}_0|} V(\mathbf{r}_0)\psi(\mathbf{r}_0)\, d^3\mathbf{r}_0,} \qquad [11.67]$$

where ψ_0 satisfies the *free*-particle Schrödinger equation,

$$(\nabla^2 + k^2)\psi_0 = 0. \qquad [11.68]$$

Equation 11.67 is the **integral form of the Schrödinger equation**; it is entirely equivalent to the more familiar differential form. At first glance it *looks* like an explicit *solution* to the Schrödinger equation (for any potential)—which is too good to be true. Don't be deceived: There's a ψ under the integral sign on the right hand side, so you can't do the integral unless you already know the solution! Nevertheless, the integral form can be very powerful, and it is particularly well suited to scattering problems, as we'll see in the following section.

Problem 11.8 Check that Equation 11.65 satisfies Equation 11.52, by direct substitution. *Hint:* $\nabla^2(1/r) = -4\pi\delta^3(\mathbf{r})$.[12]

****Problem 11.9** Show that the ground state of hydrogen (Equation 4.80) satisfies the integral form of the Schrödinger equation, for the appropriate V and E (note that E is *negative*, so $k = i\kappa$, where $\kappa \equiv \sqrt{-2mE}/\hbar$).

11.4.2 The First Born Approximation

Suppose $V(\mathbf{r}_0)$ is localized about $\mathbf{r}_0 = 0$ (that is, the potential drops to zero outside some finite region, as is typical for a scattering problem), and we want to calculate $\psi(\mathbf{r})$ at points *far away* from the scattering center. Then $|\mathbf{r}| \gg |\mathbf{r}_0|$ for all points that contribute to the integral in Equation 11.67, so

$$|\mathbf{r}-\mathbf{r}_0|^2 = r^2 + r_0^2 - 2\mathbf{r}\cdot\mathbf{r}_0 \cong r^2\left(1 - 2\frac{\mathbf{r}\cdot\mathbf{r}_0}{r^2}\right), \qquad [11.69]$$

and hence

$$|\mathbf{r}-\mathbf{r}_0| \cong r - \hat{r}\cdot\mathbf{r}_0. \qquad [11.70]$$

---

[12] See, for example, D. Griffiths, *Introduction to Electrodynamics*, 3rd ed. (Prentice Hall, Upper Saddle River, NJ, 1999), Section 1.5.3.

Let

$$\mathbf{k} \equiv k\hat{r}; \tag{11.71}$$

then

$$e^{ik|\mathbf{r}-\mathbf{r}_0|} \cong e^{ikr}e^{-i\mathbf{k}\cdot\mathbf{r}_0}, \tag{11.72}$$

and therefore

$$\frac{e^{ik|\mathbf{r}-\mathbf{r}_0|}}{|\mathbf{r}-\mathbf{r}_0|} \cong \frac{e^{ikr}}{r}e^{-i\mathbf{k}\cdot\mathbf{r}_0} \tag{11.73}$$

(In the *denominator* we can afford to make the more radical approximation $|\mathbf{r}-\mathbf{r}_0| \cong r$; in the *exponent* we need to keep the next term. If this puzzles you, try writing out the next term in the expansion of the denominator. What we are doing is expanding in powers of the small quantity (r_0/r), and dropping all but the lowest order.)

In the case of scattering, we want

$$\psi_0(\mathbf{r}) = Ae^{ikz} \tag{11.74}$$

representing an incident plane wave. For large r, then,

$$\psi(\mathbf{r}) \cong Ae^{ikz} - \frac{m}{2\pi\hbar^2}\frac{e^{ikr}}{r}\int e^{-i\mathbf{k}\cdot\mathbf{r}_0}V(\mathbf{r}_0)\psi(\mathbf{r}_0)\,d^3\mathbf{r}_0. \tag{11.75}$$

This is in the standard form (Equation 11.12), and we can read off the scattering amplitude:

$$f(\theta,\phi) = -\frac{m}{2\pi\hbar^2 A}\int e^{-i\mathbf{k}\cdot\mathbf{r}_0}V(\mathbf{r}_0)\psi(\mathbf{r}_0)\,d^3\mathbf{r}_0. \tag{11.76}$$

So far, this is *exact*. Now we invoke the **Born approximation**: Suppose the incoming plane wave is *not substantially altered by the potential*; then it makes sense to use

$$\psi(\mathbf{r}_0) \approx \psi_0(\mathbf{r}_0) = Ae^{ikz_0} = Ae^{i\mathbf{k}'\cdot\mathbf{r}_0}, \tag{11.77}$$

where

$$\mathbf{k}' \equiv k\hat{z}, \tag{11.78}$$

inside the integral. (This would be the *exact* wave function, if V were zero; it is essentially a *weak potential* approximation.[13]) In the Born approximation, then,

$$\boxed{f(\theta,\phi) \cong -\frac{m}{2\pi\hbar^2}\int e^{i(\mathbf{k}'-\mathbf{k})\cdot\mathbf{r}_0}V(\mathbf{r}_0)\,d^3\mathbf{r}_0.} \tag{11.79}$$

[13]Generally, partial wave analysis is useful when the incident particle has low energy, for then only the first few terms in the series contribute significantly; the Born approximation applies when the potential is weak, compared to the incident energy, so the deflection is small.

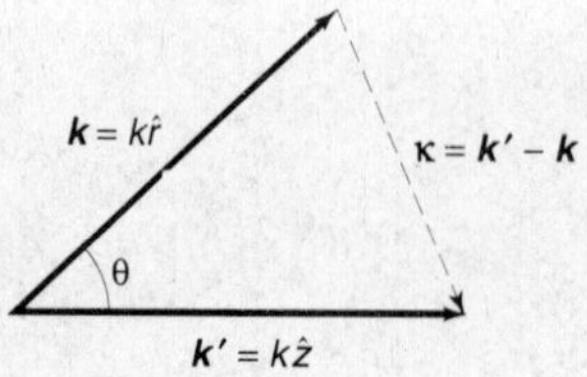

FIGURE 11.11: Two wave vectors in the Born approximation: $\mathbf{k}'$ points in the *incident* direction, $\mathbf{k}$ in the *scattered* direction.

(In case you have lost track of the definitions of $\mathbf{k}'$ and $\mathbf{k}$, they both have magnitude k, but the former points in the direction of the incident beam, while the latter points toward the detector—see Figure 11.11; $\hbar(\mathbf{k} - \mathbf{k}')$ is the **momentum transfer** in the process.)

In particular, for **low energy** (long wavelength) **scattering**, the exponential factor is essentially constant over the scattering region, and the Born approximation simplifies to

$$f(\theta, \phi) \cong -\frac{m}{2\pi\hbar^2}\int V(\mathbf{r})\, d^3\mathbf{r}, \quad \text{(low energy).} \tag{11.80}$$

(I dropped the subscript on $\mathbf{r}$, since there is no likelihood of confusion at this point.)

Example 11.4 Low-energy soft-sphere scattering.[14] Suppose

$$V(\mathbf{r}) = \begin{cases} V_0, & \text{if } r \le a, \\ 0, & \text{if } r > a. \end{cases} \tag{11.81}$$

In this case the low-energy scattering amplitude is

$$f(\theta, \phi) \cong -\frac{m}{2\pi\hbar^2} V_0 \left(\frac{4}{3}\pi a^3\right) \tag{11.82}$$

(independent of θ and ϕ), the differential cross-section is

$$\frac{d\sigma}{d\Omega} = |f|^2 \cong \left(\frac{2mV_0 a^3}{3\hbar^2}\right)^2 \tag{11.83}$$

and the total cross-section is

$$\sigma \cong 4\pi \left(\frac{2mV_0 a^3}{3\hbar^2}\right)^2 \tag{11.84}$$

---

[14] You can't apply the Born approximation to *hard*-sphere scattering ($V_0 = \infty$)—the integral blows up. The point is that we assumed the potential is *weak*, and doesn't change the wave function much in the scattering region. But a *hard* sphere changes it *radically*—from Ae^{ikz} to *zero*.

For a **spherically symmetrical potential**, $V(\mathbf{r}) = V(r)$—but *not* necessarily at low energy—the Born approximation again reduces to a simpler form. Define

$$\boldsymbol{\kappa} \equiv \mathbf{k}' - \mathbf{k}, \qquad [11.85]$$

and let the polar axis for the $\mathbf{r}_0$ integral lie along $\boldsymbol{\kappa}$, so that

$$(\mathbf{k}' - \mathbf{k}) \cdot \mathbf{r}_0 = \kappa r_0 \cos\theta_0. \qquad [11.86]$$

Then

$$f(\theta) \cong -\frac{m}{2\pi\hbar^2}\int e^{i\kappa r_0 \cos\theta_0} V(r_0) r_0^2 \sin\theta_0 \, dr_0 \, d\theta_0 \, d\phi_0. \qquad [11.87]$$

The ϕ_0 integral is trivial (2π), and the θ_0 integral is one we have encountered before (see Equation 11.59). Dropping the subscript on r, we are left with

$$\boxed{f(\theta) \cong -\frac{2m}{\hbar^2\kappa}\int_0^\infty r V(r)\sin(\kappa r)\, dr, \quad \text{(spherical symmetry)}.} \qquad [11.88]$$

The angular dependence of f is carried by κ; in Figure 11.11 we see that

$$\kappa = 2k\sin(\theta/2). \qquad [11.89]$$

Example 11.5 **Yukawa scattering.** The **Yukawa potential** (which is a crude model for the binding force in an atomic nucleus) has the form

$$V(r) = \beta\frac{e^{-\mu r}}{r}, \qquad [11.90]$$

where β and μ are constants. The Born approximation gives

$$f(\theta) \cong -\frac{2m\beta}{\hbar^2\kappa}\int_0^\infty e^{-\mu r}\sin(\kappa r)\, dr = -\frac{2m\beta}{\hbar^2(\mu^2+\kappa^2)}. \qquad [11.91]$$

(You get to work out the integral for yourself, in Problem 11.11.)

Example 11.6 **Rutherford scattering.** If we put in $\beta = q_1 q_2/4\pi\epsilon_0$, $\mu = 0$, the Yukawa potential reduces to the Coulomb potential, describing the electrical interaction of two point charges. Evidently the scattering amplitude is

$$f(\theta) \cong -\frac{2m q_1 q_2}{4\pi\epsilon_0\hbar^2\kappa^2}, \qquad [11.92]$$

or (using Equations 11.89 and 11.51):

$$f(\theta) \cong -\frac{q_1 q_2}{16\pi\epsilon_0 E \sin^2(\theta/2)}. \qquad [11.93]$$

The differential cross-section is the square of this:

$$\frac{d\sigma}{d\Omega} = \left[\frac{q_1 q_2}{16\pi\epsilon_0 E \sin^2(\theta/2)}\right]^2, \qquad [11.94]$$

which is precisely the Rutherford formula (Equation 11.11). It happens that for the Coulomb potential, classical mechanics, the Born approximation, and quantum field theory all yield the same result. As they say in the computer business, the Rutherford formula is amazingly "robust."

∗**Problem 11.10** Find the scattering amplitude, in the Born approximation, for soft-sphere scattering at arbitrary energy. Show that your formula reduces to Equation 11.82 in the low-energy limit.

Problem 11.11 Evaluate the integral in Equation 11.91, to confirm the expression on the right.

∗∗**Problem 11.12** Calculate the total cross-section for scattering from a Yukawa potential, in the Born approximation. Express your answer as a function of E.

∗**Problem 11.13** For the potential in Problem 11.4,

(a) calculate $f(\theta)$, $D(\theta)$, and σ, in the low-energy Born approximation;

(b) calculate $f(\theta)$ for arbitrary energies, in the Born approximation;

(c) show that your results are consistent with the answer to Problem 11.4, in the appropriate regime.

11.4.3 The Born Series

The Born approximation is similar in spirit to the **impulse approximation** in classical scattering theory. In the impulse approximation we begin by pretending

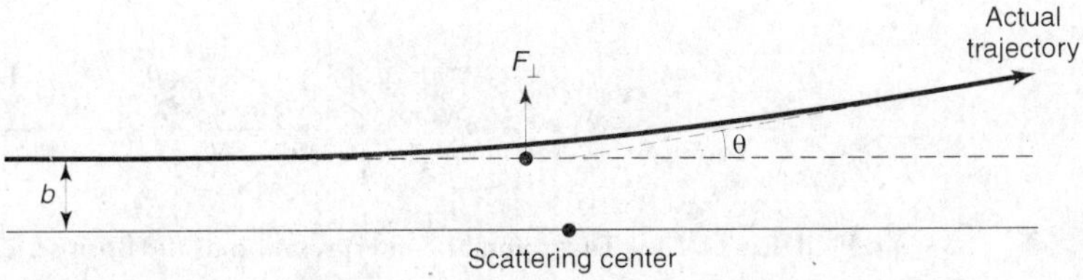

FIGURE 11.12: The impulse approximation assumes that the particle continues undeflected, and calculates the transverse momentum delivered.

that the particle keeps going in a straight line (Figure 11.12), and compute the transverse impulse that would be delivered to it in that case:

$$I = \int F_\perp \, dt. \qquad [11.95]$$

If the deflection is relatively small, this should be a good approximation to the transverse momentum imparted to the particle, and hence the scattering angle is

$$\theta \cong \tan^{-1}(I/p), \qquad [11.96]$$

where p is the incident momentum. This is, if you like, the "first-order" impulse approximation (the *zeroth*-order is what we *started* with: no deflection at all). Likewise, in the zeroth-order Born approximation the incident plane wave passes by with no modification, and what we explored in the previous section is really the first-order correction to this. But the same idea can be iterated to generate a series of higher-order corrections, which presumably converge to the exact answer.

The integral form of the Schrödinger equation reads

$$\psi(\mathbf{r}) = \psi_0(\mathbf{r}) + \int g(\mathbf{r} - \mathbf{r}_0) V(\mathbf{r}_0) \psi(\mathbf{r}_0) \, d^3\mathbf{r}_0, \qquad [11.97]$$

where ψ_0 is the incident wave,

$$g(\mathbf{r}) \equiv -\frac{m}{2\pi\hbar^2} \frac{e^{ikr}}{r} \qquad [11.98]$$

is the Green's function (into which I have now incorporated the factor $2m/\hbar^2$, for convenience), and V is the scattering potential. Schematically,

$$\psi = \psi_0 + \int gV\psi. \qquad [11.99]$$

Suppose we take this expression for ψ, and plug it in under the integral sign:

$$\psi = \psi_0 + \int gV\psi_0 + \iint gVgV\psi. \qquad [11.100]$$

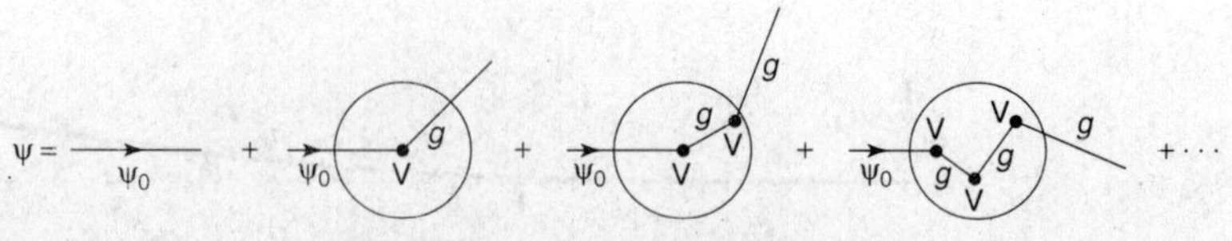

FIGURE 11.13: Diagrammatic interpretation of the Born series (Equation 11.101).

Iterating this procedure, we obtain a formal series for ψ:

$$\psi = \psi_0 + \int gV\psi_0 + \int\int gVgV\psi_0 + \int\int\int gVgVgV\psi_0 + \cdots. \quad [11.101]$$

In each integrand only the *incident* wavefunction (ψ_0) appears, together with more and more powers of gV. The *first* Born approximation truncates the series after the second term, but it is pretty clear how one generates the higher-order corrections.

The Born series can be represented diagrammatically as shown in Figure 11.13. In zeroth order ψ is untouched by the potential; in first order it is "kicked" once, and then "propagates" out in some new direction; in second order it is kicked, propagates to a new location, is kicked again, and then propagates out; and so on. In this context the Green's function is sometimes called the **propagator**—it tells you how the disturbance propagates between one interaction and the next. The Born series was the inspiration for Feynman's formulation of relativistic quantum mechanics, which is expressed entirely in terms of **vertex factors** (V) and propagators (g), connected together in **Feynman diagrams**.

Problem 11.14 Calculate θ (as a function of the impact parameter) for Rutherford scattering, in the impulse approximation. Show that your result is consistent with the exact expression (Problem 11.1(a)), in the appropriate limit.

∗ ∗ ∗**Problem 11.15** Find the scattering amplitude for low-energy soft-sphere scattering in the *second* Born approximation. *Answer:* $-(2mV_0a^3/3\hbar^2)[1 - (4mV_0a^2/5\hbar^2)]$.

FURTHER PROBLEMS FOR CHAPTER 11

∗ ∗ ∗**Problem 11.16** Find the Green's function for the *one*-dimensional Schrödinger equation, and use it to construct the integral form (analogous to Equation 11.67). *Answer:*

$$\psi(x) = \psi_0(x) - \frac{im}{\hbar^2 k}\int_{-\infty}^{\infty} e^{ik|x-x_0|}V(x_0)\psi(x_0)\,dx_0. \quad [11\ 102]$$

****Problem 11.17** Use your result in Problem 11.16 to develop the Born approximation for one-dimensional scattering (on the interval $-\infty < x < \infty$, with no "brick wall" at the origin). That is, choose $\psi_0(x) = Ae^{ikx}$, and assume $\psi(x_0) \cong \psi_0(x_0)$ to evaluate the integral. Show that the reflection coefficient takes the form:

$$R \cong \left(\frac{m}{\hbar^2 k}\right)^2 \left|\int_{-\infty}^{\infty} e^{2ikx} V(x)\, dx\right|^2 \qquad [11.103]$$

Problem 11.18 Use the one-dimensional Born approximation (Problem 11.17) to compute the transmission coefficient ($T = 1 - R$) for scattering from a delta function (Equation 2.114) and from a finite square well (Equation 2.145). Compare your results with the exact answers (Equations 2.141 and 2.169).

Problem 11.19 Prove the **optical theorem**, which relates the total cross-section to the imaginary part of the forward scattering amplitude:

$$\sigma = \frac{4\pi}{k} \mathrm{Im}(f(0)). \qquad [11.104]$$

Hint: Use Equations 11.47 and 11.48.

CHAPTER 12

AFTERWORD

Now that you have (I hope) a sound understanding of what quantum mechanics *says*, I would like to return to the question of what it *means*—continuing the story begun in Section 1.2. The source of the problem is the indeterminacy associated with the statistical interpretation of the wave function. For Ψ (or, more generally, the *quantum state*—it could be a spinor, for example) does not uniquely determine the outcome of a measurement; all it provides is the statistical distribution of possible results. This raises a profound question: Did the physical system "actually have" the attribute in question *prior* to the measurement (the so-called **realist** viewpoint), or did the act of measurement itself "create" the property, limited only by the statistical constraint imposed by the wave function (the **orthodox** position)—or can we duck the question entirely, on the grounds that it is "metaphysical" (the **agnostic** response)?

According to the realist, quantum mechanics is an *incomplete* theory, for even if you know *everything quantum mechanics has to tell you* about the system (to wit: its wave function), still you cannot determine all of its features. Evidently there is some *other* information, external to quantum mechanics, which (together with Ψ) is required for a complete description of physical reality.

The orthodox position raises even more disturbing problems, for if the act of measurement forces the system to "take a stand," helping to *create* an attribute that was not there previously,[1] then there is something very peculiar about the

[1] This may be *strange*, but it is not *mystical*, as some popularizers would like to suggest. The so-called **wave-particle duality**, which Niels Bohr elevated to the status of a cosmic principle (**complementarity**), makes electrons sound like unpredictable adolescents, who sometimes behave like

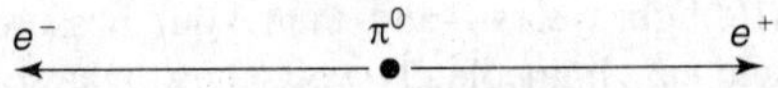

FIGURE 12.1: Bohm's version of the EPR experiment: A π^0 at rest decays into an electron-positron pair.

measurement process. Moreover, in order to account for the fact that an immediately repeated measurement yields the same result, we are forced to assume that the act of measurement **collapses** the wave function, in a manner that is difficult, at best, to reconcile with the normal evolution prescribed by the Schrödinger equation.

In light of this, it is no wonder that generations of physicists retreated to the agnostic position, and advised their students not to waste time worrying about the conceptual foundations of the theory.

12.1 THE EPR PARADOX

In 1935, Einstein, Podolsky, and Rosen[2] published the famous **EPR paradox**, which was designed to prove (on purely theoretical grounds) that the realist position is the only sustainable one. I'll describe a simplified version of the EPR paradox, introduced by David Bohm. Consider the decay of the neutral pi meson into an electron and a positron:

$$\pi^0 \rightarrow e^- + e^+.$$

Assuming the pion was at rest, the electron and positron fly off in opposite directions (Figure 12.1). Now, the pion has spin zero, so conservation of angular momentum requires that the electron and positron are in the singlet configuration:

$$\frac{1}{\sqrt{2}}(\uparrow_-\downarrow_+ - \downarrow_-\uparrow_+). \qquad [12.1]$$

If the electron is found to have spin up, the positron must have spin down, and vice versa. Quantum mechanics can't tell you *which* combination you'll get, in any particular pion decay, but it does say that the measurements will be *correlated*, and you'll get each combination half the time (on average). Now suppose

adults, and sometimes, for no particular reason, like children. I prefer to avoid such language. When I say that a particle does not have a particular attribute before its measurement, I have in mind, for example, an electron in the spin state $\chi = \begin{pmatrix} 1 \\ 0 \end{pmatrix}$; a measurement of the x-component of its angular momentum could return the value $\hbar/2$, or (with equal probability) the value $-\hbar/2$, but until the measurement is made it simply *does not have* a well-defined value of S_x.

[2]A. Einstein, B. Podolsky, and N. Rosen, *Phys. Rev.* **47**, 777 (1935).

we let the electron and positron fly *way* off—10 meters, in a practical experiment, or, in principle, 10 light years—and then you measure the spin of the electron. Say you get spin up. Immediately you know that someone 20 meters (or 20 light years) away will get spin down, if that person examines the positron.

To the realist, there's nothing surprising in this—the electron *really had* spin up (and the positron spin down) from the moment they were created ... it's just that quantum mechanics didn't know about it. But the "orthodox" view holds that neither particle had either spin up *or* spin down until the act of measurement intervened: Your measurement of the electron collapsed the wave function, and instantaneously "produced" the spin of the positron 20 meters (or 20 light years) away. Einstein, Podolsky, and Rosen considered such "spooky action-at-a-distance" (Einstein's words) preposterous. They concluded that the orthodox position is untenable; the electron and positron must have had well-defined spins all along, whether quantum mechanics can calculate them or not.

The fundamental assumption on which the EPR argument rests is that no influence can propagate faster than the speed of light. We call this the principle of **locality**. You might be tempted to propose that the collapse of the wave function is *not* instantaneous, but "travels" at some finite velocity. However, this would lead to violations of angular momentum conservation, for if we measured the spin of the positron before the news of the collapse had reached it, there would be a fifty-fifty probability of finding *both* particles with spin up. Whatever one might think of such a theory in the abstract, the experiments are unequivocal: No such violation occurs—the (anti-)correlation of the spins is perfect. Evidently the collapse of the wave function—whatever its ontological status—is instantaneous.

Problem 12.1 Entangled states. The singlet spin configuration (Equation 12.1) is the classic example of an *entangled state*—a two-particle state that cannot be expressed as the product of two one-particle states, and for which, therefore, one cannot really speak of "the state" of either particle separately. You might wonder whether this is somehow an artifact of bad notation—maybe some linear combination of the one-particle states would disentangle the system. Prove the following theorem:

Consider a two-level system, $|\phi_a\rangle$ and $|\phi_b\rangle$, with $\langle\phi_i|\phi_j\rangle = \delta_{ij}$. (For example, $|\phi_a\rangle$ might represent spin up and $|\phi_b\rangle$ spin down.) The two-particle state

$$\alpha|\phi_a(1)\rangle|\phi_b(2)\rangle + \beta|\phi_b(1)\rangle|\phi_a(2)\rangle$$

(with $\alpha \neq 0$ and $\beta \neq 0$) *cannot* be expressed as a product

$$|\psi_r(1)\rangle|\psi_s(2)\rangle,$$

for *any* one-particle states $|\psi_r\rangle$ and $|\psi_s\rangle$.

Hint: Write $|\psi_r\rangle$ and $|\psi_s\rangle$ as linear combinations of $|\phi_a\rangle$ and $|\phi_b\rangle$.

12.2 BELL'S THEOREM

Einstein, Podolsky, and Rosen did not doubt that quantum mechanics is *correct*, as far as it goes; they only claimed that it is an *incomplete* description of physical reality: The wave function is not the whole story—some *other* quantity, λ, is needed, in addition to Ψ, to characterize the state of a system fully. We call λ the "hidden variable" because, at this stage, we have no idea how to calculate or measure it.[3] Over the years, a number of hidden variable theories have been proposed, to supplement quantum mechanics;[4] they tend to be cumbersome and implausible, but never mind—until 1964 the program seemed eminently worth pursuing. But in that year J. S. Bell proved that *any* local hidden variable theory is *incompatible* with quantum mechanics.[5]

Bell suggested a generalization of the EPR/Bohm experiment: Instead of orienting the electron and positron detectors along the *same* direction, he allowed them to be rotated independently. The first measures the component of the electron spin in the direction of a unit vector **a**, and the second measures the spin of the positron along the direction **b** (Figure 12.2). For simplicity, let's record the spins in units of $\hbar/2$; then each detector registers the value $+1$ (for spin up) or -1 (spin down), along the direction in question. A table of results, for many π^0 decays, might look like this:

electron	positron	product
+1	−1	−1
+1	+1	+1
−1	+1	−1
+1	−1	−1
−1	−1	+1
⋮	⋮	⋮

[3]The hidden variable could be a single number, or it could be a whole *collection* of numbers; perhaps λ is to be calculated in some future theory, or maybe it is for some reason of principle incalculable. It hardly matters. All I am asserting is that there must be *something*—if only a *list* of the outcomes of every possible experiment—associated with the system prior to a measurement.

[4]D. Bohm, *Phys. Rev.* **85**, 166, 180 (1952).

[5]Bell's original paper (*Physics* **1**, 195 (1964)) is a gem: brief, accessible, and beautifully written.

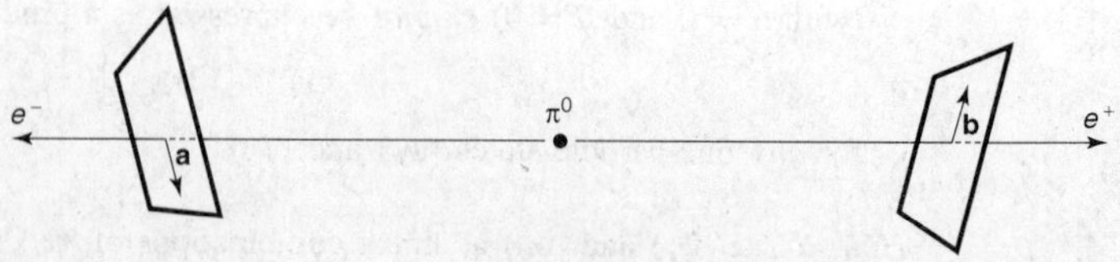

FIGURE 12.2: Bell's version of the EPR-Bohm experiment: detectors independently oriented in directions a and b.

Bell proposed to calculate the *average* value of the *product* of the spins, for a given set of detector orientations. Call this average $P(\mathbf{a}, \mathbf{b})$. If the detectors are parallel ($\mathbf{b} = \mathbf{a}$), we recover the original EPRB configuration; in this case one is spin up and the other spin down, so the product is always -1, and hence so too is the average:

$$P(\mathbf{a}, \mathbf{a}) = -1. \quad [12.2]$$

By the same token, if they are *anti*-parallel ($\mathbf{b} = -\mathbf{a}$), then every product is $+1$, so

$$P(\mathbf{a}, -\mathbf{a}) = +1. \quad [12.3]$$

For arbitrary orientations, quantum mechanics predicts

$$\boxed{P(\mathbf{a}, \mathbf{b}) = -\mathbf{a} \cdot \mathbf{b}} \quad [12.4]$$

(see Problem 4.50). What Bell discovered is that *this result is incompatible with any local hidden variable theory*.

The argument is stunningly simple. Suppose that the "complete" state of the electron/positron system is characterized by the hidden variable(s) λ (λ varies, in some way that we neither understand nor control, from one pion decay to the next). Suppose further that the outcome of the *electron* measurement is independent of the orientation ($\mathbf{b}$) of the *positron* detector—which may, after all, be chosen by the experimenter at the positron end just before the electron measurement is made, and hence far too late for any subluminal message to get back to the electron detector. (This is the locality assumption.) Then there exists some function $A(\mathbf{a}, \lambda)$ which gives the result of an electron measurement, and some other function $B(\mathbf{b}, \lambda)$ for the positron measurement. These functions can only take on the values ± 1:[6]

$$A(\mathbf{a}, \lambda) = \pm 1; \quad B(\mathbf{b}, \lambda) = \pm 1. \quad [12.5]$$

[6]This already concedes far more than a *classical* determinist would be prepared to allow, for it abandons any notion that the particles could have well-defined angular momentum vectors with simultaneously determinate components. But never mind—the point of Bell's argument is to demonstrate that quantum mechanics is incompatible with *any* local deterministic theory—even one that bends over backwards to be accommodating.

When the detectors are aligned, the results are perfectly (anti)-correlated:

$$A(\mathbf{a}, \lambda) = -B(\mathbf{a}, \lambda), \tag{12.6}$$

for all λ.

Now, the average of the product of the measurements is

$$P(\mathbf{a}, \mathbf{b}) = \int \rho(\lambda) A(\mathbf{a}, \lambda) B(\mathbf{b}, \lambda)\, d\lambda, \tag{12.7}$$

where $\rho(\lambda)$ is the probability density for the hidden variable. (Like any probability density, it is nonnegative, and satisfies the normalization condition $\int \rho(\lambda)\, d\lambda = 1$, but beyond this we make no assumptions about $\rho(\lambda)$; different hidden variable theories would presumably deliver quite different expressions for ρ.) In view of Equation 12.6, we can eliminate B:

$$P(\mathbf{a}, \mathbf{b}) = -\int \rho(\lambda) A(\mathbf{a}, \lambda) A(\mathbf{b}, \lambda)\, d\lambda, \tag{12.8}$$

If $\mathbf{c}$ is any *other* unit vector,

$$P(\mathbf{a}, \mathbf{b}) - P(\mathbf{a}, \mathbf{c}) = -\int \rho(\lambda) \left[A(\mathbf{a}, \lambda) A(\mathbf{b}, \lambda) - A(\mathbf{a}, \lambda) A(\mathbf{c}, \lambda) \right] d\lambda. \tag{12.9}$$

Or, since $[A(\mathbf{b}, \lambda)]^2 = 1$:

$$P(\mathbf{a}, \mathbf{b}) - P(\mathbf{a}, \mathbf{c}) = -\int \rho(\lambda) \left[1 - A(\mathbf{b}, \lambda) A(\mathbf{c}, \lambda) \right] A(\mathbf{a}, \lambda) A(\mathbf{b}, \lambda)\, d\lambda \tag{12.10}$$

But it follows from Equation 12.5 that $-1 \le [A(\mathbf{a}, \lambda) A(\mathbf{b}, \lambda)] \le +1$; moreover $\rho(\lambda)[1 - A(\mathbf{b}, \lambda) A(\mathbf{c}, \lambda)] \ge 0$, so

$$|P(\mathbf{a}, \mathbf{b}) - P(\mathbf{a}, \mathbf{c})| \le \int \rho(\lambda) \left[1 - A(\mathbf{b}, \lambda) A(\mathbf{c}, \lambda) \right] d\lambda, \tag{12.11}$$

or, more simply:

$$\boxed{|P(\mathbf{a}, \mathbf{b}) - P(\mathbf{a}, \mathbf{c})| \le 1 + P(\mathbf{b}, \mathbf{c}).} \tag{12.12}$$

This is the famous **Bell inequality**. It holds for *any* local hidden variable theory (subject only to the minimal requirements of Equations 12.5 and 12.6), for we have made no assumptions whatever as to the nature or number of the hidden variable(s), or their distribution (ρ).

But it is easy to show that the quantum mechanical prediction (Equation 12.4) is incompatible with Bell's inequality. For example, suppose all three vectors lie in

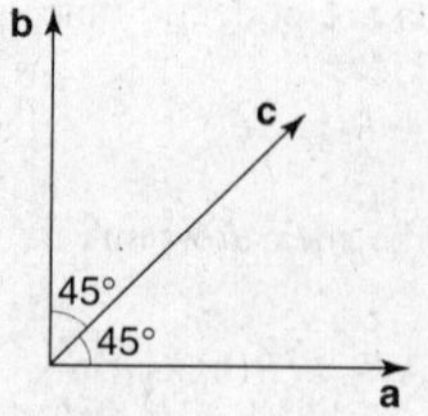

FIGURE 12.3: An orientation of the detectors that demonstrates quantum violations of Bell's inequality.

a plane, and **c** makes a 45° angle with **a** and **b** (Figure 12.3); in this case quantum mechanics says

$$P(\mathbf{a}, \mathbf{b}) = 0, \quad P(\mathbf{a}, \mathbf{c}) = P(\mathbf{b}, \mathbf{c}) = -0.707,$$

which is patently inconsistent with Bell's inequality:

$$0.707 \not\leq 1 - 0.707 = 0.293.$$

With Bell's modification, then, the EPR paradox proves something far more radical than its authors imagined: If they are right, then not only is quantum mechanics *incomplete*, it is downright *wrong*. On the other hand, if quantum mechanics is right, then *no* hidden variable theory is going to rescue us from the nonlocality Einstein considered so preposterous. Moreover, we are provided with a very simple experiment to settle the issue once and for all.

Many experiments to test Bell's inequality were performed in the '60's and '70's, culminating in the work of Aspect, Grangier, and Roger.[7] The detaiis do not concern us here (they actually used two-photon atomic transitions, not pion decays). To exclude the remote possibility that the positron detector might somehow "sense" the orientation of the electron detector, both orientations were set quasi-randomly *after* the photons were already in flight. The results were in excellent agreement with the predictions of quantum mechanics, and clearly incompatible with Bell's inequality.[8]

Ironically, the experimental confirmation of quantum mechanics came as something of a shock to the scientific community. But not because it spelled the demise of "realism"—most physicists had long since adjusted to this (and for those who could not, there remained the possibility of *nonlocal* hidden variable theories,

---

[7] A. Aspect, P. Grangier, and G. Roger, *Phys. Rev. Lett.* **49**, 91 (1982). For more recent experiments see G. Weihs *et al.*, *Phys. Rev. Lett.* **81**, 5039 (1998).

[8] Bell's theorem involves *averages* and it is conceivable that an apparatus such as Aspect's contains some secret bias which selects out a nonrepresentative sample, thus distorting the average. In 1989, an improved version of Bell's theorem was proposed, in which a *single measurement* suffices to distinguish between the quantum prediction and that of any local hidden variable theory. See D. Greenberger, M. Horne, A. Shimony, and A. Zeilinger, *Am. J. Phys.* **58**, 1131 (1990) and N. David Mermin, *Am. J. Phys.* **58**, 731 (1990).

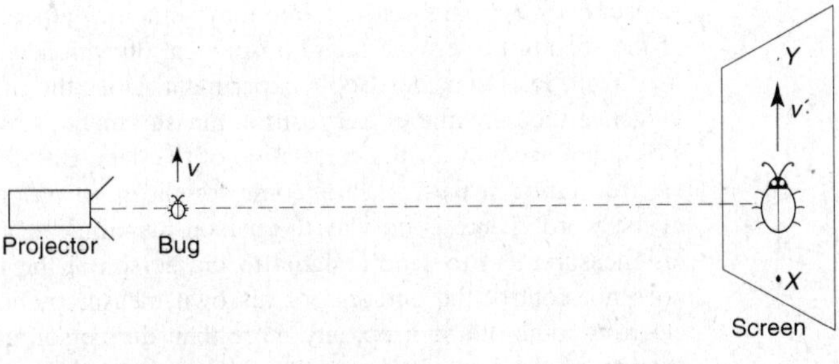

FIGURE 12.4: The shadow of the bug moves across the screen at a velocity v' greater than c, provided the screen is far enough away.

to which Bell's theorem does not apply[9]). The real shock was the demonstration that *nature itself is fundamentally nonlocal.* Nonlocality, in the form of the instantaneous collapse of the wave function (and for that matter also in the symmetrization requirement for identical particles) had always been a feature of the orthodox interpretation, but before Aspect's experiment it was possible to hope that quantum nonlocality was somehow a nonphysical artifact of the formalism, with no detectable consequences. That hope can no longer be sustained, and we are obliged to reexamine our objection to instantaneous action-at-a-distance.

Why *are* physicists so squeamish about superluminal influences? After all, there are many things that travel faster than light. If a bug flies across the beam of a movie projector, the speed of its shadow is proportional to the distance to the screen; in principle, that distance can be as large as you like, and hence the *shadow* can move at arbitrarily high velocity (Figure 12.4). However, the shadow does not carry any *energy*; nor can it transmit a *message* from one point to another on the screen. A person at point X cannot *cause anything to happen* at point Y by manipulating the passing shadow.

On the other hand, a *causal* influence that propagated faster than light would carry unacceptable implications. For according to special relativity there exist inertial frames in which such a signal propagates *backward in time*—the effect preceding the cause—and this leads to inescapable logical anomalies. (You could, for example, arrange to kill your infant grandfather. Not a good idea!) The question is, are the superluminal influences predicted by quantum mechanics and detected by

[9]It is a curious twist of fate that the EPR paradox, which *assumed* locality in order to *prove* realism, led finally to the demise of locality and left the issue of realism undecided—the outcome (as Mermin put it) Einstein would have liked *least.* Most physicists today consider that if they can't have *local* realism, there's not much point in realism at *all,* and for this reason nonlocal hidden variable theories occupy a rather peripheral place. Still, some authors—notably Bell himself, in *Speakable and Unspeakable in Quantum Mechanics* (Cambridge University Press, 1987)—argue that such theories offer the best hope of bridging the conceptual gap between the measured system and the measuring apparatus, and for supplying an intelligible mechanism for the collapse of the wave function.

Aspect *causal*, in this sense, or are they somehow ethereal enough (like the motion of the shadow) to escape the philosophical objection?

Well, let's consider Bell's experiment. Does the measurement of the electron *influence* the outcome of the positron measurement? Assuredly it *does*—otherwise we cannot account for the correlation of the data. But does the measurement of the electron *cause* a particular outcome for the positron? Not in any ordinary sense of the word. There is no way the person manning the electron detector could use his measurement to send a signal to the person at the positron detector, since he does not control the outcome of his own measurement (he cannot *make* a given electron come out spin up, any more than the person at X can affect the passing shadow of the bug). It is true that he can decide *whether to make a measurement at all*, but the positron monitor, having immediate access only to data at his end of the line, cannot tell whether the electron was measured or not, for the lists of data compiled at the two ends, considered separately, are completely random. It is only when we *compare* the two lists later that we discover the remarkable correlations. In another reference frame the positron measurements occur *before* the electron measurements, and yet this leads to no logical paradox—the observed correlation is entirely symmetrical in its treatment, and it is a matter of indifference whether we say the observation of the electron influenced the measurement of the positron, or the other way around. This is a wonderfully delicate kind of influence whose only manifestation is a subtle correlation between two lists of otherwise random data.

We are led, then, to distinguish two types of influence: the "causal" variety, which produce actual changes in some physical property of the receiver, detectable by measurements on that subsystem alone, and an "ethereal" kind, which do not transmit energy or information, and for which the only evidence is a correlation in the data taken on the two separate subsystems—a correlation which by its nature cannot be detected by examining either list alone. Causal influences *cannot* propagate faster than light, but there is no compelling reason why ethereal ones should not. The influences associated with the collapse of the wave function are of the latter type, and the fact that they "travel" faster than light may be surprising, but it is not, after all, catastrophic.[10]

12.3 THE NO-CLONE THEOREM

Quantum measurements are typically **destructive**, in the sense that they alter the state of the system measured. This is how the uncertainty principle is enforced in the laboratory. You might wonder why we don't just make a bunch of identical copies (**clones**) of the original state, and measure *them*, leaving the system itself

[10]An enormous amount has been written about Bell's theorem. My favorite is an inspired essay by David Mermin in *Physics Today* (April 1985, page 38). An extensive bibliography will be found in L. E. Ballentine, *Am. J. Phys.* **55**, 785 (1987).

unscathed. It can't be done. Indeed, if you could build a cloning device (a "quantum Xerox machine"), quantum mechanics would be out the window.

For example, it would then be possible to send superluminal messages using the EPRB experiment. Say the message to be transmitted, from the operator of the positron detector to the operator of the electron detector, is either "yes" or "no." If the message is "yes," the sender measures S_z (of the positron). Never mind what result she gets—all that matters is that she makes the measurement, for this means that the electron is now in the definite state $\uparrow$ or $\downarrow$ (never mind which). The receiver immediately makes a million clones of the electron, and measures S_z on each of them. If they all yield the same answer (never mind *which* answer), we can be pretty sure that the electron *was* in fact measured, so the message is "yes." If half of them are spin up, and half spin down, then the electron was definitely *not* measured, and the message is "no."

But you *can't* make a quantum Xerox machine, as Wootters, Zurek, and Dieks proved in 1982.[11] Schematically, we want the machine to take as input a particle in state $|\psi\rangle$ (the one to be copied), plus a second particle in state $|X\rangle$ (the "blank sheet of paper"), and spit out *two* particles in the state $|\psi\rangle$ (original plus copy):

$$|\psi\rangle|X\rangle \rightarrow |\psi\rangle|\psi\rangle. \qquad [12.13]$$

Suppose we have made a device that successfully clones the state $|\psi_1\rangle$:

$$|\psi_1\rangle|X\rangle \rightarrow |\psi_1\rangle|\psi_1\rangle, \qquad [12.14]$$

and also works for state $|\psi_2\rangle$:

$$|\psi_2\rangle|X\rangle \rightarrow |\psi_2\rangle|\psi_2\rangle \qquad [12.15]$$

($|\psi_1\rangle$ and $|\psi_2\rangle$ might be spin up and spin down, for example, if the particle is an electron). So far, so good. But what happens when we feed in a linear combination $|\psi\rangle = \alpha|\psi_1\rangle + \beta|\psi_2\rangle$? Evidently we get[12]

$$|\psi\rangle|X\rangle \rightarrow \alpha|\psi_1\rangle|\psi_1\rangle + \beta|\psi_2\rangle|\psi_2\rangle, \qquad [12.16]$$

which is not at all what we wanted—what we *wanted* was

$$\begin{aligned} |\psi\rangle|X\rangle \rightarrow |\psi\rangle|\psi\rangle &= [\alpha|\psi_1\rangle + \beta|\psi_2\rangle][\alpha|\psi_1\rangle + \beta|\psi_2\rangle] \\ &= \alpha^2|\psi_1\rangle|\psi_1\rangle + \beta^2|\psi_2\rangle|\psi_2\rangle + \alpha\beta[|\psi_1\rangle|\psi_2\rangle + |\psi_2\rangle|\psi_1\rangle]. \end{aligned} \qquad [12.17]$$

[11] W. K. Wootters and W. H. Zurek, *Nature* **299**, 802 (1982); D. Dieks, *Phys. Lett. A* **92**, 271 (1982).

[12] This assumes that the device acts *linearly* on the state $|\psi\rangle$, as it must, since the time-dependent Schrödinger equation (which presumably governs the process) is linear.

You can make a machine to clone spin-up electrons and spin-down electrons, but it's going to fail for any nontrivial linear combinations. It's as though you bought a Xerox machine that copies vertical lines perfectly, and also horizontal lines, but completely distorts diagonals.

12.4 SCHRÖDINGER'S CAT

The measurement process plays a mischievous role in quantum mechanics: It is here that indeterminacy, nonlocality, the collapse of the wave function, and all the attendant conceptual difficulties arise. Absent measurement, the wave function evolves in a leisurely and deterministic way, according to the Schrödinger equation, and quantum mechanics looks like a rather ordinary field theory (much simpler than classical electrodynamics, for example, since there is only *one* field (Ψ), instead of *two* (**E** and **B**), and it's a *scalar*). It is the bizarre role of the measurement process that gives quantum mechanics its extraordinary richness and subtlety. But what, exactly, *is* a measurement? What makes it so different from other physical processes?[13] And how can we tell when a measurement has occurred?

Schrödinger posed the essential question most starkly, in his famous **cat paradox**:[14]

> A cat is placed in a steel chamber, together with the following hellish contraption.... In a Geiger counter there is a tiny amount of radioactive substance, so tiny that maybe within an hour one of the atoms decays, but equally probably none of them decays. If one decays then the counter triggers and via a relay activates a little hammer which breaks a container of cyanide. If one has left this entire system for an hour, then one would say the cat is living if no atom has decayed. The first decay would have poisoned it. The wave function of the entire system would express this by containing equal parts of the living and dead cat.

At the end of the hour, the wave function of the cat has the schematic form

$$\psi = \frac{1}{\sqrt{2}}(\psi_{\text{alive}} + \psi_{\text{dead}}). \qquad [12.18]$$

[13] There is a school of thought that rejects this distinction, holding that the system and the measuring apparatus should be described by one great big wave function which itself evolves according to the Schrödinger equation. In such theories there is no collapse of the wave function, but one must typically abandon any hope of describing individual events—quantum mechanics (in this view) applies only to *ensembles* of identically prepared systems. See, for example, Philip Pearle *Am. J. Phys.* **35**, 742 (1967), or Leslie E. Ballentine, *Quantum Mechanics: A Modern Development*, 2nd ed., World Scientific, Singapore (1998).

[14] E. Schrödinger, *Naturwiss.* **48**, 52 (1935); translation by Josef M. Jauch, *Foundations of Quantum Mechanics*, Addison-Wesley, Reading (1968), p. 185.

The cat is neither alive nor dead, but rather a linear combination of the two, until a measurement occurs—until, say, you peek in the window to check. At that moment your observation forces the cat to "take a stand": dead or alive. And if you find him to be dead, then it's really *you* who killed him, by looking in the window.

Schrödinger regarded this as patent nonsense, and I think most physicists would agree with him. There is something absurd about the very idea of a *macroscopic* object being in a linear combination of two palpably different states. An electron can be in a linear combination of spin up and spin down, but a cat simply cannot *be* in a linear combination of alive and dead. How are we to reconcile this with the orthodox interpretation of quantum mechanics?

The most widely accepted answer is that the triggering of the Geiger counter constitutes the "measurement," in the sense of the statistical interpretation, not the intervention of a human observer. It is the essence of a measurement that some *macroscopic* system is affected (the Geiger counter, in this instance). The measurement occurs at the moment when the microscopic system (described by the laws of quantum mechanics) interacts with the macroscopic system (described by the laws of classical mechanics) in such a way as to leave a permanent record. The macroscopic system itself is not permitted to occupy a linear combination of distinct states.[15]

I would not pretend that this is an entirely satisfactory resolution, but at least it avoids the stultifying solipsism of Wigner and others, who persuaded themselves that it is the involvement of human consciousness that constitutes a measurement in quantum mechanics. Part of the problem is the word "measurement" itself, which certainly carries a suggestion of human participation. Heisenberg proposed the word "event," which might be preferable. But I'm afraid "measurement" is so ingrained by now that we're stuck with it. And, in the end, no manipulation of the terminology can completely exorcise this mysterious ghost.

12.5 THE QUANTUM ZENO PARADOX

The collapse of the wave function is undoubtedly the *most* peculiar feature of this whole bizarre story. It was introduced on purely theoretical grounds, to account for the fact that an immediately repeated measurement reproduces the same value. But surely such a radical postulate must carry directly observable consequences. In

[15]Of course, in some ultimate sense the macroscopic system is *itself* described by the laws of quantum mechanics. But wave functions, in the first instance, describe individual elementary particles; the wave function of a macroscopic object would be a monstrously complicated composite, built out of all the wave functions of its 10^{23} constituent particles. Presumably somewhere in the statistics of large numbers macroscopic linear combinations become extremely improbable. Indeed, if you *were* able somehow to get a damped pendulum (say) into a linear combination of macroscopically distinct quantum states, it would, in a tiny fraction of the damping time, revert to an ordinary classical state. This phenomenon is called **decoherence**. See, for example, R. Omnes, *The Interpretation of Quantum Mechanics* (Princeton, 1994), Chapter 7

1977 Misra and Sudarshan[16] proposed what they called the **quantum Zeno effect** as a dramatic experimental demonstration of the collapse of the wave function. Their idea was to take an unstable system (an atom in an excited state, say), and subject it to repeated measurements. Each observation collapses the wave function, resetting the clock, and it is possible by this means to delay indefinitely the expected transition to the lower state.[17]

Specifically, suppose a system starts out in the excited state ψ_2, which has a natural lifetime τ for transition to the ground state ψ_1. Ordinarily, for times substantially less than τ, the probability of a transition is proportional to t (see Equation 9.42); in fact, since the transition rate is $1/\tau$,

$$P_{2\to 1} = \frac{t}{\tau}. \quad [12.19]$$

If we make a measurement after a time t, then, the probability that the system is still in the *upper* state is

$$P_1(t) = 1 - \frac{t}{\tau}. \quad [12.20]$$

Suppose we *do* find it to be in the upper state. In that case the wave function collapses back to ψ_2, and the process starts all over again. If we make a *second* measurement, at $2t$, the probability that the system is *still* in the upper state is evidently

$$\left(1 - \frac{t}{\tau}\right)^2 \approx 1 - \frac{2t}{\tau}, \quad [12.21]$$

which is the same as it would have been had we never made the first measurement at t. This is what one would naively expect; if it were the whole story there would be nothing gained by repeatedly observing the system, and there would be no quantum Zeno effect.

However, for *extremely* short times, the probability of a transition is *not* proportional to t, but rather to t^2 (see Equation 9.39):[18]

$$P_{2\to 1} = \alpha t^2 \quad [12.22]$$

In this case the probability that the system is still in the upper state after the two measurements is

$$\left(1 - \alpha t^2\right)^2 \approx 1 - 2\alpha t^2, \quad [12.23]$$

---

[16] B. Misra and E. C. G. Sudarshan, *J. Math. Phys.* **18**, 756 (1977).

[17] This effect doesn't have much to do with Zeno, but it *is* reminiscent of the old adage, "a watched pot never boils," so it is sometimes called the **watched pot phenomenon**.

[18] In the argument leading to linear time dependence, we assumed that the function $\sin^2(\Omega t/2)/\Omega^2$ in Equation 9.39 was a sharp spike. However, the *width* of the "spike" is of order $\Delta\omega = 4\pi/t$, and for *extremely* short t this approximation fails, and the integral becomes $(t^2/4)\int \rho(\omega)d\omega$.

whereas if we had never made the first measurement it would have been

$$1-\alpha(2t)^2 \approx 1-4\alpha t^2. \qquad [12.24]$$

Evidently our observation of the system after time t decreased the net probability of a transition to the lower state!

Indeed, if we examine the system at n regular intervals, from $t = 0$ out to $t = T$ (that is, we make measurements at T/n, $2T/n$, $3T/n$, ..., T), the probability that the system is still in the upper state at the end is

$$\left(1-\alpha(T/n)^2\right)^n \approx 1-\frac{\alpha}{n}T^2, \qquad [12.25]$$

which goes to 1 in the limit $n \rightarrow \infty$: A *continuously* observed unstable system never decays at all! Some authors regard this as an absurd conclusion, and a proof that the collapse of the wave function is fallacious. However, their argument hinges on a rather loose interpretation of what constitutes "observation." If the track of a particle in a bubble chamber amounts to "continuous observation," then the case is closed, for such particles certainly do decay (in fact, their lifetime is not measurably extended by the presence of the detector). But such a particle is only intermittently interacting with the atoms in the chamber, and for the quantum Zeno effect to occur the successive measurements must be made *extremely* rapidly, in order to catch the system in the t^2 regime.

As it turns out, the experiment is impractical for spontaneous transitions, but it can be done using *induced* transitions, and the results are in excellent agreement with the theoretical predictions.[19] Unfortunately, this experiment is not as compelling a confirmation of the collapse of the wave function as its designers hoped; the observed effect can be accounted for in other ways.[20]

****

In this book I have tried to tell a consistent and coherent story: The wave function (Ψ) represents the state of a particle (or system); particles do not in general possess specific dynamical properties (position, momentum, energy, angular momentum, etc.) until an act of measurement intervenes; the probability of getting a particular value in any given experiment is determined by the statistical interpretation of Ψ; upon measurement the wave function collapses, so that an immediately repeated measurement is certain to yield the same result. There are other possible interpretations—nonlocal hidden variable theories, the "many worlds" picture, "consistent histories," ensemble models, and others—but I believe this one

[19] W. M. Itano, D. J. Heinzen, J. J. Bollinger, and D. J. Wineland, *Phys. Rev. A* **41**, 2295 (1990).

[20] L. E. Ballentine, *Found. Phys.* **20**, 1329 (1990); T. Petrosky, S. Tasaki, and I. Prigogine, *Phys. Lett. A* **151**, 109 (1990).

is conceptually the *simplest*, and certainly it is the one shared by most physicists today.[21] It has stood the test of time, and emerged unscathed from every experimental challenge. But I cannot believe this is the end of the story; at the very least, we have much to learn about the nature of measurement and the mechanism of collapse. And it is entirely possible that future generations will look back, from the vantage point of a more sophisticated theory, and wonder how we could have been so gullible.

[21] See Daniel Styer *et al.*, *Am. J. Phys.* **70**, 288 (2002).

APPENDIX

LINEAR ALGEBRA

Linear algebra abstracts and generalizes the arithmetic of ordinary vectors, such as those we encounter in first-year physics. The generalization is in two directions: (1) We allow the scalars to be *complex* numbers, and (2) we do not restrict ourselves to three dimensions.

A.1 VECTORS

A **vector space** consists of a set of **vectors** ($|\alpha\rangle$, $|\beta\rangle$, $|\gamma\rangle$, ...), together with a set of **scalars** (a, b, c, ...),[1] which is **closed**[2] under two operations: vector addition and scalar multiplication.

- **Vector Addition**

  The "sum" of any two vectors is another vector:

$$|\alpha\rangle + |\beta\rangle = |\gamma\rangle. \qquad \text{[A.1]}$$

  Vector addition is **commutative**:

$$|\alpha\rangle + |\beta\rangle = |\beta\rangle + |\alpha\rangle, \qquad \text{[A.2]}$$

[1] For our purposes, the scalars will be ordinary complex numbers. Mathematicians can tell you about vector spaces over more exotic fields, but such objects play no role in quantum mechanics. Note that α, β, γ ... are *not* (ordinarily) numbers; they are *names* (labels)—"Charlie," for instance, or "F43A-9GL," or whatever you care to use to identify the vector in question.

[2] That is to say, these operations are always well-defined, and will never carry you outside the vector space.

and **associative**:

$$|\alpha\rangle + (|\beta\rangle + |\gamma\rangle) = (|\alpha\rangle + |\beta\rangle) + |\gamma\rangle. \qquad [A.3]$$

There exists a **zero** (or **null**) **vector**,[3] $|0\rangle$, with the property that

$$|\alpha\rangle + |0\rangle = |\alpha\rangle, \qquad [A.4]$$

for every vector $|\alpha\rangle$. And for every vector $|\alpha\rangle$ there is an associated **inverse vector** $(|-\alpha\rangle)$,[4] such that

$$|\alpha\rangle + |-\alpha\rangle = |0\rangle. \qquad [A.5]$$

- **Scalar Multiplication**

  The "product" of any scalar with any vector is another vector:

$$a|\alpha\rangle = |\gamma\rangle. \qquad [A.6]$$

  Scalar multiplication is **distributive** with respect to vector addition:

$$a(|\alpha\rangle + |\beta\rangle) = a|\alpha\rangle + a|\beta\rangle, \qquad [A.7]$$

  and with respect to scalar addition:

$$(a+b)|\alpha\rangle = a|\alpha\rangle + b|\alpha\rangle. \qquad [A.8]$$

  It is also **associative** with respect to the ordinary multiplication of scalars:

$$a(b|\alpha\rangle) = (ab)|\alpha\rangle. \qquad [A.9]$$

  Multiplication by the scalars 0 and 1 has the effect you would expect:

$$0|\alpha\rangle = |0\rangle; \quad 1|\alpha\rangle = |\alpha\rangle. \qquad [A.10]$$

  Evidently $|-\alpha\rangle = (-1)|\alpha\rangle$ (which we write more simply as $-|\alpha\rangle$).

There's a lot less here than meets the eye—all I have done is to write down in abstract language the familiar rules for manipulating vectors. The virtue of such abstraction is that we will be able to apply our knowledge and intuition about the behavior of ordinary vectors to other systems that happen to share the same formal properties.

---

[3] It is customary, where no confusion can arise, to write the null vector without the adorning bracket: $|0\rangle \to 0$.

[4] This is funny notation, since α is not a number. I'm simply adopting the name "−Charlie" for the inverse of the vector whose name is "Charlie." More natural terminology will suggest itself in a moment.

A **linear combination** of the vectors $|\alpha\rangle$, $|\beta\rangle$, $|\gamma\rangle$, ..., is an expression of the form

$$a|\alpha\rangle + b|\beta\rangle + c|\gamma\rangle + \cdots . \qquad [A.11]$$

A vector $|\lambda\rangle$ is said to be **linearly independent** of the set $|\alpha\rangle$, $|\beta\rangle$, $|\gamma\rangle$, ..., if it cannot be written as a linear combination of them. (For example, in three dimensions the unit vector $\hat{k}$ is linearly independent of $\hat{\imath}$ and $\hat{\jmath}$, but any vector in the xy plane is linearly *dependent* on $\hat{\imath}$ and $\hat{\jmath}$.) By extension, a *set* of vectors is "linearly independent" if each one is linearly independent of all the rest. A collection of vectors is said to **span** the space if *every* vector can be written as a linear combination of the members of this set.[5] A set of *linearly independent* vectors that spans the space is called a **basis**. The number of vectors in any basis is called the **dimension** of the space. For the moment we shall assume that the dimension (n) is *finite*.

With respect to a prescribed basis

$$|e_1\rangle, \ |e_2\rangle, \ldots , |e_n\rangle, \qquad [A.12]$$

any given vector

$$|\alpha\rangle = a_1|e_1\rangle + a_2|e_2\rangle + \cdots + a_n|e_n\rangle, \qquad [A.13]$$

is uniquely represented by the (ordered) n-tuple of its **components**:

$$|\alpha\rangle \leftrightarrow (a_1, a_2, \ldots , a_n). \qquad [A.14]$$

It is often easier to work with the components than with the abstract vectors themselves. To add vectors, you add their corresponding components:

$$|\alpha\rangle + |\beta\rangle \leftrightarrow (a_1 + b_1, a_2 + b_2, \ldots , a_n + b_n); \qquad [A.15]$$

to multiply by a scalar you multiply each component:

$$c|\alpha\rangle \leftrightarrow (ca_1, ca_2, \ldots , ca_n); \qquad [A.16]$$

the null vector is represented by a string of zeroes:

$$|0\rangle \leftrightarrow (0, 0, \ldots , 0); \qquad [A.17]$$

and the components of the inverse vector have their signs reversed:

$$|-\alpha\rangle \leftrightarrow (-a_1, -a_2, \ldots , -a_n). \qquad [A.18]$$

[5]A set of vectors that spans the space is also called **complete**, though I personally reserve that word for the infinite-dimensional case, where subtle questions of convergence may arise.

The only *dis*advantage of working with components is that you have to commit yourself to a particular basis, and the same manipulations will look very different to someone working in a different basis.

Problem A.1 Consider the ordinary vectors in 3 dimensions $(a_x\hat{\imath} + a_y\hat{\jmath} + a_z\hat{k})$, with complex components.

(a) Does the subset of all vectors with $a_z = 0$ constitute a vector space? If so, what is its dimension; if not, why not?

(b) What about the subset of all vectors whose z component is 1? *Hint:* Would the sum of two such vectors be in the subset? How about the null vector?

(c) What about the subset of vectors whose components are all equal?

∗**Problem A.2** Consider the collection of all polynomials (with complex coefficients) of degree less than N in x.

(a) Does this set constitute a vector space (with the polynomials as "vectors")? If so, suggest a convenient basis, and give the dimension of the space. If not, which of the defining properties does it lack?

(b) What if we require that the polynomials be *even* functions?

(c) What if we require that the leading coefficient (i.e., the number multiplying x^{N-1}) be 1?

(d) What if we require that the polynomials have the value 0 at $x = 1$?

(e) What if we require that the polynomials have the value 1 at $x = 0$?

Problem A.3 Prove that the components of a vector with respect to a given basis are *unique*.

A.2 INNER PRODUCTS

In three dimensions we encounter two kinds of vector products: the dot product and the cross product. The latter does not generalize in any natural way to n-dimensional vector spaces, but the former *does*—in this context it is usually called the **inner product**. The inner product of two vectors ($|\alpha\rangle$ and $|\beta\rangle$) is a complex number, which we write as $\langle\alpha|\beta\rangle$, with the following properties:

$$\langle\beta|\alpha\rangle = \langle\alpha|\beta\rangle^*, \tag{A.19}$$

$$\langle\alpha|\alpha\rangle \geq 0, \quad \text{and} \quad \langle\alpha|\alpha\rangle = 0 \Leftrightarrow |\alpha\rangle = |0\rangle, \tag{A.20}$$

$$\langle\alpha|(b|\beta\rangle + c|\gamma\rangle) = b\langle\alpha|\beta\rangle + c\langle\alpha|\gamma\rangle. \tag{A.21}$$

Apart from the generalization to complex numbers, these axioms simply codify the familiar behavior of dot products. A vector space with an inner product is called an **inner product space**.

Because the inner product of any vector with itself is a nonnegative number (Equation A.20), its square root is *real*—we call this the **norm** of the vector:

$$\|\alpha\| \equiv \sqrt{\langle\alpha|\alpha\rangle}; \tag{A.22}$$

it generalizes the notion of "length." A **unit vector** (one whose norm is 1) is said to be **normalized** (the word should really be "normal," but I guess that sounds too anthropomorphic). Two vectors whose inner product is zero are called **orthogonal** (generalizing the notion of "perpendicular"). A collection of mutually orthogonal normalized vectors,

$$\langle\alpha_i|\alpha_j\rangle = \delta_{ij}, \tag{A.23}$$

is called an **orthonormal set**. It is always possible (see Problem A.4), and almost always convenient, to choose an *orthonormal basis*; in that case the inner product of two vectors can be written very neatly in terms of their components:

$$\langle\alpha|\beta\rangle = a_1^* b_1 + a_2^* b_2 + \cdots + a_n^* b_n, \tag{A.24}$$

the norm (squared) becomes

$$\langle\alpha|\alpha\rangle = |a_1|^2 + |a_2|^2 + \cdots + |a_n|^2, \tag{A.25}$$

and the components themselves are

$$a_i = \langle e_i|\alpha\rangle. \tag{A.26}$$

(These results generalize the familiar formulas $\mathbf{a} \cdot \mathbf{b} = a_x b_x + a_y b_y + a_z b_z$, $|\mathbf{a}|^2 = a_x^2 + a_y^2 + a_z^2$, and $a_x = \hat{\imath} \cdot \mathbf{a}$, $a_y = \hat{\jmath} \cdot \mathbf{a}$, $a_z = \hat{k} \cdot \mathbf{a}$, for the three-dimensional orthonormal basis $\hat{\imath}$, $\hat{\jmath}$, $\hat{k}$.) From now on we shall *always* work in orthonormal bases, unless it is explicitly indicated otherwise.

Another geometrical quantity one might wish to generalize is the *angle* between two vectors. In ordinary vector analysis $\cos\theta = (\mathbf{a} \cdot \mathbf{b})/|\mathbf{a}||\mathbf{b}|$. But because the inner product is in general a complex number, the analogous formula (in an arbitrary inner product space) does not define a (real) angle θ. Nevertheless, it is still true that the *absolute value* of this quantity is a number no greater than 1,

$$|\langle\alpha|\beta\rangle|^2 \le \langle\alpha|\alpha\rangle\langle\beta|\beta\rangle. \qquad [A.27]$$

(This important result is known as the **Schwarz inequality**; the proof is given in Problem A.5.) So you can, if you like, define the angle between $|\alpha\rangle$ and $|\beta\rangle$ by the formula

$$\cos\theta = \sqrt{\frac{\langle\alpha|\beta\rangle\langle\beta|\alpha\rangle}{\langle\alpha|\alpha\rangle\langle\beta|\beta\rangle}}. \qquad [A.28]$$

***Problem A.4** Suppose you start out with a basis $(|e_1\rangle, |e_2\rangle, \ldots, |e_n\rangle)$ that is *not* orthonormal. The **Gram-Schmidt procedure** is a systematic ritual for generating from it an orthonormal basis $(|e'_1\rangle, |e'_2\rangle, \ldots, |e'_n\rangle)$. It goes like this:

(i) Normalize the first basis vector (divide by its norm):

$$|e'_1\rangle = \frac{|e_1\rangle}{\|e_1\|}.$$

(ii) Find the projection of the second vector along the first, and subtract it off:

$$|e_2\rangle - \langle e'_1|e_2\rangle|e'_1\rangle.$$

This vector is orthogonal to $|e'_1\rangle$; normalize it to get $|e'_2\rangle$.

(iii) Subtract from $|e_3\rangle$ its projections along $|e'_1\rangle$ and $|e'_2\rangle$:

$$|e_3\rangle - \langle e'_1|e_3\rangle|e'_1\rangle - \langle e'_2|e_3\rangle|e'_2\rangle.$$

This is orthogonal to $|e'_1\rangle$ and $|e'_2\rangle$; normalize it to get $|e'_3\rangle$. And so on.

Use the Gram-Schmidt procedure to orthonormalize the 3-space basis $|e_1\rangle = (1+i)\hat{\imath} + (1)\hat{\jmath} + (i)\hat{k}$, $|e_2\rangle = (i)\hat{\imath} + (3)\hat{\jmath} + (1)\hat{k}$, $|e_3\rangle = (0)\hat{\imath} + (28)\hat{\jmath} + (0)\hat{k}$.

Problem A.5 Prove the Schwarz inequality (Equation A.27). *Hint:* Let $|\gamma\rangle = |\beta\rangle - (\langle\alpha|\beta\rangle/\langle\alpha|\alpha\rangle)|\alpha\rangle$, and use $\langle\gamma|\gamma\rangle \ge 0$.

Problem A.6 Find the angle (in the sense of Equation A.28) between the vectors $|\alpha\rangle = (1+i)\hat{\imath} + (1)\hat{\jmath} + (i)\hat{k}$ and $|\beta\rangle = (4-i)\hat{\imath} + (0)\hat{\jmath} + (2-2i)\hat{k}$.

Problem A.7 Prove the **triangle inequality**: $\|(|\alpha\rangle + |\beta\rangle)\| \le \|\alpha\| + \|\beta\|$.

A.3 MATRICES

Suppose you take every vector (in 3-space) and multiply it by 17, or you rotate every vector by 39° about the z-axis, or you reflect every vector in the xy plane—these are all examples of **linear transformations**. A linear transformation[6] ($\hat{T}$) takes each vector in a vector space and "transforms" it into some other vector ($|\alpha\rangle \to |\alpha'\rangle = \hat{T}|\alpha\rangle$), subject to the condition that the operation be *linear*:

$$\hat{T}(a|\alpha\rangle + b|\beta\rangle) = a(\hat{T}|\alpha\rangle) + b(\hat{T}|\beta\rangle), \qquad \text{[A.29]}$$

for any vectors $|\alpha\rangle$, $|\beta\rangle$ and any scalars a, b.

If you know what a particular linear transformation does to a set of *basis* vectors, you can easily figure out what it does to *any* vector. For suppose that

$$\begin{aligned}
\hat{T}|e_1\rangle &= T_{11}|e_1\rangle + T_{21}|e_2\rangle + \cdots + T_{n1}|e_n\rangle, \\
\hat{T}|e_2\rangle &= T_{12}|e_1\rangle + T_{22}|e_2\rangle + \cdots + T_{n2}|e_n\rangle, \\
&\cdots \\
\hat{T}|e_n\rangle &= T_{1n}|e_1\rangle + T_{2n}|e_2\rangle + \cdots + T_{nn}|e_n\rangle,
\end{aligned}$$

or, more compactly,

$$\hat{T}|e_j\rangle = \sum_{i=1}^{n} T_{ij}|e_i\rangle, \quad (j = 1, 2, \ldots, n). \qquad \text{[A.30]}$$

If $|\alpha\rangle$ is an arbitrary vector,

$$|\alpha\rangle = a_1|e_1\rangle + a_2|e_2\rangle + \cdots + a_n|e_n\rangle = \sum_{j=1}^{n} a_j|e_j\rangle, \qquad \text{[A.31]}$$

then

$$\hat{T}|\alpha\rangle = \sum_{j=1}^{n} a_j\left(\hat{T}|e_j\rangle\right) = \sum_{j=1}^{n}\sum_{i=1}^{n} a_j T_{ij}|e_i\rangle = \sum_{i=1}^{n}\left(\sum_{j=1}^{n} T_{ij}a_j\right)|e_i\rangle. \qquad \text{[A.32]}$$

Evidently $\hat{T}$ takes a vector with components $a_1, a_2, \ldots, a_n$ into a vector with components[7]

$$a_i' = \sum_{j=1}^{n} T_{ij}a_j. \qquad \text{[A.33]}$$

[6]In this chapter I'll use a hat (ˆ) to denote linear transformations; this is not inconsistent with my convention in the text (putting hats on operators), for (as we shall see) quantum operators *are* linear transformations.

[7]Notice the reversal of indices between Equations A.30 and A.33. This is not a typographical error. Another way of putting it (switching $i \leftrightarrow j$ in Equation A.30) is that if the *components* transform with T_{ij}, the *basis* vectors transform with T_{ji}.

Thus the n^2 **elements** T_{ij} uniquely characterize the linear transformation $\hat{T}$ (with respect to a given basis), just as the n components a_i uniquely characterize the vector $|\alpha\rangle$ (with respect to the same basis):

$$\hat{T} \leftrightarrow (T_{11}, T_{12}, \ldots, T_{nn}). \qquad \text{[A.34]}$$

If the basis is orthonormal, it follows from Equation A.30 that

$$T_{ij} = \langle e_i|\hat{T}|e_j\rangle. \qquad \text{[A.35]}$$

It is convenient to display these complex numbers in the form of a **matrix**:[8]

$$\mathsf{T} = \begin{pmatrix} T_{11} & T_{12} & \ldots & T_{1n} \\ T_{21} & T_{22} & \ldots & T_{2n} \\ \vdots & \vdots & & \vdots \\ T_{n1} & T_{n2} & \ldots & T_{nn} \end{pmatrix} \qquad \text{[A.36]}$$

The study of linear transformations reduces then to the theory of matrices. The *sum* of two linear transformations $(\hat{S} + \hat{T})$ is defined in the natural way:

$$(\hat{S} + \hat{T})|\alpha\rangle = \hat{S}|\alpha\rangle + \hat{T}|\alpha\rangle; \qquad \text{[A.37]}$$

this matches the usual rule for adding matrices (you add their corresponding elements):

$$\mathsf{U} = \mathsf{S} + \mathsf{T} \Leftrightarrow U_{ij} = S_{ij} + T_{ij}. \qquad \text{[A.38]}$$

The *product* of two linear transformations $(\hat{S}\hat{T})$ is the net effect of performing them in succession—first $\hat{T}$, then $\hat{S}$:

$$|\alpha'\rangle = \hat{T}|\alpha\rangle; \quad |\alpha''\rangle = \hat{S}|\alpha'\rangle = \hat{S}(\hat{T}|\alpha\rangle) = \hat{S}\hat{T}|\alpha\rangle. \qquad \text{[A.39]}$$

What matrix U represents the combined transformation $\hat{U} = \hat{S}\hat{T}$? It's not hard to work it out:

$$a_i'' = \sum_{j=1}^{n} S_{ij}a_j' = \sum_{j=1}^{n} S_{ij}\left(\sum_{k=1}^{n} T_{jk}a_k\right) = \sum_{k=1}^{n}\left(\sum_{j=1}^{n} S_{ij}T_{jk}\right)a_k = \sum_{k=1}^{n} U_{ik}a_k.$$

Evidently

$$\mathsf{U} = \mathsf{S}\mathsf{T} \Leftrightarrow U_{ik} = \sum_{j=1}^{n} S_{ij}T_{jk}. \qquad \text{[A.40]}$$

[8] I'll use boldface capital letters, sans serif, to denote square matrices.

This is the standard rule for matrix multiplication—to find the ikth element of the product **ST**, you look at the ith row of **S**, and the kth column of **T**, multiply corresponding entries, and add. The same prescription allows you to multiply *rectangular* matrices, as long as the number of columns in the first matches the number of rows in the second. In particular, if we write the n-tuple of components of $|\alpha\rangle$ as an $n \times 1$ **column matrix** (or "column vector"):[9]

$$\mathbf{a} \equiv \begin{pmatrix} a_1 \\ a_2 \\ \vdots \\ a_n \end{pmatrix} \qquad [A.41]$$

the transformation rule (Equation A.33) can be expressed as a matrix product:

$$\mathbf{a}' = \mathbf{T}\mathbf{a}. \qquad [A.42]$$

Now some matrix terminology:

- The **transpose** of a matrix (which we shall write with a tilde: $\tilde{\mathbf{T}}$) is the same set of elements, but with rows and columns interchanged. In particular, the transpose of a *column* matrix is a **row matrix**:

$$\tilde{\mathbf{a}} = \begin{pmatrix} a_1 & a_2 & \dots & a_n \end{pmatrix}. \qquad [A.43]$$

For a *square* matrix taking the transpose amounts to reflecting in the **main diagonal** (upper left to lower right):

$$\tilde{\mathbf{T}} = \begin{pmatrix} T_{11} & T_{21} & \dots & T_{n1} \\ T_{12} & T_{22} & \dots & T_{n2} \\ \vdots & \vdots & & \vdots \\ T_{1n} & T_{2n} & \dots & T_{nn} \end{pmatrix} \qquad [A.44]$$

A (square) matrix is **symmetric** if it is equal to its transpose; it is **antisymmetric** if this operation reverses the sign:

$$\text{symmetric} : \tilde{\mathbf{T}} = \mathbf{T}; \quad \text{antisymmetric} : \tilde{\mathbf{T}} = -\mathbf{T}. \qquad [A.45]$$

- The (complex) **conjugate** of a matrix (which we denote, as usual, with an asterisk, $\mathbf{T}^*$), consists of the complex conjugate of every element:

$$\mathbf{T}^* = \begin{pmatrix} T_{11}^* & T_{12}^* & \dots & T_{1n}^* \\ T_{21}^* & T_{22}^* & \dots & T_{2n}^* \\ \vdots & \vdots & & \vdots \\ T_{n1}^* & T_{n2}^* & \dots & T_{nn}^* \end{pmatrix}; \quad \mathbf{a}^* = \begin{pmatrix} a_1^* \\ a_2^* \\ \vdots \\ a_n^* \end{pmatrix}. \qquad [A.46]$$

[9] I'll use boldface lowercase letters, sans serif, for row and column matrices.

A matrix is **real** if all its elements are real, and **imaginary** if they are all imaginary:

$$\text{real} : \mathbf{T}^* = \mathbf{T}; \quad \text{imaginary} : \mathbf{T}^* = -\mathbf{T}. \qquad [A.47]$$

- The **hermitian conjugate** (or **adjoint**) of a matrix (indicated by a dagger, $\mathbf{T}^\dagger$) is the transpose conjugate:

$$\mathbf{T}^\dagger \equiv \tilde{\mathbf{T}}^* = \begin{pmatrix} T_{11}^* & T_{21}^* & \dots & T_{n1}^* \\ T_{12}^* & T_{22}^* & \dots & T_{n2}^* \\ \vdots & \vdots & & \vdots \\ T_{1n}^* & T_{2n}^* & \dots & T_{nn}^* \end{pmatrix}; \quad \mathbf{a}^\dagger \equiv \tilde{\mathbf{a}}^* = \begin{pmatrix} a_1^* & a_2^* & \dots & a_n^* \end{pmatrix}. \qquad [A.48]$$

A square matrix is **hermitian** (or **self-adjoint**) if it is equal to its hermitian conjugate; if hermitian conjugation introduces a minus sign, the matrix is **skew hermitian** (or **anti-hermitian**):

$$\text{hermitian} : \mathbf{T}^\dagger = \mathbf{T}; \quad \text{skew hermitian} : \mathbf{T}^\dagger = -\mathbf{T}. \qquad [A.49]$$

In this notation the inner product of two vectors (with respect to an orthonormal basis—Equation A.24), can be written very neatly as a matrix product:

$$\langle\alpha|\beta\rangle = \mathbf{a}^\dagger\mathbf{b}. \qquad [A.50]$$

Notice that each of the three operations defined in this paragraph, if applied twice, returns you to the original matrix.

Matrix multiplication is not, in general, commutative ($\mathbf{ST} \neq \mathbf{TS}$); the *difference* between the two orderings is called the **commutator**:[10]

$$[\mathbf{S}, \mathbf{T}] \equiv \mathbf{ST} - \mathbf{TS}. \qquad [A.51]$$

The transpose of a product is the product of the transposes *in reverse order*:

$$(\widetilde{\mathbf{ST}}) = \tilde{\mathbf{T}}\tilde{\mathbf{S}}, \qquad [A.52]$$

(see Problem A.11), and the same goes for hermitian conjugates:

$$(\mathbf{ST})^\dagger = \mathbf{T}^\dagger\mathbf{S}^\dagger \qquad [A.53]$$

The **unit matrix** (representing a linear transformation that carries every vector into itself) consists of ones on the main diagonal, and zeroes everywhere else:

$$\mathbf{I} \equiv \begin{pmatrix} 1 & 0 & \dots & 0 \\ 0 & 1 & \dots & 0 \\ \vdots & \vdots & & \vdots \\ 0 & 0 & \dots & 1 \end{pmatrix} \qquad [A.54]$$

---

[10] The commutator only makes sense for *square* matrices, of course; for rectangular matrices the two orderings wouldn't even be the same size.

In other words,

$$\mathbf{I}_{ij} = \delta_{ij}. \tag{A.55}$$

The **inverse** of a (square) matrix (written $\mathbf{T}^{-1}$) is defined in the obvious way:[11]

$$\mathbf{T}^{-1}\mathbf{T} = \mathbf{T}\mathbf{T}^{-1} = \mathbf{I}. \tag{A.56}$$

A matrix has an inverse if and only if its **determinant**[12] is nonzero; in fact,

$$\mathbf{T}^{-1} = \frac{1}{\det \mathbf{T}}\tilde{\mathbf{C}}, \tag{A.57}$$

where $\mathbf{C}$ is the matrix of **cofactors** (the cofactor of element T_{ij} is $(-1)^{i+j}$ times the determinant of the submatrix obtained from $\mathbf{T}$ by erasing the ith row and the jth column). A matrix that has no inverse is said to be **singular**. The inverse of a product (assuming it exists) is the product of the inverses *in reverse order*:

$$(\mathbf{ST})^{-1} = \mathbf{T}^{-1}\mathbf{S}^{-1} \tag{A.58}$$

A matrix is **unitary** if its inverse is equal to its hermitian conjugate:[13]

$$\text{unitary} : \mathbf{U}^{\dagger} = \mathbf{U}^{-1} \tag{A.59}$$

Assuming the basis is orthonormal, the columns of a unitary matrix constitute an orthonormal set, and so too do its rows (see Problem A.12). Linear transformations represented by unitary matrices preserve inner products, since (Equation A.50)

$$\langle \alpha' | \beta' \rangle = \mathbf{a}'^{\dagger}\mathbf{b} = (\mathbf{Ua})^{\dagger}(\mathbf{Ub}) = \mathbf{a}^{\dagger}\mathbf{U}^{\dagger}\mathbf{U}\mathbf{b} = \mathbf{a}^{\dagger}\mathbf{b} = \langle \alpha | \beta \rangle. \tag{A.60}$$

***Problem A.8** Given the following two matrices:

$$\mathbf{A} = \begin{pmatrix} -1 & 1 & i \\ 2 & 0 & 3 \\ 2i & -2i & 2 \end{pmatrix}, \quad \mathbf{B} = \begin{pmatrix} 2 & 0 & -i \\ 0 & 1 & 0 \\ i & 3 & 2 \end{pmatrix}$$

---

[11] Note that the left inverse is equal to the right inverse, for if $\mathbf{AT} = \mathbf{I}$ and $\mathbf{TB} = \mathbf{I}$, then (multiplying the second on the left by $\mathbf{A}$ and invoking the first) we get $\mathbf{B} = \mathbf{A}$.

[12] I assume you know how to evaluate determinants. If not, see M. Boas, *Mathematical Methods in the Physical Sciences*, 2nd ed. (John Wiley, New York, 1983), Section 3.3.

[13] In a *real* vector space (that is, one in which the scalars are real) the hermitian conjugate is the same as the transpose, and a unitary matrix is **orthogonal**: $\tilde{\mathbf{O}} = \mathbf{O}^{-1}$. For example, rotations in ordinary 3-space are represented by orthogonal matrices.

compute: (a) $\mathsf{A}+\mathsf{B}$, (b) AB, (c) $[\mathsf{A},\mathsf{B}]$, (d) $\tilde{\mathsf{A}}$, (e) A^*, (f) $\mathsf{A}^\dagger$, (g) $\text{Tr}(\mathsf{B})$, (h) $\det(\mathsf{B})$, and (i) B^{-1}. Check that $\mathsf{BB}^{-1} = \mathsf{I}$. Does A have an inverse?

∗**Problem A.9** Using the square matrices in Problem A.8, and the column matrices

$$\mathsf{a} = \begin{pmatrix} i \\ 2i \\ 2 \end{pmatrix}, \quad \mathsf{b} = \begin{pmatrix} 2 \\ (1-i) \\ 0 \end{pmatrix},$$

find: (a) Aa, (b) $\mathsf{a}^\dagger\mathsf{b}$, (c) $\tilde{\mathsf{a}}\mathsf{Bb}$, (d) $\mathsf{ab}^\dagger$

Problem A.10 By explicit construction of the matrices in question, show that any matrix T can be written

(a) as the sum of a symmetric matrix S and an antisymmetric matrix A;

(b) as the sum of a real matrix R and an imaginary matrix M;

(c) as the sum of a hermitian matrix H and a skew-hermitian matrix K.

∗**Problem A.11** Prove Equations A.52, A.53, and A.58. Show that the product of two unitary matrices is unitary. Under what conditions is the product of two hermitian matrices hermitian? Is the sum of two unitary matrices unitary? Is the sum of two hermitian matrices hermitian?

Problem A.12 Show that the rows and columns of a unitary matrix constitute orthonormal sets.

Problem A.13 Noting that $\det(\tilde{\mathsf{T}}) = \det(\mathsf{T})$, show that the determinant of a hermitian matrix is real, the determinant of a unitary matrix has modulus 1 (hence the name), and the determinant of an orthogonal matrix is either $+1$ or -1.

A.4 CHANGING BASES

The components of a vector depend, of course, on your (arbitrary) choice of basis, and so do the elements of the matrix representing a linear transformation. We might inquire how these numbers change when we switch to a different basis.

The old basis vectors, $|e_i\rangle$ are—like *all* vectors—linear combinations of the new ones, $|f_i\rangle$:

$$\begin{aligned}|e_1\rangle &= S_{11}|f_1\rangle + S_{21}|f_2\rangle + \cdots + S_{n1}|f_n\rangle,\\ |e_2\rangle &= S_{12}|f_1\rangle + S_{22}|f_2\rangle + \cdots + S_{n2}|f_n\rangle,\\ &\cdots\\ |e_n\rangle &= S_{1n}|f_1\rangle + S_{2n}|f_2\rangle + \cdots + S_{nn}|f_n\rangle,\end{aligned}$$

(for some set of complex numbers S_{ij}), or, more compactly,

$$|e_j\rangle = \sum_{i=1}^{n} S_{ij}|f_i\rangle, \quad (j = 1, 2, \ldots, n). \qquad [A.61]$$

This is *itself* a linear transformation (compare Equation A.30),[14] and we know immediately how the components transform:

$$a_i^f = \sum_{j=1}^{n} S_{ij} a_j^e, \qquad [A.62]$$

(where the superscript indicates the basis). In matrix form

$$\mathbf{a}^f = \mathbf{S}\mathbf{a}^e. \qquad [A.63]$$

What about the matrix representing a linear transformation $\hat{T}$—how is *it* modified by a change of basis? Well, in the old basis we had (Equation A.42)

$$\mathbf{a}^{e'} = \mathbf{T}^e \mathbf{a}^e,$$

and Equation A.63—multiplying both sides by $\mathbf{S}^{-1}$—entails[15] $\mathbf{a}^e = \mathbf{S}^{-1}\mathbf{a}^f$, so

$$\mathbf{a}^{f'} = \mathbf{S}\mathbf{a}^{e'} = \mathbf{S}(\mathbf{T}^e\mathbf{a}^e) = \mathbf{S}\mathbf{T}^e\mathbf{S}^{-1}\mathbf{a}^f$$

Evidently

$$\mathbf{T}^f = \mathbf{S}\mathbf{T}^e\mathbf{S}^{-1} \qquad [A.64]$$

In general, two matrices ($\mathbf{T}_1$ and $\mathbf{T}_2$) are said to be **similar** if $\mathbf{T}_2 = \mathbf{S}\mathbf{T}_1\mathbf{S}^{-1}$ for some (nonsingular) matrix $\mathbf{S}$. What we have just found is that *matrices representing*

[14]Notice, however, the radically different perspective: In this case we're talking about one and the same *vector*, referred to two completely different *bases*, whereas before we were thinking of a completely *different* vector, referred to the *same* basis.

[15]Note that $\mathbf{S}^{-1}$ certainly exists—if $\mathbf{S}$ were singular, the $|f_i\rangle$'s would not span the space, so they wouldn't constitute a basis.

the same linear transformation, with respect to different bases, are similar. Incidentally, if the first basis is orthonormal, the second will also be orthonormal if and only if the matrix $\mathbf{S}$ is *unitary* (see Problem A.16). Since we always work in orthonormal bases, we are interested mainly in *unitary* similarity transformations.

While the *elements* of the matrix representing a given linear transformation may look very different in the new basis, two numbers associated with the matrix are unchanged: the determinant and the **trace**. For the determinant of a product is the product of the determinants, and hence

$$\det(\mathbf{T}^f) = \det(\mathbf{S}\mathbf{T}^e\mathbf{S}^{-1}) = \det(\mathbf{S})\det(\mathbf{T}^e)\det(\mathbf{S}^{-1}) = \det \mathbf{T}^e. \qquad [A.65]$$

And the trace, which is the *sum of the diagonal elements*,

$$\mathrm{Tr}(\mathbf{T}) \equiv \sum_{i=1}^{m} T_{ii}, \qquad [A.66]$$

has the property (see Problem A.17) that

$$\mathrm{Tr}(\mathbf{T}_1\mathbf{T}_2) = \mathrm{Tr}(\mathbf{T}_2\mathbf{T}_1), \qquad [A.67]$$

(for any two matrices $\mathbf{T}_1$ and $\mathbf{T}_2$), so

$$\mathrm{Tr}(\mathbf{T}^f) = \mathrm{Tr}(\mathbf{S}\mathbf{T}^e\mathbf{S}^{-1}) = \mathrm{Tr}(\mathbf{T}_e\mathbf{S}^{-1}\mathbf{S}) = \mathrm{Tr}(\mathbf{T}^e). \qquad [A.68]$$

Problem A.14 Using the standard basis $(\hat{\imath}, \hat{\jmath}, \hat{k})$ for vectors in three dimensions:

(a) Construct the matrix representing a rotation through angle θ (counterclockwise, looking down the axis toward the origin) about the z-axis.

(b) Construct the matrix representing a rotation by 120° (counterclockwise, looking down the axis) about an axis through the point (1,1,1).

(c) Construct the matrix representing reflection through the xy-plane.

(d) Check that all these matrices are orthogonal, and calculate their determinants.

Problem A.15 In the usual basis $(\hat{\imath}, \hat{\jmath}, \hat{k})$, construct the matrix $\mathbf{T}_x$ representing a rotation through angle θ about the x-axis, and the matrix $\mathbf{T}_y$ representing a rotation through angle θ about the y-axis. Suppose now we change bases, to $\hat{\imath}' = \hat{\jmath}$, $\hat{\jmath}' = -\hat{\imath}$, $\hat{k}' = \hat{k}$. Construct the matrix $\mathbf{S}$ that effects this change of basis, and check that $\mathbf{S}\mathbf{T}_x\mathbf{S}^{-1}$ and $\mathbf{S}\mathbf{T}_y\mathbf{S}^{-1}$ are what you would expect.

Problem A.16 Show that similarity preserves matrix multiplication (that is, if $\mathbf{A}^e\mathbf{B}^e = \mathbf{C}^e$, then $\mathbf{A}^f\mathbf{B}^f = \mathbf{C}^f$). Similarity does *not*, in general, preserve symmetry, reality, or hermiticity; show, however, that if $\mathbf{S}$ is *unitary*, and $\mathbf{H}^e$ is hermitian, then $\mathbf{H}^f$ is hermitian. Show that $\mathbf{S}$ carries an orthonormal basis into another orthonormal basis if and only if it is unitary.

$*$**Problem A.17** Prove that $\mathrm{Tr}(\mathbf{T}_1\mathbf{T}_2) = \mathrm{Tr}(\mathbf{T}_2\mathbf{T}_1)$. It follows immediately that $\mathrm{Tr}(\mathbf{T}_1\mathbf{T}_2\mathbf{T}_3) = \mathrm{Tr}(\mathbf{T}_2\mathbf{T}_3\mathbf{T}_1)$, but is it the case that $\mathrm{Tr}(\mathbf{T}_1\mathbf{T}_2\mathbf{T}_3) = \mathrm{Tr}(\mathbf{T}_2\mathbf{T}_1\mathbf{T}_3)$, in general? Prove it, or disprove it. *Hint:* The best disproof is always a counterexample—the simpler the better!

A.5 EIGENVECTORS AND EIGENVALUES

Consider the linear transformation in three-space consisting of a rotation, about some specified axis, by an angle θ. Most vectors will change in a rather complicated way (they ride around on a cone about the axis), but vectors that happen to lie *along* the axis have very simple behavior: They don't change at all ($\hat{T}|\alpha\rangle = |\alpha\rangle$). If θ is 180°, then vectors which lie in the "equatorial" plane reverse signs ($\hat{T}|\alpha\rangle = -|\alpha\rangle$). In a complex vector space[16] *every* linear transformation has "special" vectors like these, which are transformed into scalar multiples of themselves:

$$\hat{T}|\alpha\rangle = \lambda|\alpha\rangle; \qquad [A.69]$$

they are called **eigenvectors** of the transformation, and the (complex) number λ is their **eigenvalue**. (The *null* vector doesn't count, even though in a trivial sense it obeys Equation A.69 for *any* $\hat{T}$ and *any* λ; technically, an eigenvector is any *nonzero* vector satisfying Equation A.69.) Notice that any (nonzero) *multiple* of an eigenvector is still an eigenvector, with the same eigenvalue.

With respect to a particular basis, the eigenvector equation assumes the matrix form

$$\mathbf{Ta} = \lambda\mathbf{a}, \qquad [A.70]$$

(for nonzero $\mathbf{a}$), or

$$(\mathbf{T} - \lambda\mathbf{I})\mathbf{a} = \mathbf{0}. \qquad [A.71]$$

(Here $\mathbf{0}$ is the **zero matrix**, whose elements are all zero.) Now, if the matrix $(\mathbf{T} - \lambda\mathbf{I})$ had an *inverse*, we could multiply both sides of Equation A.71 by $(\mathbf{T} - \lambda\mathbf{I})^{-1}$, and

[16]This is *not* always true in a *real* vector space (where the scalars are restricted to real values). See Problem A.18.

conclude that $\mathbf{a} = \mathbf{0}$. But by assumption $\mathbf{a}$ is *not* zero, so the matrix $(\mathbf{T} - \lambda\mathbf{I})$ must in fact be singular, which means that its determinant is zero:

$$\det(\mathbf{T} - \lambda\mathbf{I}) = \begin{vmatrix} (T_{11} - \lambda) & T_{12} & \dots & T_{1n} \\ T_{21} & (T_{22} - \lambda) & \dots & T_{2n} \\ \vdots & \vdots & & \vdots \\ T_{n1} & T_{n2} & \dots & (T_{nn} - \lambda) \end{vmatrix} = 0. \qquad [A.72]$$

Expansion of the determinant yields an algebraic equation for λ:

$$C_n\lambda^n + C_{n-1}\lambda^{n-1} + \cdots + C_1\lambda + C_0 = 0, \qquad [A.73]$$

where the coefficients C_i depend on the elements of $\mathbf{T}$ (see Problem A.20). This is called the **characteristic equation** for the matrix; its solutions determine the eigenvalues. Notice that it's an nth-order equation, so (by the **fundamental theorem of algebra**) it has n (complex) roots.[17] However, some of these may be multiple roots, so all we can say for certain is that an $n \times n$ matrix has *at least one* and *at most n* distinct eigenvalues. The collection of all the eigenvalues of a matrix is called its **spectrum**; if two (or more) linearly independent eigenvectors share the same eigenvalue, the spectrum is said to be **degenerate**.

To construct the eigen*vectors* it is generally easiest simply to plug each λ back into Equation A.70 and solve "by hand" for the components of $\mathbf{a}$. I'll show you how it goes by working out an example.

Example A.1 Find the eigenvalues and eigenvectors of the following matrix:

$$\mathbf{M} = \begin{pmatrix} 2 & 0 & -2 \\ -2i & i & 2i \\ 1 & 0 & -1 \end{pmatrix}. \qquad [A.74]$$

Solution: The characteristic equation is

$$\begin{vmatrix} (2 - \lambda) & 0 & -2 \\ -2i & (i - \lambda) & 2i \\ 1 & 0 & (-1 - \lambda) \end{vmatrix} = -\lambda^3 + (1 + i)\lambda^2 - i\lambda = 0, \qquad [A.75]$$

and its roots are 0, 1, and i. Call the components of the first eigenvector (a_1, a_2, a_3); then

$$\begin{pmatrix} 2 & 0 & -2 \\ -2i & i & 2i \\ 1 & 0 & -1 \end{pmatrix} \begin{pmatrix} a_1 \\ a_2 \\ a_3 \end{pmatrix} = 0 \begin{pmatrix} a_1 \\ a_2 \\ a_3 \end{pmatrix} = \begin{pmatrix} 0 \\ 0 \\ 0 \end{pmatrix},$$

[17] It is here that the case of *real* vector spaces becomes more awkward, because the characteristic equation need not have any (real) solutions at all. See Problem A.18.

which yields three equations:

$$2a_1 - 2a_3 = 0,$$
$$-2ia_1 + ia_2 + 2ia_3 = 0,$$
$$a_1 - a_3 = 0.$$

The first determines a_3 (in terms of a_1): $a_3 = a_1$; the second determines a_2: $a_2 = 0$; and the third is redundant. We may as well pick $a_1 = 1$ (since any multiple of an eigenvector is still an eigenvector):

$$\mathbf{a}^{(1)} = \begin{pmatrix} 1 \\ 0 \\ 1 \end{pmatrix}, \quad \text{for } \lambda_1 = 0. \qquad [A.76]$$

For the second eigenvector (recycling the same notation for the components) we have

$$\begin{pmatrix} 2 & 0 & -2 \\ -2i & i & 2i \\ 1 & 0 & -1 \end{pmatrix} \begin{pmatrix} a_1 \\ a_2 \\ a_3 \end{pmatrix} = 1 \begin{pmatrix} a_1 \\ a_2 \\ a_3 \end{pmatrix} = \begin{pmatrix} a_1 \\ a_2 \\ a_3 \end{pmatrix},$$

which leads to the equations

$$2a_1 - 2a_3 = a_1,$$
$$-2ia_1 + ia_2 + 2ia_3 = a_2,$$
$$a_1 - a_3 = a_3,$$

with the solution $a_3 = (1/2)a_1$, $a_2 = [(1-i)/2]a_1$; this time I'll pick $a_1 = 2$, so

$$\mathbf{a}^{(2)} = \begin{pmatrix} 2 \\ 1-i \\ 1 \end{pmatrix}, \quad \text{for } \lambda_2 = 1. \qquad [A.77]$$

Finally, for the third eigenvector,

$$\begin{pmatrix} 2 & 0 & -2 \\ -2i & i & 2i \\ 1 & 0 & -1 \end{pmatrix} \begin{pmatrix} a_1 \\ a_2 \\ a_3 \end{pmatrix} = i \begin{pmatrix} a_1 \\ a_2 \\ a_3 \end{pmatrix} = \begin{pmatrix} ia_1 \\ ia_2 \\ ia_3 \end{pmatrix},$$

which gives the equations

$$2a_1 - 2a_3 = ia_1,$$
$$-2ia_1 + ia_2 + 2ia_3 = ia_2,$$
$$a_1 - a_3 = ia_3,$$

whose solution is $a_3 = a_1 = 0$, with a_2 undetermined. Choosing $a_2 = 1$, we conclude

$$\mathbf{a}^{(3)} = \begin{pmatrix} 0 \\ 1 \\ 0 \end{pmatrix}, \quad \text{for } \lambda_3 = i. \qquad \text{[A.78]}$$

If the eigenvectors span the space (as they do in the preceding example), we are free to use *them* as a basis:

$$\begin{aligned} \hat{T}|f_1\rangle &= \lambda_1|f_1\rangle, \\ \hat{T}|f_2\rangle &= \lambda_2|f_2\rangle, \\ &\dots \\ \hat{T}|f_n\rangle &= \lambda_n|f_n\rangle. \end{aligned}$$

In this basis the matrix representing $\hat{T}$ takes on a very simple form, with the eigenvalues strung out along the main diagonal, and all other elements zero:

$$\mathbf{T} = \begin{pmatrix} \lambda_1 & 0 & \dots & 0 \\ 0 & \lambda_2 & \dots & 0 \\ \vdots & \vdots & & \vdots \\ 0 & 0 & \dots & \lambda_n \end{pmatrix}, \qquad \text{[A.79]}$$

and the (normalized) eigenvectors are

$$\begin{pmatrix} 1 \\ 0 \\ 0 \\ \vdots \\ 0 \end{pmatrix}, \begin{pmatrix} 0 \\ 1 \\ 0 \\ \vdots \\ 0 \end{pmatrix}, \dots, \begin{pmatrix} 0 \\ 0 \\ 0 \\ \vdots \\ 1 \end{pmatrix}. \qquad \text{[A.80]}$$

A matrix that can be brought to **diagonal form** (Equation A.79) by a change of basis is said to be **diagonalizable** (evidently a matrix is diagonalizable if and only if its eigenvectors span the space). The similarity matrix that effects the diagonalization can be constructed by using the normalized eigenvectors (in the old basis) as the columns of $\mathbf{S}^{-1}$:

$$(\mathbf{S}^{-1})_{ij} = (\mathbf{a}^{(j)})_i. \qquad \text{[A.81]}$$

Example A.2 In Example A.1,

$$\mathbf{S}^{-1} = \begin{pmatrix} 1 & 2 & 0 \\ 0 & (1-i) & 1 \\ 1 & 1 & 0 \end{pmatrix},$$

so (using Equation A.57)

$$\mathbf{S} = \begin{pmatrix} -1 & 0 & 2 \\ 1 & 0 & -1 \\ (i-1) & 1 & (1-i) \end{pmatrix}$$

and you can check for yourself that

$$\mathbf{Sa}^{(1)} = \begin{pmatrix} 1 \\ 0 \\ 0 \end{pmatrix}, \quad \mathbf{Sa}^{(2)} = \begin{pmatrix} 0 \\ 1 \\ 0 \end{pmatrix}, \quad \mathbf{Sa}^{(3)} = \begin{pmatrix} 0 \\ 0 \\ 1 \end{pmatrix},$$

and

$$\mathbf{SMS}^{-1} = \begin{pmatrix} 0 & 0 & 0 \\ 0 & 1 & 0 \\ 0 & 0 & i \end{pmatrix}.$$

There's an obvious advantage in bringing a matrix to diagonal form: It's much easier to work with. Unfortunately, not every matrix *can* be diagonalized—the eigenvectors have to span the space. If the characteristic equation has n distinct roots, then the matrix is certainly diagonalizable, but it *may* be diagonalizable even if there are multiple roots. (For an example of a matrix that *cannot* be diagonalized, see Problem A.19.) It would be handy to know in advance (before working out all the eigenvectors) whether a given matrix is diagonalizable. A useful sufficient (though not necessary) condition is the following: A matrix is said to be **normal** if it commutes with its hermitian conjugate:

$$\text{normal}: \ [\mathbf{N}^\dagger, \mathbf{N}] = \mathbf{0}. \qquad [A.82]$$

Every normal matrix is diagonalizable (its eigenvectors span the space). In particular, every hermitian matrix, and every unitary matrix, is diagonalizable.

Suppose we have *two* diagonalizable matrices; in quantum applications the question often arises: Can they be **simultaneously diagonalized** (by the *same* similarity matrix **S**)? That is to say, does there exist a basis in which they are *both* diagonal? The answer is yes *if and only if the two matrices commute* (see Problem A.22).

∗**Problem A.18** The 2×2 matrix representing a rotation of the xy plane is

$$\mathbf{T} = \begin{pmatrix} \cos\theta & -\sin\theta \\ \sin\theta & \cos\theta \end{pmatrix}. \qquad [A.83]$$

Show that (except for certain special angles—what are they?) this matrix has no real eigenvalues. (This reflects the geometrical fact that no vector in the plane is carried into itself under such a rotation; contrast rotations in *three* dimensions.) This matrix *does*, however, have *complex* eigenvalues and eigenvectors. Find them. Construct a matrix $\mathbf{S}$ that diagonalizes $\mathbf{T}$. Perform the similarity transformation ($\mathbf{STS}^{-1}$) explicitly, and show that it reduces $\mathbf{T}$ to diagonal form.

Problem A.19 Find the eigenvalues and eigenvectors of the following matrix:

$$\mathbf{M} = \begin{pmatrix} 1 & 1 \\ 0 & 1 \end{pmatrix}.$$

Can this matrix be diagonalized?

Problem A.20 Show that the first, second, and last coefficients in the characteristic equation (Equation A.73) are:

$$C_n = (-1)^n, \quad C_{n-1} = (-1)^{n-1}\mathrm{Tr}(\mathbf{T}), \quad \text{and} \quad C_0 = \det(\mathbf{T}). \qquad [A.84]$$

For a 3×3 matrix with elements T_{ij}, what is C_1?

Problem A.21 It's obvious that the trace of a *diagonal* matrix is the sum of its eigenvalues, and its determinant is their product (just look at Equation A.79). It follows (from Equations A.65 and A.68) that the same holds for any *diagonalizable* matrix. Prove that in fact

$$\det(\mathbf{T}) = \lambda_1\lambda_2\cdots\lambda_n, \quad \mathrm{Tr}(\mathbf{T}) = \lambda_1 + \lambda_2 + \cdots + \lambda_n. \qquad [A.85]$$

for *any* matrix. (The λ's are the n solutions to the characteristic equation—in the case of multiple roots, there may be fewer linearly independent eigen*vectors* than there are solutions, but we still count each λ as many times as it occurs.) *Hint:* Write the characteristic equation in the form

$$(\lambda_1 - \lambda)(\lambda_2 - \lambda)\cdots(\lambda_n - \lambda) = 0,$$

and use the result of Problem A.20.

Problem A.22

(a) Show that if two matrices commute in *one* basis, then they commute in any basis. That is:

$$[\mathbf{T}_1^e, \mathbf{T}_2^e] = \mathbf{0} \Rightarrow [\mathbf{T}_1^f, \mathbf{T}_2^f] = \mathbf{0}. \qquad [A.86]$$

Hint: Use Equation A.64.

(b) Show that if two matrices are simultaneously diagonalizable, they commute.[18]

Problem A.23 Consider the matrix

$$\mathbf{M} = \begin{pmatrix} 1 & 1 \\ 1 & i \end{pmatrix}$$

(a) Is it normal?

(b) Is it diagonalizable?

A.6 HERMITIAN TRANSFORMATIONS

In Equation A.48 I defined the hermitian conjugate (or "adjoint") of a *matrix* as its transpose-conjugate: $\mathbf{T}^\dagger = \tilde{\mathbf{T}}^*$. Now I want to give you a more fundamental definition for the hermitian conjugate of a *linear transformation*: It is that transformation $\hat{T}^\dagger$ which, when applied to the *first* member of an inner product, gives the same result as if $\hat{T}$ itself had been applied to the *second* vector:

$$\langle \hat{T}^\dagger \alpha | \beta \rangle = \langle \alpha | \hat{T} \beta \rangle, \qquad \text{[A.87]}$$

(for all vectors $|\alpha\rangle$ and $|\beta\rangle$).[19] I have to warn you that although everybody uses it, this is lousy notation. For α and β are not *vectors* (the *vectors* are $|\alpha\rangle$ and $|\beta\rangle$), they are *names*. In particular, they are endowed with no mathematical properties at all, and the expression "$\hat{T}\beta$" is literally *nonsense*: Linear transformations act on *vectors*, not *labels*. But it's pretty clear what the notation *means*: $\hat{T}\beta$ is the name of the vector $\hat{T}|\beta\rangle$, and $\langle \hat{T}^\dagger \alpha | \beta \rangle$ is the inner product of the vector $\hat{T}^\dagger |\alpha\rangle$ with the vector $|\beta\rangle$. Notice in particular that

$$\langle \alpha | c\beta \rangle = c \langle \alpha | \beta \rangle, \qquad \text{[A.88]}$$

whereas

$$\langle c\alpha | \beta \rangle = c^* \langle \alpha | \beta \rangle, \qquad \text{[A.89]}$$

for any scalar c.

---

[18] Proving the converse (that if two diagonalizable matrices commute then they are simultaneously diagonalizable) is not so simple. See for example Eugen Merzbacher, *Quantum Mechanics*, 3rd ed., Wiley, New York (1998), Section 10.4.

[19] You may wonder whether such a transformation necessarily exists. Good question! The answer is "yes." See, for instance, P. R. Halmos, *Finite Dimensional Vector Spaces*, 2nd ed., van Nostrand, Princeton (1958), Section 44.

If you're working in an orthonormal basis (as we always shall), the hermitian conjugate of a linear transformation is represented by the hermitian conjugate of the corresponding matrix; for (using Equations A.50 and A.53),

$$\langle\alpha|\hat{T}\beta\rangle = \mathbf{a}^\dagger\mathbf{T}\mathbf{b} = (\mathbf{T}^\dagger\mathbf{a})^\dagger\mathbf{b} = \langle\hat{T}^\dagger\alpha|\beta\rangle. \qquad [A.90]$$

So the terminology is consistent, and we can speak interchangeably in the language of transformations or of matrices.

In quantum mechanics, a fundamental role is played by **hermitian transformations** ($\hat{T}^\dagger = \hat{T}$). The eigenvectors and eigenvalues of a hermitian transformation have three crucial properties:

1. **The eigenvalues of a hermitian transformation are real.**

   Proof: Let λ be an eigenvalue of $\hat{T}$: $\hat{T}|\alpha\rangle = \lambda|\alpha\rangle$, with $|\alpha\rangle \neq |0\rangle$. Then

   $$\langle\alpha|\hat{T}\alpha\rangle = \langle\alpha|\lambda\alpha\rangle = \lambda\langle\alpha|\alpha\rangle.$$

   Meanwhile, if $\hat{T}$ is hermitian, then

   $$\langle\alpha|\hat{T}\alpha\rangle = \langle\hat{T}\alpha|\alpha\rangle = \langle\lambda\alpha|\alpha\rangle = \lambda^*\langle\alpha|\alpha\rangle.$$

   But $\langle\alpha|\alpha\rangle \neq 0$ (Equation A.20), so $\lambda = \lambda^*$, and hence λ is real. QED

2. **The eigenvectors of a hermitian transformation belonging to distinct eigenvalues are orthogonal.**

   Proof: Suppose $\hat{T}|\alpha\rangle = \lambda|\alpha\rangle$ and $\hat{T}|\beta\rangle = \mu|\beta\rangle$, with $\lambda \neq \mu$. Then

   $$\langle\alpha|\hat{T}\beta\rangle = \langle\alpha|\mu\beta\rangle = \mu\langle\alpha|\beta\rangle,$$

   and if $\hat{T}$ is hermitian,

   $$\langle\alpha|\hat{T}\beta\rangle = \langle\hat{T}\alpha|\beta\rangle = \langle\lambda\alpha|\beta\rangle = \lambda^*\langle\alpha|\beta\rangle.$$

   But $\lambda = \lambda^*$ (from **1**), and $\lambda \neq \mu$, by assumption, so $\langle\alpha|\beta\rangle = 0$. QED

3. **The eigenvectors of a hermitian transformation span the space.** As we have seen, this is equivalent to the statement that any hermitian matrix can be diagonalized (see Equation A.82). This rather technical fact is, in a sense, the mathematical support on which much of quantum mechanics leans. It turns out to be a thinner reed than one might have hoped, because the proof does not carry over to infinite-dimensional vector spaces.

Problem A.24 A hermitian linear transformation must satisfy $\langle\alpha|\hat{T}\beta\rangle = \langle\hat{T}\alpha|\beta\rangle$ for all vectors $|\alpha\rangle$ and $|\beta\rangle$. Prove that it is (surprisingly) sufficient that $\langle\gamma|\hat{T}\gamma\rangle = \langle\hat{T}\gamma|\gamma\rangle$ for all vectors $|\gamma\rangle$. *Hint:* First let $|\gamma\rangle = |\alpha\rangle + |\beta\rangle$, and then let $|\gamma\rangle = |\alpha\rangle + i|\beta\rangle$.

$*$**Problem A.25** Let

$$\mathbf{T} = \begin{pmatrix} 1 & 1-i \\ 1+i & 0 \end{pmatrix}.$$

(a) Verify that $\mathbf{T}$ is hermitian.

(b) Find its eigenvalues (note that they are real).

(c) Find and normalize the eigenvectors (note that they are orthogonal).

(d) Construct the unitary diagonalizing matrix $\mathbf{S}$, and check explicitly that it diagonalizes $\mathbf{T}$.

(e) Check that det($\mathbf{T}$) and Tr($\mathbf{T}$) are the same for $\mathbf{T}$ as they are for its diagonalized form.

$**$**Problem A.26** Consider the following hermitian matrix:

$$\mathbf{T} = \begin{pmatrix} 2 & i & 1 \\ -i & 2 & i \\ 1 & -i & 2 \end{pmatrix}.$$

(a) Calculate det($\mathbf{T}$) and Tr($\mathbf{T}$).

(b) Find the eigenvalues of $\mathbf{T}$. Check that their sum and product are consistent with (a), in the sense of Equation A.85. Write down the diagonalized version of $\mathbf{T}$.

(c) Find the eigenvectors of $\mathbf{T}$. Within the degenerate sector, construct two linearly independent eigenvectors (it is this step that is always possible for a *hermitian* matrix, but not for an *arbitrary* matrix—contrast Problem A.19). Orthogonalize them, and check that both are orthogonal to the third. Normalize all three eigenvectors.

(d) Construct the unitary matrix $\mathbf{S}$ that diagonalizes $\mathbf{T}$, and show explicitly that the similarity transformation using $\mathbf{S}$ reduces $\mathbf{T}$ to the appropriate diagonal form.

Problem A.27 A *unitary transformation* is one for which $\hat{U}^\dagger\hat{U} = 1$.

(a) Show that unitary transformations preserve inner products, in the sense that $\langle\hat{U}\alpha|\hat{U}\beta\rangle = \langle\alpha|\beta\rangle$, for all vectors $|\alpha\rangle$, $|\beta\rangle$.

(b) Show that the eigenvalues of a unitary transformation have modulus 1.

(c) Show that the eigenvectors of a unitary transformation belonging to distinct eigenvalues are orthogonal.

*****Problem A.28** *Functions* of matrices are defined by their Taylor series expansions; for example,

$$e^{\mathbf{M}} \equiv \mathbf{I} + \mathbf{M} + \frac{1}{2}\mathbf{M}^2 + \frac{1}{3!}\mathbf{M}^3 + \cdots \qquad \text{[A.91]}$$

(a) Find $\exp(\mathbf{M})$, if

$$\text{(i) } \mathbf{M} = \begin{pmatrix} 0 & 1 & 3 \\ 0 & 0 & 4 \\ 0 & 0 & 0 \end{pmatrix}; \quad \text{(ii) } \mathbf{M} = \begin{pmatrix} 0 & \theta \\ -\theta & 0 \end{pmatrix}$$

(b) Show that if $\mathbf{M}$ is diagonalizable, then

$$\det\left(e^{\mathbf{M}}\right) = e^{\mathrm{Tr}(\mathbf{M})} \qquad \text{[A.92]}$$

Comment: This is actually *true* even if $\mathbf{M}$ is *not* diagonalizable, but it's harder to prove in the general case.

(c) Show that if the matrices $\mathbf{M}$ and $\mathbf{N}$ *commute*, then

$$e^{\mathbf{M}+\mathbf{N}} = e^{\mathbf{M}}e^{\mathbf{N}} \qquad \text{[A.93]}$$

Prove (with the simplest counterexample you can think up) that Equation A.93 is *not* true, in general, for *non*-commuting matrices.

(d) If $\mathbf{H}$ is hermitian, show that $e^{i\mathbf{H}}$ is unitary.

INDEX

I

J

K

L

T